Frontiers of Intelligent Control and Information Processing

Frontiers of Intelligent Control and Information Processing

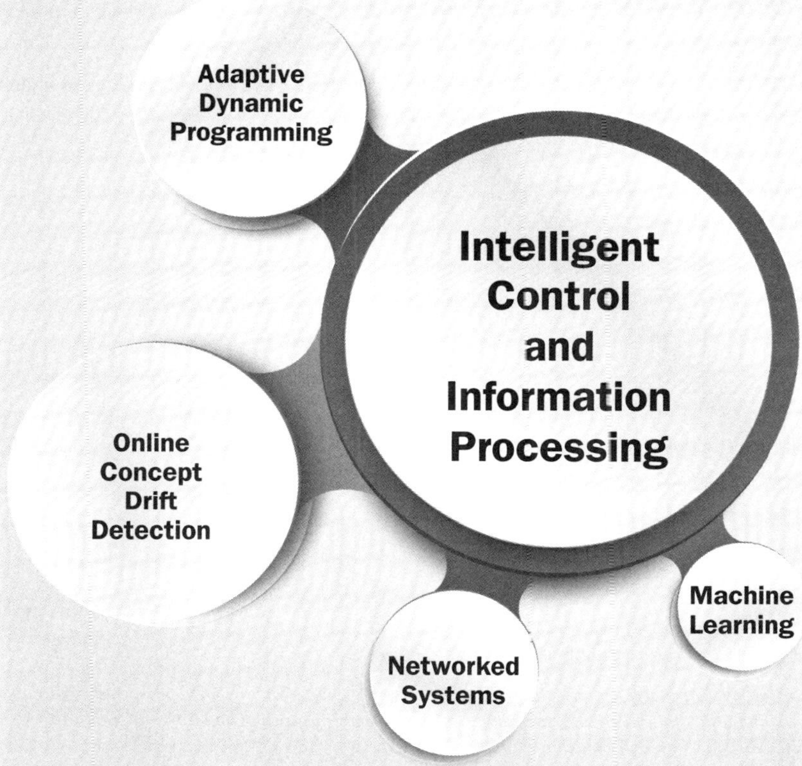

Editors

Derong Liu (Chinese Academy of Sciences, China)

Cesare Alippi (Politecnico di Milano, Italy)

Dongbin Zhao (Chinese Academy of Sciences, China)

Huaguang Zhang (Northeastern University, China)

World Scientific

NEW JERSEY • LONDON • SINGAPORE • BEIJING • SHANGHAI • HONG KONG • TAIPEI • CHENNAI

Published by

World Scientific Publishing Co. Pte. Ltd.
5 Toh Tuck Link, Singapore 596224
USA office: 27 Warren Street, Suite 401-402, Hackensack, NJ 07601
UK office: 57 Shelton Street, Covent Garden, London WC2H 9HE

Library of Congress Cataloging-in-Publication Data
Frontiers of intelligent control and information processing / edited by Derong Liu, University of Illinois at Chicago USA, Cesare Alippi, Politecnico di Milano, Italy, Dongbin Zhao, The Institute of Automation, Chinese Academy of Sciences, China, Huaguang Zhang, Institute of Electric Automation, Northeastern University, Shenyang, China.
pages cm
Includes bibliographical references.
ISBN 978-9814616874 (hardcover : alk. paper)
1. Automatic control. 2. Information technology. I. Liu, Derong, 1963–
TJ216.F76 2014
629.8--dc23
2014015264

British Library Cataloguing-in-Publication Data
A catalogue record for this book is available from the British Library.

Printed in Singapore by Mainland Press Pte Ltd.

Preface

Intelligence and cognition mechanisms represent the frontiers in control and information processing and define a mainstream where young researchers are invited to join the effort. In this direction the book aims to promote research in *Intelligent Control and Information Processing* by identifying some up-to-date research directions and challenges.

The book is composed of 17 chapters that cover hot research areas from intelligent control to intelligent information processing with a special focus on intelligent optimal control with adaptive/approximate dynamic programming and reinforcement learning, intelligent control and information processing of networked systems and machine learning techniques for big data.

The first six chapters focus on recent advances in intelligent optimal control with adaptive/approximate dynamic programming and reinforcement learning.

Chapter 1 by Abouheaf and Lewis, brings together discrete Hamiltonian mechanics, distributed multi-agent control, optimal control theory, game theory, and reinforcement learning techniques to formulate and solve the multi-agent graphical games problem. Ideas from cooperative control are used to attain synchronization among the agents' dynamics to leader's dynamics. The chapter lays the mathematical foundation for developing adaptive learning solutions for dynamic graphical games.

In Chapter 2, Yang, Liu, Wei, and Wang develop a reinforcement learning-based online adaptive control to deliver a desired tracking performance for a class of discrete-time (DT) nonlinear systems in the presence of unknown bounded disturbances. The research investigates multi-input-multi-output unknown nonaffine nonlinear DT systems. An action neural network is employed for both generating the optimal control signal and cancelling the nonlinearity of unknown DT systems. A critic neural network is then built to estimate the cost function, which satisfies the recursive equations derived from heuristic dynamic programming.

In Chapter 3, Ni, He, and Zhong investigate a data-driven heuristic dynamic programming architecture to tackle the partially observed Markov decision process. Specifically, they include a state estimator neural network to recover the full system information for the action network, so that the optimal control policy can still be achieved under partially observed environment.

In Chapter 4, Zhu and Zhao focus on online reinforcement learning for continuous state-action systems. Continuous state-space problems are more difficult than finite state ones since the number of states is infinite. Both the recursive least-square policy iteration method and the multi-samples in each cell approach have been studied and their performance is illustrated with examples.

In Chapter 5, He, Qin, and Wu propose a novel adaptive iterative learning algorithm for a class of constrained robotic manipulators with uncertainties and external disturbances. The controller design is based on a reduced form of the robot model. The uncertain parameters are estimated in time domain whereas the repetitive disturbances are compensated in the iterations. With the adoption of a saturated learning method, all signals in the closed loop are guaranteed to be bounded.

In Chapter 6, Xu, Sahoo, and Jagannathan develop a novel time-based finite horizon neural dynamic programming scheme for nonlinear networked control systems. Neural networks are used to generate the identifier, the critic and the action networks yielding optimal control policies in the presence of uncertain system dynamics due to communication network imperfections such as network-induced delays and packet losses.

A second group of chapters bridges between methodologies of intelligent control and information processing.

In particular, Chapter 7 by Yan and Wang investigates model predictive control (MPC) solutions which generate control signals by means of a real-time optimization of a performance index over a finite moving horizon of predicted future. Novel MPC approaches are developed for nonlinear systems with un-modeled dynamics based on neural networks.

In Chapter 8, Zhao and Liu propose a natural approach for packet-based communication and control co-design framework for networked control systems, which enables a sequence of control signals to be sent over the network simultaneously within one data packet. In turn, this makes it possible to actively compensate for the communication constraints in networked control systems with specially designed compensation mechanisms.

In Chapter 9, DasGupta and Srinivasan focus on privacy preserving computational models which are evidenced by distributed computing applications and game-theoretic settings, due to the increasingly widespread usage of sensitive data in networked environments. Since perfect privacy is often either impossible or too costly to achieve (e.g., it requires impractically extensive communication steps) they review various notions of approximate privacy.

The last group of chapters addresses some challenging topics in machine learning, deep learning, pattern recognition, dynamic optimization, and classification, and describes real world applications such as electromyography, smart grid, image encryption, and big data.

Chapter 10 by Alippi, Boracchi, Bu, and Zhao focuses on online detection of the occurrence of concept drift (i.e., structural changes) in datastreams, which represents a hot research topic with striking implications, e.g., network traffic monitoring and credit-card fraud detection, monitoring-based systems and big data applications. The chapter describes a change-detection method based on encoding-decoding mechanism, which is computationally simple, yet effective method for detecting concept drift in datastreams.

In Chapter 11, Ju, Ouyang, Wilamowska-Korsak, and Liu focus on the challenging problems of achieving a satisfactory rate for the surface electromyography pattern recognition, which is becoming a main issue in on-going research in rehabilitation and prosthetics. This chapter introduces nonlinear feature extractions and nonlinear classification solutions to efficiently identify different human hand manipulations based on surface electromyography signals.

In Chapter 12, Severini, Squartini, and Piazza investigate dynamic pricing policies as an effective asset at a micro-grid level, and propose a hybrid energy management scheme. With the nonlinear nature of a micro grid, neural-network forecasting abilities can provide a sustainable support under realistic operating conditions. Based on the forecast of solar energy production and grid energy prices and outdoor temperature, the optimization of tasks allocation is aimed to lower both the user costs and the grid burden while accounting for user's thermal comfort.

In Chapter 13, He and Yen introduce an ensemble method to compare six state-of-the-art multi-objective evolutionary algorithms designed specifically for many-objective optimization problems under a number of carefully crafted benchmark problems, by combining a number of performance metrics using double elimination tournament selection. The double elimination design allows characteristically poor performance of a quality algorithm to still be able to win it all.

In Chapter 14, Wen and Zeng concern about the problems of fuzzy modeling and synchronization of memristive Lorenz circuits with Chua's circuits. Considering the state-dependent properties of memristor, a new fuzzy model employing parallel distributed compensation gives a new way to analyze the complicated memristive circuits with only two subsystems. Several examples are also given to illustrate the effectiveness and potential applications in image encryption of the results.

In Chapter 15, Chen and Yang extend the original twin support vector machine and propose a novel graph embedded total margin twin support vector machine (GTM-TSVM). The central idea of GTM-TSVM is the plane of one class is required to be far away from the other class of samples. Moreover the intra-class and inter-class graphs which respectively characterize the proximity relationships between samples of within and between classes are embedded into GTM-TSVM formulation so as to exploit the underlying manifold structure of data.

In Chapter 16, Zhou, Guo, and Chen investigate extensions of Kullback Leibler (KL) information measure. The proposed method goes under the framework of minimum description length principle. In the developed methods, regularization parameters are selected by the criterion of minimization the KL divergence and approximated efficiently by second-order Taylor expansion.

In Chapter 17, Guo, Wang, and Wu study the impact of big data on evolutionary algorithms and propose a new design for evolutionary algorithms. Different from traditional ideas, the processing for big data should fully take account of data itself. An example of correlation analysis for big data in business intelligence is presented based on the evolutionary algorithms which demonstrates that the evolutionary algorithm is useful and helpful for e-business to make correct judgements for commodities storage, transfer and sales.

At last but not least, we wish to thank both authors and editors for having allowed us to finish the book within the set deadline. We would also like to thank Mr. Xiong Yang for his help in preparing the LaTeX manuscript. We believe readers will take advantage of the book content since it sets the frontiers of research. It requires everyone's effort to take the next steps so as to move the research frontier further.

Derong Liu
Beijing, China

Cesare Alippi
Milano, Italy

Dongbin Zhao
Beijing, China

Huaguang Zhang
Shenyang, China
2014

Contents

Preface v

1. Dynamic Graphical Games: Online Adaptive Learning Solutions Using Approximate Dynamic Programming 1

Mohammed I. Abouheaf and Frank L. Lewis

1.1 Introduction . . . 2
1.2 Graphs and Synchronization of Multi-Agent Dynamical Systems . 4
1.2.1 Graphs . . . 4
1.2.2 Synchronization and tracking error dynamics . . . 5
1.3 Multiple Player Cooperative Games on Graphs . . . 7
1.3.1 Graphical games . . . 7
1.3.2 Comparison of graphical games with standard dynamic games . . . 8
1.3.3 Nash equilibrium for graphical games . . . 9
1.3.4 Hamiltonian equation for dynamic graphical games . . . 9
1.3.5 Bellman equation for dynamic graphical games . . . 12
1.3.6 Discrete Hamilton-Jacobi theory: Equivalence of Bellman and discrete-time Hamilton Jacobi equations . . . 14
1.3.7 Stability and Nash solution of the graphical games . . . 19
1.4 Approximate Dynamic Programming for Graphical Games . . . 21
1.4.1 Heuristic dynamic programming for graphical games . . . 21
1.4.2 Dual heuristic programming for graphical games . . . 27
1.5 Coupled Riccati Recursions . . . 33
1.6 Graphical Game Solutions by Actor-Critic Learning . . . 35
1.6.1 Actor-critic networks and tuning . . . 35
1.6.2 Actor-critic offline tuning with exploration . . . 37
1.6.3 Actor-critic online tuning in real-time . . . 38
1.7 Graphical Game Example and Simulation Results . . . 38
1.7.1 Riccati recursion offline solution . . . 39
1.7.2 Simulation results using offline actor-critic tuning . . . 40
1.7.3 Simulation results using online actor-critic tuning . . . 41

1.8 Conclusions 44

2. Reinforcement-Learning-Based Online Learning Control for Discrete-Time Unknown Nonaffine Nonlinear Systems 49

Xiong Yang, Derong Liu, Qinglai Wei, and Ding Wang

2.1 Introduction 49
2.2 Problem Statement and Preliminaries 53
2.2.1 Dynamics of nonaffine nonlinear discrete-time systems . . . 53
2.2.2 A single-hidden layer neural network 54
2.3 Controller Design via Reinforcement Learning 55
2.3.1 A basic controller design approach 55
2.3.2 Critic neural network and weight update law 60
2.3.3 Action neural network and weight update law 61
2.4 Stability Analysis and Performance of the Closed-Loop System . . 63
2.5 Numerical Examples 70
2.5.1 Example 1 70
2.5.2 Example 2 74
2.6 Conclusions 76

3. Experimental Studies on Data-Driven Heuristic Dynamic Programming for POMDP 83

Zhen Ni, Haibo He and Xiangnan Zhong

3.1 Introduction 83
3.2 Markov Decision Process and Partially Observable Markov Decision Process 86
3.2.1 Markov decision process 86
3.2.2 Partially observable Markov decision process 88
3.3 Problem Formulation with the State Estimator 89
3.4 Data-Driven HDP Algorithm for POMDP 91
3.4.1 Learning in the state estimator network 92
3.4.2 Learning in the critic and the action network 93
3.5 Simulation Study 94
3.5.1 Case study one 95
3.5.2 Case study two 96
3.5.3 Case study three 98
3.6 Conclusions and Discussion 100

4. Online Reinforcement Learning for Continuous-State Systems 107

Yuanheng Zhu and Dongbin Zhao

4.1 Introduction 107
4.2 Background of Reinforcement Learning 109

4.3 RLSPI Algorithm 110
4.3.1 Policy iteration 110
4.3.2 RLSPI 111
4.4 Examples of RLSPI 114
4.4.1 Linear discrete-time system 114
4.4.2 Nonlinear discrete-time system 118
4.5 MSEC Algorithm 120
4.5.1 State aggregation 121
4.5.2 MSEC 121
4.6 Examples of MSEC 124
4.6.1 Inverted pendulum 125
4.6.2 Results of MSEC 126
4.7 Conclusions 129

5. Adaptive Iterative Learning Control of Robot Manipulators in the Presence of Environmental Constraints 133

Xiongxiong He, Zhenhua Qin and Xianqing Wu

5.1 Introduction 133
5.2 Problem Statement and Preliminaries 135
5.2.1 Dynamics of constrained robot systems 135
5.3 Adaptive Iterative Learning Control Design 140
5.3.1 The dynamic model conversion and assumption 140
5.3.2 Adaptive iterative learning control design 142
5.4 Stability and Convergence Analysis of the Closed-Loop System . . 143
5.5 Numerical Examples 145
5.6 Conclusions 149

6. Neural Network Control of Nonlinear Discrete-Time Systems in Affine Form in the Presence of Communication Network 151

Hao Xu, Avimanyu Sahoo and Sarangapani Jagannathan

6.1 Introduction 151
6.2 NNCS Background and Traditional Optimal Control 154
6.2.1 Nonlinear networked control systems representation 154
6.2.2 Traditional stochastic optimal control 155
6.3 Stochastic Optimal Controller Design for NNCS 157
6.3.1 Online NN-identifier design 157
6.3.2 Stochastic value function setup and critic NN design . . . 160
6.3.3 Action NN estimation of optimal control input 163
6.3.4 Closed-loop stability 164
6.3.5 Simulation results 166
6.4 Extension to Infinite Horizon 171

6.4.1 Approximation of the optimal stochastic value function and control policy using NN . . . 171
6.4.2 Simulation results . . . 175
6.5 Event-Triggered Control of NNCS . . . 178
6.5.1 Event-triggered control design of uncertain continuous time nonlinear system . . . 178
6.5.2 Closed-loop system stability and event condition design . . 182
6.5.3 Simulation results . . . 184
6.6 Conclusions . . . 185
6.7 Appendix . . . 186
6.7.1 The proof of Theorem 6.3 . . . 186
6.7.2 The proof of Theorem 6.4 . . . 187

7. Nonlinear and Robust Model Predictive Control of Systems with Unmodeled Dynamics Based on Supervised Learning and Neurodynamic Optimization 193

Zheng Yan and Jun Wang

7.1 Introduction . . . 193
7.2 Problem Formulation . . . 196
7.2.1 Nominal stability . . . 197
7.2.2 Quadratic programming formulation . . . 197
7.2.3 Minimax problem formulation . . . 200
7.3 Neural Network Approaches . . . 203
7.3.1 Extreme learning machine . . . 203
7.3.2 Simplified dual network . . . 204
7.3.3 A two-layer recurrent neural network . . . 205
7.3.4 Overall MPC approach . . . 205
7.4 Simulation Results . . . 206
7.4.1 Example 1 . . . 206
7.4.2 Example 2 . . . 211
7.5 Conclusions . . . 215

8. Packet-Based Communication and Control Co-Design for Networked Control Systems 221

Yun-Bo Zhao and Guo-Ping Liu

8.1 Introduction . . . 221
8.2 Packet-Based Transmission in Networked Control Systems . . . 224
8.3 Packet-Based Control for Networked Control Systems . . . 226
8.3.1 Packet-based control for NCSs: A unified model . . . 226
8.3.2 Design of the packet-based control scheme . . . 228
8.4 Stability of Packet-Based Networked Control Systems . . . 230

8.4.1 A switched system theory approach 231
8.4.2 A delay dependent analysis approach 232
8.5 Packet-Based Controller Design: A GPC-Based Approach 234
8.6 Numerical and Experimental Examples 236
8.6.1 Numerical examples 236
8.6.2 Experimental example 238
8.7 Conclusions 240

9. Review of Some Approximate Privacy Measures of Multi-Agent Communication Protocols 245

Bhaskar DasGupta and Venkatakumar Srinivasan

9.1 Introduction 245
9.1.1 Perfect vs. approximately perfect privacy 248
9.1.2 Privacy analysis in other environments 248
9.2 Various Frameworks to Quantify Privacy of Protocols 249
9.2.1 Communication complexity based approaches 249
9.2.2 Information-theoretic approaches 249
9.2.3 Cryptographic approaches 250
9.2.4 Two-agent differential privacy framework 250
9.3 Benchmark Problems and Functions 251
9.4 Examples of Standard Communication Protocols 252
9.5 A Geometric Approach to Quantify Privacy 254
9.5.1 Tiling functions and dissection protocols 258
9.5.2 Generalization for $d > 2$ agents 260
9.6 Conclusions 260

10. Encoding-Decoding Machines for Online Concept-Drift Detection on Datastreams 263

Cesare Alippi, Giacomo Boracchi, Li Bu, and Dongbin Zhao

10.1 Introduction 263
10.2 A CDT Based on the Encoding-Decoding Machine 266
10.2.1 The observation model 266
10.2.2 The recurrent encoding-decoding machine 267
10.2.3 The encoding-decoding CDT 267
10.3 An Encoding-Decoding Machine Based on SVD 268
10.3.1 The SVD-based encoding-decoding machine 268
10.3.2 The residuals 269
10.3.3 Learning $\mathcal{M}_0$ 270
10.3.4 Extensions 271
10.4 An ICI-Based Encoding-Decoding CDT 271
10.4.1 Designing the encoding-decoding CDT 271

10.4.2 Adaptation . . . 273
10.4.3 Single vs. multiple sensors . . . 274
10.5 Experiments . . . 274
10.5.1 The "ARMA" dataset . . . 276
10.5.2 The "hairdryer" application . . . 277
10.5.3 The mountain temperature application . . . 277
10.6 Conclusions . . . 280

11. Recognizing sEMG Patterns for Interacting with Prosthetic Manipulation 283

Zhaojie Ju, Gaoxiang Ouyang, Marzena Wilamowska-Korsak, and Honghai Liu

11.1 Introduction . . . 283
11.2 Recurrence Plot and Quantification Analysis . . . 286
11.3 Nonlinear Recognition Method . . . 288
11.3.1 Gaussian mixture models . . . 288
11.3.2 Fuzzy Gaussian mixture models . . . 291
11.3.3 Recognizing sEMG signals using FGMMs . . . 293
11.4 Experimental Results . . . 294
11.4.1 Data collection and description . . . 294
11.4.2 Feature extraction and parameter setting . . . 296
11.4.3 Recognition with one single feature . . . 297
11.4.4 Recognition with multiple features . . . 298
11.4.5 Recognition with different classifiers . . . 300
11.5 Conclusions . . . 303

12. Energy Demand Management Through Uncertain Data Forecasting: A Hybrid Approach 309

Marco Severini, Stefano Squartini and Francesco Piazza

12.1 Introduction . . . 309
12.2 Home Energy Management Problem: The Model . . . 311
12.3 Modelling the Thermal Optimization Sub-Problem . . . 312
12.3.1 Notations . . . 312
12.3.2 Heat pump constraints . . . 313
12.4 Neural Networks for Uncertain Data Forecasting . . . 314
12.5 Optimization Algorithms . . . 316
12.6 Smart Home Energy Management: Case Studies and Simulation Results . . . 317
12.6.1 Thermal model characterization . . . 317
12.6.2 Task scheduling and energy cost accounting . . . 319
12.6.3 Solar production . . . 320

12.6.4 Data forecasting 320
12.6.5 Energy management performance 321
12.7 Conclusions 331

13. Many-Objective Evolutionary Algorithms and Hybrid Performance Metrics 335

Zhenan He and Gary G. Yen

13.1 Introduction 335
13.2 Literature Review on Performance Metrics 340
13.2.1 Metrics assessing the number of Pareto optimal solutions in the set 340
13.2.2 Metrics measuring the closeness of the solutions to the true Pareto front 341
13.2.3 Metrics focusing on distribution of the solutions 342
13.2.4 Metrics concerning spread of the solutions 343
13.2.5 Metrics considering both closeness and diversity 344
13.3 Literature Review on Many-Objective Evolutionary Algorithms . . 344
13.4 Performance Metrics Ensemble 346
13.4.1 The proposed framework 346
13.4.2 Double elimination tournament 347
13.5 Experimental Results 352
13.5.1 Selected benchmark test problems 352
13.5.2 Selected performance metrics 354
13.5.3 Parameter setting in experiment 354
13.5.4 Experiment results 354
13.5.5 Observations and insights 358
13.6 Conclusions 360

14. Synchronization Control of Memristive Chaotic Circuits and Their Applications in Image Encryptions 365

Shiping Wen and Zhigang Zeng

14.1 Introduction 365
14.2 Synchronization of Memristive Chua's Circuits 366
14.2.1 Memristive Chua's circuits 367
14.2.2 Adaptive synchronization control of memristive Chua's circuits 368
14.2.3 Numerical simulations 371
14.2.4 Conclusions 374
14.3 Synchronization of Memristive Lorenz Circuits 375
14.3.1 Modeling and fuzzy synchronization of memristive Lorenz circuits 375

14.3.2 Synchronization of fuzzy memristive Lorenz circuits with memristive Chua's circuits . . . 380
14.3.3 Simulation results . . . 381
14.3.4 Conclusions . . . 383

15. Graph Embedded Total Margin Twin Support Vector Machine and Its Applications 385

Xiaobo Chen and Jian Yang

15.1 Introduction . . . 385
15.2 A Brief Review on TSVM . . . 388
15.3 Graph Embedded Total Margin TSVM . . . 390
15.3.1 Model formulation of GTM-TSVM . . . 390
15.3.2 Algorithm derivation . . . 392
15.3.3 Weighting factors for GTM-TSVM . . . 394
15.4 Experimental Results and Analysis . . . 396
15.4.1 Experimental specification . . . 396
15.4.2 A toy example . . . 397
15.4.3 Evaluations on UCI benchmark data sets . . . 397
15.4.4 Experiments on USPS and MIT CBCL database . . . 402
15.5 Conclusions . . . 404

16. Regularized Covariance Matrix Estimation Based on MDL Principle 407

Xiuling Zhou, Ping Guo and C. L. Philip Chen

16.1 Introduction . . . 407
16.2 Theoretical Backgrounds . . . 409
16.2.1 Gaussian mixture model classifier . . . 409
16.2.2 Classical discriminant analysis . . . 410
16.2.3 KLIM method . . . 412
16.3 Covariance Matrix Estimation with Multi-Regularization Parameters . . . 414
16.3.1 KLIM_L method with multi-regularization parameters . . . 414
16.3.2 Multi-regularization parameters estimation . . . 415
16.3.3 Comparison of KLIM_L with regularization methods . . . 418
16.3.4 Experiments and discussions . . . 418
16.4 Covariance Matrix Estimation with Variable Regularization Parameters . . . 422
16.4.1 Covariance matrix estimation with variable regularization parameters . . . 422
16.4.2 Variable regularization matrix estimation . . . 423
16.4.3 Comparisons . . . 425
16.4.4 Experiments and discussions . . . 426

16.5 Conclusions 428

17. An Evolution of Evolutionary Algorithms with Big Data 431

Weian Guo, Lei Wang and Qidi Wu

17.1 Introduction 431
17.2 Design of Hyper-Heuristic Framework for Evolutionary Algorithms 434
17.2.1 Genetic algorithm 434
17.2.2 Particle swarm optimization 435
17.2.3 Biogeography-based optimization 435
17.2.4 Design of framework of hyper-heuristic algorithm 438
17.2.5 Simulations for the novel framework of evolutionary algorithms 439
17.3 Performances Analysis of Evolutionary Algorithms for High-Dimensional Benchmarks 442
17.3.1 Uniform blended migration operator 442
17.3.2 Heuristic migration operator 443
17.3.3 Extended migration operator 443
17.3.4 Numerical simulations for high dimensional benchmarks . . 444
17.4 Correlation Analysis for Business Intelligence Based on Evolutionary Algorithms 446
17.4.1 Correlation model 451
17.4.2 Simulation for correlation analysis 452
17.5 Conclusions 454

Index 457

Chapter 1

Dynamic Graphical Games: Online Adaptive Learning Solutions Using Approximate Dynamic Programming

Mohammed I. Abouheaf and Frank L. Lewis

Abstract The chapter develops a new class of multi-agent discrete-time dynamic games, known as dynamic graphical games. In graphical games, the information flow between agents is prescribed by a communication graph structure. For that reason a local performance index is defined for each agent that depends only on the local information available to this agent. This chapter brings together Discrete Hamiltonian mechanics, distributed multi-agent control, optimal control theory, game theory, and reinforcement learning techniques to formulate and solve these multi-agent graphical games. Ideas from cooperative control are used to attain synchronization among the agents' dynamics to the leader dynamics. The chapter lays the mathematical foundation to develop adaptive learning solutions for the dynamic graphical games. Novel coupled Hamilton Jacobi equations and coupled Bellman equations are developed for this class of the dynamic graphical games. The relationship between the coupled Hamilton Jacobi equations and the coupled Bellman equations is shown. The graphical game Bellman equations are shown to be equivalent to certain graphical game Hamilton Jacobi Bellman equations developed herein. Nash solutions are given in terms of solutions to a set of coupled discrete-time Hamilton-Jacobi-Bellman equations. Novel coupled Riccati recursions are developed for the dynamic graphical game. Reinforcement learning value iteration algorithms are developed to solve the dynamic graphical games online in real-time. The developed multi-agent value iteration algorithms learn the Nash solution online in real-time. Value iteration convergence proofs are provided. Approximate Dynamic Programming techniques namely Heuristic Dynamic Programming and Dual Heuristic Dynamic Programming are developed and extended to solve the dynamic graphical games. Online and offline adaptive learning structures are implemented using actor-critic network structures to solve these dynamic graphical games.

1.1 Introduction

This chapter introduces a new class of multi-agent discrete-time games known as dynamic graphical games, where the information flow between agents is restricted by a communication graph structure. The agents' error dynamics are coupled dynamical systems driven by the local control input of each agent and its neighbors. Cooperative control ideas are used to synchronize the agents' dynamics to the leader dynamics. Discrete mechanics is employed to show the relation between the coupled Hamilton Jacobi equations and Bellman equations for the dynamic graphical games. Optimal control theory is used to develop novel coupled discrete-time Hamilton Jacobi Bellman (HJB) equations for the dynamic graphical games. Game theory provides a theoretical frame work to solve the set of coupled HJB equations. Nash equilibrium policies are given in terms of solutions to a set of coupled HJB equations. Approximate Dynamic Programming (ADP) techniques [Werbos (1989, 1992)] are used to solve the coupled HJB equations. Value iteration algorithms are developed to solve the dynamic graphical games. Convergence of the value iteration algorithm is shown under mild assumptions about the graph interconnectivity properties. Offline and real-time actor-critic network structures are developed to verify the proper performance of the developed value iteration algorithms.

Cooperative control techniques are applied to the distributed multi-agent systems. Here the dynamical multi-agent systems are interconnected by a communication graph structure. The cooperative control applications involve unmanned vehicles, autonomous teams, mobile robots, and sensor networks, etc. [Beard and Stepanyan (2003); Mu *et al.* (2005)]. The cooperative control problems are classified into consensus control problems and synchronization control problems [Fax and Murray (2004); Jadbabaie *et al.* (2003); Olfati-Saber and Murray (2004); Olfati-Saber *et al.* (2007); Qu (2009); Ren *et al.* (2005); Ren and Beard (2005, 2008); Sun *et al.* (2006); Tsitsiklis (1984)]. In the consensus control problems, the agents converge to uncontrollable node dynamics or to the leader dynamics. In the synchronization control problems, control protocols are properly designed such that each agent reaches the same state [Hong *et al.* (2006); Li *et al.* (2004); Ren *et al.* (2007); Wang and Chen (2002)].

Static graphical games have been studied in the computational intelligence community. The work on static graphical games has focused on developing efficient algorithms to compute standard Nash equilibrium for payoffs generally given in terms of matrices. Such algorithms are simplified in that they only have complexity on the order of the maximum node degree in the graph, not on the order of the number of players N. This work has mostly been limited to undirected connected graphs, normally trees [Kakade *et al.* (2003); Kearns *et al.* (2001); Shoham and Leyton-Brown (2008)]. Multi-player decision control problems are formulated using game theory. The optimal policy for each agent is achieved through optimizing the performance index. Thus the agents' optimal policies form Nash equilibrium solution. This is equivalent to solving the Hamilton Jacobi (HJ) equations of the multi-player cooperative games or the coupled Riccati equations in the LQR case [Basar and Olsder (1999); Freiling *et al.* (1996); Gajic and Li (1988); Kakade *et al.* (2003); Kearns *et al.* (2001); Shoham and Leyton-Brown (2008)]. These equations are difficult to solve. The standard dynamic games have a centralized global information structure [Abou-Kandil *et al.* (2003); Basar and Olsder (1999)]. Continuous-time graphical games are in traduced in Vamvoudakis *et al.* (2012).

Discrete forms of the Euler–Lagrange and Hamiltonian equations can be obtained using discrete versions of variational principles [Marsden and West (2001); Suris (2003, 2004)]. The formulation for the discrete Lagrangian mechanics is introduced in Marsden and West (2001). Direct approaches are used to formulate the discrete Hamiltonian mechanics in Gonzalez (1996) and McLachlan *et al.* (1999). The Euclidean relationship between symplectic integrators, discrete-time optimal control [Lewis *et al.* (2012)], and distributed network optimization [Lall and West (2006)] emphasize the importance of the Hamiltonian discrete-time forms. The optimal control theory employed the classical mechanics to solve the optimal control problem [Bellman (1957); Bryson (1996); Lewis *et al.* (2012)]. The classical mechanics extended the Hamilton Jacobi equation to the HJB equation, where the optimal cost-to-go function is the solution to this HJB equation [Bellman (1957); Bryson (1996); Lewis *et al.* (2012)].

ADP techniques are developed to solve the optimal control problem online in Werbos (1992) and offline in Bertsekas and Tsitsiklis (1996). The offline techniques are referred to as neuro-dynamic programming. ADP techniques are used in adaptive control, computational intelligence, applied mathematics, and operation research, etc. [Prokhorov and Wunsch (1997); Si *et al.* (2004); Werbos (1992, 2007, 2009)]. ADP techniques are classified into four main schemes: Heuristic Dynamic Programming (HDP), Dual Heuristic Dynamic Programming (DHDP), Action Dependent Heuristic Dynamic Programming (ADHDP), and Action Dependent Dual Heuristic Dynamic Programming (ADDHDP) [Werbos (1992)].

Reinforcement Learning (RL) is concerned with learning from interaction in a dynamic environment [Sutton and Barto (1998)]. The agent picks its action in such a way that minimizes the sum of cumulative reward [Weiss (1999); Sutton and Barto (1998)]. RL techniques involve Value Iteration (VI) or Policy Iteration (PI) algorithms [Bertsekas and Tsitsiklis (1996)]. These algorithms are used to solve the optimal control problem for dynamic systems in real-time [Al-Tamimi *et al.* (2008); Vamvoudakis and Lewis (2010); Vrabie *et al.* (2009); Werbos (1974, 1992)]. RL algorithms are used to solve multi-player games for finite-state systems in Busoniu *et al.* (2008); Littman (2001); Vrancx *et al.* (2008) and to learn online in real-time the solutions for the optimal control problems of the differential games in Dierks and Jagannathan (2010); Johnson *et al.* (2010); Vamvoudakis and Lewis (2010, 2011); Vrabie *et al.* (2009).

Actor-critic networks are Temporal Difference methods with separate structures that explicitly represent the policies apart, from the value structures [Widrow *et al.* (1973)]. These structures involve forward-in-time algorithms for computing optimal decisions that are implemented online in real-time. The actor component applies control policies to their environment, while the critic rewards some decisions and punishes other decisions [Sutton and Barto (1998)]. Based on this assessment, the actor policy is updated at each learning step [Sutton and Barto (1998); Widrow *et al.* (1973)]. A general analytical model for learning with a critic is formulated and analyzed in Widrow *et al.* (1973). Online actor-critic network structures based on value iteration are used to solve the optimal control problems in real-time in Werbos (1974, 1989, 1992).

The chapter is organized as follows. In Section 1.2, the synchronization problem in multi-agent systems is reviewed and the error dynamics for each agent are derived. These dynamics are driven by the controls of the agent and its neighbors. In Section 1.3, a dynamic graphical game is formulated through defining a performance index that likewise depends on information from the agent and its neighbors. The graph game is unlike standard multi-player dynamical games in the literature, which have a centralized state. The graphical game Hamiltonian and Bellman equations are developed. The relation between these Bellman equations and the graph games HJB equations is elucidated using discrete mechanics. Furthermore, Nash equilibrium and stability proofs are given in terms of the solutions to the coupled HJB equations. In Section 1.4, value iteration algorithms based on HDP and DHP techniques are developed to solve the coupled graphical game discrete-time HJB equations along with the convergence proof for the HDP algorithm. In Section 1.5, coupled Riccati equations are used to solve the dynamic graphical games that are equivalent to solving the developed DHP algorithm. In Section 1.6, offline and online adaptive learning structures are formulated using actor-critic networks. These networks are motivated by the graph game DHP algorithm developed in Section 1.4. In Section 1.7, a graphical game simulation example is used to verify the proper performance of the derived coupled Riccati recursions and the developed DHP Algorithms. The simulation is conducted using data obtained along the system trajectory both online in real-time and offline.

1.2 Graphs and Synchronization of Multi-Agent Dynamical Systems

1.2.1 *Graphs*

The directed graph or digraph Gr is defined as the pair $Gr = (V, \mathcal{E})$ with a nonempty finite set of N nodes or vertices $V = \{v_1, \ldots, v_N\}$ and a set of edges or arcs $\mathcal{E} \subseteq V \times V$. General directed graphs with fixed topology are considered in this chapter. The graph is assumed to be simple, e.g., no repeated edges and no self-loops $(v_i, v_i) \notin \mathcal{E}, \forall i$. The connectivity matrix E is defined such that $E = [e_{ij}]$ with $e_{ij} > 0$ if

$(v_j, v_i) \in \mathcal{E}$ and $e_{ij} = 0$ otherwise. The set of neighbors of every node v_i is $N_i = \{v_j : (v_j, v_i) \in \mathcal{E}\}$. Define the in-degree matrix as a diagonal matrix $D = \text{diag}\{d_i\}$, with $d_i = \sum_{j \in N_i} e_{ij}$ the weighted in-degree of node i. The graph Laplacian matrix L is defined as $L = D - E$.

A directed path from node v_0 to node v_r is defined as a sequence of nodes $v_0, v_1, \ldots, v_r$ such that $(v_i, v_{i+1}) \in \mathcal{E},\ i \in \{0, 1, \ldots, r-1\}$. A directed graph is strongly connected if there is a directed path from v_i to v_j for all distinct nodes $v_i, v_j \in V$. A directed tree is a digraph where each node has in-degree equal to one except for a single node which has in-degree equal to zero. The node with zero in-degree is known as the root node. A spanning tree of a digraph is a directed tree formed by a subset of graph edges that connects all the nodes of the graph. A strongly connected digraph contains a spanning tree.

1.2.2 *Synchronization and tracking error dynamics*

Consider the communication graph $Gr = (V, \mathcal{E})$ having N agents, each with dynamics given by

$$x_i(k+1) = Ax_i(k) + B_i u_i(k) \tag{1.1}$$

where $x_i(k) \in \mathbb{R}^n$ is the state vector of node i, and $u_i(k) \in \mathbb{R}^{m_i}$ is the control input vector for node i. A control or leader node v_0 has command generator dynamics [Lewis (1992)] $x_0(k) \in \mathbb{R}^n$ given by

$$x_0(k+1) = Ax_0(k). \tag{1.2}$$

The leader is connected to a small percentage of the nodes in the graph. That is, there is an edge (v_0, v_i) only for a small number of nodes i.

The control objective is to design the control inputs $u_i(k)$, using information only from neighbor nodes, so that all agent states synchronize to the leader state, that is, $\lim\limits_{k \to \infty} \|x_i(k) - x_0(k)\| = 0, \forall i$.

To study synchronization on graphs, define the local neighborhood tracking error [Khoo *et al.* (2009)] $\varepsilon_i(k) \in R^n$ for each node i as

$$\varepsilon_i(k) = \sum_{j \in N_i} e_{ij}\big(x_j(k) - x_i(k)\big) + g_i\big(x_0(k) - x_i(k)\big) \tag{1.3}$$

where g_i is the pinning gain of node i, which is nonzero $g_i > 0$ if the control node x_0 is connected to node i [Li *et al.* (2004)]. Assume that $g_i > 0$ for at least one node.

The global tracking error vector $\varepsilon = \begin{bmatrix} \varepsilon_1^{\mathsf{T}} & \varepsilon_2^{\mathsf{T}} & \cdots & \varepsilon_N^{\mathsf{T}} \end{bmatrix}^{\mathsf{T}} \in R^{nN}$ for all nodes is given by

$$\varepsilon(k) = -\big((L+G) \otimes I_n\big)x(k) + \big((L+G) \otimes I_n\big)\underline{x}_0(k) \tag{1.4}$$

where the global node state vector is $x = \begin{bmatrix} x_1^{\mathsf{T}} & x_2^{\mathsf{T}} & \cdots & x_N^{\mathsf{T}} \end{bmatrix}^{\mathsf{T}} \in R^{nN}$. This can be written as

$$\varepsilon(k) = -\big((L+G)\otimes I_n\big)\eta(k) \tag{1.5}$$

where the global disagreement vector or the synchronization error [Olfati-Saber and Murray (2004)] is

$$\eta(k) = \big(x(k) - \underline{x}_0(k)\big) \in R^{nN} \tag{1.6}$$

with $\underline{x}_0 = \underline{I}x_0 \in R^{nN}$, $\underline{I} = \underline{1}\otimes I_n \in R^{nN\times n}$ and $\underline{1}$ the N-vector of ones. $G = \mathrm{diag}\{g_i\} \in R^{N\times N}$is a diagonal matrix of pinning gains.

If the graph contains a spanning tree and at least one pinning gain into at least one root node is nonzero, then $(L+G)$ is nonsingular. Then, the next result shows that the disagreement vector can be made arbitrarily small by making the local neighborhood tracking errors small. The maximum and minimum singular values of a matrix are denoted as $\bar{\sigma}(\cdot)$ and $\underline{\sigma}(\cdot)$, respectively.

Lemma 1.1. *(Graph Synchronization Error) Let the graph have a spanning tree and the pinning gain into at least one root node be nonzero. Then the synchronization error is bounded by*

$$\big\|\eta(k)\big\| \le \big\|\varepsilon(k)\big\|/\underline{\sigma}(L+G). \tag{1.7}$$

Proof. Under the hyposthesis $(L+G)$ is nonsingular. Then $\underline{\sigma}(L+G)\neq 0$ and (1.5) implies (1.7), with $\varepsilon(k)\equiv 0$ if and only if the nodes synchronize, that is

$$x(k) = \underline{I}x_0(k). \tag{1.8}$$

This completes the proof. □

The objective is to minimize the local neighborhood tracking errors $\varepsilon_i(k)$, which in view of Lemma 1.1 will guarantee approximate synchronization. For ease of notation x_{ik} is written as $x_i(k)$, and so on, when the time index k is clear. The dynamics of the local neighborhood tracking error for node i are given by

$$\varepsilon_{i(k+1)} = \sum_{j\in N_i} e_{ij}\big(x_{j(k+1)} - x_{i(k+1)}\big) + g_i\big(x_{0(k+1)} - x_{i(k+1)}\big).$$

Then,

$$\begin{aligned}\varepsilon_{i(k+1)} = {} & A\Big(\sum_{j\in N_i} e_{ij}\big(x_{jk} - x_{ik}\big) + g_i\big(x_{0k} - x_{ik}\big)\Big)\\ & + \sum_{j\in N_i} e_{ij}\big(B_j u_{jk} - B_i u_{ik}\big) - g_i B_i u_{ik}\end{aligned}$$

or

$$\varepsilon_{i(k+1)} \equiv f_i(\varepsilon_{ik}, u_{ik}, u_{-ik}) = A\varepsilon_{ik} - (d_i + g_i)B_i u_{ik} + \sum_{j\in N_i} e_{ij} B_j u_{jk}. \tag{1.9}$$

These error dynamics are interacting dynamical systems driven by the control actions of agent i and all of its neighbors.

1.3 Multiple Player Cooperative Games on Graphs

In this section, multi-player games are defined on graphs by introducing local performance indices. This captures both competitive and cooperative team aspects among the nodes. Principles of optimal control [Lewis *et al.* (2012)] are used to develop Hamiltonian functions and costates. The Bellman equations are derived for the dynamic graphical games and the Discrete Hamilton-Jacobi Theory [Lall and West (2006)] is used to show the relation between the Hamiltonian functions and the Bellman functions. These results lay the mathematical foundation to solve the dynamic graphical games in subsequent sections.

Graphical games are based on the responses of each agent i to the other players in graph. Define the control actions of the neighbors of each agent i as

$$u_{-i} = \{u_j | j \in N_i\}. \tag{1.10}$$

Also define the group of control actions of the neighbors of each agent i and the control actions of the neighbors to the neighbors of each agent i as

$$u_{-i,-\{-i\}} = \{u_j | j \in N_i, N_{\{-i\}}\} \tag{1.11}$$

and the actions of all the agents in the graph excluding i as

$$u_{\bar{i}} = \{u_j | j \in N, j \neq i\}. \tag{1.12}$$

1.3.1 *Graphical games*

The local neighborhood dynamics (1.9) are driven by the control inputs of node i and all of its neighbors. In order to define a dynamic graphical game, the local performance index for each node is written as

$$J_i = \sum_{k=0}^{\infty} U_i(\varepsilon_{ik}, u_{ik}, u_{-ik}) = \frac{1}{2}\sum_{k=0}^{\infty}\Big(\varepsilon_{ik}^{\mathsf{T}} Q_{ii}\varepsilon_{ik} + u_{ik}^{\mathsf{T}} R_{ii} u_{ik} + \sum_{j\in N_i} u_{jk}^{\mathsf{T}} R_{ij} u_{jk}\Big) \tag{1.13}$$

where $Q_{ii} > 0, R_{ii} > 0, R_{ij} > 0$ are symmetric time-invariant weighting matrices. This performance index depends on the state and control of node i and the controls of all its neighbors. The dynamics (1.9) and the performance indices (1.13) depend on the graph topology $Gr = (V, \mathcal{E})$.

Definition 1.1. The control sequence $\tilde{u}_i = \{u_{ik}\}_{k=0}^{\infty}, \forall i \in N$ is said to be admissible if it locally stabilizes (1.9) and guarantees that J_i is finite [Zhang *et al.* (2009)].

Definition 1.2. The dynamic graphical game with local dynamics (1.9) and performance indices (1.13) is well-formed if $R_{ij} \neq 0 \rightleftharpoons e_{ij} \in E$.

1.3.2 Comparison of graphical games with standard dynamic games

Dynamic graphical games as introduced in this chapter are a special case of the dynamic games normally discussed in the literature. The standard N-player dynamic game as defined in Basar and Olsder (1999), has centralized dynamics

$$\delta(k+1) = F\delta(k) + \sum_{i=1}^{N} B^i u_i(k) \tag{1.14}$$

where $\delta(k) \in R^{\tilde{N}}$ is the centralized state and $u_i(k) \in R^{m_i}$ is the control input for each player i. In these standard games, the performance index for each player i, is defined as

$$\tilde{J}_i(\{\delta(k), u_i(k)\}_{k\geq 0}) = \frac{1}{2}\sum_{k=0}^{\infty}\Big(\delta^{\mathsf{T}}(k)\hat{Q}_{ii}\delta(k) + \sum_{j\in N} u_j^{\mathsf{T}}(k)\hat{R}_{ij}u_j(k)\Big) \tag{1.15}$$

where $\hat{Q}_{ii} > 0 \in R^{\tilde{N}\times\tilde{N}}$, $\hat{R}_{ij} > 0 \in R^{m_j\times m_j}$.

Both the dynamics (1.14) and the performance indices (1.15) depend on the control actions of all other players.

To compare these standard games to the graphical games defined in this chapter, write the global error dynamics for the graphical game as

$$\varepsilon(k+1) = (I_N \otimes A)\varepsilon(k) - \big((L+G)\otimes I_n\big)\bar{B}u(k) \tag{1.16}$$

with $\bar{B} = \operatorname{diag}\{B_1, \ldots, B_i, \ldots, B_N\}$ and $u(k) = \begin{bmatrix} u_1^{\mathsf{T}} \; u_2^{\mathsf{T}} \cdots u_N^{\mathsf{T}} \end{bmatrix}^{\mathsf{T}}$ is the global vector of control inputs. To find the relation between the standard centralized game (1.14) and the dynamic graphical game (1.16), define $F = (I_N \otimes A)$ and write (1.14) as

$$\delta(k+1) = (I_N \otimes A)\delta(k) + \begin{bmatrix} B^1 \cdots B^N \end{bmatrix} u(k). \tag{1.17}$$

Now, define l_{ij} as the ij^{th} element of $(L+G)$ and write (1.16) as

$$\varepsilon(k+1) = (I_N \otimes A)\varepsilon(k) - [l_{ij}B_j]u(k) \tag{1.18}$$

where $[l_{ij}B_j]$ is a matrix whose ij^{th} block is $l_{ij}B_j$. (1.17) and (1.18) are same if one defines the block column matrix B^j as $B^j = \begin{bmatrix} l_{1j}B_j^{\mathsf{T}} \; l_{2j}B_j^{\mathsf{T}} \cdots l_{Nj}B_j^{\mathsf{T}} \end{bmatrix}^{\mathsf{T}}$.

It is easily observed that the graphical game local performance index (1.13) is a special case of the standard game performance index (1.15) with appropriate definition of $\hat{Q}_{ij}, \hat{R}_{ij}$. Therefore, the graphical game is a special case of the standard game which explicitly displays the graph topology through $(L+G)$ in (1.18).

The graphical game formulation explicitly captures the structure of the communication graph. Therefore, its analysis clearly reveals the interplay of individual node dynamics and the graph topology within a multi-player game. Moreover, as seen in Section 1.6, it allows the solution of the game in a distributed fashion. Note that the coupled game Riccati equations of each agent presented in Basar and Olsder (1999), depend on the policies of all other agents, and so provide a centralized solution for the game.

1.3.3 *Nash equilibrium for graphical games*

In this section, the definition of Nash equilibrium is introduced. The concept of interactive Nash equilibrium for graphical games is briefly mentioned. The interactive Nash equilibrium requires that the graph be strongly connected.

Definition 1.3. (Global Nash Equilibrium [Basar and Olsder (1999)]) The N-player graphical game with N-tuple of optimal control policies $\{u_1^*, u_2^*, \ldots, u_N^*\}$ is said to have a global Nash equilibrium solution if for all $i \in N$

$$J_i^* \triangleq J_i(u_i^*, u_{\bar{i}}^*) \leq J_i(u_i\ u_{\bar{i}}^*). \tag{1.19}$$

The N-tuple $\{J_1^*, J_2^*, \ldots, J_N^*\}$ is called the Nash equilibrium outcome of the N-player game.

Remark 1.1. In a disconnected graph, the agents can be in Nash equilibrium, yet have no influence in each other. Therefore, the stronger definition of interactive Nash equilibrium is introduced in Vamvoudakis *et al.* (2012) to rule out such undesirable situations. All agents are in interactive Nash equlibrium if they are all in Nash equilibrium and the graph is strongly connected. This concept is not used in this chapter.

1.3.4 *Hamiltonian equation for dynamic graphical games*

The optimal control problem uses the Lagrange multiplier approach to find the optimal control sequence by minimizing the associated performance index [Lewis *et al.* (2012)]. The optimal control problem is used to solve the dynamic graphical games by following the procedure in Lewis *et al.* (2012).

Using the optimal control theory to solve the dynamic graphical game requires a constraint function for each agent i $\psi_i(\bar{\varepsilon}_{ik}, u_{ik}, u_{-ik,-\{-ik\}})$ specified at each time k so that

$$\bar{\varepsilon}_{i(k+1)} = \begin{bmatrix} \varepsilon_{i(k+1)} \\ \vdots \\ \varepsilon_{-i(k+1)} \end{bmatrix} \equiv \psi_i(\bar{\varepsilon}_{ik}, u_{ik}, u_{-ik,-\{-ik\}}) \in R^{nN_{i,j}} \tag{1.20}$$

where $\bar{\varepsilon}_{ik}$ is a vector of the agent's state ε_{ik} and the states of its neighbors ε_{-ik}. $N_{i,j}$ is the total number of each agent i and its neighbors.

Each constraint function $\psi_i(\bar{\varepsilon}_{ik}, u_{ik}, u_{-ik,-\{-ik\}})$ has an associated Lagrange multiplier $\lambda_{i(k+1)}$. It is noticed that the constraint function will depend only on the local information available to each agent i.

Let $\lambda_{i(k+1)} \in R^{nN_{i,j}}$ and define the augmented performance index $J_i^{'}$ using the constraint function (1.20) and the performance index (1.13) so that

$$\begin{aligned} J_i^{'} &= \Phi(\Pi, \bar{\varepsilon}_{i\Pi}) + \sum_{k=l}^{\Pi-1} U_i(\varepsilon_{ik}, u_{ik}, u_{-ik}) \\ &= \Phi(\Pi, \bar{\varepsilon}_{i\Pi}) + \sum_{k=l}^{\Pi-1} \Big(U_i(\varepsilon_{ik}, u_{ik}, u_{-ik}) \\ &\quad + \lambda_{i(k+1)}^{\mathsf{T}} (\psi_i(\bar{\varepsilon}_{ik}, u_{ik}, u_{-ik,-\{-ik\}}) - \bar{\varepsilon}_{i(k+1)}) \Big) \end{aligned} \tag{1.21}$$

where $\Phi(\Pi, \bar{\varepsilon}_{i\Pi})$ is a function of the states $\bar{\varepsilon}_{i\Pi}$ at time Π, $\lambda_{ik} \equiv \lambda_i(k)$ is the costate variable of each agent i.

Define the Hamiltonian function for each agent i as

$$\begin{aligned} H_i(\bar{\varepsilon}_{ik}, \lambda_{i(k+1)}, u_{ik}, u_{-ik}) = \lambda_{i(k+1)}^{\mathsf{T}} \bar{\varepsilon}_{i(k+1)} + \frac{1}{2} \Big(\varepsilon_{ik}^{\mathsf{T}} Q_{ii} \varepsilon_{ik} \\ + u_{ik}^{\mathsf{T}} R_{ii} u_{ik} + \sum_{j \in N_i} u_{jk}^{\mathsf{T}} R_{ij} u_{jk} \Big). \end{aligned} \tag{1.22}$$

(1.21) can be written as

$$\begin{aligned} J_i^{'} &= \Phi(\Pi, \bar{\varepsilon}_{i\Pi}) - \lambda_{i\Pi}^{\mathsf{T}} \bar{\varepsilon}_{i\Pi} + H_i(\bar{\varepsilon}_{il}, \lambda_{i(l+1)}, u_{il}, u_{-il}) \\ &\quad + \sum_{k=l+1}^{\Pi-1} \Big(H_i(\bar{\varepsilon}_{ik}, \lambda_{i(k+1)}, u_{ik}, u_{-ik}) - \lambda_{ik}^{\mathsf{T}} \bar{\varepsilon}_{ik} \Big). \end{aligned} \tag{1.23}$$

The increment in the augmented performance index $J_i^{'}$ (1.23) due to increments in the variables $\bar{\varepsilon}_{ik}$, $\lambda_{i(k+1)}$, u_{-ik} and u_{ik} is given by

$$\begin{aligned} \partial J_i^{'} &= \left(\frac{\partial \Phi(\Pi, \bar{\varepsilon}_{i\Pi})}{\partial \bar{\varepsilon}_{i\Pi}} - \lambda_{i\Pi} \right)^{\mathsf{T}} \partial \bar{\varepsilon}_{i\Pi} + \frac{\partial H_i\big(\bar{\varepsilon}_{il}, \lambda_{i(l+1)}, u_{il}, u_{-il}\big)}{\partial \bar{\varepsilon}_{il}}^{\mathsf{T}} \partial \bar{\varepsilon}_{il} \\ &\quad + \frac{\partial H_i\big(\bar{\varepsilon}_{il}, \lambda_{i(l+1)}, u_{il}, u_{-il}\big)}{\partial u_{il}}^{\mathsf{T}} \partial u_{il} + \sum_{j \in N_i} \left(\frac{\partial H_i\big(\bar{\varepsilon}_{il}, \lambda_{i(l+1)}, u_{il}, u_{-il}\big)}{\partial u_{jl}}^{\mathsf{T}} \partial u_{jl} \right) \\ &\quad + \sum_{k=l+1}^{\Pi-1} \left(\left(\frac{\partial H_i\big(\bar{\varepsilon}_{ik}, \lambda_{i(k+1)}, u_{ik}, u_{-ik}\big)}{\partial \bar{\varepsilon}_{ik}} - \lambda_{ik} \right)^{\mathsf{T}} \partial \bar{\varepsilon}_{ik} \right. \\ &\quad + \frac{\partial H_i\big(\bar{\varepsilon}_{ik}, \lambda_{i(k+1)}, u_{ik}, u_{-ik}\big)}{\partial u_{ik}}^{\mathsf{T}} \partial u_{ik} \\ &\quad \left. + \sum_{j \in N_i} \left(\frac{\partial H_i\big(\bar{\varepsilon}_{ik}, \lambda_{i(k+1)}, u_{ik}, u_{-ik}\big)}{\partial u_{jk}}^{\mathsf{T}} \partial u_{jk} \right) \right) \\ &\quad + \sum_{k=l+1}^{\Pi} \left(\frac{\partial H_i\big(\bar{\varepsilon}_{i(k-1)}, \lambda_{ik}, u_{i(k-1)}, u_{-i(k-1)}\big)}{\partial \lambda_{ik}} - \bar{\varepsilon}_{ik} \right)^{\mathsf{T}} \partial \lambda_{ik}. \end{aligned} \tag{1.24}$$

According to the Lagrange multiplier theory, the increment in the augmented performance index $\partial J_i'$ is zero at a constrained minimum [Lewis *et al.* (2012)]. Then the necessary conditions for a constrained minimum are given by

$$\bar{\varepsilon}_{i(k+1)} = \frac{\partial H_i\big(\bar{\varepsilon}_{ik}, \lambda_{i(k+1)}, u_{ik}, u_{-ik}\big)}{\partial \lambda_{i(k+1)}}, \tag{1.25}$$

$$\lambda_{ik} = \frac{\partial H_i\big(\bar{\varepsilon}_{ik}, \lambda_{i(k+1)}, u_{ik}, u_{-ik}\big)}{\partial \bar{\varepsilon}_{ik}}, \tag{1.26}$$

$$\frac{\partial H_i\big(\bar{\varepsilon}_{ik}, \lambda_{i(k+1)}, u_{ik}, u_{-ik}\big)}{\partial u_{ik}} = 0, \tag{1.27}$$

$$\sum_{j \in N_i} \left(\frac{\partial H_i\big(\bar{\varepsilon}_{ik}, \lambda_{i(k+1)}, u_{ik}, u_{-ik}\big)}{\partial u_{jk}}^{\mathsf{T}} \partial u_{jk} \right) = 0. \tag{1.28}$$

It is shown that the Lagrange multiplier or the agent's costate λ_{ik} is a variable that is determined by its own dynamical equation (1.26). The state vector (1.25) and the costate (1.26) are coupled difference equations. The optimal control policy is found by (1.27), and the costate λ_{ik} represents an intermediate step to find the optimal control solution.

The best response solution of the dynamic graphical game finds the optimal control policy for each agent, given fixed policies for its neighbors, i.e., $\partial u_{jk} = 0,\ u_{jk} \in \{u_{-ik}\}$. Therefore, the coupled stationarity condition (1.28) of each agent holds regardless of the value of the gradient $\partial H_i(\bar{\varepsilon}_{ik}, \lambda_{i(k-1)}, u_{ik}, u_{-ik})/\partial u_{-ik}$.

The boundary conditions are given by the non-recursive equations

$$\frac{\partial H_i\big(\bar{\varepsilon}_{il}, \lambda_{i(l+1)}, u_{il}, u_{-il}\big)}{\partial \bar{\varepsilon}_{il}}^{\mathsf{T}} \partial \bar{\varepsilon}_{il} = 0, \tag{1.29}$$

$$\left(\frac{\partial \Phi(\Pi, \bar{\varepsilon}_{i\Pi})}{\partial \bar{\varepsilon}_{i\Pi}} - \lambda_{i\Pi} \right)^{\mathsf{T}} \partial \bar{\varepsilon}_{i\Pi} = 0, \tag{1.30}$$

$$\sum_{j \in N_i} \left(\frac{\partial H_i\big(\bar{\varepsilon}_{il}, \lambda_{i(l+1)}, u_{il}, u_{-il}\big)}{\partial u_{jl}}^{\mathsf{T}} \partial u_{jl} \right) = 0. \tag{1.31}$$

The initial state vector $\bar{\varepsilon}_{il}$ is fixed, thus, (1.29) holds and the gradient $\partial H_i\big(\bar{\varepsilon}_{il}, \lambda_{i(l-1)}, u_{il}, u_{-il}\big)/\partial \bar{\varepsilon}_{il}$ doesn't impose any additional constraints. Similarly for the fixed final state case, (1.30) holds regardless of the value $(\partial \Phi(\Pi, \bar{\varepsilon}_{i\Pi})/\partial \bar{\varepsilon}_{i\Pi}) - \lambda_{i\Pi}$. On the other hand, if the free final state case is considered, then the following condition needs to be satisfied

$$\lambda_{i\Pi} = \frac{\partial \Phi(\Pi, \bar{\varepsilon}_{i\Pi})}{\partial \bar{\varepsilon}_{i\Pi}}. \tag{1.32}$$

Similarly the best response solution of the dynamic graphical game results in zero change in initial neighbors' policies u_{-il}, i.e., $\partial u_{jl} = 0,\ u_{jl} \in \{u_{-il}\}$. Then, (1.31) holds regardless of the value of the gradient $\partial H_i\big(\bar{\varepsilon}_{il}, \lambda_{i(l+1)}, u_{il}, u_{-il}\big)/\partial u_{-il}$.

Remark 1.2. The objective of minimizing the local performance index for each agent i, is to obtain the best response solution for this agent with fixed neighbors' policies, i.e., $(\partial u_{jk}) = 0,\ \forall j \in N_i$. This means that the increment in the augmented performance index due to the neighbors' policies is zero $\partial j_i^{'} = 0$.

The Hamiltonian (1.22) for each agent i optimize the performance index (1.13) along the equality constraints $\bar{\varepsilon}_{i(k+1)}$ (1.20), where $\bar{\varepsilon}_{ik}$ is a vector of the agent's state ε_{ik} and the states of its neighbors ε_{-ik}.

The costate equation for each agent i is found using the stationarity condition [Lewis *et al.* (2012)] so that

$$\lambda_{ik} = \bar{A}_i^{\mathsf{T}} \lambda_{i(k+1)} + \bar{Q}_i \bar{\varepsilon}_{ik} \tag{1.33}$$

where

$$\bar{A}_i^{\mathsf{T}} = (I_{N_{i,j}} \otimes A^{\mathsf{T}}) \in R^{nN_{i,j} \times nN_{i,j}}, \quad \bar{Q}_i = \mathrm{diag}\{0, \ldots, Q_{ii}, \ldots, 0\} \in R^{nN_{i,j} \times nN_{i,j}}.$$

The optimal control policy is given by applying the stationarity condition [Lewis *et al.* (2012)] $\partial H_i / \partial u_{ik} = 0$, so that

$$u_{ik}^* = R_{ii}^{-1} \Big(\big[\cdots (g_i + d_i) \cdots - e_{ji} \cdots \big] \otimes B_i^{\mathsf{T}} \Big) \lambda_{i(k+1)}. \tag{1.34}$$

Or,

$$u_{ik}^* = M_i \lambda_{i(k+1)} \tag{1.35}$$

where $M_i = R_{ii}^{-1} \Big(\big[\cdots (g_i + d_i) \cdots - e_{ji} \cdots \big] \otimes B_i^{\mathsf{T}} \Big)$.

Substituting the optimal policy (1.34) for each agent i into

$$H_i\big(\bar{\varepsilon}_{ik}, \lambda_{i(k+1)}, u_{ik}, u_{-ik}\big) = 0$$

yields the graph game Discrete-Time Hamilton optimality equations

$$\begin{aligned} H_i\big(\bar{\varepsilon}_{ik}, \lambda_{i(k+1)}, u_{ik}^*, u_{-ik}^*\big) &= \lambda_{i(k+1)}^{\mathsf{T}} \bar{\varepsilon}_{i(k+1)} + \frac{1}{2} \Bigg(\varepsilon_{ik}^{\mathsf{T}} Q_{ii} \varepsilon_{ik} \\ &\quad + u_{ik}^{*\mathsf{T}} R_{ii} u_{ik}^* + \sum_{j \in N_i} u_{jk}^{*\mathsf{T}} R_{ij} u_{jk}^* \Bigg) = 0. \end{aligned} \tag{1.36}$$

1.3.5 *Bellman equation for dynamic graphical games*

Given fixed admissible policies of node i and its neighbors, the value function for each agent i is given as

$$V_i(\bar{\varepsilon}_{ik}) = \sum_{l=k}^{\infty} U_i\big(\varepsilon_{il}, u_{il}, u_{-il}\big) \tag{1.37}$$

where $\bar{\varepsilon}_{ik}$ is a vector of the agent's state ε_{ik} and the states of its neighbors ε_{-ik}.

Taking the first difference of (1.37) yields the graphical game Bellman equation

$$V_i(\bar{\varepsilon}_{ik}) = \frac{1}{2}\left(\varepsilon_{ik}^{\mathsf{T}} Q_{ii}\varepsilon_{ik} + u_{ik}^{\mathsf{T}} R_{ii}u_{ik} + \sum_{j\in N_i} u_{jk}^{\mathsf{T}} R_{ij}u_{jk}\right) + V_i(\bar{\varepsilon}_{i(k+1)}) \tag{1.38}$$

with initial conditions $V_i(\mathbf{0}) = 0$. These equations depend on the graph structure.

Define the first difference of the value function $V_i(\bar{\varepsilon}_{ik})$ as

$$\Delta V_i(\bar{\varepsilon}_{ik}) = V_i(\bar{\varepsilon}_{i(k+1)}) - V_i(\bar{\varepsilon}_{ik}) \tag{1.39}$$

and its gradient as

$$\nabla V_i(\bar{\varepsilon}_{i(k+1)}) = \partial V_i(\bar{\varepsilon}_{i(k+1)})/\partial \bar{\varepsilon}_{i(k+1)}. \tag{1.40}$$

The relation between the first difference of the value function and its gradient is of prime interest in the upcoming development.

The objective of the graphical games optimization problem is to find for each i the optimal value

$$V_i^o(\bar{\varepsilon}_{ik}) = \min_{\tilde{u}_i}(V_i(\bar{\varepsilon}_{ik})) = \min_{\tilde{u}_i}\left(\sum_{l=k}^{\infty} U_i\left(\varepsilon_{il}, u_{il}, u_{-il}\right)\right).$$

According to the Bellman optimality principle

$$V_i^o(\bar{\varepsilon}_{ik}) = \min_{u_{ik}}\left(\frac{1}{2}\left(\varepsilon_{ik}^{\mathsf{T}} Q_{ii}\varepsilon_{ik} + u_{ik}^{\mathsf{T}} R_{ii}u_{ik} + \sum_{j\in N_i} u_{jk}^{\mathsf{T}} R_{ij}u_{jk}\right) + V_i^o(\bar{\varepsilon}_{i(k+1)})\right). \tag{1.41}$$

Consequently, the optimal control policy for each agent i is

$$u_{ik}^o = R_{ii}^{-1}\left(\left[\cdots(g_i+d_i)\cdots - e_{ji}\cdots\right]\otimes B_i^{\mathsf{T}}\right)\nabla V_i^o(\bar{\varepsilon}_{i(k+1)}). \tag{1.42}$$

Or,

$$u_{ik}^o = M_i \nabla V_i^o\left(\bar{\varepsilon}_{i(k+1)}\right). \tag{1.43}$$

Substituting (1.43) into (1.41) yields the graph game Bellman optimality equations

$$\begin{aligned} V_i^o(\bar{\varepsilon}_{ik}) = \frac{1}{2}\Big(&\varepsilon_{ik}^{\mathsf{T}} Q_{ii}\varepsilon_{ik} + \nabla V_i^o\left(\bar{\varepsilon}_{i(k+1)}\right)^{\mathsf{T}} M_i^{\mathsf{T}} R_{ii} M_i \nabla V_i^o\left(\bar{\varepsilon}_{i(k+1)}\right) \\ &+ \sum_{j\in N_i} \nabla V_j^o\left(\bar{\varepsilon}_{i(k+1)}\right)^{\mathsf{T}} M_j^{\mathsf{T}} R_{ij} M_j \nabla V_j^o\left(\bar{\varepsilon}_{i(k+1)}\right)\Big) \\ &+ V_i^o\left(\bar{\varepsilon}_{i(k+1)}\right) \end{aligned} \tag{1.44}$$

with initial conditions $V_i^o(\mathbf{0}) = 0$.

1.3.6 *Discrete Hamilton-Jacobi theory: Equivalence of Bellman and discrete-time Hamilton Jacobi equations*

The next result (c.f. [Lall and West (2006)]) relates the costate variable in the Hamiltonian function (1.22) to the value function $V_i(\bar{\varepsilon}_{ik})$ through the Bellman equation (1.38). It also introduces the Discrete-Time Hamilton Jacobi equation which relates the gradient of the value function $\nabla V_i(\bar{\varepsilon}_{i(k+1)})$ and its first difference $\Delta V_i(\bar{\varepsilon}_{ik}) = V_i(\bar{\varepsilon}_{i(k+1)}) - V_i(\bar{\varepsilon}_{ik})$.

Theorem 1.1. *(Discrete-Time Hamilton Jacobi Equation) Consider the Hamiltonian equations* (1.22) *and define the value functions* $V_i(\bar{\varepsilon}_{ik})$ *by* (1.37). *Then,* $V_i(\bar{\varepsilon}_{ik})$ *satisfies the discrete-time Hamilton Jacobi (DTHJ) equation*

$$\Delta V_i(\bar{\varepsilon}_{ik}) - \nabla V_i(\bar{\varepsilon}_{i(k+1)})^{\mathsf{T}} \bar{\varepsilon}_{i(k+1)} + H_i\Big(\bar{\varepsilon}_{ik}, \nabla V_i(\bar{\varepsilon}_{i(k+1)}), u_{ik}, u_{-ik}\Big) = 0. \quad (1.45)$$

Proof. The objective of the optimization problem is to minimize the performance index for each agent i (1.13), which is subject to the constraints

$$\bar{\varepsilon}_{i(k+1)} = \psi_i(\bar{\varepsilon}_{ik}, u_{ik}, u_{-ik,-\{-ik\}}). \quad (1.46)$$

The augmented value function is defined as follows

$$V_i(\bar{\varepsilon}_{ik}) = \sum_{l=k}^{\infty} \Big\{ U_i(\varepsilon_{il}, u_{il}, u_{-il}) + \lambda_{i(l+1)}^{\mathsf{T}} \left[\psi_i(\bar{\varepsilon}_{il}, u_{il}, u_{-il,-\{-il\}}) - \bar{\varepsilon}_{i(l+1)} \right] \Big\}. \quad (1.47)$$

The corresponding Hamiltonian function of agent i at time index l is,

$$H_i(\bar{\varepsilon}_{il}, \lambda_{i(l+1)}, u_{il}, u_{-il}) = U_i(\varepsilon_{il}, u_{il}, u_{-il}) + \lambda_{i(l+1)}^{\mathsf{T}} \psi_i(\bar{\varepsilon}_{il}, u_{il}, u_{-il,-\{-il\}}). \quad (1.48)$$

(1.47) and (1.48) yield

$$\Delta V_i(\bar{\varepsilon}_{ik}) + H_i(\bar{\varepsilon}_{ik}, \lambda_{i(k+1)}, u_{ik}, u_{-ik}) - \lambda_{i(k+1)}^{\mathsf{T}} \bar{\varepsilon}_{i(k+1)} = 0. \quad (1.49)$$

Take the derivative of (1.49) with respect to $\bar{\varepsilon}_{i(k+1)}$ as

$$\begin{aligned} &\nabla V_i(\bar{\varepsilon}_{i(k+1)}) + \Big(\partial(\lambda_{i(k+1)})/\partial\bar{\varepsilon}_{i(k+1)}\Big)^{\mathsf{T}} \Big(\partial H_i(\bar{\varepsilon}_{ik}, \lambda_{i(k+1)}, u_{ik}, u_{-ik})/\partial\lambda_{i(k+1)}\Big) \\ &\quad - \Big[\big(\partial(\lambda_{i(k+1)})/\partial\bar{\varepsilon}_{i(k+1)}\big)^{\mathsf{T}} \bar{\varepsilon}_{i(k+1)} + \lambda_{i(k+1)}\Big] = 0. \end{aligned}$$

Rearranging the last equation, yields

$$\begin{aligned} \nabla V_i(\bar{\varepsilon}_{i(k+1)}) &= \lambda_{i(k+1)} + \Big(\partial(\lambda_{i(k+1)})/\partial\bar{\varepsilon}_{i(k+1)}\Big)^{\mathsf{T}} \\ &\quad \times \Big(\bar{\varepsilon}_{i(k+1)} - \partial H_i(\bar{\varepsilon}_{ik}, \lambda_{i(k+1)}, u_{ik}, u_{-ik})/\partial\lambda_{i(k+1)}\Big). \end{aligned} \quad (1.50)$$

(1.50) and $\bar{\varepsilon}_{i(k+1)} = \partial H_i(\bar{\varepsilon}_{ik}, \lambda_{i(k+1)}, u_{ik}, u_{-ik})/\partial\lambda_{i(k+1)}$ from (1.22) yields $\lambda_{i(k+1)} = \nabla V_i(\bar{\varepsilon}_{i(k+1)})$. Substituting this into (1.49) yields (1.45). □

This proof provides motivation for henceforth defining the costate in terms of the value function as

$$\lambda_{i(k+1)} = \nabla V_i\big(\bar{\varepsilon}_{i(k+1)}\big). \tag{1.51}$$

Using (1.51) in (1.36) yields the Discrete-Time HJB equation

$$H_i\Big(\bar{\varepsilon}_{ik}, \nabla V_i\big(\bar{\varepsilon}_{i(k+1)}\big), u_{ik}^*, u_{-ik}^*\Big) = \nabla V_i\big(\bar{\varepsilon}_{i(k+1)}\big)^{\mathsf{T}} \bar{\varepsilon}_{i(k+1)} + \frac{1}{2}\Bigg(\varepsilon_{ik}^{\mathsf{T}} Q_{ii} \varepsilon_{ik} + u_{ik}^{*\mathsf{T}} R_{ii} u_{ik}^* + \sum_{j \in N_i} u_{jk}^{*\mathsf{T}} R_{ij} u_{jk}^*\Bigg) = 0 \tag{1.52}$$

where $u_{ik}^* = R_{ii}^{-1}\big(\big[\cdots (g_i + d_i) \cdots - e_{ji} \cdots\big] \otimes B_i^{\mathsf{T}}\big) \nabla V_i\big(\bar{\varepsilon}_{i(k+1)}\big)$.

The next result shows the relation between the optimal control policy (1.34) based on the Hamiltonian equation (1.22) and the optimal control policy (1.42) based on Bellman equation (1.38). It also relates the Hamiltonian equation (1.22) along the optimal trajectories and Bellman optimality equation (1.44).

Theorem 1.2. *(Discrete-Time HJB Equation)*

(*a*) *Let* $0 < V_i^*\big(\bar{\varepsilon}_{ik}\big) \in C^2$, $\forall i$ *satisfy the discrete-time Hamilton Jacobi Bellman (DTHJB) equation*

$$\begin{aligned} H_i\Big(\bar{\varepsilon}_{ik}, \nabla V_i^*\big(\bar{\varepsilon}_{i(k+1)}\big), u_{ik}^*, u_{-ik}^*\Big) &= \nabla V_i^*\big(\bar{\varepsilon}_{i(k+1)}\big)^{\mathsf{T}} \bar{\varepsilon}_{i(k+1)} + \frac{1}{2}\Bigg(\varepsilon_{ik}^{\mathsf{T}} Q_{ii} \varepsilon_{ik} \\ &\quad + u_{ik}^{*\mathsf{T}} R_{ii} u_{ik}^* + \sum_{j \in N_i} u_{jk}^{*\mathsf{T}} R_{ij} u_{jk}^*\Bigg) \\ &= 0 \end{aligned} \tag{1.53}$$

with initial condition $V_i^*(\mathbf{0}) = 0$, *where*

$$u_{ik}^* = R_{ii}^{-1}\Big(\big[\cdots (g_i + d_i) \cdots - e_{ji} \cdots\big] \otimes B_i^{\mathsf{T}}\Big) \nabla V_i^*\big(\bar{\varepsilon}_{i(k+1)}\big). \tag{1.54}$$

Then, $V_i^*\big(\bar{\varepsilon}_{ik}\big)$ *satisfies the Bellman optimality equation* (1.44)

(*b*) *Let* (A, B_i) $\forall i$ *be reachable. Let* $0 < V_i^*\big(\bar{\varepsilon}_{ik}\big) \in C^2$ $\forall i$ *satisfy* (1.44). *Then* $V_i^*\big(\bar{\varepsilon}_{ik}\big)$ *satisfies* (1.53).

Proof. (a) If $V_i^*\big(\bar{\varepsilon}_{ik}\big)$ satisfies (1.53) and u_{ik}^* is given by (1.54), then

$$H_i\Big(\bar{\varepsilon}_{ik}, \nabla V_i^*\big(\bar{\varepsilon}_{i(k+1)}\big), u_{ik}^*, u_{-ik}^*\Big) = 0.$$

Then, by Theorem 1.1 $\Delta V_i^*\big(\bar{\varepsilon}_{ik}\big) = \nabla V_i^*\big(\bar{\varepsilon}_{i(k+1)}\big)^{\mathsf{T}} \bar{\varepsilon}_{i(k+1)}$. Therefore, $V_i^*\big(\bar{\varepsilon}_{ik}\big)$ satisfies (1.44).

(b) Completing the squares on the Hamiltonian equation (1.22) for an arbitrary smooth function $V_i(\bar{\varepsilon}_{ik})$ yields

$$\begin{aligned} H_i\Big(\bar{\varepsilon}_{ik}, \nabla V_i(\bar{\varepsilon}_{i(k+1)}), u_{ik}, u_{-ik}\Big) &= H_i\Big(\bar{\varepsilon}_{ik}, \nabla V_i(\bar{\varepsilon}_{i(k+1)}), u_{ik}^*, u_{-ik}^*\Big) \\ &\quad + \frac{1}{2}(u_{ik} - u_{ik}^*)^\mathsf{T} R_{ii}(u_{ik} - u_{ik}^*) \\ &\quad + \frac{1}{2}\sum_{j\in N_i}(u_{jk} - u_{jk}^*)^\mathsf{T} R_{ij}(u_{jk} - u_{jk}^*) \\ &\quad + \sum_{j\in N_i} u_{jk}^{*\mathsf{T}} R_{ij}(u_{jk} - u_{jk}^*) + \nabla_i V_i^\mathsf{T}(\bar{\varepsilon}_{i(k+1)}) \\ &\quad \times \sum_{j\in N_i} e_{ij} B_j (u_{jk} - u_{jk}^*) + \nabla_{-i} V_i^\mathsf{T}(\bar{\varepsilon}_{i(k+1)}) \\ &\quad \times \bar{g}_i \bar{B}_i(\bar{u}_{-ik} - \bar{u}_{-ik}^*) \end{aligned} \tag{1.55}$$

where

$$\bar{g}_i = \begin{bmatrix} -(g_j + d_j) \cdot e_{jc} \cdot e_{jc'} \cdot & e_{jj'} \\ \vdots \quad . \;\vdots\; . \;\vdots\; . & \vdots \\ e_{j'j} \quad \cdot e_{j'c} \cdot e_{j'c'} \cdot & -(g_{j'} + d_{j'}) \end{bmatrix} \otimes I_n,$$

$$\bar{u}_{-ik} = \left[u_j^\mathsf{T} \;.\; u_c^\mathsf{T} \;.\; u_{c'}^\mathsf{T} \;.\; u_{j'}^\mathsf{T}\right]^\mathsf{T} \qquad \bar{B}_i = \mathrm{diag}\{B_j, .B_c, .B_{c'}, ., B_{j'}\},$$

$\{j, j'\} \in N_i$, $c \in N_j, c' \in N_{j'}, \{c, c'\} \neq i$. $\nabla_i V_i(\bar{\varepsilon}_{i(k+1)}) = \partial V_i(\bar{\varepsilon}_{i(k+1)})/\partial\varepsilon_{i(k+1)}$, $\nabla_{-i} V_i(\bar{\varepsilon}_{i(k+1)}) = \partial V_i(\bar{\varepsilon}_{i(k+1)})/\partial\varepsilon_{-i(k+1)}$, and $u_{ik}^* = M_i \nabla V_i(\bar{\varepsilon}_{i(k+1)})$.

Now, let $V_i(\bar{\varepsilon}_{ik}) \in C^2$, $\forall i$ satisfy the Bellman equation (1.38). The Hamiltonian function at $V_i^*(\bar{\varepsilon}_{ik})$ for any arbitrary control policies is given by

$$\begin{aligned} H_i\Big(\bar{\varepsilon}_{ik}, \nabla V_i^*(\bar{\varepsilon}_{i(k+1)}), u_{ik}, u_{-ik}\Big) &= \Big(\nabla V_i^*(\bar{\varepsilon}_{i(k+1)})^\mathsf{T} \bar{\varepsilon}_{i(k+1)}\Big) + \frac{1}{2}\Big(\varepsilon_{ik}^\mathsf{T} Q_{ii}\varepsilon_{ik} \\ &\quad + u_{ik}^\mathsf{T} R_{ii} u_{ik} + \sum_{j\in N_i} u_{jk}^\mathsf{T} R_{ij} u_{jk}\Big) \\ &= \frac{1}{2}(u_{ik} - u_{ik}^*)^\mathsf{T} R_{ii}(u_{ik} - u_{ik}^*) \\ &\quad + \frac{1}{2}\sum_{j\in N_i}(u_{jk} - u_{jk}^*)^\mathsf{T} R_{ij}(u_{jk} - u_{jk}^*) \\ &\quad + \sum_{j\in N_i} u_{jk}^{*T} R_{ij}(u_{jk} - u_{jk}^*) \\ &\quad + \nabla_i V_i^\mathsf{T}(\bar{\varepsilon}_{i(k+1)}) \sum_{j\in N_i} e_{ij} B_j(u_{jk} - u_{jk}^*) \\ &\quad + \nabla_{-i} V_i^\mathsf{T}(\bar{\varepsilon}_{i(k+1)}) \bar{g}_i \bar{B}_i \\ &\quad \times (\bar{u}_{-ik} - \bar{u}_{-ik}^*). \end{aligned} \tag{1.56}$$

Add zero to the Bellman equation (1.38) to obtain

$$V_i(\bar{\varepsilon}_{ik}) = \frac{1}{2}\Big(\varepsilon_{ik}^{\mathsf{T}} Q_{ii}\varepsilon_{ik} + u_{ik}^{\mathsf{T}} R_{ii} u_{ik} + \sum_{j\in N_i} u_{jk}^{\mathsf{T}} R_{ij} u_{jk}\Big) \\ + \nabla V_i^*(\bar{\varepsilon}_{i(k+1)})^{\mathsf{T}} \bar{\varepsilon}_{i(k+1)} - \nabla V_i^*(\bar{\varepsilon}_{i(k+1)})^{\mathsf{T}} \bar{\varepsilon}_{i(k+1)} \\ + V_i(\bar{\varepsilon}_{i(k+1)}). \tag{1.57}$$

Introducing the Hamiltonian equation (1.56) into this equation yields

$$V_i(\bar{\varepsilon}_{ik}) = \frac{1}{2}(u_{ik} - u_{ik}^*)^{\mathsf{T}} R_{ii}(u_{ik} - u_{ik}^*) \\ + \frac{1}{2}\sum_{j\in N_i}(u_{jk} - u_{jk}^*)^{\mathsf{T}} R_{ij}(u_{jk} - u_{jk}^*) \\ + \sum_{j\in N_i} u_{jk}^{*\mathsf{T}} R_{ij}(u_{jk} - u_{jk}^*) + \nabla_i V_i^{\mathsf{T}}(\bar{\varepsilon}_{i(k+1)}) \\ \times \sum_{j\in N_i} e_{ij} B_j (u_{jk} - u_{jk}^*) + \nabla_{-i} V_i^{\mathsf{T}}(\bar{\varepsilon}_{i(k+1)}) \bar{g}_i \bar{B}_i \\ \times (\bar{u}_{-ik} - \bar{u}_{-ik}^*) - \nabla V_i^*(\bar{\varepsilon}_{i(k+1)})^{\mathsf{T}} \bar{\varepsilon}_{i(k+1)} + V_i(\bar{\varepsilon}_{i(k+1)}). \tag{1.58}$$

Applying Bellman's optimality principle, yields that $V_i^o(\bar{\varepsilon}_{ik})$ satisfies the following equation

$$V_i^o(\bar{\varepsilon}_{ik}) = \min_{u_{ik}} \Big(V_i^o(\bar{\varepsilon}_{i(k+1)}) - \nabla V_i^*(\bar{\varepsilon}_{i(k+1)})^{\mathsf{T}} \bar{\varepsilon}_{i(k+1)} \\ + \frac{1}{2}(u_{ik} - u_{ik}^*)^{\mathsf{T}} R_{ii}(u_{ik} - u_{ik}^*) + \sum_{j\in N_i} u_{jk}^{*\mathsf{T}} R_{ij}(u_{jk} - u_{jk}^*) \\ + \frac{1}{2}\sum_{j\in N_i}(u_{jk} - u_{jk}^*)^{\mathsf{T}} R_{ij}(u_{jk} - u_{jk}^*) \\ + \nabla_i V_i^{\mathsf{T}}(\bar{\varepsilon}_{i(k+1)}) \sum_{j\in N_i} e_{ij} B_j(u_{jk} - u_{jk}^*) \\ + \nabla_{-i} V_i^{\mathsf{T}}(\bar{\varepsilon}_{i(k+1)}) \bar{g}_i \bar{B}_i (\bar{u}_{-ik} - \bar{u}_{-ik}^*)\Big). \tag{1.59}$$

Applying the stationarity conditions $\partial V_i^o(\bar{\varepsilon}_{ik})/\partial u_{ik} = 0$ yields the control policy u_{ik}^o given by solving

$$-\big([\cdots (g_i + d_i) \cdots - e_{ji} \cdots] \otimes B_i^{\mathsf{T}}\big)\big(\nabla V_i^o(\bar{\varepsilon}_{i(k+1)}) - \nabla V_i^*(\bar{\varepsilon}_{i(k+1)})\big) \\ + R_{ii}(u_{ik}^o - u_{ik}^*) = 0.$$

Then,

$$u_{ik}^o - u_{ik}^* = R_{ii}^{-1}\big([\cdots (g_i + d_i) \cdots - e_{ji} \cdots] \otimes B_i^{\mathsf{T}}\big) \\ \times \big(\nabla V_i^o(\bar{\varepsilon}_{i(k+1)}) - \nabla V_i^*(\bar{\varepsilon}_{i(k+1)})\big). \tag{1.60}$$

The Hessians of the Hamiltonian equations (1.22) and Bellman equations (1.38) with respect to all control policies are positive definite values so that $\nabla^2_{u_{ik}}(H_i) = \partial^2 H_i/\partial u_{ik}^2 = R_{ii} > 0$ and $\nabla^2_{u_{ik}}(V_i) = \partial^2 V_i/\partial u_{ik}^2 = R_{ii} > 0$. Therefore, the optimal control policy is unique and $u_{ik}^* = u_{ik}^o, \forall k$.

Now, (1.60) and the costate equation (1.33) show that

$$R_{ii}^{-1}\Big([\cdots(g_i+d_i)\cdots - e_{ji}\cdots]\otimes B_i^{\mathsf{T}}\Big)(A^{\mathsf{T}})^p \\ \times\Big(\nabla V_i^o\big(\varepsilon_{i(k+1)}\big) - \nabla V_i^*\big(\varepsilon_{i(k+1)}\big)\Big) = 0,\ \forall i, \forall k,\ p = 0,\ldots,n-1. \tag{1.61}$$

The reachability matrix

$$\tilde{U}_i = [B_i\ AB_i\ A^2B_i\ \cdots\ A^{n-1}B_i] \tag{1.62}$$

under the hypothesis has full rank. Therefore, since $V_i^*(\mathbf{0}) = 0$ and $V_i^o(\mathbf{0}) = 0$ then,

$$V_i^*\left(\bar{\varepsilon}_{ik}\right) = V_i^o\left(\bar{\varepsilon}_{ik}\right),\ \forall k. \tag{1.63}$$

This completes the proof. □

Remark 1.3. The solution of the dynamic graphical game using optimal control theory depends on optimizing the performance index for each agent i (1.13). This performance index arises directly from the dynamics of each agent i (1.9). It is worth to note that the utility function (1.13) used to solve the optimal control problem using the Hamiltonian or Bellman structures must be the same [Basar and Olsder (1999); Lewis *et al.* (2012)]. The optimal control sequence using these structures depends on the choice of the intermediate costate variable $\lambda_{i(k+1)}$ or the gradient of the value function $V_i(\bar{\varepsilon}_{i(k+1)})$ respectively.

Remark 1.4. The results of Theorem 1.1 and Theorem 1.2 justify the dependence of the value function $V_i(\cdot)$ given by Bellman equation (1.38) on the constraints function $\bar{\varepsilon}_{i(k+1)}$ of the Hamiltonian problem (1.20). Theorem 1.1 showed that the costate variable $\lambda_{i(k+1)}$ is related to the gradient of the value function for each agent i $V_i(\cdot)$ so that $\lambda_{i(k+1)} = \partial V_i\big(\bar{\varepsilon}_{i(k+1)}\big)/\partial\bar{\varepsilon}_{i(k+1)}$. Moreover, Theorem 1.2 showed the equivalence between the Discrete-time HJB equation (1.53) and the Bellman optimality equation (1.44), which infers the relation between the intermediate solution costate variable $\lambda_{i(k+1)}$ and the gradient of the value function $\nabla V_i\big(\bar{\varepsilon}_{i(k+1)}\big)$ for each agent. As implied by Theorem 1.2, the choice of value function structure $V_i\big(\bar{\varepsilon}_{i(k+1)}\big)$ affects this equivalence. Similarly, the choice of the value function structure $V_i\big(\bar{\varepsilon}_{i(k+1)}\big)$ affects the equivalence between the optimal control policies (1.34) and (1.42) given by the stationarity conditions applied to the Hamiltonian and Bellman equations respectively. The dynamics of the intermediate solution costate variable $\lambda_{i(k+1)}$ and the gradient of the value function $\nabla V_i\big(\bar{\varepsilon}_{i(k+1)}\big)$ are used to obtain the optimal control policies (1.34) and (1.42).

1.3.7 *Stability and Nash solution of the graphical games*

It is now proven that the solutions to the coupled Bellman optimality equations (1.44) provide a stable Nash solution for dynamic graphical games. Each agent has optimal control policies given by (1.42).

Theorem 1.3. *(Stability and Solution for Cooperative Nash Equilibrium)*

(*a*) *Let* $0 < V_i^*(\bar{\varepsilon}_{ik}) \in C^2$ *satisfy DTHJB* (1.53), *or equivalently the Bellman optimality equation*(1.44). *Let all agents use control policies*(1.54). *Let the graph contain a spanning tree with the pinning gain into at least one root node nonzero. Then, the error dynamics* (1.9) *are asymptotically stable, and all agents synchronize to the leader node dynamics* (1.2). *Moreover,* $V_i^*(\bar{\varepsilon}_{ik})$ *is a Lyapunov function for* (1.9).
(*b*) $J_i^*(u_{ik}^*, u_{\bar{i}k}^*) = V_i^*(\bar{\varepsilon}_{ik})$.
(*c*) *All agents are in Nash equilibrium.*

Proof. (a) $V_i^*(\bar{\varepsilon}_{ik})$ satisfies the Bellman optimality equation so that

$$V_i^*(\bar{\varepsilon}_{i(k+1)}) - V_i^*(\bar{\varepsilon}_{ik}) = -U_i^*\left(\varepsilon_{ik}, u_{ik}^*, u_{-ik}^*\right) < 0. \tag{1.64}$$

Therefore, V_i^* serves as a Lyapunov function, and the error system is asymptotically stable. If there is a spanning tree, according to Lemma 1.1, all agents synchronize to the leaders node dynamics.

(b) Using Theorem 1.2 and DTHJB (1.53), then the Hamiltonian equation (1.55) for arbitrary control policies is given by

$$\begin{aligned} H_i\Big(\bar{\varepsilon}_{ik}, \nabla V_i^*\left(\bar{\varepsilon}_{i(k+1)}\right), u_{ik}, u_{-ik}\Big) &= \nabla V_i^*\left(\bar{\varepsilon}_{i(k+1)}\right)^{\mathsf{T}} \bar{\varepsilon}_{i(k+1)} + U_i\left(\varepsilon_{ik}, u_{ik}, u_{jk}\right) \\ &= \frac{1}{2}(u_{ik} - u_{ik}^*)^{\mathsf{T}} R_{ii}(u_{ik} - u_{ik}^*) \\ &\quad + \frac{1}{2}\sum_{j\in N_i}(u_{jk} - u_{jk}^*)^{\mathsf{T}} R_{ij}(u_{jk} - u_{jk}^*) \\ &\quad + \sum_{j\in N_i} u_{jk}^{*\mathsf{T}} R_{ij}(u_{jk} - u_{jk}^*) \\ &\quad + \nabla V_i^{\mathsf{T}}\left(\bar{\varepsilon}_{i(k+1)}\right) \sum_{j\in N_i} e_{ij} B_j (u_{jk} - u_{jk}^*) \\ &\quad + \nabla_{-i} V_i^{\mathsf{T}}\left(\bar{\varepsilon}_{i(k+1)}\right) \bar{g}_i \\ &\quad \times \bar{B}_i(\bar{u}_{-ik} - \bar{u}_{-ik}^*). \end{aligned} \tag{1.65}$$

Using the result in part a yields $V_i^*(\bar{\varepsilon}_i(\infty)) = 0$ and

$$J_i(u_{il}, u_{-il}) = V_i^*\left(\bar{\varepsilon}_i(\infty)\right) + \sum_{k=l}^{\infty} U_i\left(\varepsilon_{ik}, u_{ik}, u_{-ik}\right) \tag{1.66}$$

by adding zero,

$$J_i(u_{il}, u_{-il}) = V_{il}^*(\bar{\varepsilon}_{il}) + \sum_{k=l}^{\infty} \Big(U_i(\varepsilon_{ik}, u_{ik}, u_{-ik}) - U_i^*(\varepsilon_{ik}, u_{ik}^*, u_{-ik}^*) \Big). \tag{1.67}$$

The Hamiltonian function for arbitrary control inputs is given by

$$\begin{aligned} H_i\Big(\bar{\varepsilon}_{ik}, \nabla V_i^*\big(\bar{\varepsilon}_{i(k+1)}\big), u_{ik}, u_{-ik}\Big) &= \nabla V_i^*\big(\bar{\varepsilon}_{i(k+1)}\big)^{\mathsf{T}} \bar{\varepsilon}_{i(k+1)}\big|_{u_{ik}, u_{-ik}} \\ &\quad + U_i\big(\varepsilon_{ik}, u_{ik}, u_{-ik}\big). \end{aligned} \tag{1.68}$$

The Hamiltonian function for optimal control inputs is given by

$$\begin{aligned} H_i\Big(\bar{\varepsilon}_{ik}, \nabla V_i^*\big(\bar{\varepsilon}_{i(k+1)}\big), u_{ik}^*, u_{-ik}^*\Big) &= \nabla V_i^*\big(\bar{\varepsilon}_{i(k+1)}\big)^{\mathsf{T}} \bar{\varepsilon}_{i(k+1)}\big|_{u_{ik}^*, u_{-ik}^*} \\ &\quad + U_i^*\big(\varepsilon_{ik}, u_{ik}^*, u_{-ik}^*\big). \end{aligned} \tag{1.69}$$

Then,

$$\begin{aligned} U_i\big(\varepsilon_{ik}, u_{ik}, u_{-ik}\big) - U_i^*\big(\varepsilon_{ik}, u_{ik}^*, u_{-ik}^*\big) &\\ &= H_i\Big(\bar{\varepsilon}_{ik}, \nabla V_i^*\big(\bar{\varepsilon}_{i(k+1)}\big), u_{ik}, u_{-ik}\Big) \\ &\quad - \nabla V_i^*\big(\bar{\varepsilon}_{i(k+1)}\big)^{\mathsf{T}} \bar{\varepsilon}_{i(k+1)}\big|_{u_{ik}, u_{-ik}} \\ &\quad + \nabla V_i^*\big(\bar{\varepsilon}_{i(k+1)}\big)^{\mathsf{T}} \bar{\varepsilon}_{i(k+1)}\big|_{u_{ik}^*, u_{-ik}^*}. \end{aligned} \tag{1.70}$$

Then, the performance index is given by

$$\begin{aligned} J_i(u_{il}, u_{-il}) = V_i^*\big(\bar{\varepsilon}_{il}\big) + \sum_{k=l}^{\infty} \bigg(&\frac{1}{2}(u_{ik} - u_{ik}^*)^{\mathsf{T}} R_{ii}(u_{ik} - u_{ik}^*) \\ &+ \frac{1}{2} \sum_{j \in N_i} (u_{jk} - u_{jk}^*)^{\mathsf{T}} R_{ij}(u_{jk} - u_{jk}^*) \\ &+ u_{ik}^{*\mathsf{T}} R_{ii}(u_{ik} - u_{ik}^*) + \sum_{j \in N_i} u_{jk}^{*\mathsf{T}} R_{ij}(u_{jk} - u_{jk}^*) \bigg). \end{aligned} \tag{1.71}$$

From (1.71) at optimal control policies (equilibrium), the performance index is given by the unique value $V_i^*\big(\bar{\varepsilon}_{il}\big)$

$$J_i(u_{il}^*, u_{-il}^*) = V_i^*\big(\bar{\varepsilon}_{il}\big). \tag{1.72}$$

(c) Given that the summation of the performance index is always positive for arbitrary control policies so that

$$\sum_{k=l}^{\infty} U_i\big(\varepsilon_{ik}, u_{ik}, u_{-ik}^*\big) - U_i^*\big(\varepsilon_{ik}, u_{ik}^*, u_{-ik}^*\big) > 0. \tag{1.73}$$

Then, according to (1.73) the argument of the performance index (1.71) is positive for arbitrary control policy. So, it is now straightforward that

$$J_{il}(u_{il}^*, u_{\bar{il}}^*) \leq J_{il}(u_{il}, u_{\bar{il}}^*). \tag{1.74}$$

So that Nash equilibrium exists according to Definition 1.3. □

1.4 Approximate Dynamic Programming for Graphical Games

VI is a form of RL [Bertsekas and Tsitsiklis (1996); Sutton and Barto (1998)]. It will be used here to obtain solution methods for multi-player cooperative graphical games. In this section, VI algorithms are proposed to solve DTHJB equation (1.53). Two algorithms are given that generalize HDP and DHP [Werbos (1992)] to the case of graph games. VI convergence is proven. The standard derivation of DHP from the HDP algorithm used for the single agent case does not work for graph games because it omits the interactions due to the neighbors.

1.4.1 *Heuristic dynamic programming for graphical games*

HDP was developed by Werbos (1992) for the single-agent case. Here, HDP is extended to the case of multi-player graph games. Under the hypotheses of Theorem 1.3 and Lemma 1.1, the next algorithm extends HDP or VI to graphical games. It is based on coupled Bellman equation (1.38).

Algorithm 1 (HDP for Graphical Games)

Step 1: Start with arbitrary initial policies u_{ik}^0 and values $\tilde{V}_i^0(\bar{\varepsilon}_{ik})$.
Step 2: (Value Update) Solve for $\tilde{V}_i^{l+1}$ using Bellman equations

$$\tilde{V}_i^{l+1}(\bar{\varepsilon}_{ik}) = \frac{1}{2}\Big(\varepsilon_{ik}^{\mathsf{T}} Q_{ii}\varepsilon_{ik} + u_{ik}^{l\mathsf{T}} R_{ii} u_{ik}^l + \sum_{j\in N_i} u_{jk}^{l\mathsf{T}} R_{ij} u_{jk}^l\Big) + \tilde{V}_i^l(\bar{\varepsilon}_{i(k+1)}) \tag{1.75}$$

where l is the iteration index.
Step 3: (Policy Improvement) Update the control policies using

$$u_{ik}^{l+1} = R_{ii}^{-1}\Big(\big[\cdots(g_i+d_i)\cdots - e_{ji}\cdots\big] \otimes B_i^{\mathsf{T}}\Big)\nabla\tilde{V}_i\big(\bar{\varepsilon}_{i(k+1)}\big)^{l+1}. \tag{1.76}$$

Step 4: Iterate steps 2 and 3 until convergence is verified, in the sense that $\left\|\tilde{V}_i(\bar{\varepsilon}_{ik})^{l+1} - \tilde{V}_i(\bar{\varepsilon}_{ik})^{l}\right\|$ is small. End.

Remark 1.5. The appearance of the edge weights e_{ji} coming out of node i in (1.76) means that this Algorithm can only be implemented using distributed local neighborhood information on undirected graphs.

Theorem 1.4 'to follow' provides the convergence proof for Algorithm 1 when all agents update their policies at each iteration. The following notations and technical lemmas are needed.

The optimal control policy sequence $\{L_i^l\}_{l=0}^{\infty} \in R^{m_i}$ for each agent i at iteration l in Algorithm 1 is given by

$$L_i^l = \arg\min_{u_{ik}} \left(\frac{1}{2}\left(\varepsilon_{ik}^{\mathsf{T}} Q_{ii}\varepsilon_{ik} + u_{ik}^{\mathsf{T}} R_{ii} u_{ik} + \sum_{j\in N_i} u_{jk}^{\mathsf{T}} R_{ij} u_{jk}\right) + V_i^l\left(\bar{\varepsilon}_{i(k+1)}\right)\right) \tag{1.77}$$

and the associated value function sequence is given by

$$V_i^{l+1}\left(\bar{\varepsilon}_{ik}\right) = \frac{1}{2}\left(\varepsilon_{ik}^{\mathsf{T}} Q_{ii}\varepsilon_{ik} + L_i^{l\mathsf{T}} R_{ii} L_i^l + \sum_{j\in N_i} L_j^{l\mathsf{T}} R_{ij} L_j^l\right) + V_i^l\left(\bar{\varepsilon}_{i(k+1)}\right). \tag{1.78}$$

Consider arbitrary admissible policies for the agents $\{M_i^l\}_{l=0}^{\infty} \in R^{m_i}$, with associated value function sequences

$$Z_i^{l+1}\left(\bar{\varepsilon}_{ik}\right) = \frac{1}{2}\left(\varepsilon_{ik}^{\mathsf{T}} Q_{ii}\varepsilon_{ik} + M_i^{l\mathsf{T}} R_{ii} M_i^l + \sum_{j\in N_i} M_j^{l\mathsf{T}} R_{ij} M_j^l\right) + Z_i^l\left(\bar{\varepsilon}_{i(k+1)}\right). \tag{1.79}$$

The next two technical lemmas are required for the proof of Theorem 1.4. They are motivated by Al-Tamimi *et al.* (2008) and Lancaster and Rodman (1995).

Lemma 1.2. *Given arbitrary stabilizing control policies $\{M_i^l\}_{l=0}^{\infty} \in R^{m_i}$ and $\{M_j^l\}_{l=0}^{\infty} \in R^{m_j}$ for agent i and its neighbor j, let the associated value sequence be $\{Z_i^l\}_{l=0}^{\infty} \in R^1, Z_i^0 \geq 0$. Define the sequences of control policies generated by Algorithm 1 agent i and its neighbor j as $\{L_i^l\}_{l=0}^{\infty} \in R^{m_i}$ and $\{L_j^l\}_{l=0}^{\infty} \in R^{m_j}$ respectively, with associated value sequences $\{V_i^l\}_{l=0}^{\infty} \in R^1$. Suppose that $\bar{\sigma}(R_{jj}^{-1}R_{ij})$ is small. Then, starting with $0 \leq V_i^0 \leq Z_i^0$, one has*

$$0 \leq V_i^l \leq Z_i^l. \tag{1.80}$$

Proof. The value function sequences for each agent i are given by (1.78) and (1.79), the arbitrary stabilizing control sequences for each agent i and its neighbor j are given by $M_i^l = \left(L_i^l + (M_i^l - L_i^l)\right)$ and $M_j^l = \left(L_j^l + (M_j^l - L_j^l)\right)$, respectively. The arbitrary value sequence $\{Z_i^l\}_{l=0}^{\infty}$ for each agent i is given as

$$\begin{aligned}
Z_i^{l+1}(M_i^l, M_j^l) &\equiv F_i(Z_i^l, M_i^l, M_j^l) \\
&= Z_i^l(M_i^l, M_j^l) + \frac{1}{2}\Big(\varepsilon_i^{\mathsf{T}} Q_{ii}\varepsilon_i + \left(L_i^l + (M_i^l - L_i^l)\right)^{\mathsf{T}} \\
&\quad \times R_{ii}\left(L_i^l + (M_i^l - L_i^l)\right) + \sum_{j\in N_i}\left(L_j^l + (M_j^l - L_j^l)\right)^{\mathsf{T}} \\
&\quad \times R_{ij}\left(L_j^l + (M_j^l - L_j^l)\right)\Big).
\end{aligned}$$

Rearranging this equation yields

$$\begin{aligned}Z_i^{l+1}(M_i^l, M_j^l) \equiv{}& F_i(Z_i^l, L_i^l, L_j^l) + \frac{1}{2}\sum_{j\in N_i}(M_j^l - L_j^l)^\top R_{ij}(M_j^l - L_j^l)\\&+ \frac{1}{2}(M_i^l - L_i^l)^\top R_{ii}(M_i^l - L_i^l) + Z_i^l(M_i^l, M_j^l)\\&+ M_i^{l\top} R_{ii} L_i^l + \sum_{j\in N_i} M_j^{l\top} R_{ij} L_j^l - \Big(Z_i^l(L_i^l, L_j^l)\\&+ L_i^{l\top} R_{ii} L_i^l + \sum_{j\in N_i} L_j^{l\top} R_{ij} L_j^l\Big)\end{aligned} \tag{1.81}$$

where

$$F_i(Z_i^l, L_i^l, L_j^l) = Z_i^l(L_i^l, L_j^l) + \frac{1}{2}\Big(\varepsilon_i^\top Q_{ii}\varepsilon_i + L_i^{l\top} R_{ii} L_i + \sum_{j\in N_i} L_j^{l\top} R_{ij} L_j^l\Big).$$

(1.81) yields

$$\begin{aligned}Z_i^{l+1}(M_i^l, M_j^l) \equiv{}& F_i(Z_i^l, L_i^l, L_j^l) + \frac{1}{2}\sum_{j\in N_i}(M_j^l - L_j^l)^\top R_{ij}(M_j^l - L_j^l)\\&+ \frac{1}{2}(M_i^l - L_i^l)^\top R_{ii}(M_i^l - L_i^l) + \sum_{j\in N_i}(M_j^l - L_j^l)^\top R_{ij} L_j^l\\&+ Z_i^l(M_i^l, M_j^l) + M_i^{l\top} R_{ii} L_i^l - \Big(Z_i^l(L_i^l, L_j^l) + L_i^{l\top} R_{ii} L_i^l\Big).\end{aligned} \tag{1.82}$$

Since

$$\begin{aligned}\sum_{k=0}^{\infty}\Big(& Z_i^l(M_i^l, M_j^l) + M_i^{l\top} R_{ii} L_i^l - (Z_i^l(L_i^l, L_j^l) + L_i^{l\top} R_{ii} L_i^l)\\&+ \frac{1}{2}(M_i^l - L_i^l)^\top R_{ii}(M_i^l - L_i^l)\Big) > 0\end{aligned} \tag{1.83}$$

and

$$\frac{1}{2}\sum_{j\in N_i}(M_j^l - L_j^l)^\top R_{ij}(M_j^l - L_j^l) + \sum_{j\in N_i}(M_j^l - L_j^l)^\top R_{ij} L_j^l > 0.$$

Under the condition

$$\frac{1}{2}(M_j^l - L_j^l)^\top R_{ij}(M_j^l - L_j^l) > (L_j^l - M_j^l)^\top R_{ij} L_j^l, \ \forall j. \tag{1.84}$$

Considering the optimal policy given by

$$L_j^l = R_{jj}^{-1}\Big([\cdots (g_j + d_j)\cdots - e_{rj}\cdots] \otimes E_j^\top\Big)\nabla V_j(\bar{\varepsilon}_{j(k+1)})^l, \ r \in N_j.$$

Then inequality (1.84) yields

$$\sum_{j\in N_i} \frac{1}{2}\underline{\sigma}(R_{ij})\,\|\Delta E_{jk}\| > \sum_{j\in N_i} \bar{\sigma}\big(R_{jj}^{-1}R_{ij}\big)\Big((g_j+d_j)\,\big\|\nabla_j V_j^l(\bar{\varepsilon}_{j(k+1)})\big\| + \sum_{r\in N_j}\Big(e_{rj}\,\big\|\nabla_r V_j^l(\bar{\varepsilon}_{j(k+1)})\big\|\Big)\Big)\,\|B_j\|$$

where $\|\Delta E_{jk}\| = \big\|L_j^l - M_j^l\big\|$.

(1.83) and under the assumption that $\bar{\sigma}\big(R_{jj}^{-1}R_{ij}\big)$ is small, then (1.81) yields

$$Z_i^{l+1}\big(M_i^l, M_j^l\big) = F_i\big(Z_i^l, M_i^l, M_j^l\big) \ge \hat{Z}_i^{l+1} = F_i\big(Z_i^l, L_i^l, L_j^l\big). \tag{1.85}$$

Similarly,

$$V_i^{l+1}\big(M_i^l, M_j^l\big) = F_i\big(V_i^l, M_i^l, M_j^l\big) \ge V_i^{l+1}\big(L_i^l, L_j^l\big) = F_i\big(V_i^l, L_i^l, L_j^l\big). \tag{1.86}$$

Using the initial sequence $0 \le V_i^0\big(M_i^l, M_j^l\big) \le Z_i^0\big(M_i^l, M_j^l\big)$, then by induction $0 \le V_i^l\big(M_i^l, M_j^l\big) \le Z_i^l\big(M_i^l, M_j^l\big)$. Using (1.85) which gives the lower bound on the arbitrary value sequence $Z_i^{l+1}\big(M_i^l, M_j^l\big)$ yields

$$\begin{aligned} 0 \le V_i^{l+1}(M_i^l) &= F_i\big(V_i^l, M_i^l\big) \\ &\le \hat{Z}_i^{l+1} = F_i\big(Z_i^l, L_i^l\big) \\ &\le Z_i^{l+1}\big(M_i^l\big) = F_i\big(Z_i^l, M_i^l\big). \end{aligned} \tag{1.87}$$

Then,

$$0 \le V_i^l \le \hat{Z}_i^l \le Z_i^l. \tag{1.88}$$

This completes the proof. □

Lemma 1.3. *Define the sequences of control policies generated by Algorithm 1 for agent i and its neighbor j as $\{L_i^l\}_{l=0}^{\infty} \in R^{m_i}$ and $\{L_j^l\}_{l=0}^{\infty} \in R^{m_j}$ respectively, with associated value sequences $\{V_i^l\}_{l=0}^{\infty} \in R^1$. Suppose that $\bar{\sigma}(R_{jj}^{-1}R_{ij})$ is small. Then there exists a finite upper bound $\bar{\bar{U}}$ such that*

$$0 \le V_i^l \le \bar{\bar{U}}. \tag{1.89}$$

Proof. M_i and M_j are stabilizing control policies for each agent i and its neighbor j. Using (1.79) and the sequence $V_i^0 = Z_i^0 = 0$ yields

$$\begin{aligned} Z_i^{l+1}\big(M_i^l, M_j^l\big) - Z_i^l\big(M_i^l, M_j^l\big) &= F_i\big(Z_i^l, M_i^l, M_j^l\big) - F_i\big(Z_i^{l-1}, M_i^l, M_j^l\big) \\ &= Z_i^l\big(M_i^l, M_j^l\big) + \frac{1}{2}\Big(\varepsilon_i^{\mathsf{T}} Q_{ii}\varepsilon_i + M_i^{l\mathsf{T}} R_{ii} M_i^l \\ &\quad + \sum_{j\in N_i} M_j^{l\mathsf{T}} R_{ij} M_j^l\Big) - Z_i^{l-1}\big(M_i^l, M_j^l\big) \\ &\quad - \frac{1}{2}\Big(\varepsilon_i^{\mathsf{T}} Q_{ii}\varepsilon_i + M_i^{l\mathsf{T}} R_{ii} M_i^l + \sum_{j\in N_i} M_j^{l\mathsf{T}} R_{ij} M_j^l\Big). \end{aligned}$$

Rearranging this equation yields

$$\begin{aligned} Z_i^{l+1}(M_i^l, M_j^l, k) - Z_i^l(M_i^l, M_j^l, k) &= Z_i^l(M_i^l, M_j^l, k+1) - Z_i^{l-1}(M_i^l, M_j^l, k+1) \\ &= Z_i^{l-1}(M_i^l, M_j^l, k+2) - Z_i^{l-2}(M_i^l, M_j^l, k+2) \\ &\vdots \\ &= Z_i^1(M_i^l, M_j^l, k+l) - Z_i^0(M_i^l, M_j^l, k+l) \end{aligned} \tag{1.90}$$

with $Z_i^0(M_i^l, M_j^l, k+l) = 0$.

Rearranging (1.90) yields

$$\begin{aligned} Z_i^{l+1}(M_i^l, M_j^l, k) &= Z_i^1(M_i^l, M_j^l, k+l) + Z_i^l(M_i^l, M_j^l, k) \\ &= Z_i^1(M_i^l, M_j^l, k+l) + Z_i^1(M_i^l, M_j^l, k+l-1) \\ &\quad + Z_i^{l-1}(M_i^l, M_j^l, k) \\ &= Z_i^1(M_i^l, M_j^l, k+l) + Z_i^1(M_i^l, M_j^l, k+l-1) \\ &\quad + \cdots + Z_i^1(M_i^l, M_j^l, k). \end{aligned} \tag{1.91}$$

Then, (1.91) is written as

$$\begin{aligned} Z_i^{l+1}(M_i^l, M_j^l, k) &= \sum_{n=0}^{l} Z_i^1(M_i^l, M_j^l, k+n) \\ &= \sum_{n=0}^{l} \frac{1}{2}\Big(\varepsilon_i^{\mathsf{T}}(k+n) Q_{ii} \varepsilon_i(k+n) \\ &\quad + M_i^{l\mathsf{T}}(k+n) R_{ii} M_i^l(k+n) \\ &\quad + \sum_{j \in N_i} M_j^{l\mathsf{T}}(k+n) R_{ij} M_j^l(k+n)\Big). \end{aligned} \tag{1.92}$$

Since the used policies are stabilizable policies, then

$$Z_i^{l+1}(M_i^l, M_j^l, k) \leq \sum_{n=0}^{\infty} Z_i^1(M_i^l, M_j^l, k+n) = \bar{\bar{U}}. \tag{1.93}$$

This inequality satisfies (1.89). □

The next result proves convergence of Algorithm 1 when all agents update their policies simultaneously.

Theorem 1.4. *(Convergence of HDP Algorithm 1) Let all agents update their policies simultaneously using Algorithm 1. Suppose that $\bar{\sigma}(R_{jj}^{-1} R_{ij})$ is small. Then, the solution sequences V_i^l converge to the unique solution $V_i^*(\bar{\varepsilon}_{ik})$ $\forall i$ of (1.44).*

Proof. Using the optimal control policy (1.77), then the results from Lemma 1.2 and Lemma 1.3 yield

$$0 \leq V_i^l \leq Z_i^l \leq \bar{\bar{U}}. \tag{1.94}$$

Using the value function sequence (1.78) and (1.79) yields

$$V_i^{l+1}(L_i^l, L_j^l) = \frac{1}{2}\Big(\varepsilon_i^\top Q_{ii}\varepsilon_i + L_i^{l\top} R_{ii} L_i^l + \sum_{j\in N_i} L_j^{l\top} R_{ij} L_j^l\Big) + V_i^l(L_i^l, L_j^l) \quad (1.95)$$

$$Z_i^l(L_i^l, L_j^l) = \frac{1}{2}\Big(\varepsilon_i^\top Q_{ii}\varepsilon_i + L_i^{l\top} R_{ii} L_i^l + \sum_{j\in N_i} L_j^{l\top} R_{ij} L_j^l\Big) + Z_i^{l-1}(L_i^l, L_j^l). \quad (1.96)$$

After subtracting (1.95) and (1.96), one has

$$V_i^{l+1}(L_i^l, L_j^l) - Z_i^l(L_i^l, L_j^l) = V_i^l(L_i^l, L_j^l) - Z_i^{l-1}(L_i^l, L_j^l). \quad (1.97)$$

By setting $V_i^0 = Z_i^0 = 0$ and $V_i^1(L_i^1, L_j^1) = \varepsilon_{i0}^\top Q_{ii}\varepsilon_{i0} \geq 0$, in (1.95) and (1.96), one has,

$$V_i^1(L_i^1, L_j^1) - Z_i^0(L_i^1, L_j^1) \geq 0. \quad (1.98)$$

Similarly with previous developments we will make the following induction hypothesis $V_i^l(L_i^l, L_j^l) - Z_i^{l-1}(L_i^l, L_j^l) \geq 0$. Because of that hypothesis it is straightforward that (1.97) yields $V_i^{l+1}(L_i^l, L_j^l) - Z_i^l(L_i^l, L_j^l) \geq 0$.

Similarly, the value function sequences (1.95) and (1.96) with policy given by L_i^{l+1} yield

$$V_i^{l+1}(L_i^{l+1}, L_j^l) - Z_i^l(L_i^{l+1}, L_j^l) = V_i^l(L_i^{l+1}, L_j^l) - Z_i^{l-1}(L_i^{l+1}, L_j^l). \quad (1.99)$$

After setting $V_i^0 = Z_i^0 = 0$ and $V_i^1(L_i^2, L_j^1) = \varepsilon_{i0}^\top Q_{ii}\varepsilon_{i0} \geq 0$, then (1.95), (1.96), and (1.99) yield

$$V_i^1(L_i^2, L_j^1) - Z_i^0(L_i^2, L_j^1) \geq 0. \quad (1.100)$$

Now the induction hypothesis becomes $V_i^{l+1}(L_i^{l+1}, L_j^l) - Z_i^l(L_i^{l+1}, L_j^l) \geq 0$. Therefore, for all the policies $L_i^{l+1}, \forall l$ the following monotonic sequence holds

$$V_i^{l+1} > Z_i^l > V_i^l > \cdots \geq 0. \quad (1.101)$$

Since $Z_i^l(L_i^l, L_j^l)$ is the lower bound of $V_i^{l+1}(L_i^l, L_j^l)$ and Lemma 1.3 sets $\bar{\bar{U}}$ as an upper value of V_i^{l+1}, then (1.94) and (1.101) yield

$$0 \leq V_i^l \leq Z_i^l \leq V_i^{l+1} \leq \bar{\bar{U}}. \quad (1.102)$$

From (1.102), the sequence of V_i^l is increasing and it has an upper bound $\bar{\bar{U}}$, which means that V_i^l will converge to the optimal solution $V_i^*(\bar{\varepsilon}_{ik})$ monotonically by satisfying

$$V_i^*(\bar{\varepsilon}_{ik}) = \frac{1}{2}\Big(\varepsilon_{ik}^\top Q_{ii}\varepsilon_{ik} + L_i^{*\top} R_{ii} L_i^* + \sum_{j\in N_i} L_j^{*\top} R_{ij} L_j^*\Big) + V_i^*(\bar{\varepsilon}_{i(k+1)}) \quad (1.103)$$

where

$$L_i^* = R_{ii}^{-1}\Big(\big[\cdots (g_i + d_i)\cdots - e_{ji}\cdots\big] \otimes B_i^\top\Big)\nabla V_i^*(\bar{\varepsilon}_{i(k+1)}). \quad (1.104)$$

This completes the proof. □

1.4.2 Dual heuristic programming for graphical games

According to (1.37), the ith value function and its gradient depend on information about agent i and all of its neighbors. This leads to the appearance in HDP Algorithm 1 of the weights e_{ji} of edges leaving node i. Information about the out-neighbors of node i is not available to node i unless the graph is undirected. This is a restriction on HDP Algorithm 1. In this section a DHP Algorithm is developed for graphical games on directed graphs. In graphical games, the value function has a special structure that allows the formulation of DHP only in terms of information from the in-neighbors. This structure is given in (1.105) and used to define a local Hamiltonian (1.106) that only depends on the costate of node i. Theorem 1.5 will relate the local Hamiltonian (1.106) to the Bellman equation (1.38). It is noted that standard approaches to deriving DHP algorithms used for single agents do not extend to graphical games because they do not capture the neighbors information.

Write the ith value function as $V_i(\bar{\varepsilon}_{ik}) = \varepsilon_k^{\mathsf{T}} O_i S \bar{O}_i \varepsilon_k$, where $S \in R^{nN \times nN}$ is a global intermediate solution matrix. The existence of S is shown in Section 1.5. In the development here, this structure of the value function is used to develop the structure in (1.105). Left mapping matrix O_i selects block rows of the global matrix S corresponding to the value $V_i(\bar{\varepsilon}_{ik})$ of node i. Right mapping matrix $\bar{O}_i$ selects the states of the nodes that influence the value $V_i(\bar{\varepsilon}_{ik})$ of node i.

The left and right mappings can be selected to preserve the graph structure as follows. Let

$$V_i(\bar{\varepsilon}_{ik}) = \varepsilon_k^{\mathsf{T}} Z_i^{\mathsf{T}} Z_i S \bar{Z}_i \varepsilon_k \tag{1.105}$$

where $Z_i = [0 \cdots [I_n]_{ii} \cdots 0] \in R^{n \times nN}$ and $\bar{Z}_i = \operatorname{diag}\{a_{ij}\} \in R^{nN \times nN}$ where

$$a_{ij} = \begin{cases} I_n, & j \in \{i, N_i\}, \\ 0_n, & \text{else.} \end{cases}$$

In the simulation examples, this distributed structure is observed to suffice.

Given the value function (1.105), the node dynamics (1.9), and the performance indices (1.13), define the reduced Hamiltonian function [Lewis *et al.* (2012)] of each agent i on the trajectory of agent i in the local form

$$\begin{aligned}\tilde{H}_i(\varepsilon_{ik}, \tilde{\lambda}_{i(k+1)}, u_{ik}, u_{-ik}) = \tilde{\lambda}_{i(k+1)}^{\mathsf{T}} \varepsilon_{i(k+1)} + \frac{1}{2}\Big(\varepsilon_{ik}^{\mathsf{T}} Q_{ii} \varepsilon_{ik} \\ + u_{ik}^{\mathsf{T}} R_{ii} u_{ik} + \sum_{j \in N_i} u_{jk}^{\mathsf{T}} R_{ij} u_{jk}\Big)\end{aligned} \tag{1.106}$$

where $\tilde{\lambda}_{ik} \equiv \tilde{\lambda}_i(k)$ is the costate variable of each agent i. Then, the local costate equations are found using $\partial \tilde{H}_i / \partial \varepsilon_{ik} = \tilde{\lambda}_{ik}$ to be

$$\tilde{\lambda}_{ik} = A^{\mathsf{T}} \tilde{\lambda}_{i(k+1)} + Q_{ii} \varepsilon_{ik}. \tag{1.107}$$

The optimal control policies are found using $u_{ik}^{o} = \arg\min_{u_{ik}} \left(\tilde{H}_i\right)$ to be

$$\tilde{u}_{ik}^{*} = (g_i + d_i)R_{ii}^{-1}B_i^{\mathsf{T}}\tilde{\lambda}_{i(k+1)}. \tag{1.108}$$

The following theorem relates the costate variable in the Hamiltonian equation (1.106) to the value (1.105) through the Bellman equation (1.38). It also introduces the Discrete-Time HJB equation which relates the gradient of the value function $\nabla_i V_i\left(\bar{\varepsilon}_{i(k+1)}\right)$ and its first difference $\Delta V_i\left(\bar{\varepsilon}_{ik}\right)$.

Theorem 1.5. *(Discrete-Time Hamilton Jacobi Equation) Consider the Hamiltonian equations* (1.106) *and define the value functions* $V_i\left(\bar{\varepsilon}_{ik}\right)$ *by* (1.37)*, with the structure* (1.105)*. Then,* $V_i\left(\bar{\varepsilon}_{ik}\right)$ *satisfies the discrete-time Hamilton Jacobi (DTHJ) equation*

$$\begin{aligned}\Delta V_i\left(\bar{\varepsilon}_{ik}\right) &- Z_i\nabla V_i\left(\bar{\varepsilon}_{i(k+1)}\right)^{\mathsf{T}}\varepsilon_{i(k+1)}\\ &+ \tilde{H}_i\Big(\varepsilon_{ik}, Z_i\nabla V_i\left(\bar{\varepsilon}_{i(k+1)}\right), u_{ik}, u_{-ik}\Big) = 0.\end{aligned} \tag{1.109}$$

Proof. The objective of the optimization problem is to minimize the performance index for each agent i (1.13), which is subject to the constraints function

$$\varepsilon_{i(k+1)} = f_i\left(\varepsilon_{ik}, u_{ik}, u_{-ik}\right). \tag{1.110}$$

The augmented value function is defined as follows

$$V_i\left(\bar{\varepsilon}_{ik}\right) = \sum_{l=k}^{\infty}\left\{U_i\left(\varepsilon_{il}, u_{il}, u_{-il}\right) + \tilde{\lambda}_{i(l+1)}^{\mathsf{T}}\left[f_i\left(\varepsilon_{il}, u_{il}, u_{-il}\right) - \varepsilon_{i(l+1)}\right]\right\}. \tag{1.111}$$

The corresponding Hamiltonian function of agent i at time index l is,

$$\tilde{H}_i\left(\varepsilon_{il}, \lambda_{i(l+1)}, u_{il}, u_{-il}\right) = U_i\left(\varepsilon_{il}, u_{il}, u_{-il}\right) + \tilde{\lambda}_{i(l+1)}^{\mathsf{T}} f_i\left(\varepsilon_{il}, u_{il}, u_{-il}\right). \tag{1.112}$$

(1.111) and (1.112) yield

$$\Delta V_i\left(\bar{\varepsilon}_{ik}\right) + \tilde{H}_i\left(\varepsilon_{ik}, \tilde{\lambda}_{i(k+1)}, u_{ik}, u_{-ik}\right) - \tilde{\lambda}_{i(k+1)}^{\mathsf{T}}\varepsilon_{i(k+1)} = 0. \tag{1.113}$$

Take the derivative of (1.113) with respect to $\varepsilon_{i(k+1)}$ as

$$\begin{aligned}Z_i\nabla V_i\left(\bar{\varepsilon}_{i(k+1)}\right) &+ \left(\partial\left(\tilde{\lambda}_{i(k+1)}\right)/\partial\varepsilon_{i(k+1)}\right)^{\mathsf{T}}\left(\partial\tilde{H}_i\left(\varepsilon_{ik}, \tilde{\lambda}_{i(k+1)}, u_{ik}, u_{-ik}\right)/\partial\tilde{\lambda}_{i(k+1)}\right)\\ &- \left[\left(\partial\left(\tilde{\lambda}_{i(k+1)}\right)/\partial\varepsilon_{i(k+1)}\right)^{\mathsf{T}}\varepsilon_{i(k+1)} + \tilde{\lambda}_{i(k+1)}\right] = 0.\end{aligned}$$

Rearranging the last equation, yields

$$\begin{aligned}Z_i\nabla V_i\left(\bar{\varepsilon}_{i(k+1)}\right) &= \tilde{\lambda}_{i(k+1)} + \left(\partial\left(\tilde{\lambda}_{i(k+1)}\right)/\partial\varepsilon_{i(k+1)}\right)^{\mathsf{T}}\\ &\times\left(\varepsilon_{i(k+1)} - \left(\partial\tilde{H}_i\left(\varepsilon_{ik}, \tilde{\lambda}_{i(k+1)}, u_{ik}, u_{-ik}\right)/\partial\tilde{\lambda}_{i(k+1)}\right)\right).\end{aligned} \tag{1.114}$$

(1.114) and $\varepsilon_{i(k+1)} = \left(\partial\tilde{H}_i(\varepsilon_{ik}, \tilde{\lambda}_{i(k+1)}, u_{ik}, u_{-ik})/\partial\tilde{\lambda}_{i(k+1)}\right)$ from (1.106) yields $\tilde{\lambda}_{i(k+1)} = Z_i\nabla V_i(\bar{\varepsilon}_{i(k+1)}) = \nabla_i V_i(\bar{\varepsilon}_{i(k+1)})$. Substituting this into (1.113) yields (1.109). □

Remark 1.6. By assuming the value function structure (1.105), the Hamiltonian equation (1.22) is reduced to the local form (1.106) or

$$H_i\left(\varepsilon_{ik}, \nabla\tilde{V}_i(\varepsilon_{i(k+1)}), u_{ik}, u_{-ik}\right) = \nabla\tilde{V}_i(\varepsilon_{i(k+1)})^{\mathsf{T}}\varepsilon_{i(k+1)} + \frac{1}{2}\left(\varepsilon_{ik}^{\mathsf{T}}Q_{ii}\varepsilon_{ik} + u_{ik}^{\mathsf{T}}R_{ii}u_{ik} + \sum_{j\in N_i} u_{jk}^{\mathsf{T}}R_{ij}u_{jk}\right) = 0 \quad (1.115)$$

with reduced value function

$$\tilde{V}_i(\bar{\varepsilon}_{ik}) = \varepsilon_{ik}^{\mathsf{T}}Z_i\{Z_i^{\mathsf{T}}Z_iS\,\bar{Z}_i\}\varepsilon_k. \quad (1.116)$$

Remark 1.7. Assuming the value function with the structure (1.105) or (1.106) then the Bellman equation (1.38) is reduced to

$$\tilde{V}_i(\bar{\varepsilon}_{ik}) = \frac{1}{2}\left(\varepsilon_{ik}^{\mathsf{T}}Q_{ii}\varepsilon_{ik} + u_{ik}^{\mathsf{T}}R_{ii}u_{ik} + \sum_{j\in N_i} u_{jk}^{\mathsf{T}}R_{ij}u_{jk}\right) + \tilde{V}_i(\bar{\varepsilon}_{i(k+1)}). \quad (1.117)$$

Taking the gradient with respect to ε_{ik} of the both sides of the Bellman equation (1.117) yields the costate equation

$$\nabla\tilde{V}_i(\bar{\varepsilon}_{ik}) = A^{\mathsf{T}}\nabla\tilde{V}_i(\bar{\varepsilon}_{i(k+1)}) + Q_{ii}\varepsilon_{ik},\ \tilde{V}_i(\mathbf{0}) = 0. \quad (1.118)$$

Algorithm 1, HDP for graph games, is based on the Bellman equation (1.38) or equivalently (1.117). The DHP Algorithm is based on the costate (1.107). To derive a DHP Algorithm for graph games, for policies $u_{ik}, \forall i$, consider the Hamiltonian equation (1.115), with initial condition $\tilde{V}_i(\mathbf{0}) = 0$. The next results show the equivalence between the Bellman equation (1.38) or equivalently (1.117) and the Hamiltonian equation (1.115) for arbitrary admissible control policies.

Theorem 1.6. *(Discrete-Time Hamiltonian Equation) Let* $0 < \tilde{V}_i(\bar{\varepsilon}_{ik}) \in C^2$, $\forall i$ *satisfy the Hamiltonian equation* (1.115), *given arbitrary admissible control policies. Then* $\tilde{V}_i(\bar{\varepsilon}_{ik})$ *satisfies the Bellman equation* (1.38) *or equivalently* (1.117).

Proof. Let the Hamiltonian equation (1.115) hold. Then, Theorem 1.1, Theorem 1.5, Remark 1.6, and Remark 1.7 yield

$$\Delta\tilde{V}_i(\bar{\varepsilon}_{ik}) = \nabla\tilde{V}_i(\bar{\varepsilon}_{i(k+1)})^{\mathsf{T}}\bar{\varepsilon}_{i(k+1)}. \quad (1.119)$$

Therefore, $\tilde{V}_i(\bar{\varepsilon}_{ik})$ satisfies (1.38) or equivalently (1.117). □

Theorem 1.7. *(Discrete-Time Bellman Equation) Let* (A, B_i) , $\forall i$ *be reachable. Let* $0 < V_i^{'}(\bar{\varepsilon}_{ik}) \in C^2$, $\forall i$ *satisfy the discrete-time Bellman equation* (1.38) *or equivalently* (1.117), *given arbitrary admissible control policies. Then* $V_i^{'}(\bar{\varepsilon}_{ik})$ *satisfies the Hamiltonian equation* (1.115).

Proof. Applying the stationarity conditions on the Hamiltonian and the Bellman equations (1.115) and (1.38) or equivalently (1.117), yields the optimal control inputs $u_{ik}^o = (d_i + g_i)R_{ii}^{-1}B_i^{\mathsf{T}}\nabla\tilde{V}_i(\bar{\varepsilon}_{i(k+1)})$ and $u_{ik}^{'} = (d_i + g_i)R_{ii}^{-1}B_i^{\mathsf{T}}\nabla V_i^{'}(\bar{\varepsilon}_{i(k+1)})$ respectively [Lewis *et al.* (2012)].

The utility function $U_i(\varepsilon_{ik}, u_{ik}, u_{-ik})$ is the same for both the Hamiltonian equation (1.115) and the Bellman equation (1.38) or equivalently (1.117). Using Theorem 1.1 or Theorem 1.5 yields

$$H_i\Big(\varepsilon_{ik}, \nabla\tilde{V}_i(\bar{\varepsilon}_{i(k+1)}), u_{ik}, u_{-ik}\Big) = \nabla\tilde{V}_i(\bar{\varepsilon}_{i(k+1)})^{\mathsf{T}}\varepsilon_{i(k+1)} - V_i^{'}(\bar{\varepsilon}_{i(k+1)}) + V_i^{'}(\bar{\varepsilon}_{ik}). \tag{1.120}$$

Applying the stationarity conditions [Lewis *et al.* (2012)] on (1.120) yields

$$\partial H_i/\partial u_{ik} = 0 \Rightarrow \partial H_i/\partial u_{ik} = \big(\partial\varepsilon_{i(k+1)}/\partial u_{ik}\big)^{\mathsf{T}}\nabla\tilde{V}_i(\bar{\varepsilon}_{i(k+1)}) - \big(\partial\varepsilon_{i(k+1)}/\partial u_{ik}\big)^{\mathsf{T}}\nabla V_i^{'}(\bar{\varepsilon}_{i(k+1)}). \tag{1.121}$$

This equation yields

$$\frac{\partial H_i}{\partial u_{ik}} = 0, \;\; (d_i + g_i)B_i^{\mathsf{T}}\Big(\nabla\tilde{V}_i(\bar{\varepsilon}_{i(k+1)}) - \nabla V_i^{'}(\bar{\varepsilon}_{i(k+1)})\Big) = 0. \tag{1.122}$$

Then,

$$(d_i + g_i)B_i^{\mathsf{T}}(\nabla\tilde{V}_i(\bar{\varepsilon}_{ik}) - \nabla V_i^{'}(\bar{\varepsilon}_{ik})) = 0, \; \forall k. \tag{1.123}$$

Therefore, (1.123) and the costate equations (1.107), (1.118) show that

$$(g_i + d_i)R_{ii}^{-1}B_i^{\mathsf{T}}(A^{\mathsf{T}})^p\Big(\nabla\tilde{V}_i(\bar{\varepsilon}_{i(k+1)}) - \nabla V_i^{'}(\bar{\varepsilon}_{i(k+1)})\Big) = 0 \tag{1.124}$$

$\forall k, \; p = 0, \ldots, n-1.$

The reachability matrix

$$\tilde{U}_i = \big[B_i \; AB_i \; A^2B_i \; \cdots \; A^{n-1}B_i\big] \tag{1.125}$$

under the hypothesis has full rank. Given the initial conditions $V_i^{'}(\mathbf{0}) = 0$ and $\tilde{V}_i(\mathbf{0}) = 0$. Therefore,

$$V_i^{'}(\bar{\varepsilon}_{ik}) = \tilde{V}_i(\bar{\varepsilon}_{ik}), \; \forall k. \tag{1.126}$$

This completes the proof. □

A DHP Algorithm is derived for graph games that properly includes the neighbors' policies. Taking the gradient with respect to ε_{ik} of the both sides of the Hamiltonian equation (1.115) and using the definition of the costate (1.107) yields

$$\nabla\tilde{V}_i(\bar{\varepsilon}_{ik}) = A^{\mathsf{T}}\nabla\tilde{V}_i(\bar{\varepsilon}_{i(k+1)}) + Q_{ii}\varepsilon_{ik}, \;\; \tilde{V}_i(\mathbf{0}) = 0. \tag{1.127}$$

This is the costate equation (1.33) with costate $\lambda_{i(k+1)} = \nabla\tilde{V}_i(\bar{\varepsilon}_{i(k+1)})$.

The next result develops an alternative Bellman equation that allows the definition of a DHP Algorithm for graphical games.

Theorem 1.8. *(Bellman Equation for Costate-State Product)*

(*a*) *Let* $\tilde{V}_i(\bar{\varepsilon}_{ik})$ *satisfy* (1.115). *Then it also satisfies*

$$\begin{aligned}\varepsilon_{ik}^{\mathsf{T}}\nabla\tilde{V}_i(\bar{\varepsilon}_{ik}) &= (g_i+d_i)\nabla\tilde{V}_i(\bar{\varepsilon}_{i(k+1)})^{\mathsf{T}}B_iu_{ik}\\ &\quad-\sum_{j\in N_i}e_{ij}\nabla\tilde{V}_i(\bar{\varepsilon}_{i(k+1)})^{\mathsf{T}}B_ju_{jk}\\ &\quad+\frac{1}{2}\Big(\varepsilon_{ik}^{\mathsf{T}}Q_{ii}\varepsilon_{ik}-u_{ik}^{\mathsf{T}}R_{ii}u_{ik}-\sum_{j\in N_i}u_{jk}^{\mathsf{T}}R_{ij}u_{jk}\Big).\end{aligned} \tag{1.128}$$

(*b*) *Let* $\tilde{V}_i(\bar{\varepsilon}_{ik})$ *satisfy* (1.128). *Then, it also satisfies* (1.115).

Proof. (a) Multiply (1.127) by $\varepsilon_{ik}^{\mathsf{T}}$ so that

$$\varepsilon_{ik}^{\mathsf{T}}\nabla\tilde{V}_i(\bar{\varepsilon}_{ik})=\varepsilon_{ik}^{\mathsf{T}}Q_{ii}\varepsilon_{ik}+\varepsilon_{ik}^{\mathsf{T}}A^{\mathsf{T}}\nabla\tilde{V}_i(\bar{\varepsilon}_{i(k+1)}). \tag{1.129}$$

The Hamiltonian equation (1.115) and the error dynamics (1.9) yield

$$\begin{aligned}H_i\Big(\varepsilon_{ik},\nabla\tilde{V}_i(\bar{\varepsilon}_{i(k+1)}),u_{ik},u_{-ik}\Big) &= \nabla\tilde{V}_i(\bar{\varepsilon}_{i(k+1)})^{\mathsf{T}}A\varepsilon_{ik}-(d_i+g_i)\\ &\quad\times\nabla\tilde{V}_i(\bar{\varepsilon}_{i(k+1)})^{\mathsf{T}}B_iu_{ik}\\ &\quad+\sum_{j\in N_i}e_{ij}\nabla\tilde{V}_i(\bar{\varepsilon}_{i(k+1)})^{\mathsf{T}}B_ju_{jk}\\ &\quad+\frac{1}{2}\Big(\varepsilon_{ik}^{\mathsf{T}}Q_{ii}\varepsilon_{ik}+u_{ik}^{\mathsf{T}}R_{ii}u_{ik}\\ &\quad+\sum_{j\in N_i}u_{jk}^{\mathsf{T}}R_{ij}u_{jk}\Big)=0.\end{aligned} \tag{1.130}$$

Using (1.129) in (1.130), yields

$$\begin{aligned}&\varepsilon_{ik}^{\mathsf{T}}\nabla\tilde{V}_i(\bar{\varepsilon}_{ik})-\varepsilon_{ik}^{\mathsf{T}}Q_{ii}\varepsilon_{ik}-(d_i+g_i)\nabla\tilde{V}_i(\bar{\varepsilon}_{i(k+1)})^{\mathsf{T}}B_iu_{ik}\\ &\quad+\sum_{j\in N_i}e_{ij}\nabla\tilde{V}_i(\bar{\varepsilon}_{i(k+1)})^{\mathsf{T}}B_ju_{jk}+\frac{1}{2}\Big(\varepsilon_{ik}^{\mathsf{T}}Q_{ii}\varepsilon_{ik}\\ &\quad+u_{ik}^{\mathsf{T}}R_{ii}u_{ik}+\sum_{j\in N_i}u_{jk}^{\mathsf{T}}R_{ij}u_{jk}\Big)=0.\end{aligned} \tag{1.131}$$

Rearranging (1.131) yields (1.128).

(b) Adding and subtracting $\nabla\tilde{V}_i(\bar{\varepsilon}_{i(k+1)})^{\mathsf{T}}A\varepsilon_{ik}+\frac{1}{2}\varepsilon_{ik}^{-}Q_{ii}\varepsilon_{ik}$ on the right-hand side of equation (1.128) yields

$$\begin{aligned}\varepsilon_{ik}^{\mathsf{T}}\nabla\tilde{V}_i(\bar{\varepsilon}_{ik}) &= \nabla\tilde{V}_i(\bar{\varepsilon}_{i(k+1)})^{\mathsf{T}}A\varepsilon_{ik} - \nabla\tilde{V}_i(\bar{\varepsilon}_{i(k+1)})^{\mathsf{T}}A\varepsilon_{ik} \\ &\quad + (g_i+d_i)\nabla\tilde{V}_i(\bar{\varepsilon}_{i(k+1)})^{\mathsf{T}}B_iu_{ik} \\ &\quad - \sum_{j\in N_i} e_{ij}\nabla\tilde{V}_i(\bar{\varepsilon}_{i(k+1)})^{\mathsf{T}}B_ju_{jk} + \frac{1}{2}\varepsilon_{ik}^{\mathsf{T}}Q_{ii}\varepsilon_{ik} \\ &\quad - \frac{1}{2}\varepsilon_{ik}^{\mathsf{T}}Q_{ii}\varepsilon_{ik} + \frac{1}{2}\Big(\varepsilon_{ik}^{\mathsf{T}}Q_{ii}\varepsilon_{ik} - u_{ik}^{\mathsf{T}}R_{ii}u_{ik} \\ &\quad - \sum_{j\in N_i} u_{jk}^{\mathsf{T}}R_{ij}u_{jk}\Big). \end{aligned} \tag{1.132}$$

Or,

$$\begin{aligned}\varepsilon_{ik}^{\mathsf{T}}\nabla\tilde{V}_i(\bar{\varepsilon}_{ik}) &= \nabla\tilde{V}_i(\bar{\varepsilon}_{i(k+1)})^{\mathsf{T}}A\varepsilon_{ik} - \nabla\tilde{V}_i(\bar{\varepsilon}_{i(k+1)})^{\mathsf{T}}A\varepsilon_{ik} \\ &\quad + \varepsilon_{ik}^{\mathsf{T}}Q_{ii}\varepsilon_{ik} + (g_i+d_i)\nabla\tilde{V}_i(\bar{\varepsilon}_{i(k+1)})^{\mathsf{T}}B_iu_{ik} \\ &\quad - \sum_{j\in N_i} e_{ij}\nabla\tilde{V}_i(\bar{\varepsilon}_{i(k+1)})^{\mathsf{T}}B_ju_{jk} \\ &\quad - \frac{1}{2}\Big(\varepsilon_{ik}^{\mathsf{T}}Q_{ii}\varepsilon_{ik} + u_{ik}^{\mathsf{T}}R_{ii}u_{ik} + \sum_{j\in N_i} u_{jk}^{\mathsf{T}}R_{ij}u_{jk}\Big). \end{aligned} \tag{1.133}$$

Using (1.129) in this equation yields (1.115). □

Note that both sides of (1.128) are scalars, since they are inner products of the state and costate. This equation relates the state-costate product of node i at time k to the state-costate of node i at time $k+1$ and the policies of the neighbors. As such, (1.128) can be used to write down value iteration algorithms based on the state-costate inner product.

Based on (1.128), a Dual Heuristic Programming (DHP) Algorithm for graph games is now given.

Algorithm 2 (DHP for Graphical Games)

Step 1: Start with arbitrary initial policies u_{ik}^0.

Step 2: (Costate-State Product Update) Solve for $\nabla\tilde{V}_i(\bar{\varepsilon}_{ik})^{l+1}$ using

$$\begin{aligned}\nabla\tilde{V}_i^{(l+1)\mathsf{T}}(\bar{\varepsilon}_{ik})\varepsilon_{ik} &= \frac{1}{2}\Big(\varepsilon_{ik}^{\mathsf{T}}Q_{ii}\varepsilon_{ik} - u_{ik}^{l\mathsf{T}}R_{ii}u_{ik}^l - \sum_{j\in N_i} u_{jk}^{l\mathsf{T}}R_{ij}u_{jk}^l\Big) \\ &\quad + (d_i+g_i)\nabla\tilde{V}_i^{l\mathsf{T}}(\bar{\varepsilon}_{i(k+1)})B_iu_{ik}^l \\ &\quad - \sum_{j\in N_i} e_{ij}\nabla\tilde{V}_i^{l\mathsf{T}}(\bar{\varepsilon}_{i(k+1)})B_ju_{jk}^l \end{aligned} \tag{1.134}$$

where l is the value iteration step.

Step 3: (Policy Improvement) Update the control policies using

$$u_{ik}^{l+1} = (d_i+g_i)R_{ii}^{-1}B_i^{\mathsf{T}}\nabla\tilde{V}_i(\bar{\varepsilon}_{i(k+1)})^{l+1}, \forall i=1,\ldots,N. \tag{1.135}$$

Step 4: Iterate steps 2 and 3 until convergence is verified in the sense that $\left\|\nabla\tilde{V}_i(\bar{\varepsilon}_{ik})^{l+1} - \nabla\tilde{V}_i(\bar{\varepsilon}_{ik})^{l}\right\|$ is small. End.

Note that, if this Algorithm converges, then it converges to the solution of (1.128) with control policy (1.135). By Theorem 1.8, this is equivalent to (1.115) with the control policy (1.135) and the value (1.116). This is DTHJB equation (1.53) with the value (1.105) or equivalently (1.116). According to Theorem 1.2, DTHJB equation (1.53) is equivalent to the Bellman optimality equation (1.44). Therefore, if Algorithm 2 converges, it converges to the optimal value and the optimal control.

Algorithm 2 is in contrast to the standard single-player DHP Algorithm [Werbos (1992)] which propagates the costate, which is a vector. Algorithm 2 on the other hand, propagates the state-costate inner product, which is a scalar. Therefore, its implementation is simplified. See Section 1.6.

1.5 Coupled Riccati Recursions

In this section, coupled Riccati recursions for graphical games are derived that are equivalent to performing Algorithm 2. These Riccati recursions can be used to solve the graphical game offline. This development justifies the value function structure in (1.105) or equivalently (1.116).

The graphical game (1.9), (1.13) has a linear state equation and quadratic cost functions with infinite horizon. Therefore, assume that the global costate variable $\lambda_k = [\lambda_{1k}^{\mathsf{T}} \ \cdots \ \lambda_{Nk}^{\mathsf{T}}]^{\mathsf{T}}$ is linear in the global state $\varepsilon_k = [\varepsilon_{1k}^{\mathsf{T}} \ \cdots \ \varepsilon_{Nk}^{\mathsf{T}}]^{\mathsf{T}}$ so that

$$\lambda_k = S\varepsilon_k. \tag{1.136}$$

This assumption is justified in the statement of Theorem 1.9 below, where coupled Riccati recursions are given that are equivalent to (1.134) with controls (1.135).

Based on the structure (1.105) or equivalently (1.116), one can derive coupled Riccati recursions that are equivalent to (1.134) as given in the following theorem.

Theorem 1.9. *(Coupled Riccati Recursions) Assume* $\tilde{V}_i^l(\bar{\varepsilon}_{ik}) = \varepsilon_k^{\mathsf{T}} Z_i^{\mathsf{T}} Z_i S^l \bar{Z}_i \varepsilon_k$ *in Algorithm 2. Then,* $S_i^l = Z_i S^l \bar{Z}_i$ *satisfies the coupled Riccati recursions*

$$\begin{aligned}\hat{S}_i^{l+1} &= (\bar{A} - \bar{E}\bar{B}K^l)^{\mathsf{T}}(-\bar{Z}_i^{\mathsf{T}} S^{l\mathsf{T}} Z_i^{\mathsf{T}} \bar{\epsilon}_i \bar{B}\bar{g}\bar{R}^{-1}\bar{B}^{\mathsf{T}} Z\bar{S}^l\bar{Z}) \\ &\quad \times (\bar{A} - \bar{E}\bar{B}K^l) + \bar{Q}_i + K^{l\mathsf{T}}\bar{R}_i K^l \end{aligned} \tag{1.137}$$

which are equivalent to (1.134) for each l. Here $K^l = \bar{g}\bar{R}^{-1}\bar{B}^{\mathsf{T}} Z\bar{S}^l\bar{Z}(I + \bar{E}\bar{B}\bar{g}\bar{R}^{-1}\bar{B}^{\mathsf{T}} Z\bar{S}^l\bar{Z})^{-1}\bar{A}$ *and other matrices are defined in the proof.*

Proof. According to Theorem 1.5, the costate variable is given by

$$\tilde{\lambda}_{ik}^l = \nabla\tilde{V}_i^l(\bar{\varepsilon}_{ik}), \ \ \forall i \in N.$$

The costate variable $\nabla \tilde{V}_i^l\left(\bar{\varepsilon}_{i(k+1)}\right)$ for each agent i is given by

$$\nabla \tilde{V}_i^l\left(\bar{\varepsilon}_{i(k+1)}\right) = Z_i S^l \bar{Z}_i \varepsilon_{k+1}^l. \tag{1.138}$$

Error dynamics (1.9), (1.138) and optimal control policies (1.108) for each agent i yield the closed-loop systems

$$\begin{aligned} \varepsilon_{i(k+1)}^l &= A\varepsilon_{ik} - (g_i+d_i)^2 B_i R_{ii}^{-1} B_i^{\mathsf{T}} Z_i S^l \bar{Z}_i \varepsilon_{k+1}^l \\ &\quad + \sum_{j\in N_i}\left(e_{ij}(g_j+d_j) B_j R_{jj}^{-1} B_j^{\mathsf{T}} Z_j S^l \bar{Z}_j \varepsilon_{k+1}^l\right). \end{aligned} \tag{1.139}$$

This yields the global dynamics

$$\varepsilon_{k+1}^l = \bar{A}\varepsilon_k - \bar{E}\bar{B}\bar{g}\bar{R}^{-1}\bar{B}^{\mathsf{T}} Z \bar{S}^l \bar{Z}\varepsilon_{k+1}^l \tag{1.140}$$

where $\bar{E} = (L+G)\otimes I_n$, $\bar{A} = \left(I_N \otimes A\right) \in R^{nN\times nN}$, $\bar{B} = \operatorname{diag}\{B_1 \ldots, B_i, \ldots, B_N\} \in R^{nN\times mN}$, $\bar{g} = \operatorname{diag}\{(g_1+d_1), \ldots, (g_i+d_i), \ldots, (g_N+d_N)\} \otimes I_m$, $\bar{R} = \operatorname{diag}\{R_{11}, \ldots, R_{ii}, \ldots, R_{NN}\}$, $Z = \operatorname{diag}\{Z_1, \ldots, Z_i, \ldots, Z_N\} \in R^{nN\times nN^2}$, $\bar{S}^l = (I_N \otimes S^l) \in R^{nN^2\times nN^2}$, $\bar{Z} = [\bar{Z}_1^{\mathsf{T}} \ldots \bar{Z}_i^{\mathsf{T}} \ldots, \bar{Z}_N^{\mathsf{T}}]^{\mathsf{T}} \in R^{nN^2\times nN}$.

Rearranging (1.140) yields the global error dynamics

$$\varepsilon_{k+1}^l = \left(I + \bar{E}\bar{B}\bar{g}\bar{R}^{-1}\bar{B}^{\mathsf{T}} Z \bar{S}^l \bar{Z}\right)^{-1} \bar{A}\varepsilon_k. \tag{1.141}$$

Using the optimal control policy (1.108), the global optimal control policy is

$$u_k^l = \bar{g}\bar{R}^{-1}\bar{B}^{\mathsf{T}} Z \bar{S}^l \bar{Z}\varepsilon_{k+1}^l.$$

Using (1.141) in this equation yields

$$u_k^l = \bar{g}\bar{R}^{-1}\bar{B}^{\mathsf{T}} Z \bar{S}^l \bar{Z}(I + \bar{E}\bar{B}\bar{g}\bar{R}^{-1}\bar{B}^{\mathsf{T}} Z \bar{S}^l \bar{Z})^{-1}\bar{A}\varepsilon_k. \tag{1.142}$$

Using optimal control policy (1.108) in (1.134) yields

$$\begin{aligned} \varepsilon_{ik}^{\mathsf{T}} \nabla \tilde{V}_i^{l+1}\left(\bar{\varepsilon}_{ik}\right) &= \frac{1}{2}\left(\varepsilon_{ik}^{\mathsf{T}} Q_{ii}\varepsilon_{ik} + u_{ik}^{l\mathsf{T}} R_{ii} u_{ik}^l - \sum_{j\in N_i} u_{jk}^{l\mathsf{T}} R_{ij} u_{jk}^l\right) \\ &\quad - \sum_{j\in N_i} e_{ij} \nabla \tilde{V}_i^{l\mathsf{T}}\left(\bar{\varepsilon}_{i(k+1)}\right) B_j u_{jk}^l. \end{aligned}$$

Using (1.138) in this equation yields

$$\varepsilon_k^{\mathsf{T}} Z_i^{\mathsf{T}} Z_i S^{l+1} \bar{Z}_i \varepsilon_k = \varepsilon_k^{\mathsf{T}} \bar{Q}_i \varepsilon_k + u_k^{l\mathsf{T}} \bar{R}_i u_k^l - \varepsilon_{(k+1)}^{l\mathsf{T}} \bar{Z}_i^{\mathsf{T}} S^{l\mathsf{T}} Z_i^{\mathsf{T}} \bar{e}_i \bar{B} u_k^l$$

where $\bar{Q}_i = \dfrac{1}{2}\operatorname{diag}\{0, \ldots, Q_{ii}, \ldots, 0\}$, $\bar{R}_i = \dfrac{1}{2}\operatorname{diag}\{-R_{i1}, \ldots, R_{ii}, \ldots, -R_{iN}\}$, $\bar{e}_i = I_n \otimes \{e_{ij}\}, j \in N$.

Rearranging this equation yields,

$$\varepsilon_k^{\mathsf{T}} \hat{S}_i^{l+1} \varepsilon_k = \varepsilon_k^{\mathsf{T}} \bar{Q}_i \varepsilon_k + u_k^{l\mathsf{T}} \bar{R}_i u_k^l - \varepsilon_{(k+1)}^{l\mathsf{T}} \bar{Z}_i^{\mathsf{T}} S^{l\mathsf{T}} Z_i^{\mathsf{T}} \bar{e}_i \bar{B} u_k^l \tag{1.143}$$

where $\hat{S}_i^{l+1} \equiv Z_i^{\mathsf{T}} Z_i S^{l+1} \bar{Z}_i \in R^{nN \times nN}$. Using (1.141) and (1.142) into (1.143) yields the coupled Riccati recursions

$$\begin{aligned}\hat{S}_i^{l+1} = \bar{Q}_i + \bar{A}^{\mathsf{T}}\Big(I + \bar{E}\bar{B}\bar{g}\bar{R}^{-1}\bar{B}^{\mathsf{T}} Z\bar{S}^l \bar{Z}\Big)^{-\mathsf{T}} \\ \times \Big(\bar{Z}^{\mathsf{T}}\bar{S}^{l\mathsf{T}} Z^{\mathsf{T}}\bar{B}\bar{R}^{-1}\bar{g}\bar{R}_i - \bar{Z}_i^{\mathsf{T}} S^{l\mathsf{T}} Z_i^{\mathsf{T}}\bar{\epsilon}_i\bar{B}\Big) \\ \times \bar{g}\bar{R}^{-1}\bar{B}^{\mathsf{T}} Z\bar{S}^l\bar{Z}\Big(I + \bar{E}\bar{B}\bar{g}\bar{R}^{-1}\bar{B}^{\mathsf{T}} Z\bar{S}^l\bar{Z}\Big)^{-1}\bar{A}. \qquad (1.144)\end{aligned}$$

(1.143) can be written in Joseph Stabilized Form [Lewis *et al.* (2012)] so that

$$\begin{aligned}\hat{S}_i^{l+1} = (\bar{A} - \bar{E}\bar{B}K^l)^{\mathsf{T}}\Big(-\bar{Z}_i^{\mathsf{T}} S^{l\mathsf{T}} Z_i^{\mathsf{T}}\bar{e}_i\bar{B}\bar{g}\bar{R}^{-1}\bar{B}^{\mathsf{T}} Z\bar{S}^l\bar{Z}\Big) \\ \times (\bar{A} - \bar{E}\bar{B}K^l) + \bar{Q}_i + K^{l\mathsf{T}}\bar{R}_i K^l \qquad (1.145)\end{aligned}$$

where $K^l = \bar{g}\bar{R}^{-1}\bar{B}^{\mathsf{T}} Z\bar{S}^l\bar{Z}(I + \bar{E}\bar{B}\bar{g}\bar{R}^{-1}\bar{B}^{\mathsf{T}} Z\bar{S}^l\bar{Z})^{-1}\bar{A}$. The global matrix S^l is constructed from each recursion so that

$$S^{l+1} = \Big[(Z_1\hat{S}_1^{l+1}\bar{Z}_1)^{\mathsf{T}} \cdots (Z_i\hat{S}_i^{l+1}\bar{Z}_i)^{\mathsf{T}} \cdots (Z_N\hat{S}_N^{l+1}\bar{Z}_N)^{\mathsf{T}}\Big]^{\mathsf{T}}.$$

□

1.6 Graphical Game Solutions by Actor-Critic Learning

This section develops actor-critic network structures based on value function approximation [Werbos (1992)] that can be used to solve the graphical games. Both online and offline actor-critic learning algorithms are given. These actor-critic structures are motivated by graph games DHP Algorithm 2, where each agent i has its own critic to perform the evaluation update (1.134) and its own actor to perform the policy update (1.135). Here, a critic network is used to approximate the costate for each agent i and an actor network is used to approximate the control policy for each agent i. The critic and actor structures depend on the neighbors, or local information.

1.6.1 *Actor-critic networks and tuning*

The costate $\nabla\tilde{V}_i(\bar{\varepsilon}_{i(k+1)})$ for each agent i is approximated by a critic network $\nabla\hat{V}_i(\cdot|W_{ic})$, and the control policy is approximated by an actor network $\hat{u}_i(\cdot|W_{ia})$ so that

$$\nabla\hat{V}_{ik}(W_{ic}) = W_{ic}^{\mathsf{T}} Z_{ik} \qquad (1.146)$$

$$\hat{u}_{ik}(W_{ia}) = W_{ia}^{\mathsf{T}} Z_{ik} \qquad (1.147)$$

where W_{ic} and W_{ia} are the critic and actor weights, respectively; Z_{ik} is a vector of the state ε_{ik} of node i and the states of its neighbors ε_{-ik}.

Let $\zeta_{u_{ik}}^{\tilde{V}_i(\bar{\varepsilon}_{ik})}$ be the approximation error of the actor network so that

$$\zeta_{u_{ik}}^{\tilde{V}_i(\bar{\varepsilon}_{ik})} = \hat{u}_{ik}(W_{ia}) - \bar{u}_{ik} = W_{ia}^{\mathsf{T}} Z_{ik} - \bar{u}_{ik} \tag{1.148}$$

where based on (1.135), the target control policy $\bar{u}_{ik}$ is given in terms of the critic network by

$$\bar{u}_{ik} = (g_i + d_i) R_{ii}^{-1} B_i^{\mathsf{T}} \nabla \hat{V}_{i(k+1)}. \tag{1.149}$$

The norm squared of the actor approximation error is

$$\mathrm{err}_{actor} = \frac{1}{2}\left(\zeta_{u_{ik}}^{\tilde{V}_i(\bar{\varepsilon}_{ik})}\right)^{\mathsf{T}} \zeta_{u_{ik}}^{\tilde{V}_i(\bar{\varepsilon}_{ik})}. \tag{1.150}$$

The change in the actor weights is given by gradient descent on this function whose gradient is

$$\begin{aligned} -\Delta W_{ia}^{l\mathsf{T}} &= \frac{\partial \mathrm{err}_{actor}}{\partial \zeta_{u_{ik}}^{\tilde{V}_i(\bar{\varepsilon}_{ik})}} \frac{\partial \zeta_{u_{ik}}^{\tilde{V}_i(\bar{\varepsilon}_{ik})}}{\partial W_{ia}^{\mathsf{T}}}\bigg|_{W_{ia}^{\mathsf{T}} = W_{ia}^{l\mathsf{T}}} \\ &= \left(W_{ia}^{\mathsf{T}} Z_{ik} - \bar{u}_{ik}\right) Z_{ik}^{\mathsf{T}}\Big|_{W_{ia}^{\mathsf{T}} = W_{ia}^{l\mathsf{T}}}. \end{aligned}$$

The update rule for the actor weights is therefore given by

$$W_{ia}^{(l+1)\mathsf{T}} = W_{ia}^{l\mathsf{T}} - \mu_{ia}\left(W_{ia}^{l\mathsf{T}} Z_{ik} - \bar{u}_{ik}\right)(Z_{ik})^{\mathsf{T}} \tag{1.151}$$

where $0 < \mu_{ia} < 1$ is the actor network learning rate.

The costate-state product update equation is given by (1.134). Let $\chi_{\bar{\varepsilon}_{ik}}^{\tilde{V}_i(\bar{\varepsilon}_{ik})}$ be the target value of the critic network at step l so that

$$\begin{aligned} \chi_{\bar{\varepsilon}_{ik}}^{\tilde{V}_i(\bar{\varepsilon}_{ik})} &= \frac{1}{2}\Big(\varepsilon_{ik}^{\mathsf{T}} Q_{ii} \varepsilon_{ik} - \hat{u}_{ik}^{l\mathsf{T}} R_{ii} \hat{u}_{ik}^{l} - \sum_{j \in N_i} \hat{u}_{jk}^{l\mathsf{T}} R_{ij} \hat{u}_{jk}^{l}\Big) \\ &+ (d_i + g_i) \nabla \hat{V}_{i(k+1)}^{l\mathsf{T}} B_i \hat{u}_{jk}^{l} \\ &- \sum_{j \in N_i} e_{ij} \nabla \hat{V}_{i(k+1)}^{l\mathsf{T}} B_j \hat{u}_{jk}^{l}. \end{aligned} \tag{1.152}$$

The critic network approximation error at step l is given by

$$\zeta_{\bar{\varepsilon}_{ik}}^{\tilde{V}_i(\bar{\varepsilon}_{ik})} = \chi_{\bar{\varepsilon}_{ik}}^{\tilde{V}_i(\bar{\varepsilon}_{ik})} - \nabla \hat{V}_{ik}^{\mathsf{T}}(W_{ic}) \varepsilon_{ik}. \tag{1.153}$$

The square sum of the approximation error for the critic networks is given by

$$\mathrm{err}_{critic} = \frac{1}{2}\left(\zeta_{\bar{\varepsilon}_{ik}}^{\tilde{V}_i(\bar{\varepsilon}_{ik})}\right)^{\mathsf{T}} \zeta_{\bar{\varepsilon}_{ik}}^{\tilde{V}_i(\bar{\varepsilon}_{ik})} = \frac{1}{2}\left\|\chi_{\bar{\varepsilon}_{ik}}^{\tilde{V}_i(\bar{\varepsilon}_{ik})} - \left(W_{ic}^{\mathsf{T}} Z_{ik}\right)^{\mathsf{T}} \varepsilon_{ik}\right\|_2^2. \tag{1.154}$$

The change in the critic weights is given by gradient descent on this function whose gradient is

$$\begin{aligned} -\Delta W_{ic}^{l\mathsf{T}} &= \left(\partial \mathrm{err}_{critic} / \partial \zeta_{\bar{\varepsilon}_{ik}}^{\tilde{V}_i(\bar{\varepsilon}_{ik})}\right)\left(\partial \zeta_{\bar{\varepsilon}_{ik}}^{\tilde{V}_i(\bar{\varepsilon}_{ik})} / \partial W_{ic}^{\mathsf{T}}\right)\Big|_{W_{ic}^{\mathsf{T}} = W_{ic}^{l\mathsf{T}}} \\ &= \left(\chi_{\bar{\varepsilon}_{ik}}^{\tilde{V}_i(\bar{\varepsilon}_{ik})} - (W_{ic}^{l\mathsf{T}} Z_{ik})^{\mathsf{T}} \varepsilon_{ik}\right) \varepsilon_{ik} Z_{ik}^{\mathsf{T}}. \end{aligned} \tag{1.155}$$

Therefore, the update rule for the critic weights is given by

$$W_{ic}^{(l+1)T} = W_{ic}^{l\mathsf{T}} - \mu_{ic}\left(\chi_{\bar{\varepsilon}_{ik}}^{\tilde{V}_i(\bar{\varepsilon}_{ik})} - \left(W_{ic}^{l\mathsf{T}} Z_{ik}\right)^{\mathsf{T}} \varepsilon_{ik}\right)\varepsilon_{ik} Z_{ik}^{\mathsf{T}} \tag{1.156}$$

where $0 < \mu_{ic} < 1$ is the critic network learning rate.

Actor and Critic tuning algorithms (1.151) and (1.156) are derived from graph games DHP Algorithm 2. Using these tuning algorithms, the actor-critic weights can be tuned either offline or online, as detailed next.

1.6.2 *Actor-critic offline tuning with exploration*

The following Algorithm is used for solving the graphical game by offline tuning of the actor-critic network structures. It is similar to an algorithm in Zhang *et al.* (2009). This offline Algorithm is useful because random points are selected in the state space, which guarantees sufficient exploration [Sutton and Barto (1998)] in computing the actor and critic network weights.

Algorithm 3 (Actor-critic network offline tuning)

Initialize the actor weights W_{ia}^0 randomly and initialize the critic weights W_{ic}^0 with zero values.
Do loop 1 (q Iterations) {Start with random initial state ε_{i0}, $\forall i$.}
Do Loop 2 (l Iterations) {

2.1 Calculate $\hat{u}_{ik}^l$, $\forall i$ using (1.147).
2.2 Calculate the dynamics $\varepsilon_{i(k+1)}^l$, $\forall i$ using (1.9).
2.3 Calculate costate variable $\nabla\hat{V}_{i(k+1)}^l$, $\forall i$ in terms of dynamics $\varepsilon_{i(k+1)}^l$, $\forall i$ using (1.146).
2.4 Critic update rule

$$W_{ic}^{(l+1)\mathsf{T}} = W_{ic}^{l\mathsf{T}} - \mu_{ic}\left(\chi_{\bar{\varepsilon}_{ik}}^{\tilde{V}_i(\bar{\varepsilon}_{ik})} - \left(W_{ic}^{l\mathsf{T}} Z_{ik}\right)^{\mathsf{T}} \varepsilon_{ik}\right)\varepsilon_{ik} Z_{ik}^{\mathsf{T}}$$

where $\chi_{\bar{\varepsilon}_{ik}}^{\tilde{V}_i(\bar{\varepsilon}_{ik})}$ is given by (1.152).
2.5 Actor update rule

$$W_{ia}^{(l+1)T} = W_{ia}^{l\mathsf{T}} - \mu_{ia}\left(W_{ic}^{l\mathsf{T}} Z_{ik} - \bar{u}_{ik}^l\right) Z_{ik}^{\mathsf{T}}$$

where $\bar{u}_{ik}^l = (g_i + d_i) R_{ii}^{-1} B_i^{\mathsf{T}} \nabla\hat{V}_{i(k+1)}^l$.
2.6 On convergence of the actor-critic weights end Loop 2}.

Transfer the actor-critic weights to the next iteration $(q+1)$.
Do steps 2.1 to 2.6.
On convergence of $\left\|\nabla\hat{V}_i(\bar{\varepsilon}_{ik})^{q+1} - \nabla\hat{V}_i(\bar{\varepsilon}_{ik})^q\right\|$ End Loop 2}.
End Loop 1.

1.6.3 *Actor-critic online tuning in real-time*

The following Algorithm is used for solving the graphical game by online tuning of the actor-critic network structures in real-time using data measured along the system trajectories.

Algorithm 4 (Actor-Critic Network Online Tuning)

Initialize the actor weights W_{ia}^{0} randomly and initialize the critic weights W_{ic}^{0} with zero values.
Do Loop (l iterations) {

2.1 Start with given initial state ε_{i0}, $\forall i$ on the system trajectory.
2.2 Calculate $\hat{u}_{ik}^{l}$, $\forall i$ using (1.147).
2.3 Calculate the dynamics $\varepsilon_{i(k+1)}^{l}$, $\forall i$ using (1.9).
2.4 Calculate costate variable $\nabla\hat{V}_{i(k+1)}^{l}$, $\forall i$ in terms of dynamics $\varepsilon_{i(k+1)}^{l}$, $\forall i$ using (1.146).
2.5 Critic update rule

$$W_{ic}^{(l+1)T} = W_{ic}^{l\mathsf{T}} - \mu_{ic}\left(\chi_{\bar{\varepsilon}_{ik}}^{\tilde{V}_i(\bar{\varepsilon}_{ik})} - \left(W_{ic}^{l\mathsf{T}} Z_{ik}\right)^{\mathsf{T}} \varepsilon_{ik}\right)\varepsilon_{ik} Z_{ik}^{\mathsf{T}}$$

where $\chi_{\bar{\varepsilon}_{ik}}^{\tilde{V}_i\left(\bar{\varepsilon}_{ik}\right)}$ is given by (1.152).
2.6 Actor update rule

$$W_{ia}^{(l+1)T} = W_{ia}^{l\mathsf{T}} - \mu_{ia}\left(W_{ia}^{l\mathsf{T}} Z_{ik} - \bar{u}_{ik}^{l}\right)\left(Z_{ik}\right)^{\mathsf{T}}$$

where $\bar{u}_{ik}^{l} = (g_i + d_i)R_{ii}^{-1}B_i^{\mathsf{T}}\nabla\hat{V}_{i(k+1)}^{l}$.
2.7 On convergence of $\left\|\nabla\hat{V}_i\left(\bar{\varepsilon}_{ik}\right)^{l+1} - \nabla\hat{V}_i\left(\bar{\varepsilon}_{ik}\right)^{l}\right\|$.

End Loop}.

1.7 Graphical Game Example and Simulation Results

The graphical game problem can be solved offline using coupled Riccati recursions (1.137) or Algorithm 3. It can be solved online in real-time by Algorithm 4. Algorithm 3 and Algorithm 4 are derived using graph games DHP Algorithm 2. In this section, simulation examples are performed to verify the proper performance of these algorithms. First, Algorithm 3 will be tested offline and it will be shown that it converges to the same solution as does the Riccati recursions (1.137), thereby finding the optimal value and optimal control. Next, an online simulation using Algorithm 4 is performed where the actor and critic weights are tuned in real-time using data measured along the system trajectories. Stability and synchronization of each agent to the leader node's trajectory are verified.

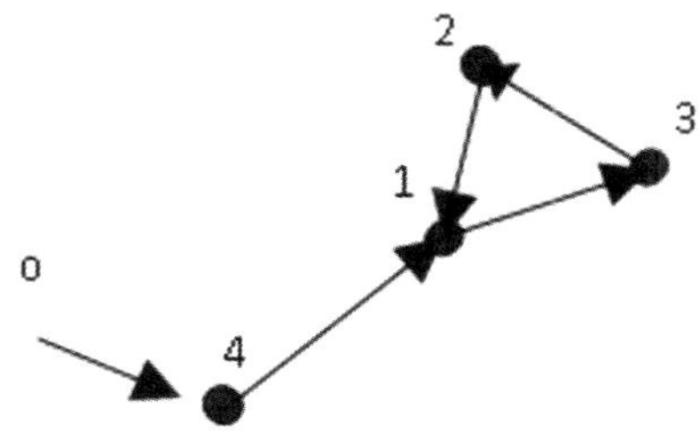

Fig. 1.1 Offline tuning: critic weights tuning of agent no. 1

Consider the directed graph with four agents shown in Fig. 1.1. The data of the graph example are given as follows: Agents' dynamics:

$$A = \begin{bmatrix} 0.995 & 0.09983 \\ -0.09983 & 0.995 \end{bmatrix}, \quad B_1 = \begin{bmatrix} 0.2047 \\ 0.08984 \end{bmatrix}, \quad B_2 = \begin{bmatrix} 0.2147 \\ 0.2895 \end{bmatrix},$$
$$B_3 = \begin{bmatrix} 0.2097 \\ 0.1897 \end{bmatrix}, \quad B_4 = \begin{bmatrix} 0.2 \\ 0.1 \end{bmatrix}.$$

Pinning gains: $g_1 = g_2 = g_3 = 0,\ g_4 = 1.$

Graph connectivity: $e_{12} = 0.8,\ e_{14} = 0.7,\ e_{23} = 0.6,\ e_{31} = 0.8.$

Weighting matrices: $Q_{11} = Q_{22} = Q_{33} = Q_{44} = I_{2\times 2},\quad R_{11} = R_{22} = R_{33} = R_{44} = 1,\ R_{12} = R_{14} = R_{23} = R_{31} = 1.$

1.7.1 *Riccati recursion offline solution*

By solving the set of coupled Riccati recursions (1.137), one converges to a global solution matrix S for the dynamic graphical game such that (1.105) satisfies (1.128) with controls (1.135). This is equivalent to solving the set of the DTHJB equations (1.53) for the dynamic graphical game. The derived Riccati structure finds the optimal value function for each agent i in terms of a sub-matrix of the global solution matrix S. The sub-matrix solution for each agent i is given using the pre-described mappings (1.105) or (1.116). Under the structure of (1.105), one can write

$$S^l = \left[\, S_1^{l\mathsf{T}}\ S_2^{l\mathsf{T}} \cdots S_N^{l\mathsf{T}} \,\right]^\mathsf{T} \in R^{nN\times nN} \tag{1.157}$$

where $S_i^l = Z_i S^l \bar{Z}_i \in R^{n\times nN}$ and $V_i^l = \varepsilon_{ik}^\mathsf{T} S_i^l \varepsilon_k \in R^{n\times nN}$. Note that $\bar{Z}_i \varepsilon_k$ selects the states of node i and its neighbors.

Two simulation cases are considered. The first case defines the global matrix S^l in each Riccati recursion (1.137) as a matrix which has entries from all the nodes or agents. The second case defines the global matrix S^l in each Riccati recursion (1.137) as a matrix which has entries only from the agent and its neighbors and zero entries otherwise. In both cases the Riccati recursions converge to the same global solution matrix S

$$S=\begin{bmatrix} 0.5100 & 0.0057 & -0.0038 & -0.0065 & 0 & 0 & -0.0064 & -0.0072 \\ 0.0057 & 0.5032 & -0.0067 & -0.0112 & 0 & 0 & -0.0074 & -0.0082 \\ 0 & 0 & 0.5015 & 0.0025 & -0.0042 & -0.0047 & 0 & 0 \\ 0 & 0 & 0.0025 & 0.5042 & -0.0046 & -0.0051 & 0 & 0 \\ -0.0109 & -0.0062 & 0 & 0 & 0.5028 & 0.0031 & 0 & 0 \\ -0.0061 & -0.0035 & 0 & 0 & 0.0031 & 0.5034 & 0 & 0 \\ 0 & 0 & 0 & 0 & 0 & 0 & 0.5047 & 0.0052 \\ 0 & 0 & 0 & 0 & 0 & 0 & 0.0052 & 0.5058 \end{bmatrix} \tag{1.158}$$

The importance of this result is showing that the derived set of coupled Riccati recursions (1.137) for each agent i depends only on the entries of that agent and its neighbors. This motivates the selection of the vector Z_{ik} in (1.146) and (1.147) to be a vector of the state ε_{ik} of each node i and the states of its neighbors ε_{-ik}.

1.7.2 *Simulation results using offline actor-critic tuning*

Algorithm 3 is now used to implement the offline actor-critic tuning. It is seen that this Algorithm converges to the Riccati recursion solution (1.158). The same graphical game example is used with learning rates ($\mu_{ic}=0.001$, $\mu_{ia}=0.001$, $\forall i$). Algorithm 3 uses random states ε_{ik}, which guarantees sufficient exploration in computing the actor and critic network weights. Fig. 1.2 shows the critic weights of agent *1* at each iteration. Fig. 1.3 shows the actor weights of agent *1* at each iteration.

The critic weights for the four agents converge to the overall weight matrix

$$W_c=\begin{bmatrix}\bar{W}_{1c}\\ \bar{W}_{2c}\\ \bar{W}_{3c}\\ \bar{W}_{4c}\end{bmatrix}$$

$$=\begin{bmatrix} 0.5096 & 0.0124 & -0.0035 & -0.0124 & 0 & 0 & -0.0105 & -0.0078 \\ 0.0124 & 0.5079 & -0.0114 & -0.0196 & 0 & 0 & -0.0090 & -0.0094 \\ 0 & 0 & 0.5037 & 0.0058 & -0.0095 & -0.0112 & 0 & 0 \\ 0 & 0 & 0.0058 & 0.5034 & -0.0073 & -0.0100 & 0 & 0 \\ -0.0239 & -0.0177 & 0 & 0 & 0.5045 & 0.0081 & 0 & 0 \\ -0.0167 & -0.0116 & 0 & 0 & 0.0081 & 0.5054 & 0 & 0 \\ 0 & 0 & 0 & 0 & 0 & 0 & 0.5041 & 0.0033 \\ 0 & 0 & 0 & 0 & 0 & 0 & 0.0033 & 0.5013 \end{bmatrix} \tag{1.159}$$

where $\bar{W}_{ic}\in R^{n\times nN}$ is a matrix that has the critic weights in W_{ic} of each agent i in the appropriate positions of the state and its neighbors, and zeros otherwise. These critic weights closely match the solution (1.158) given by solving the Riccati recursions. These results show that the offline tuned actor-critic weights from Algorithm 3 converge to the same values as the Riccati matrices of (1.137).

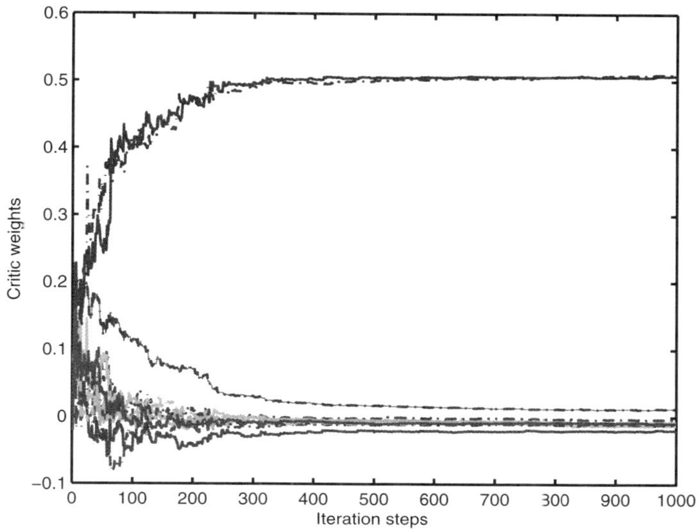

Fig. 1.2 Offline tuning: critic weights tuning of agent no. 1

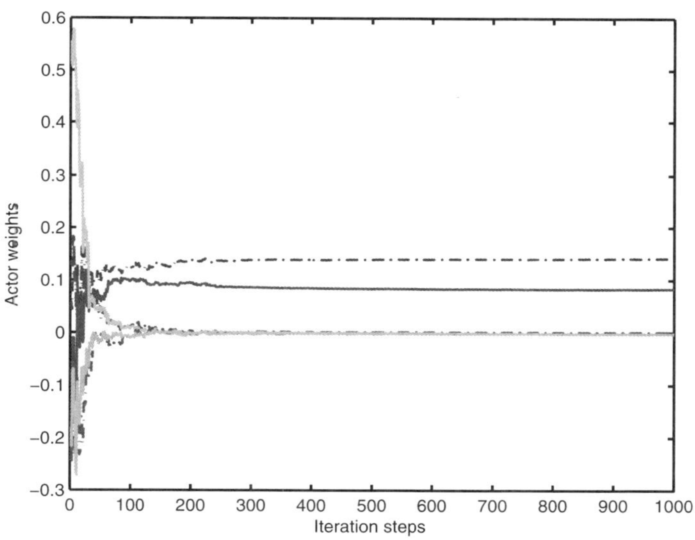

Fig. 1.3 Offline tuning: actor weights tuning of agent no. 1

1.7.3 *Simulation results using online actor-critic tuning*

Algorithm 4 is now used to implement online actor-critic tuning in real-time using data measured along the system trajectories. The same graphical game example is used with learning rates ($\mu_{ic} = 0.1$, $\mu_{ia} = 0.1$, $\forall i$). Fig. 1.4 shows the critic weights of agent *1*. Fig. 1.5 shows the actor weights of agent *1*. Fig. 1.6 shows the dynamics of all four agents. Fig. 1.7 shows the phase plane plots of all four agents including the leader node 4. Finally, Fig. 1.8 shows the neighborhood tracking error dynamics of all four agents. These figures show that Algorithm 4 yields stability and

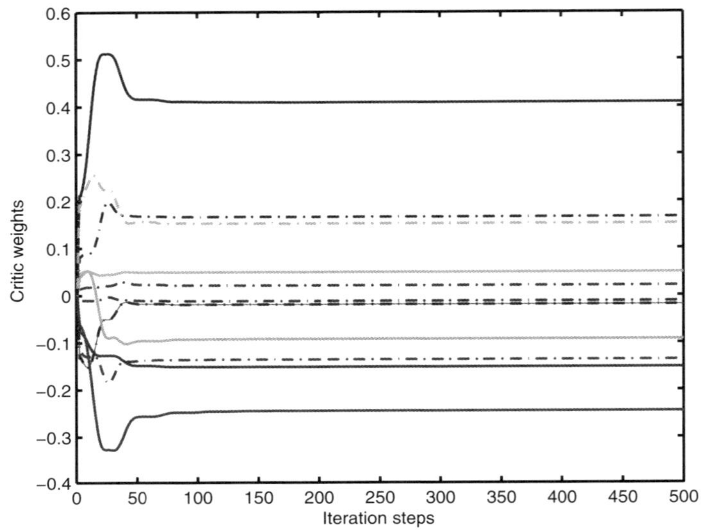

Fig. 1.4 Online tuning: critic weights tuning of agent no. 1

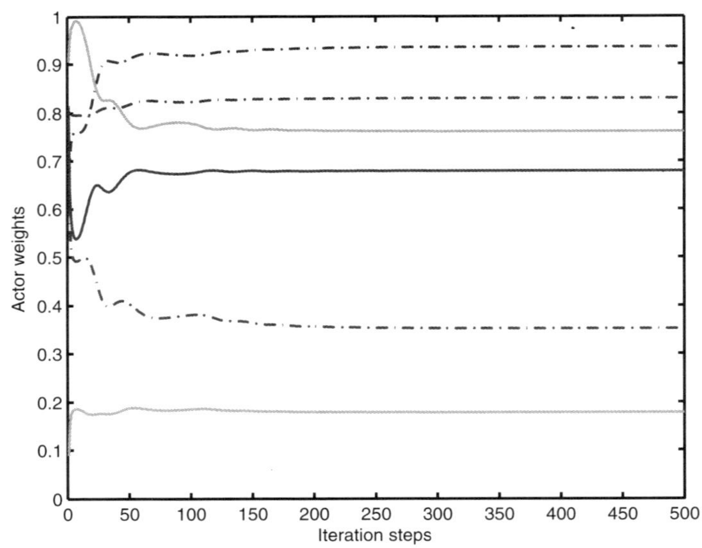

Fig. 1.5 Online tuning: actor weights tuning of agent no. 1

synchronization to the leader's state. The weights learned online using Algorithm 4 do not converge to the same value as the weights found offline using Algorithm 3 and the Riccati recursions. This is because exploration cannot be guaranteed using online methods without adding probing noise.

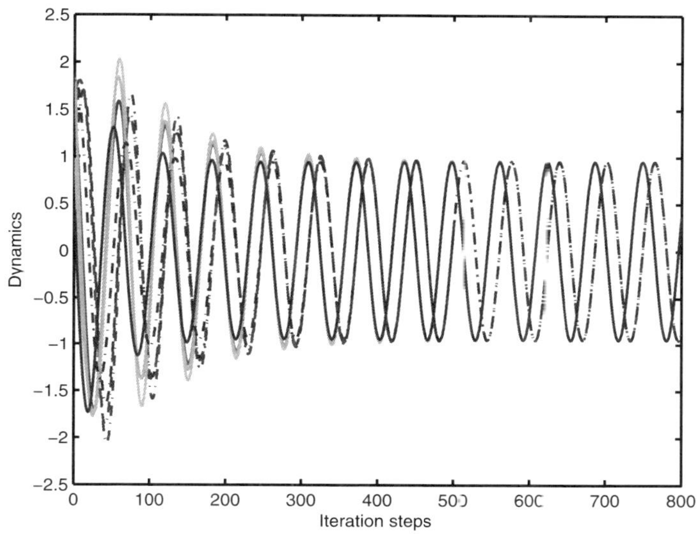

Fig. 1.6 Agents' dynamics

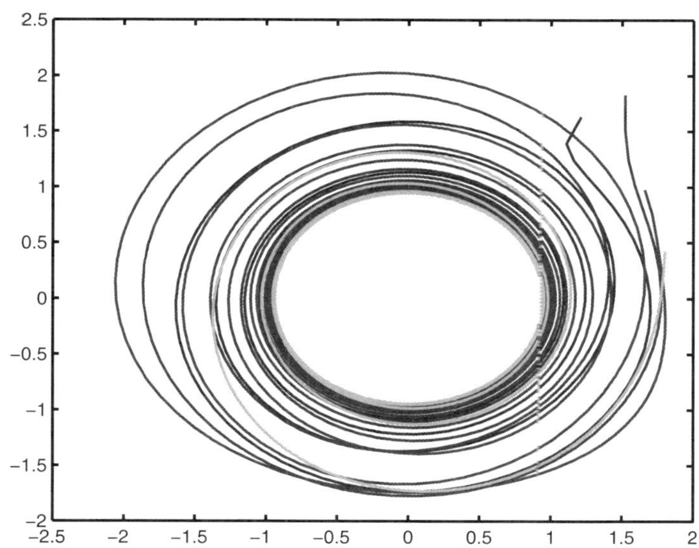

Fig. 1.7 Phase plane plot

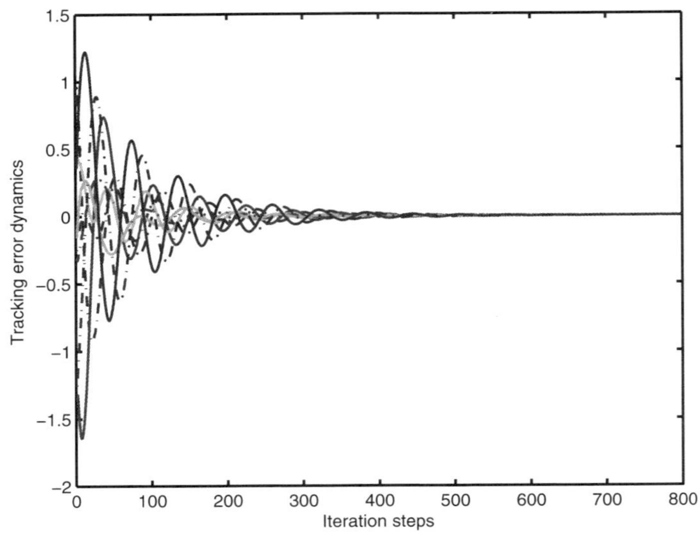

Fig. 1.8 Error tracking dynamics

1.8 Conclusions

The chapter lays the mathematical foundation to develop adaptive learning solutions for a class of discrete-time dynamic games known as dynamic graphical games, where interactions between the agents are prescribed by a communication graph structure. This chapter brings together discrete Hamiltonian mechanics, distributed multi-agent control, optimal control theory, game theory, and RL techniques to formulate and solve these multi-agent dynamic graphical games. Graphical game Bellman equations are derived and shown to be equivalent to the graphical game HJB equations developed herein. Nash solutions are given in terms of solutions to the set of coupled discrete-time HJB equations. ADP methods, namely HDP and DHP, are proposed to solve the dynamic graphical games using only local information. Novel DHP structure is derived and extended to solve the dynamic graphical game, since the standard DHP structure does not capture the neighbors' information. Novel set of coupled Riccati recursions are developed to solve the dynamic graphical game offline based on the derived DHP structure. RL value iteration algorithms are developed to solve the dynamic graphical games in an online manner along with its proof of convergence. Finally, offline and real-time adaptive learning structures based on actor-critic techniques are developed to implement DHP by using two approximator network structures.

Acknowledgement

This work was supported by NSF grant ECCS-1128050, ONR grant N00014-13-1-0562, AFOSR EOARD Grant 13-3055, ARO grant W911NF-11-D-0001, China NNSF grant 61120106011, and China Education Ministry Project 111 (No. B08015).

References

Abou-Kandil, H., Freiling, G., Ionescu, V. and Jank, G. (2003). *Matrix Riccati Equations in Control and Systems Theory* (Birkhäuser).

Al-Tamimi, A., Lewis, F. L. and Abu-Khalaf, M. (2008). Discrete-time nonlinear HJB solution using approximate dynamic programming: Convergence proof, *IEEE Transactions on Systems, Man, and Cybernetics, Part B: Cybernetics* **38**, 4, pp. 943–949.

Basar, T. and Olsder, G. J. (1999). *Dynamic Noncooperative Game Theory*, 2nd edn. (SIAM: Philadelphia).

Beard, R. W. and Stepanyan, V. (2003). Information consensus in distributed multiple vehicle coordinated control, in *Proceedings of the IEEE Conference on Decision and Control* (Maui, Hawaii), pp. 2029–2034.

Bellman, R. E. (1957). *Dynamic Programming* (Princeton University Press).

Bertsekas, D. P. and Tsitsiklis, J. N. (1996). *Neuro-Dynamic Programming* (Athena Scientific).

Bryson, A. E. (1996). Optimal control–1950 to 1985, *IEEE Control Systems* **16**, 3, pp. 26–33.

Busoniu, L., Babuska, R. and De-Schutter, B. (2008). A comprehensive survey of multiagent reinforcement learning, *IEEE Transactions on Systems, Man, and Cybernetics, Part C: Applications and Reviews* **38**, 2, pp. 156–172.

Dierks, T. and Jagannathan, S. (2010). Optimal control of affine nonlinear continuous-time systems using an online Hamilton-Jacobi-Isaacs formulation, in *49th IEEE Conference on Decision and Control* (Atlanta, USA), pp. 3048–3053.

Fax, J. A. and Murray, R. M. (2004). Information flow and cooperative control of vehicle formations, *IEEE Transactions on Automatic Control* **49**, 9, pp. 1465–1476.

Freiling, G., Jank, G. and Abou-Kandil, H. (1996). On global existence of solutions to coupled matrix Riccati equations in closed-loop Nash games, *IEEE Transactions on Automatic Control* **41**, 2, pp. 264–269.

Gajic, Z. and Li, T. Y. (1988). Simulation results for two new algorithms for solving coupled algebraic Riccati equations, in *Third International Symposium on Differential Games* (Sophia, Antipolis).

Gonzalez, O. (1996). Time integration and discrete Hamiltonian systems, *Journal of Nonlinear Science* **6**, 5, pp. 449–467.

Hong, Y., Hu, J. and Gao, L. (2006). Tracking control for multi-agent consensus with an active leader and variable topology, *Automatica* **42**, 7, pp. 1177–1182.

Jadbabaie, A., Lin, J. and Morse, A. S. (2003). Coordination of groups of mobile autonomous agents using nearest neighbor rules, *IEEE Transactions on Automatic Control* **48**, 6, pp. 988–1001.

Johnson, M., Hiramatsu, T., Fitz-Coy, N. and Dixon, W. E. (2010). Asymptotic stackelberg optimal control design for an uncertain Euler Lagrange system, in *49th IEEE Conference on Decision and Control* (Atlanta, USA), pp. 6686–6691.

Kakade, S., Kearns, M., Langford, J. and Ortiz, L. (2003). Correlated equilibria in graphical games, in *Proceedings of the 4th ACM Conference on Electronic Commerce* (New York, USA), pp. 42–47.

Kearns, M. J., Littman, M. L. and Singh, S. P. (2001). Graphical models for game theory, in *Proceedings of 17th Annual Conference on Uncertainty in Artificial Intelligence* (San Francisco, USA), pp. 253–260.

Khoo, S., Xie, L. and Man, Z. (2009). Robust finite-time consensus tracking algorithm for multirobot systems, *IEEE/ASME Transactions on Mechatronics* **14**, 2, pp. 219–228.

Lall, S. and West, M. (2006). Discrete variational Hamiltonian mechanics, *Journal of Physics A: Mathematical and General* **39**, 19, pp. 5509–5519.

Lancaster, P. and Rodman, L. (1995). *Algebraic Riccati Equations* (Oxford University Press).

Lewis, F. L. (1992). *Applied Optimal Control and Estimation: Digital Design and Implementation* (Prentice Hall).

Lewis, F. L., Vrabie, D. and Syrmos, V. L. (2012). *Optimal Control*, 3rd edn. (Wiley).

Li, X., Wang, X. and Chen, G. (2004). Pinning a complex dynamical network to its equilibrium, *IEEE Transactions on Circuits and Systems I: Regular Papers* **51**, 10, pp. 2074–2087.

Littman, M. L. (2001). Value-function reinforcement learning in Markov games, *Cognitive Systems Research* **2**, 1, pp. 55–66.

Marsden, J. E. and West, M. (2001). Discrete mechanics and variational integrators, *Acta Numerica* **10**, 5, pp. 357–514.

McLachlan, R. I., Quispel, G. R. and Robidoux, N. (1999). Geometric integration using discrete gradients, *Philosophical Transactions of the Royal Society A* **357**, 1754, pp. 1021–1045.

Mu, S., Chu, T. and Wang, L. (2005). Coordinated collective motion in a motile particle group with a leader, *Physica A: Statistical Mechanics and its Applications* **351**, 2–4, pp. 211–226.

Olfati-Saber, R., Fax, J. A. and Murray, R. M. (2007). Consensus and cooperation in networked multi-agent systems. *Proceedings of the IEEE* **95**, 1, pp. 215–233.

Olfati-Saber, R. and Murray, R. M. (2004). Consensus problems in networks of agents with switching topology and time-delays, *IEEE Transactions on Automatic Control* **49**, 9, pp. 1520–1533.

Prokhorov, D. V. and Wunsch, D. C. (1997). Adaptive critic designs, *IEEE Transactions on Neural Networks* **8**, 5, pp. 997–1007.

Qu, Z. (2009). *Cooperative Control of Dynamical Systems: Applications to Autonomous Vehicles* (Springer).

Ren, W., Beard, R. W. and Atkins, E. (2005). A survey of consensus problems in multiagent coordination, in *Proceedings of the American Control Conference* (Portland, USA), pp. 1859–1864.

Ren, W. and Beard, R. W. (2005). Consensus seeking in multiagent systems under dynamically changing interaction topologies, *IEEE Transactions on Automatic Control* **50**, 5, pp. 655–661.

Ren, W. and Beard, R. W. (2008). *Distributed Consensus in Multi-vehicle Cooperative Control* (Springer).

Ren, W., Moore, K. L. and Chen, Y. Q. (2007). High-order and model reference consensus algorithms in cooperative control of multivehicle systems, *Journal of Dynamic Systems, Measurement, and Control* **129**, 5, pp. 678–688

Weiss, G. (ed.) (1999). *Multiagent Systems: A Modern Approach to Distributed Artificial Intelligence* (MIT Press).

Shoham, Y. and Leyton-Brown, K. (2009). *Multiagent Systems: Algorithmic, Game-Theoretic, and Logical Foundations* (Cambridge University Press).

Si, J., Barto, A. G., Powell, W. B. and Wunsch, D. (2004). *Handbook of Learning and Approximate Dynamic Programming* (Wiley-IEEE Press).

Sun, Y. G., Wang, L. and Xie, G. (2006). Average consensus in directed networks of dynamic agents with time-varying communication delays, in *45th IEEE Conference on Decision and Control* (San Diego, USA), pp. 3393–3398.

Suris, Y. B. (2003). *The Problem of Integrable Discretization: Hamiltonian Approach* (Birkhäuser).

Suris, Y. B. (2004). Discrete Lagrangian models, in B. Grammaticos, T. Tamizhmani, and Y. Kosmann-Schwarzbach (eds.), *Discrete Integrable Systems* (Springer, Berlin), pp. 111–184.

Sutton, R. S. and Barto, A. G. (1998). *Reinforcement Learning–An Introduction* (MIT Press).

Tsitsiklis, J. (1984). *Problems in Decentralized Decision Making and Computation*, Ph.D. Dissertation, Department of Electrical Engineering and Computer Science, MIT, Cambridge, USA.

Vamvoudakis, K. G. and Lewis, F. L. (2010). Online actor-critic algorithm to solve the continuous-time infinite horizon optimal control problem, *Automatica* **46**, 5, pp. 878–888.

Vamvoudakis, K. G. and Lewis, F. L. (2011). Multi-player non-zero-sum games: Online adaptive learning solution of coupled Hamilton-Jacobi equations, *Automatica* **47**, 8, pp. 1556–1569.

Vamvoudakis, K.G., Lewis, F. L. and Hudas, G. R. (2012). Multi-agent differential graphical games: Online adaptive learning solution for synchronization with optimality, *Automatica* **48**, 8, pp. 1598–1611.

Vrabie, D., Pastravanu, O., Abu-Khalaf, M. and Lewis, F. L. (2009). Adaptive optimal control for continuous-time linear systems based on policy iteration, *Automatica* **45**, 2, pp. 477–484.

Vrancx, P., Verbeeck, K. and Nowe, A. (2008). Decentralized learning in Markov games, *IEEE Transactions on Systems, Man, and Cybernetics, Part B: Cybernetics* **38**, 4, pp. 976–981.

Wang, X. F. and Chen, G. (2002). Pinning control of scale-free dynamical networks, *Physica A: Statistical Mechanics and its Applications* **310**, 3-4, pp. 521–531.

Werbos, P. J. (1974). *Beyond Regression: New Tools for Prediction and Analysis in the Behavior Sciences*, Ph.D. Dissertation, Harward University, USA.

Werbos, P. J. (1989). Neural networks for control and system identification, in *Proceedings of the 28th IEEE Conference on Decision and Control* (Tampa, Florida), pp. 260–265.

Werbos, P. J. (1992). Approximate dynamic programming for real-time control and neural modeling, in D. A. White and D. A. Sofge (eds.), *Handbook of Intelligent Control* (Van Nostrand Reinhold, New York), pp. 493–525.

Werbos, P. J. (2007). Using ADP to understand and replicate brain intelligence: The next level design, in L. I. Perlovsky and R. Kozma (eds.), *Neurodynamics of Cognition and Consciousness* (Springer, Berlin), pp. 109–123.

Werbos, P. J. (2009). Intelligence in the brain: A theory of how it works and how to build it, *Neural Networks* **22**, 3, pp. 200–212.

Widrow, B., Gupta, N. and Maitra, S. (1973). Punish/Reward: Learning with a critic in adaptive threshold systems, *IEEE Transactions on Systems, Man and Cybernetics* **3**, 5, pp. 455–465.

Zhang, H., Luo, Y. and Liu, D. (2009). Neural-network-based near-optimal control for a class of discrete-time affine nonlinear systems with control constraints, *IEEE Transactions on Neural Networks* **20**, 9, pp. 1490–1503.

Chapter 2

Reinforcement-Learning-Based Online Learning Control for Discrete-Time Unknown Nonaffine Nonlinear Systems

Xiong Yang, Derong Liu, Qinglai Wei, and Ding Wang

Abstract In this chapter, a reinforcement-learning-based online adaptive control is developed to deliver a desired tracking performance for a class of discrete-time (DT) nonlinear systems in the presence of unknown bounded disturbances. We study multi-input-multi-output unknown nonaffine nonlinear DT systems and employ two neural networks (NNs): an action NN is applied to generate the control signal and it is also designed to cancel the nonlinearity of unknown DT systems, and a critic NN is utilized to estimate the cost function, which satisfies the recursive equations derived from heuristic dynamic programming. The weights of both the action NN and the critic NN are directly updated online instead of preliminarily offline training. Based on Lyapunov's direct method, the overall adaptive scheme guarantees that all signals involved in the closed-loop system are uniformly ultimately bounded and the tracking errors can converge to a small neighborhood of zero. Simulation results are presented to demonstrate the effectiveness of the developed approach.

2.1 Introduction

Controller designs for systems with complex nonlinear dynamics have been an active area of research for at least a quarter of the century in the control community [Chemachema (2012); Chen and Khalil (1995); Ge *et al.* (1999); Lewis *et al.* (1996); Nakanishi and Schaal (2004); Narendra and Mukhopadhyay (1994)]. Plenty of distinguished results have been obtained owing to the advance in nonlinear control theory, especially feedback linearization technique, Lyapunov direct approach, and NN control. The objective of designing a controller is generally considered as finding stable controllers for nonlinear systems. However, stability is only a bare minimum requirement in system designs. The optimality based on a prescribed cost function is often required to be taken into consideration for control problems of nonlinear systems. In other words, control schemes should be developed to guarantee the stability of the closed-loop system, while keeping the cost function as small as possible.

In order to get such a controller, many significant methods have been proposed. Among these approaches, dynamic programming (DP) has been extensively applied to generate optimal control for nonlinear systems by employing Bellman's principle of optimality [Bellman (1957)]. The method ensures to perform optimization backward-in-time. However, a serious drawback about DP is that the computation is untenable to be run with the increasing dimension of nonlinear systems, which is referred to "curse of dimensionality". In addition, the backward direction of search prohibits the wide use of DP in real-time control process. On the other hand, with considerable investigations on the field of artificial NNs, researchers find NNs can be successfully applied to intelligent control due to their properties of nonlinearity, adaptivity, self-learning, and fault tolerance [Haykin (2008); Yu (2009)]. Consequently, NNs are used for universal function approximation in adaptive dynamic programming (ADP) algorithms, which are proposed by Werbos (1991, 1992, 2007, 2008), as methods to solve optimal control problems forward-in-time. There are several synonyms used for ADP including "adaptive dynamic programming" [Liu *et al.* (2013a,b); Murray *et al.* (2002); Wang *et al.* (2012a,b); Wei *et al.* (2010); Wang *et al.* (2009, 2011); Wei and Liu (2012); Zhang *et al.* (2011)], "approximate dynamic programming" [Al-Tamimi *et al.* (2008); Liu (2005)], "adaptive critic designs"(ACDs) [Prokhorov and Wunsch (1997)], "neuro-dynamic programming" (NDP) [Bertsekas and Tsitsiklis (1996)], and "neural dynamic programming" [Si and Wang (2001)]. Furthermore, according to Werbos (1992) and Prokhorov and Wunsch (1997), ADP algorithms are classified as follows: heuristic dynamic programming (HDP), dual heuristic programming (DHP), globalized dual heuristic programming (GDHP). When the action is introduced as an additional input to the critic, ACDs are referred to action dependent version of the ACDs, such as action dependent HDP (ADHDP), action dependent DHP (ADDHP), action dependent GDHP (ADGDHP).

Nevertheless, most of the ADP algorithms are implemented either by an offline process by using iterative schemes or need *a priori* knowledge of dynamics of nonlinear systems. Since the exact knowledge of nonlinear systems is generally unavailable, it brings about great challenges to implement these algorithms. For the sake of overcoming the difficulty, reinforcement learning (RL) is introduced. RL is a class of approaches used in machine learning to methodically revise the actions of an agent based on responses from its environment [Sutton and Barto (1998)]. A distinct difference between the traditional supervised NN learning and RL is that, there is no prescribed behavior or training model proposed to RL schemes. If the cost function is viewed as the reinforcement signal, then ADP algorithms become RL approaches. Consequently, ADP algorithms are actually a class of RL methods [Lewis and Vrabie (2009); Lewis and Vamvoudakis (2011)]. Since RL shares considerable common features with ADP algorithms, it is often employed for feedback controller designs.

RL-based feedback controllers for a class of nonlinear systems have been investigated in many literature [Bhasin *et al.* (2013); He and Jagannathan (2005); Lewis and Vrabie (2009); Lewis and Vamvoudakis (2011); Vamvoudakis and Lewis (2010, 2011); Yang *et al.* (2009); Yang and Jagannathan (2012); Yang *et al.* (2013)]. In He and Jagannathan (2005), an RL-based output feedback control was developed for multi-input-multi-output (MIMO) unknown affine nonlinear DT systems. By using Lyapunov's direct approach, the estimated state errors, the tracking errors and the estimated weights were all guaranteed to be uniformly ultimately bounded (UUB). After that, in Yang *et al.* (2009), a direct heuristic dynamic programming (DHDP) method was proposed to obtain online learning controller for MIMO unknown affine nonlinear DT systems. Using Lyapunov's direct method, the uniform ultimate boundedness of both the closed-loop tracking errors and the NN estimated weights was derived. Just as mentioned above, the authors took the cost function as the reinforcement signal. Recently, in Vamvoudakis and Lewis (2010), an online algorithm based on RL for affine nonlinear continuous-time (CT) systems was investigated. By using the proposed algorithm, both the optimal cost and the optimal control were well approximated in real time, while guaranteeing the uniform ultimate boundedness of the closed-loop system. In addition, the NN estimated weights were guaranteed to be UUB through Lyapunov's direct method. Later, in Vamvoudakis and Lewis (2011), RL methods were applied to multi-player differential games for nonlinear CT systems. Based on Lyapunov's direct method, the NN estimated weights were guaranteed to be UUB, while keeping the closed-loop system stable in the sense of uniform ultimate boundedness. After that, Bhasin *et al.* (2013) presented a projection algorithm based on RL to derive the optimal control of uncertain nonlinear CT systems. Based on the algorithm, the requirement of the prior knowledge of nonlinear dynamics was relaxed. Meanwhile, the actor, the critic, and the identifier were all simultaneously tuned. More recently, Yang *et al.* (2013) developed an identifer-critic architecture based on RL to investigate the optimal control of partially unknown affine nonlinear CT systems. Based on the presented method, the weights of the identifier NN and the critic NN were tuned simultaneously and were guaranteed to be UUB, while keeping the closed-loop system stable in the sense of uniform ultimate boundedness.

However, all of aforementioned literature deal with feedback control problems of RL methods for *affine* nonlinear systems. To the best of our knowledge, there are rather few investigations on feedback control of RL approaches for *nonaffine* nonlinear systems, especially MIMO unknown nonaffine nonlinear DT systems. Though there exist some researches about nonaffine nonlinear DT systems [Deng *et al.* (2008); Noriega and Wang (1998); Yang *et al.* (2008)], most of them focus on feedback control problems of nonlinear autoregressive moving average with exogenous inputs (NARMAX) systems. It is less convenient than the state-form of nonaffine nonlinear systems for purpose of adaptive control using NNs. On the other hand, since the output of affine nonlinear systems is linear with respect to the control input, it is easy to design a controller to follow prescribed trajectories by using feedback linearization methods. Nevertheless, feedback linearization approaches cannot be implemented for nonaffine nonlinear systems, for the output of this type of systems depends nonlinearly on the control signal. It gives rise to great difficulties for researchers to design an efficient controller of such a nonaffine nonlinear system, which aims at achieving desired trajectories. Furthermore, in real engineering, control approaches of affine nonlinear systems do not always hold and control methods for nonaffine nonlinear systems are necessary. Therefore, control problems of RL methods for unknown nonaffine nonlinear systems are very significant in both theory and applications.

The objective of this chapter is to develop an online direct adaptive control based on RL methods by delivering a desired tracking performance for MIMO unknown nonaffine nonlinear DT systems with unknown bounded disturbances. Two NNs are employed in controller designs. An action NN is to generate the control signal. Meanwhile, by using Implicit Function Theorem, the action NN approximation is well designed to cancel the nonlinearity of unknown nonlinear DT systems, for purpose of utilizing feedback linearization methods. A critic NN is used to estimate the cost function, which satisfies the recursive equations derived from HDP. The weights of both the action NN and the critic NN are directly updated online instead of preliminary offline training. By employing Lyapunov's direct method, the overall adaptive scheme guarantees that all signals involved are UUB and the tracking errors can converge to a small neighborhood of zero.

It is worth pointing out that a distinct difference between above literature [Bhasin *et al.* (2013); He and Jagannathan (2005); Vamvoudakis and Lewis (2010, 2011); Yang *et al.* (2009); Yang and Jagannathan (2012)] and the present chapter is that, in our case, we deal with MIMO unknown nonaffine nonlinear DT systems, which is a rather complicated case as mentioned before. In addition, the technique applied to design controllers for nonaffine nonlinear systems is considered as the feedback linearization approach, however, it is quite different from the method applied to the classic affine nonlinear systems. For affine nonlinear systems, we can get the concrete form of the controller by directly solving the system equations with respect to the input signal. While for nonaffine nonlinear systems, the controller can be only derived as the abstract form by applying the Implicit Function Theorem to system equations.

The chapter is organized as follows. Section 2.2 provides the problem statement and preliminaries. Section 2.3 develops an online adaptive control by using RL approaches. Section 2.4 shows the stability analysis and the performance of the closed-loop system. Section 2.5 provides two simulation results to verify the effectiveness of the developed theory. Finally, Section 2.6 gives several concluding remarks.

For convenience, notations used in this chapter are listed as follows:

- $\mathbb{R}$ denotes the set of the real numbers, $\mathbb{R}^m$ and $\mathbb{R}^{m\times n}$ denote the set of the real m-vectors and the set of the real $m \times n$ matrices, respectively. T is a transpose signal.
- Ω is a compact set of $\mathbb{R}^m$, $C^m(\Omega) = \{f^{(m)} \in C^1 | f : \Omega \to \mathbb{R}^m\}$. When α is a vector, $\|\alpha\|$ denotes the Euclidean norm of α. When A is a matrix, $\|A\|$ denotes the 2-norm of A.
- I_m, I_{mn}, and I_{s_1} $(s_1 \in \mathbb{N}_+)$ represent the set of $m \times m$ identity matrices, the set of $mn \times mn$ identity matrices, and the set of $s_1 \times s_1$ identity matrices, respectively.

2.2 Problem Statement and Preliminaries

2.2.1 *Dynamics of nonaffine nonlinear discrete-time systems*

Consider an mnth-order MIMO nonaffine nonlinear DT plant of the form

$$\begin{aligned} x_1(k+1) &= x_2(k) \\ &\vdots \\ x_{n-1}(k+1) &= x_n(k) \\ x_n(k+1) &= h\big(x(k), u(x(k))\big) + d(k) \\ y(k) &= x_1(k) \end{aligned} \tag{2.1}$$

where the state $x(k) = \left[x_1^{\mathsf{T}}(k), x_2^{\mathsf{T}}(k), \ldots, x_n^{\mathsf{T}}(k)\right]^{\mathsf{T}} \in \mathbb{R}^{mn}$ with $x_i(k) \in \mathbb{R}^m$ for $i = 1, \ldots, n$, and $u(x(k)) \in \mathbb{R}^m$ is the control vector which is a continuous function with respect to $x(k)$. For convenience, we denote $v(k) = u(x(k))$. $d(k) \in \mathbb{R}^m$ is an unknown disturbance bounded by a known positive constant d_M, i.e., $\|d(k)\| \leq d_M$, $h\big(x(k), v(k)\big) \in \mathbb{R}^m$ is an unknown nonaffine nonlinear function with $h(0,0) = 0$, and $y(k)$ is the output vector. It is assumed that the state $x(k)$ is available at the k-th step for the state-feedback controller. Meanwhile, we provide the following assumption, which is made for the controllability of the system.

Assumption 2.1. The $m \times m$ matrix $\partial h\big(x(k), v(k)\big)/\partial v(k)$ is positive definite. It implies

$$\det\left[\frac{\partial h\big(x(k), v(k)\big)}{\partial v(k)}\right] \neq 0 \tag{2.2}$$

for $\forall\ \big(x(k), v(k)\big) \in \Omega \times \mathbb{R}^m$ with a compact region $\Omega \subset \mathbb{R}^{mn}$.

Before continuing our discussion, we provide the following required definition. The definition is introduced in Lewis *et al.* (1999), and readers can refer to the monograph for more details.

Definition 2.1. (Uniform Ultimate Boundedness) The equilibrium point x_e of system (2.1) is said to be UUB if there exists a compact set $\Omega \subset \mathbb{R}^{mn}$ such that $\forall x_0 \in \Omega$ ($x_0 = x(k_0)$, k_0 is the initial time), there exists a bound $\epsilon > 0$ and a positive number $N(x_0, \epsilon)$ such that $\|x(k) - x_e\| < \epsilon$ for all $k \geq k_0 + N$.

The main objective in this chapter is to design an NN-based robust adaptive controller for system (2.1) such that the following results hold:

(a) All the estimated variables in the closed-loop system are guaranteed to be UUB.
(b) The system output $y(k)$ follows a given bounded reference signal $y_d(k)$ to a small neighborhood of the origin.
(c) The cost function (to be detailed in the subsequent section) is minimized such that the approximate optimal control is obtained. The action NN and the critic NN are tuned online instead of preliminary offline training.

Assumption 2.2. The desired trajectory of system (2.1) is described as $x_d(k) = \left[x_{1d}^{\mathsf{T}}(k), x_{2d}^{\mathsf{T}}(k), \ldots, x_{nd}^{\mathsf{T}}(k)\right]^{\mathsf{T}} \in \mathbb{R}^{mn}$. $x_{id}(k)$ is arbitrarily selected and satisfies $x_{id}(k+1) = x_{(i+1)d}(k)$, $i = 1, \ldots, n-1$. The desired output trajectory $y_d(k)$ is bounded by a known smooth function over the compact set Ω.

From Assumption 2.2 and system (2.1), we can obtain $x_{(i+1)d}(k) = y_d(k+i)$, $i = 0, \ldots, n-1$. Accordingly, the tracking error can be defined as

$$\begin{aligned} e_i(k) &= y_d(k+i) - y(k+i) \\ &= x_{(i+1)d}(k) - x_{i+1}(k) \end{aligned} \tag{2.3}$$

where $i = 0, \ldots, n-1$.

2.2.2 *A single-hidden layer neural network*

The purpose of this section is to introduce a single-hidden layer feedforward NN, which is applied to our controller architecture. As we know, $u(k)$ in system (2.1) is generally a nonlinear function with respect to $x(k)$. The lack of a general structure makes the design of such a nonlinear controller rather intractable. In addition, the nonlinear controller design is actually the function approximation problem. Hence, according to the universal approximation property of NNs [Haykin (2008)], a single-hidden layer feedforward NN can be successfully utilized to deal with the problem. A general function $\mathcal{G}(x) \in \mathrm{C}^m(\Omega)$ can be written as [Haykin (2008)]:

$$\mathcal{G}(x) = W^{\mathsf{T}}\sigma(V^{\mathsf{T}}x) + \varepsilon(x) \tag{2.4}$$

with $\sigma(\cdot)$ the activation function, $\varepsilon(x)$ the NN functional reconstruction error, V and W the weight vectors for the input layer to the hidden layer and the hidden layer to the output layer, respectively. The number of hidden layer nodes is denoted as N (see Fig. 2.1). It has been shown in Igelnik and Pao (1995) that, if the hidden layer weight vector V is selected initially at random and held constant, as N is large enough, $\varepsilon(x)$ can be arbitrarily small. That is, there exists a positive number N_0 such that $N \geq N_0$ implies $\|\varepsilon(x)\| \leq \varepsilon_N$. In general, activation functions $\sigma(x)$ are bounded, measurable, nondecreasing functions from real numbers onto $[-1, 1]$ which include, for instance, hyperbolic tangent function $\sigma(x) = (e^x - e^{-x})/(e^x + e^{-x})$, etc. In this chapter, this type of NNs is applied to derive (a)–(e).

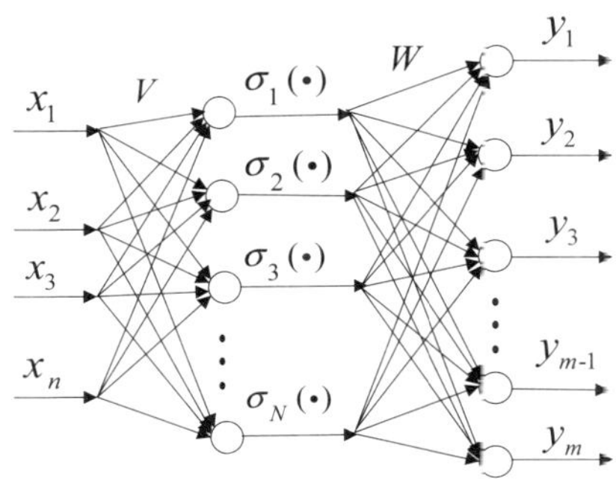

Fig. 2.1 Structure of a single-hidden layer feedforward NN

2.3 Controller Design via Reinforcement Learning

The purpose of this section is to develop an online learning controller by using RL methods. Three subsections are included in this section. First, a basic controller design approach is developed. Then, the design of the critic NN is introduced. Finally, the design of the action NN is presented.

2.3.1 *A basic controller design approach*

Since the output of nonaffine nonlinear systems depends nonlinearly on the control input, feedback linearization methods cannot be used to design the controller. In order to deal with this problem, we develop a novel control law for plant (2.1) based on Hovakimyan *et al.* (2002) and Park *et al.* (2005), which study CT adaptive control problems.

From system (2.1), we have

$$\begin{aligned} y(k+n) &= h\big(x(k), v(k)\big) + d(k) \\ &= \alpha v(k) + f\big(x(k), v(k)\big) + d(k) \end{aligned} \tag{2.5}$$

where $f\big(x(k), v(k)\big) = h\big(x(k), v(k)\big) - \alpha v(k)$, and $\alpha > 0$ is a design constant.

Define the control $v(k)$ as

$$v(k) = \frac{1}{\alpha}\big(v_s(k) - v_a(k)\big) \tag{2.6}$$

where $v_s(k)$ is a feedback controller to stabilize linearized error dynamics, $v_a(k)$

is an adaptive controller designed to approximate the unknown nonlinear term $f\big(x(k), \upsilon(k)\big)$ by using NNs.

From (2.5) and (2.6), we obtain

$$y(k+n) = f\big(x(k), \upsilon(k)\big) - \upsilon_a(k) + \upsilon_s(k) + d(k). \tag{2.7}$$

In view of the task of $\upsilon_s(k)$ and $\upsilon_a(k)$, we propose

$$\begin{cases} \upsilon_a(k) = \hat{f}\big(x(k), \upsilon(k)\big) \\ \upsilon_s(k) = y_d(k+n) + \lambda_1 e_{n-1}(k) + \cdots + \lambda_n e_0(k) \end{cases} \tag{2.8}$$

where $\hat{f}\big(x(k), \upsilon(k)\big)$ is an approximation of $f\big(x(k), \upsilon(k)\big)$, $e_{n-1}(k), \ldots, e_0(k)$ are the delayed values of $e_n(k)$, and $\lambda_1, \ldots, \lambda_n$ are constant matrices selected such that $|z^n + \lambda_1 z^{n-1} + \cdots + \lambda_n|$ is stable, i.e., the solutions of $|z^n + \lambda_1 z^{n-1} + \cdots + \lambda_n| = 0$ are located inside the unit circle centered at the origin.

Let the approximation error of $f\big(x(k), \upsilon(k)\big)$ be

$$\tilde{f}\big(x(k), \upsilon(k)\big) = \hat{f}\big(x(k), \upsilon(k)\big) - f\big(x(k), \upsilon(k)\big). \tag{2.9}$$

Then, we can derive the following lemma.

Lemma 2.1. *Assume that the tracking error $e_i(k)$ is given by (2.3) and $\upsilon_s(k)$ is proposed as in (2.8). Then, the error dynamics can be derived as*

$$\boldsymbol{e}(k+1) = \tilde{A}\boldsymbol{e}(k) + \tilde{B}\big[\tilde{f}(x(k), \upsilon(k)) - d(k)\big] \tag{2.10}$$

where

$$\begin{gathered} \boldsymbol{e}(k) = \big[e_0^{\mathsf{T}}(k), \ldots, e_{n-1}^{\mathsf{T}}(k)\big]^{\mathsf{T}} \\ A = \begin{bmatrix} 0 & 1 & \cdots & 0 \\ \vdots & \vdots & & \vdots \\ -\lambda_n & -\lambda_{n-1} & \cdots & -\lambda_1 \end{bmatrix}, \quad B = \begin{bmatrix} 0 \\ \vdots \\ 1 \end{bmatrix} \\ \tilde{A} = A \otimes I_m, \quad \tilde{B} = B \otimes I_m. \end{gathered} \tag{2.11}$$

Proof. From Assumption 2.2 and (2.3), we have

$$\begin{aligned} e_i(k) &= y_d(k+i) - y(k+i) \\ &= x_{id}(k+1) - x_i(k+1) \\ &= e_{i-1}(k+1), \quad i = 1, \ldots, n-1. \end{aligned} \tag{2.12}$$

Meanwhile, by using (2.7)–(2.9), we can obtain

$$e_n(k) = -\boldsymbol{\lambda}^{\mathsf{T}}\boldsymbol{e}(k) + \tilde{f}\big(x(k), \upsilon(k)\big) - d(k)$$

where $\boldsymbol{\lambda} = [\lambda_n, \ldots, \lambda_1]^{\mathsf{T}}$.

Noticing that $e_n(k) = e_{n-1}(k+1)$, we derive

$$e_{n-1}(k+1) = -\boldsymbol{\lambda}^{\mathsf{T}}\boldsymbol{e}(k) + \tilde{f}\big(x(k), \upsilon(k)\big) - d(k). \tag{2.13}$$

Accordingly, combining (2.12) and (2.13), we can get

$$\begin{cases} e_0(k+1) = e_1(k) \\ \quad\vdots \\ e_{n-1}(k+1) = -\boldsymbol{\lambda}^{\mathsf{T}}\boldsymbol{e}(k) + \tilde{f}\big(x(k), \upsilon(k)\big) - d(k). \end{cases} \tag{2.14}$$

Rewriting (2.14) in the vector form, and observing the tracking error $\boldsymbol{e}(k) \in \mathbb{R}^{mn}$, we derive (2.10) and (2.11). □

Remark 2.1. If there exists a feedforward control $\upsilon_a(k)$ successfully canceling the term $f\big(x(k), \upsilon(k)\big)$, i.e., $\tilde{f}\big(x(k), \upsilon(k)\big) = 0$, and ignore the disturbance term $d(k)$, i.e., $d(k) = 0$, then the closed-loop system becomes a linear system $\boldsymbol{e}(k+1) = \tilde{A}\boldsymbol{e}(k)$. Since $\lambda_1, \ldots, \lambda_n$ are constant matrices selected such that $|z^n + \lambda_1 z^{n-1} + \cdots + \lambda_n|$ is stable, it is obvious that $\tilde{A}$ can keep the linear system $\boldsymbol{e}(k+1) = \tilde{A}\boldsymbol{e}(k)$ stable (for short, $\tilde{A}$ is a stable matrix). Therefore, letting $f\big(x(k), \upsilon(k)\big) = \upsilon_a(k)$ and $d(k) = 0$, $\upsilon_s(k)$ can make the tracking error $\boldsymbol{e}(k)$ exponentially converge to zero as time increases. This shows that the definition of $\upsilon_s(k)$ in (2.8) makes sense.

Though the design of $\upsilon_s(k)$ is reasonable, one may doubt whether $\upsilon_a(k)$ exists or not. The following theorem is developed to show that the controller $\upsilon_a(k)$ does exist. For convenience, the Implicit Function Theorem for vector-valued functions is first provided, which plays a significant role in the subsequent proof.

Lemma 2.2. *(Implicit Function Theorem [Apostol (1974)]) Let $\boldsymbol{f} = (f_1, \ldots, f_n)$ be a vector-valued function defined on an open set S in $\mathbb{R}^{m+n}$ with values in $\mathbb{R}^n$. Suppose $\boldsymbol{f} \in C^1$ on S. Let $(\boldsymbol{x}_0, \boldsymbol{y}_0)$ be a point in S for which $\boldsymbol{f}(\boldsymbol{x}_0, \boldsymbol{y}_0) = 0$, and for which the $n \times n$ determinant $\det[\partial \boldsymbol{f}(\boldsymbol{x}_0, \boldsymbol{y}_0)/\partial \boldsymbol{y}_0] \neq 0$. Then there exists a n-dimensional open set T_0 containing $\boldsymbol{y}_0$ and one, and only one, vector-valued function $\boldsymbol{g}$, defined on T_0 and having values in $\mathbb{R}^n$, such that (i) $\boldsymbol{y}_0 = \boldsymbol{g}(\boldsymbol{x}_0)$; (ii) for $\forall$ $(\boldsymbol{x}_0, \boldsymbol{y}_0) \in T_0$, $\boldsymbol{f}\big(\boldsymbol{x}_0, \boldsymbol{g}(\boldsymbol{x}_0)\big) = 0$.*

By using (2.6), we can define

$$\begin{aligned} F\big(x(k), \upsilon_a(k), \upsilon_s(k)\big) &= f\big(x(k), \upsilon(k)\big) - \upsilon_a(k) \\ &= f\left(x(k), \frac{\upsilon_s(k) - \upsilon_a(k)}{\alpha}\right) - \upsilon_a(k). \end{aligned}$$

Let $F\big(x(k), \upsilon_a(k), \upsilon_s(k)\big) = 0$. Then, we have

$$F\big(x(k), \upsilon_a(k), \upsilon_s(k)\big) = f\left(x(k), \frac{\upsilon_s(k) - \upsilon_a(k)}{\alpha}\right) - \upsilon_a(k) = 0. \tag{2.15}$$

Theorem 2.1. *Assume that the following matrix inequality holds:*

$$\alpha\theta_1 I_m \leq \frac{\partial h\big(x(k), \upsilon(k)\big)}{\partial \upsilon(k)} \leq \alpha\theta_2 I_m \tag{2.16}$$

where $0 < \theta_1 < \theta_2 \leq 2$. Then, there exists a compact set $\Omega' \subset \Omega$ and a unique $v_a\big(x(k), v_s(k)\big)$ (Note: $v_a(k)$ is a function with respect to $x(k)$ and $v_s(k)$) satisfying (2.15) for $\forall\ \big(x(k), v_s(k)\big) \in \Omega' \times U$, where $U \subset \mathbb{R}^m$ is a compact set.

Proof. In order to utilize Lemma 2.2, the proof is divided into two parts. First, we show that there exists a solution of (2.15) (the solution is denoted by $v_a^*(k)$). Then, we show that $\det\big[\partial F\big(x(k), v_s(k), v_a^*(k)\big)/\partial v_a^*(k)\big] \neq 0$.

(i) The proof for showing the existence of $v_a^*(k)$.
In light of the expression of (2.15), if the conclusion is true, we have

$$v_a^*(k) = f\left(x(k), \frac{v_s(k) - v_a^*(k)}{\alpha}\right). \tag{2.17}$$

That is, $v_a^*(k)$ is the fixed point of (2.17). Consequently, we just need to prove that $f\big(x(k), \cdot\big)$ is the contracting operator on a compact set $U \in \mathbb{R}^m$. Since $x(k)$ is defined on the compact Ω and $v(k)$ is a continuous function with respect to $x(k)$, by the knowledge of Functional Analysis [Rudin (1991)], we can derive that $v(\Omega)$ is a compact set on $\mathbb{R}^m$. Hence, we can select $U = v(\Omega)$. Notice that

$$\begin{aligned}\left\|\frac{\partial f\big(x(k), v(k)\big)}{\partial v_a(k)}\right\| &= \left\|\frac{\partial f\big(x(k), v(k)\big)}{\partial v(k)}\frac{\partial v(k)}{\partial v_a(k)}\right\| \\ &= \left\|\left(\frac{\partial h\big(x(k), v(k)\big)}{\partial v(k)} - \alpha I_m\right)\left(-\frac{I_m}{\alpha}\right)\right\| \\ &= \left\|I_m - \frac{\partial h\big(x(k), v(k)\big)}{\alpha\partial v(k)}\right\|.\end{aligned} \tag{2.18}$$

By using the matrix inequality (2.16), we have

$$\begin{cases} I_m - \dfrac{\partial h\big(x(k), v(k)\big)}{\alpha\partial v(k)} \geq (1-\theta_2)I_m \\ I_m - \dfrac{\partial h\big(x(k), v(k)\big)}{\alpha\partial v(k)} \leq (1-\theta_1)I_m. \end{cases} \tag{2.19}$$

Therefore, from (2.18) and (2.19), we can get

$$\left\|\frac{\partial f\big(x(k), v(k)\big)}{\partial v_a(k)}\right\| \leq 1. \tag{2.20}$$

Noticing that $f\big(x(k), v(k)\big)$ is a continuous function and $\big(x(k), v(k)\big)$ is defined on the compact set $\Omega \times U$, we can conclude that $f\big(\Omega \times U\big)$ is a compact set on $\mathbb{R}^{nm} \times \mathbb{R}^m$. Hence, we can draw the conclusion that $f\big(x(k), \cdot\big)$ is a completely continuous operator [Zeidler (1985)]. By using Schauder's Fixed-Point Theorem [Zeidler (1985)] and from (2.20), we derive that there exists at least a fixed point for the operator $f\big(x(k), \cdot\big)$ on the compact set U. That is, there exists $v_a^*(k)$ satisfying (2.15).

(ii) The proof for $\det\left[\partial F\big(x(k), v_s(k), v_a^*(k)\big)/\partial v_a^*(k)\right] \neq 0$.
Notice that

$$\begin{aligned}
&\frac{\partial F\big(x(k), v_s(k), v_a(k)\big)}{\partial v_a(k)}\bigg|_{v_a(k)=v_a^*(k)} \\
&= \frac{\partial\Big(h\big(x(k), v(k)\big) - \alpha v(k)\Big)}{\partial v(k)} \frac{\partial v(k)}{\partial v_a(k)}\bigg|_{v_a(k)=v_a^*(k)} - I_m \\
&= -\frac{\partial h\big(x(k), v(k)\big)}{\partial v(k)}\bigg|_{v_a(k)=v_a^*(k)}.
\end{aligned}$$

By using (2.2), we can derive that $\det\left[\partial F\big(x(k), v_s(k), v_a^*(k)\big)/\partial v_a^*(k)\right] \neq 0$.

Hence, combining (i) and (ii), and using Lemma 2.2, we obtain that there exists a unique $v_a\big(x(k), v_s(k)\big)$ satisfying (2.15) for $\forall\ \big(x(k), v_s(k)\big) \in \Omega' \times U\ (\Omega' \subset \Omega)$. □

Remark 2.2. From Assumption 2.1, one shall notice that the matrix inequality (2.16) is actually one of the properties of the positive definite matrix $\partial h\big(x(k), v(k)\big)/\partial v(k)$. This technique was utilized in Lewis *et al.* (1996) and Lewis *et al.* (1999). Hence, the assumption about (2.16) makes sense.

Remark 2.3. Though, by using Schauder's Fixed-Point Theorem, we derive that there exists at least one solution $v_a^*(k)$ satisfying (2.15) in part (i), it doesn't impair the validity of the conclusion, for we just need to show the existence of $v_a^*(k)$. The uniqueness of $v_a^*(k)$ is guaranteed by Implicit Function Theorem (Lemma 2.2). Therefore, one shall find the solution of (2.15) to be unique for a given local domain. This feature satisfies the nonlinearity of the function $F\big(x(k), v_a(k), v_s(k)\big)$. Moreover, from the above analysis, we don't need $f\big(x(k), \cdot\big)$ to be a strictly contracting operator, which is a more relaxed condition than Hovakimyan *et al.* (2002) and Park *et al.* (2005).

Remark 2.4. Since there exists a unique $v_a(k)$ satisfying (2.17) on Ω', by means of the feature of $v_a(k)$, we can conclude that $f\big(x(k), v(k)\big)$ is actually a function with respect to $x(k)$ and $v_s(k)$. Moreover, from the definition of $e_i(k)$ in (2.3) and the expression of $v_s(k)$ in (2.8), $f\big(x(k), v(k)\big)$ can be represented as a function with respect to $x(k)$ and $x_d(k)$. Hence, in the remainder of this chapter, we denote

$$f\big(x(k), x_d(k)\big) = f\big(x(k), v(k)\big)$$

and

$$\tilde{f}\big(x(k), x_d(k)\big) = \tilde{f}\big(x(k), v(k)\big).$$

2.3.2 *Critic neural network and weight update law*

In this subsection, a critic NN is used to approximate the cost function $J(k)$ defined as in (2.22). The utility function [Si and Wang (2001)] depending on the tracking error $e(k)$ is described by

$$r(k) = \big[r_1(k), \dots, r_m(k)\big] \in \mathbb{R}^m \tag{2.21}$$

with

$$r_i(k) = \begin{cases} 0, & \text{if} \quad \|\tilde{e}_i(k)\| \leq \epsilon \\ 1, & \text{if} \quad \|\tilde{e}_i(k)\| > \epsilon \end{cases} \qquad i = 1, \dots, m,$$

where $\tilde{e}(k) = \boldsymbol{\lambda}^{\mathsf{T}}\boldsymbol{e}(k) \in \mathbb{R}^m$, $\tilde{e}_i(k)$ is the ith element of the vector $\tilde{e}(k)$, and $\epsilon > 0$ is a prescribed threshold. The utility function $r(k)$ is considered as the performance index, where $r_i(k) = 0$ and $r_i(k) = 1$ stand for the good and poor tracking performance, respectively.

The cost function $J(k) \in \mathbb{R}^m$ [He and Jagannathan (2005)] is defined as

$$J(k) = \tau^N r(k+1) + \tau^{N-1} r(k+2) + \cdots + \tau^{k+1} r(N) \tag{2.22}$$

with $0 < \tau \leq 1$ a design parameter, and N the final instant of time.

From (2.22), we can derive

$$J(k) = \tau J(k-1) - \tau^{N+1} r(k)$$

which is the Bellman equation.

Hence, the prediction error for the critic NN can be given by

$$e_c(k) = \hat{J}(k) - \tau\hat{J}(k-1) + \tau^{N+1} r(k) \tag{2.23}$$

where $\hat{J}(k)$ is the critic NN output, and it is an approximation of $J(k)$.

The critic NN is implemented by a single-hidden layer feedforward NN, and its output is given by

$$\hat{J}(k) = \hat{w}_c^{\mathsf{T}}(k)\sigma\big(\vartheta_c^{\mathsf{T}} x(k)\big) = \hat{w}_c^{\mathsf{T}}(k)\sigma_c\big(x(k)\big) \tag{2.24}$$

where $\vartheta_c \in \mathbb{R}^{nm \times s_1}$ is the weight vector for the input layer to the hidden layer of the critic NN, $\hat{w}_c(k) \in \mathbb{R}^{s_1 \times m}$ is the estimated weight vector for the hidden layer to the output layer of the critic NN, and s_1 is the number of nodes in the hidden layer. Since the hidden layer weights are initialized randomly and kept constant, the activation function $\sigma(\vartheta_c^{\mathsf{T}} x(k))$ is written as $\sigma_c(x(k))$ for briefly.

The objective function to be minimized by the critic NN is defined as

$$E_c(k) = \frac{1}{2} e_c^{\mathsf{T}}(k) e_c(k). \tag{2.25}$$

The weight update law for the critic NN is a gradient-based adaptation, which is described by

$$\hat{w}_c(k+1) = \hat{w}_c(k) + \Delta\hat{w}_c(k) \tag{2.26}$$

where

$$\begin{aligned}\Delta\hat{w}_c(k) &= l_c\left[-\frac{\partial E_c(k)}{\partial \hat{w}_c(k)}\right]\\ &= l_c\left[-\frac{\partial E_c(k)}{\partial e_c(k)}\frac{\partial e_c(k)}{\partial \hat{J}(k)}\frac{\partial \hat{J}(k)}{\partial \hat{w}_c(k)}\right]\end{aligned}$$

and $0 < l_c < 1$ is the learning rate of the critic NN.

From (2.23)–(2.26), we can derive the weight update law for the critic NN as

$$\begin{aligned}\hat{w}_c(k+1) &= \hat{w}_c(k) - l_c\sigma_c\big(x(k)\big)e_c^{\mathsf{T}}(k)\\ &= \hat{w}_c(k) - l_c\sigma_c(x(k))\Big[\hat{w}_c^{\mathsf{T}}(k)\sigma_c\big(x(k)\big) + \tau^{N+1}\\ &\quad \times r(k) - \tau\hat{w}_c^{\mathsf{T}}(k-1)\sigma_c\big(x(k-1)\big)\Big]^{\mathsf{T}}.\end{aligned} \tag{2.27}$$

2.3.3 *Action neural network and weight update law*

In this subsection, an action NN is employed to generate the input signal and approximate the unknown nonlinear function $f\big(x(k), v(k)\big)$. Due to the controller design described by (2.8), the error for the action NN should consist of the functional approximation error $\tilde{f}\big(x(k), v(k)\big)$ and the error between the nominal prescribed cost function $J_d(k) \in \mathbb{R}^m$ and the critic NN output $\hat{J}(k) \in \mathbb{R}^m$.

Noticing that $\tilde{f}\big(x(k), x_d(k)\big) = \tilde{f}\big(x(k), v(k)\big)$, the prediction error for the action NN can be developed by

$$e_a(k) = \hat{J}(k) - J_d(k) + \tilde{f}\big(x(k), x_d(k)\big). \tag{2.28}$$

The prescribed cost function $J_d(k)$ is generally considered to be zero, i.e., $J_d(k) = 0$, which represents that the system state can track the reference signal well [Si and Wang (2001)].

Therefore, the prediction error given by (2.28) becomes

$$e_a(k) = \hat{J}(k) + \tilde{f}\big(x(k), x_d(k)\big). \tag{2.29}$$

The action NN is also implemented by a single-hidden layer feedforward NN, and its output is given by

$$\hat{f}(k) = \hat{w}_a^{\mathsf{T}}(k)\sigma\big(\vartheta_a^{\mathsf{T}} z(k)\big) = \hat{w}_a^{\mathsf{T}}(k)\sigma_a\big(z(k)\big) \tag{2.30}$$

where $\hat{f}(k)$ stands for $\hat{f}\big(x(k), x_d(k)\big)$, $\vartheta_a \in \mathbb{R}^{(n+1)m\times s_2}$ is the weight vector for the input layer to the hidden layer of the action NN, $\hat{w}_c(k) \in \mathbb{R}^{s_2\times m}$ is the estimated weight vector for the hidden layer to the output layer of the action NN, and s_2 is the number of nodes in the hidden layer, and $z(k) = \big[x^{\mathsf{T}}(k)\ x_d^{\mathsf{T}}(k)\big]^{\mathsf{T}} \in \mathbb{R}^{(n+1)m}$. Since the hidden layer weights are initialized randomly and kept constant, for briefly, the activation function $\sigma\big(\vartheta_a^{\mathsf{T}} z(k)\big)$ is written as $\sigma_a(z(k))$.

Remark 2.5. From (2.8), we shall find that the output of the action NN is actually $\upsilon_a(k)$. Meanwhile, we have that $z(k) = \big[x^{\mathsf{T}}(k)\ x_{nd}^{\mathsf{T}}(k) + \mathbf{e}^{\mathsf{T}}(k)\boldsymbol{\lambda}\big]^{\mathsf{T}} \in \mathbb{R}^{(n+1)m}$.

By utilizing (2.4), $f(k)$ can accurately be represented by a single-hidden layer feedforward NN as

$$f(k) = w_a^{\mathsf{T}} \sigma_a\big(z(k)\big) + \varepsilon(k) \tag{2.31}$$

where $f(k)$ denotes $f\big(x(k), x_d(k)\big)$, $w_a \in \mathbb{R}^{s_2 \times m}$ is the ideal weight vector for the hidden layer to the output layer of the action NN, and $\varepsilon(k)$ is the action NN approximation error.

Define the weight approximation error of the action NN as

$$\tilde{w}_a(k) = \hat{w}_a(k) - w_a. \tag{2.32}$$

It is important to note that the relationship between $\tilde{w}_c(k)$ and $\hat{w}_c(k)$ is defined as the same as in (2.32). From (2.9) and (2.30), (2.31), and (2.32), the function approximation error $\tilde{f}(k)$ can be represented as

$$\tilde{f}(k) = \tilde{w}_a^{\mathsf{T}}(k)\sigma_a\big(z(k)\big) - \varepsilon(k). \tag{2.33}$$

The objective function to be minimized by the action NN is defined as a quadratic function of error as

$$E_a(k) = \frac{1}{2} e_a^{\mathsf{T}}(k) e_a(k). \tag{2.34}$$

The weight update law for the action NN is given by

$$\hat{w}_a(k+1) = \hat{w}_a(k) + \Delta \hat{w}_a(k) \tag{2.35}$$

where

$$\begin{aligned}
\Delta \hat{w}_a(k) &= l_a \left[-\frac{\partial E_a(k)}{\partial \hat{w}_a(k)} \right] \\
&= l_a \left[-\frac{\partial E_a(k)}{\partial e_a(k)} \frac{\partial e_a(k)}{\partial \tilde{f}(k)} \frac{\partial \tilde{f}(k)}{\partial \hat{w}_a(k)} \right]
\end{aligned}$$

and $0 < l_a < 1$ is the learning rate for the action NN.

From (2.29)–(2.35), we get the weight update law for the action NN as

$$\begin{aligned}
\hat{w}_a(k+1) &= \hat{w}_a(k) - l_a \sigma_a\big(z(k)\big) e_a^{\mathsf{T}}(k) \\
&= \hat{w}_a(k) - l_a \sigma_a\big(z(k)\big) \Big[\hat{w}_c^{\mathsf{T}}(k) \sigma_c\big(x(k)\big) \\
&\quad + \tilde{w}_a^{\mathsf{T}}(k) \sigma_a\big(z(k)\big) - \varepsilon(k) \Big]^{\mathsf{T}}.
\end{aligned} \tag{2.36}$$

From (2.10) and Remark 2.4, we have

$$\tilde{f}(k) = \tilde{B}^{\mathsf{T}}\big(e(k+1) - \tilde{A}e(k)\big) + d(k). \tag{2.37}$$

By utilizing (2.33) and (2.37), (2.36) can be rewritten as

$$\begin{aligned}\hat{w}_a(k+1) = \hat{w}_a(k) - l_a\sigma_a\big(z(k)\big)\Big[\hat{w}_c^{\mathsf{T}}(k)\sigma_c\big(x(k)\big)\\ + \tilde{B}^{\mathsf{T}}\big(e(k+1) - \tilde{A}e(k)\big) + d(k)\Big]^{\mathsf{T}}.\end{aligned} \tag{2.38}$$

Nevertheless, $d(k)$ is typically unavailable. Consequently, the ideal case of $d(k)$ is considered, i.e., $d(k) = 0$ [He and Jagannathan (2005); Yang *et al.* (2009); Yang and Jagannathan (2012)]. Then, the weight update rule (2.38) becomes

$$\begin{aligned}\hat{w}_a(k+1) = \hat{w}_a(k) - l_a\sigma_a(z(k))\Big[\hat{w}_c^{\mathsf{T}}(k)\sigma_c(x(k))\\ + \tilde{B}^{\mathsf{T}}\big(e(k+1) - \tilde{A}e(k)\big)\Big]^{\mathsf{T}}.\end{aligned} \tag{2.39}$$

The flowchart of the developed control algorithm is presented in Fig. 2.2.

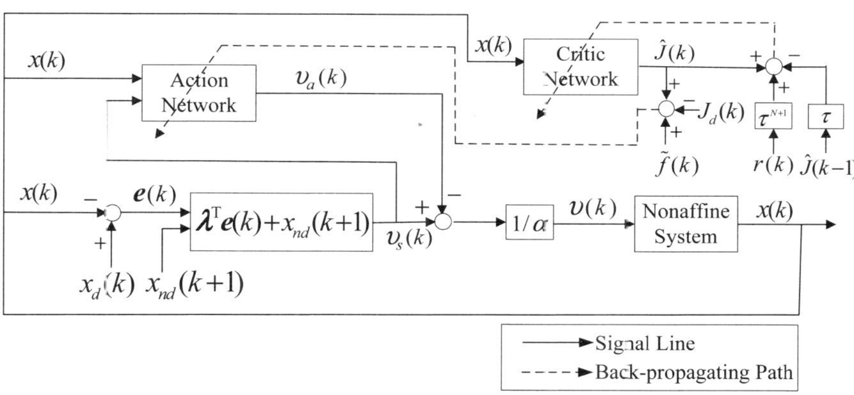

Fig. 2.2 Control algorithm for nonaffine nonlinear DT systems

2.4 Stability Analysis and Performance of the Closed-Loop System

The purpose of this section is to establish our main results by using Lyapunov's direct method. Prior to demonstrating the main theorem, we need to develop some mild assumptions and facts as follows:

Assumption 2.3. Let ideal output layer weights w_a and w_c be bounded over the compact set Ω by known positive constants w_{aM} and w_{cM}, respectively. That is,

$$\|w_a\| \le w_{aM}, \quad \|w_c\| \le w_{cM}. \tag{2.40}$$

Assumption 2.4. Let the NN approximation error $\varepsilon(k)$ be bounded by a known positive constant ε_M over the compact set Ω. That is,

$$\|\varepsilon(k)\| \le \varepsilon_M. \tag{2.41}$$

Fact 2.1. The activation functions for the action NN and the critic NN are bounded by known positive constants over the compact set Ω . That is,

$$\big\|\sigma_a\big(z(k)\big)\big\| \le \sigma_{aM}, \quad \big\|\sigma_c\big(x(k)\big)\big\| \le \sigma_{cM}.$$

Fact 2.2. From Remark 2.1, we know that $\tilde{A}$ is a stable matrix. Hence, there exists a positive definite matrix $P \in \mathbb{R}^{mn\times mn}$ satisfying the Lyapunov equation

$$\tilde{A}^{\mathsf{T}} P \tilde{A} - P = -\beta I_{mn} \tag{2.42}$$

where β is a positive constant.

Fact 2.3. By the same technique used in Lewis *et al.* (1999), there exist two known positive constants ϱ and ρ ($\varrho < \rho$) such that

$$\varrho I_m \le \tilde{B}^{\mathsf{T}} P \tilde{B} \le \rho I_m. \tag{2.43}$$

Remark 2.6. Observing that $\tilde{B} \in \mathbb{R}^{mn\times m}$ is the column full rank matrix, i.e., $\text{rank}(\tilde{B}) = m$ and $P \in \mathbb{R}^{mn\times mn}$ is positive definite, one can easily conclude that $\tilde{B}^{\mathsf{T}} P \tilde{B}$ is positive definite based on the knowledge of Matrix Analysis [Horn and Johnson (1985)]. Therefore, one can derive Fact 2.3 by using the property of the positive definite matrix, which is used in Lewis *et al.* (1999).

Next, several lemmas are developed to lay bases for deriving our main theorem.

Lemma 2.3. *Let Assumptions 2.1–2.4 hold. Taking the control input as in (2.6), and combining (2.10), (2.33), (2.42), (2.43), we can derive that the first difference of*

$$L_1(k) = \gamma_1 \boldsymbol{e}^{\mathsf{T}}(k) P \boldsymbol{e}(k)$$

as

$$\begin{aligned}\Delta L_1(k) \le\ & 2\gamma_1(\rho+\eta)\big\|\varepsilon(k)+d(k)\big\|^2 \\ & + 2\gamma_1(\rho+\eta)\big\|\xi_a(k)\big\|^2 - \gamma_1(\beta-1)\big\|\boldsymbol{e}(k)\big\|^2\end{aligned} \tag{2.44}$$

where $\eta = \big\|\tilde{A}^{\mathsf{T}} P \tilde{B}\big\|^2$, $\xi_a(k) = \tilde{w}_a^{\mathsf{T}}(k)\sigma_a\big(z(k)\big)$, *and* $\gamma_1 > 0$ *is a design parameter.*

Proof. The first difference of $L_1(k)$ is

$$\Delta L_1(k) = \gamma_1\Big[\boldsymbol{e}^{\mathsf{T}}(k+1) P \boldsymbol{e}(k+1) - \boldsymbol{e}^{\mathsf{T}}(k) P \boldsymbol{e}(k)\Big].$$

Define $Q(k) = \xi_a(k) - \big(\varepsilon(k) + d(k)\big)$. Combining (2.10), (2.33), (2.42), and

(2.43), we obtain

$$\begin{aligned}
\Delta L_1(k) &= \gamma_1\big(\tilde{A}\boldsymbol{e}(k)+\tilde{B}Q(k)\big)^{\mathsf{T}}P\big(\tilde{A}\boldsymbol{e}(k)+\tilde{B}Q(k)\big)-\gamma_1\boldsymbol{e}^{\mathsf{T}}(k)P\boldsymbol{e}(k)\\
&= \gamma_1\boldsymbol{e}^{\mathsf{T}}(k)\big(\tilde{A}^{\mathsf{T}}P\tilde{A}-P\big)\boldsymbol{e}(k)+2\gamma_1\boldsymbol{e}^{\mathsf{T}}(k)\tilde{A}^{\mathsf{T}}P\tilde{B}Q(k)\\
&\quad+\gamma_1Q^{\mathsf{T}}(k)\big(\tilde{B}^{\mathsf{T}}P\tilde{B}\big)Q(k)\\
&\le -\gamma_1\beta\|\boldsymbol{e}(k)\|^2+2\gamma_1\boldsymbol{e}^{\mathsf{T}}(k)\tilde{A}^{\mathsf{T}}P\tilde{B}Q(k)+\gamma_1\rho\|Q(k)\|^2\\
&\le \gamma_1\big(\rho+\|\tilde{A}^{\mathsf{T}}P\tilde{B}\|^2\big)\|Q(k)\|^2-\gamma_1(\beta-1)\|\boldsymbol{e}(k)\|^2.
\end{aligned}\tag{2.45}$$

Applying the Cauchy-Schwarz inequality $\|x+y\|^2\le 2\|x\|^2+2\|y\|^2$ to $\|Q(k)\|^2$ and using (2.45), we can obtain (2.44). □

Lemma 2.4. *Given that Assumptions 2.1–2.4 hold. Take the control input as in (2.6), and the utility function as defined in (2.21). Combining (2.24) and (2.27), we can obtain the first difference of*

$$L_2(k)=\frac{\gamma_2}{l_c}\mathrm{tr}\big(\tilde{w}_c^{\mathsf{T}}(k)\tilde{w}_c(k)\big)$$

as

$$\begin{aligned}
\Delta L_2(k) \le & -\gamma_2\|\xi_c(k)\|^2-\gamma_2\big(1-l_c\|\sigma_c\big(x(k)\big)\|^2\big)\\
&\times\Big\|\xi_c(k)+w_c^{\mathsf{T}}\sigma_c\big(x(k)\big)+\tau^{N+1}r(k)\\
&-\tau\hat{w}_c^{\mathsf{T}}(k-1)\sigma_c\big(x(k-1)\big)\Big\|^2\\
&+2\tau^2\gamma_2\Big\|\xi_c(k-1)\Big\|^2+2\gamma_2\Big\|w_c^{\mathsf{T}}\sigma_c\big(x(k)\big)\\
&+\tau^{N+1}r(k)-\tau w_c^{\mathsf{T}}\sigma_c\big(x(k-1)\big)\Big\|^2
\end{aligned}\tag{2.46}$$

where $\xi_c(k)=\tilde{w}_c^{\mathsf{T}}(k)\sigma_c\big(x(k)\big)$ *and* $\gamma_2>0$ *is a design parameter.*

Proof. The first difference of $L_2(k)$ is

$$\Delta L_2(k)=\frac{\gamma_2}{l_c}\mathrm{tr}\big[\tilde{w}_c^{\mathsf{T}}(k+1)\tilde{w}_c(k+1)-\tilde{w}_c^{\mathsf{T}}(k)\tilde{w}_c(k)\big].\tag{2.47}$$

Observing that $\tilde{w}_c(k)=\hat{w}_c(k)-w_c$ and using (2.27), we have

$$\begin{aligned}
\tilde{w}_c(k+1) = &\Big[I_{s_1}-l_c\sigma_c\big(x(k)\big)\sigma_c^{\mathsf{T}}\big(x(k)\big)\Big]\tilde{w}_c(k)\\
&-l_c\sigma_c\big(x(k)\big)\Big[w_c^{\mathsf{T}}\sigma_c\big(x(k)\big)+\tau^{N+1}r(k)\\
&-\tau\hat{w}_c^{\mathsf{T}}(k-1)\sigma_c\big(x(k-1)\big)\Big]^{\mathsf{T}}.
\end{aligned}\tag{2.48}$$

Combining (2.47) and (2.48), we obtain

$$\Delta L_2(k)=\frac{\gamma_2}{l_c}\mathrm{tr}\Big(\mathfrak{R}_1(k)+\mathfrak{R}_2(k)+\mathfrak{R}_3(k)\Big)\tag{2.49}$$

where

$$\begin{aligned}\mathfrak{R}_1(k) =& \left[\left(I_{s_1} - l_c\sigma_c\big(x(k)\big)\sigma_c^{\mathsf{T}}\big(x(k)\big)\right)\tilde{w}_c(k)\right]^{\mathsf{T}}\\&\times\left[\left(I_{s_1} - l_c\sigma_c\big(x(k)\big)\sigma_c^{\mathsf{T}}\big(x(k)\big)\right)\tilde{w}_c(k)\right] - \tilde{w}_c^{\mathsf{T}}(k)\tilde{w}_c(k)\\=& - l_c\left(1 - l_c\big\|\sigma_c(x(k))\big\|^2\right)\xi_c(k)\xi_c^{\mathsf{T}}(k) - l_c\xi_c(k)\xi_c^{\mathsf{T}}(k)\end{aligned}$$

$$\begin{aligned}\mathfrak{R}_2(k) =& - 2\tilde{w}_c^{\mathsf{T}}(k)\Big[I_{s_1} - l_c\sigma_c\big(x(k)\big)\sigma_c^{\mathsf{T}}\big(x(k)\big)\Big]l_c\sigma_c\big(x(k)\big)\\&\times\Big[w_c^{\mathsf{T}}\sigma_c\big(x(k)\big) + \tau^{N+1}r(k) - \tau\hat{w}_c^{\mathsf{T}}(k-1)\sigma_c\big(x(k-1)\big)\Big]^{\mathsf{T}}\\=& - 2l_c\xi_c(k)\left(1 - l_c\big\|\sigma_c(x(k))\big\|^2\right)\\&\times\Big[w_c^{\mathsf{T}}\sigma_c\big(x(k)\big) + \tau^{N+1}r(k) - \tau\hat{w}_c^{\mathsf{T}}(k-1)\sigma_c\big(x(k-1)\big)\Big]^{\mathsf{T}}\end{aligned}$$

and

$$\begin{aligned}\mathfrak{R}_3(k) =& l_c^2\Big[w_c^{\mathsf{T}}\sigma_c\big(x(k)\big) + \tau^{N+1}r(k) - \tau\hat{w}_c^{\mathsf{T}}(k-1)\sigma_c\big(x(k-1)\big)\Big]\\&\times\sigma_c^{\mathsf{T}}\big(x(k)\big)\sigma_c\big(x(k)\big)\Big[w_c^{\mathsf{T}}\sigma_c\big(x(k)\big) + \tau^{N+1}r(k)\\&- \tau\hat{w}_c^{\mathsf{T}}(k-1)\sigma_c\big(x(k-1)\big)\Big]^{\mathsf{T}}\\=& l_c^2\big\|\sigma_c(x(k))\big\|^2\Big[w_c^{\mathsf{T}}\sigma_c\big(x(k)\big) + \tau^{N+1}r(k)\\&- \tau\hat{w}_c^{\mathsf{T}}(k-1)\sigma_c\big(x(k-1)\big)\Big]\\&\times\Big[w_c^{\mathsf{T}}\sigma_c\big(x(k)\big) + \tau^{N+1}r(k) - \tau\hat{w}_c^{\mathsf{T}}(k-1)\sigma_c\big(x(k-1)\big)\Big]^{\mathsf{T}}.\end{aligned}$$

Define

$$M(k) = w_c^{\mathsf{T}}\sigma_c\big(x(k)\big) + \tau^{N+1}r(k) - \tau\hat{w}_c^{\mathsf{T}}(k-1)\sigma_c\big(x(k-1)\big). \tag{2.50}$$

From (2.49), we derive

$$\begin{aligned}\Delta L_2(k) =& \gamma_2\mathrm{tr}\Big[-\left(1 - l_c\big\|\sigma_c\big(x(k)\big)\big\|^2\right)\xi_c(k)\xi_c^{\mathsf{T}}(k)\\&- 2\xi_c(k)\left(1 - l_c\big\|\sigma_c\big(x(k)\big)\big\|^2\right)M^{\mathsf{T}}(k)\\&+ l_c\|\sigma_c\big(x(k)\big)\|^2M(k)M^{\mathsf{T}}(k) - \xi_c(k)\xi_c^{\mathsf{T}}(k)\Big]\\=& - \gamma_2\xi_c^{\mathsf{T}}(k)\xi_c(k) - \left(1 - l_c\big\|\sigma_c\big(x(k)\big)\big\|^2\right)\\&\times\gamma_2\left(\xi_c^{\mathsf{T}}(k)\xi_c(k) + 2\xi_c^{\mathsf{T}}(k)M(k)\right)\\&+ \gamma_2l_c\big\|\sigma_c\big(x(k)\big)\big\|^2M^{\mathsf{T}}(k)M(k)\end{aligned}$$

$$\begin{aligned}&= -\gamma_2\big\|\xi_c(k)\big\|^2 - \Big(1 - l_c\big\|\sigma_c(x(k))\big\|^2\Big)\\&\quad\times\gamma_2\big\|\xi_c(k) + M(k)\big\|^2 + \gamma_2\big\|M(k)\big\|^2.\end{aligned} \tag{2.51}$$

Notice that (2.50) can be rewritten as

$$\begin{aligned}M(k) = {}& w_c^{\mathsf{T}}\sigma_c\big(x(k)\big) + \tau^{N+1}r(k)\\&- \tau w_c^{\mathsf{T}}\sigma_c\big(x(k-1)\big) - \tau\xi_c\big(x(k-1)\big).\end{aligned} \tag{2.52}$$

Applying the Cauchy-Schwarz inequality $\|x+y\|^2 \le 2\|x\|^2 + 2\|y\|^2$ to $\|M(k)\|^2$, and combining (2.51) and (2.52), we can derive (2.46). □

Lemma 2.5. *Suppose that Assumptions 2.1–2.4 hold. Take the control input as in (2.6), and the utility function as in (2.21). Combining (2.24) and (2.39), we can derive the first difference of*

$$L_3(k) = \frac{\gamma_3}{l_a}\mathrm{tr}\big(\tilde{w}_a^{\mathsf{T}}(k)\tilde{w}_a(k)\big)$$

as

$$\begin{aligned}\Delta L_3(k) \le {}& 2\gamma_3\big\|w_c^{\mathsf{T}}\sigma_c\big(x(k)\big) - \varepsilon(k) - d(k)\big\|^2\\&+ 2\gamma_3\big\|\xi_c(k)\big\|^2 - \gamma_3\big\|\xi_a(k)\big\|^2\\&- \gamma_3\Big(1 - l_a\big\|\sigma_a(z(k))\big\|^2\Big)\big\|\hat{w}_c^{\mathsf{T}}(k)\sigma_c\big(x(k)\big)\\&+ \xi_a(k) - \varepsilon(k) - d(k)\big\|^2\end{aligned} \tag{2.53}$$

where $\xi_a(k) = \tilde{w}_a^{\mathsf{T}}(k)\sigma_a\big(z(k)\big)$ and $\gamma_3 > 0$ is a design parameter.

Proof. The first difference of $L_3(k)$ is

$$\Delta L_3(k) = \frac{\gamma_3}{l_a}\mathrm{tr}\Big[\tilde{w}_a^{\mathsf{T}}(k+1)\tilde{w}_a(k+1) - \tilde{w}_a^{\mathsf{T}}(k)\tilde{w}_a(k)\Big]. \tag{2.54}$$

Combining (2.10), (2.32), and (2.39), we obtain

$$\begin{aligned}\tilde{w}_a(k+1) = {}& \tilde{w}_a(k) - l_a\sigma_a\big(z(k)\big)\Big[\hat{w}_c^{\mathsf{T}}(k)\sigma_c\big(x(k)\big)\\&+ \xi_a(k) - \varepsilon(k) - d(k)\Big]^{\mathsf{T}}.\end{aligned} \tag{2.55}$$

Define

$$N(k) = \hat{w}_c^{\mathsf{T}}(k)\sigma_c\big(x(k)\big) + \xi_a(k) - \varepsilon(k) - d(k).$$

From (2.54) and (2.55), we have

$$\begin{aligned}\Delta L_3(k) = {}& \frac{\gamma_3}{l_a}\mathrm{tr}\Big[-l_a\tilde{w}_a^{\mathsf{T}}(k)\sigma_a\big(z(k)\big)N^{\mathsf{T}}(k) - l_aN(k)\\&\times\sigma_a^{\mathsf{T}}\big(z(k)\big)\Big(\tilde{w}_a(k) - l_a\sigma_a\big(z(k)\big)N^{\mathsf{T}}(k)\Big)\Big]\end{aligned}$$

$$
\begin{aligned}
&= -\gamma_3 \mathrm{tr}\Big[\xi_a(k)N^{\mathsf{T}}(k) + N(k) \\
&\quad \times \Big(\xi_a^{\mathsf{T}}(k) - l_a\big\|\sigma_a\big(z(k)\big)\big\|^2 N^{\mathsf{T}}(k)\Big)\Big] \\
&= l_a\gamma_3\big\|\sigma_a\big(z(k)\big)\big\|^2 N^{\mathsf{T}}(k)N(k) - 2\gamma_3\xi_a^{\mathsf{T}}(k)N(k) \\
&= l_a\gamma_3\big\|\sigma_a\big(z(k)\big)\big\|^2\big\|N(k)\big\|^2 + \gamma_3\big\|\xi_a(k) - N(k)\big\|^2 \\
&\quad - \gamma_3\big\|N(k)\big\|^2 - \gamma_3\big\|\xi_a(k)\big\|^2. \qquad (2.56)
\end{aligned}
$$

Notice that $N(k)$ can be rewritten as

$$N(k) = w_c^{\mathsf{T}}\sigma_c\big(x(k)\big) + \xi_a(k) - \varepsilon(k) - d(k) + \xi_c(k). \qquad (2.57)$$

Applying the Cauchy-Schwarz inequality $\|x+y\|^2 \le 2\|x\|^2 + 2\|y\|^2$ to $\|\xi_a(k) - N(k)\|^2$ and combining (2.56) and (2.57), we can get (2.53). □

With the aid of Assumptions 2.1–2.4 and Facts 2.1–2.3, our main theorem is established.

Theorem 2.2. *Consider the nonaffine nonlinear system described by (2.1) with all system states measurable. Let Assumptions 2.1–2.4 hold. Take the control input for system (2.1) as (2.6) with (2.8) and the critic NN (2.24), as well as the action NN (2.30). Moreover, let the weight update law for the critic NN and the action NN be (2.27) and (2.39), respectively. Then, the tracking error vector* $\mathbf{e}(k)$*, the action NN weight estimation error* $\tilde{w}_a(k)$*, and the critic NN weight estimation error* $\tilde{w}_c(k)$ *are UUB by positive constants* $\mathfrak{T}_1$*,* $\mathfrak{T}_2$ *and* $\mathfrak{T}_3$*, respectively, which are given by*

$$
\begin{aligned}
\mathfrak{T}_1 &= \sqrt{\frac{\mathfrak{B}_M^2}{\gamma_1(\beta-1)}} \\
\mathfrak{T}_2 &= \frac{1}{\sigma_{aM}}\sqrt{\frac{\mathfrak{B}_M^2}{\gamma_3 - 2\gamma_1(\rho+\eta)}} \\
\mathfrak{T}_3 &= \frac{1}{\sigma_{cM}}\sqrt{\frac{\mathfrak{B}_M^2}{(1-2\tau^2)\gamma_2 - 2\gamma_3}}
\end{aligned}
$$

provided that the following conditions hold:

$$
\begin{aligned}
&(a)\quad \beta > 1 \\
&(b)\quad 0 < l_c\big\|\sigma_c\big(x(k)\big)\big\|^2 < 1 \\
&(c)\quad 0 < l_a\big\|\sigma_a\big(z(k)\big)\big\|^2 < 1 \\
&(d)\quad 0 < \tau < \frac{\sqrt{2}}{2}.
\end{aligned}
\qquad (2.58)
$$

Proof. Consider the Lyapunov function candidate

$$L(k) = L_1(k) + L_2(k) + L_3(k) + L_4(k)$$

where

$$
\begin{aligned}
L_1(k) &= \gamma_1 \boldsymbol{e}^{\mathsf{T}}(k) P \boldsymbol{e}(k) \\
L_2(k) &= \frac{\gamma_2}{l_c} \operatorname{tr}\left(\tilde{w}_c^{\mathsf{T}}(k) \tilde{w}_c(k)\right) \\
L_3(k) &= \frac{\gamma_3}{l_a} \operatorname{tr}\left(\tilde{w}_a^{\mathsf{T}}(k) \tilde{w}_a(k)\right) \\
L_4(k) &= \gamma_4 \left\|\xi_c(k-1)\right\|^2.
\end{aligned}
$$

The first difference of the Lyapunov function candidate is

$$
\Delta L(k) = \Delta L_1(k) + \Delta L_2(k) + \Delta L_3(k) + \Delta L_4(k). \tag{2.59}
$$

For convenience, we denote

$$
P(k) = w_c^{\mathsf{T}} \sigma_c\big(z(k)\big) + \tau^{N+1} r(k) - \tau w_c^{\mathsf{T}} \sigma_c\big(x(k-1)\big).
$$

By using Lemmas 2.3–2.5 and (2.59), we derive

$$
\begin{aligned}
\Delta L(k) \leq & -\gamma_1(\beta-1)\left\|\boldsymbol{e}(k)\right\|^2 + \mathfrak{B}^2(k) \\
& - (\gamma_2 - 2\gamma_3 - \gamma_4)\left\|\xi_c(k)\right\|^2 \\
& - (\gamma_4 - 2\tau^2\gamma_2)\left\|\xi_c(k-1)\right\|^2 \\
& - \big(\gamma_3 - 2\gamma_1(\rho+\eta)\big)\left\|\xi_a(k)\right\|^2 \\
& - \gamma_2\Big(1 - l_c\left\|\sigma_c\big(x(k)\big)\right\|^2\Big)\left\|\xi_c(k) + M(k)\right\|^2 \\
& - \gamma_3\Big(1 - l_a\left\|\sigma_a\big(z(k)\big)\right\|^2\Big)\left\|N(k)\right\|^2
\end{aligned} \tag{2.60}
$$

where

$$
\begin{aligned}
\mathfrak{B}^2(k) = & \; 2\gamma_2\left\|P(k)\right\|^2 + 2\gamma_3\left\|w_c^{\mathsf{T}}\sigma_c(k) - \varepsilon(k) - d(k)\right\|^2 \\
& + 2\gamma_1(\rho+\eta)\left\|\varepsilon(k) + d(k)\right\|^2.
\end{aligned}
$$

By using (2.40), (2.41), and Fact 2.1, and the Cauchy-Schwarz inequality, we obtain

$$
\begin{aligned}
\mathfrak{B}^2(k) \leq & \; (12\gamma_2 + 6\gamma_3) w_{cM}^2 \sigma_{cM}^2 + 6\gamma_2 r_M^2 \\
& + \big(6\gamma_3 + 4\gamma_1(\rho+\eta)\big)(\varepsilon_M^2 + d_M^2) \\
& \triangleq \mathfrak{B}_M^2
\end{aligned} \tag{2.61}
$$

where r_M is an upper bound of $\|r(k)\|$, i.e., $\|r(k)\| \leq r_M$.

The parameters r_i $(i = 1, 2, 3)$ are selected to satisfy that

$$
\gamma_1 < \frac{\gamma_3}{2(\rho+\eta)}, \quad \gamma_2 = \frac{\gamma_4}{2\tau^2}, \quad \text{and} \quad \gamma_2 > \frac{2\gamma_3}{1-2\tau^2}. \tag{2.62}
$$

Therefore, from (2.58) and (2.62), we can conclude that (2.60) and (2.61) yield $\Delta L(k) < 0$ as long as one of the following conditions holds:

$$\|\boldsymbol{e}(k)\| > \sqrt{\frac{\mathfrak{B}_M^2}{\gamma_1(\beta - 1)}}$$

or

$$\|\xi_a(k)\| > \sqrt{\frac{\mathfrak{B}_M^2}{\gamma_3 - 2\gamma_1(\rho + \eta)}} \tag{2.63}$$

or

$$\|\xi_c(k)\| > \sqrt{\frac{\mathfrak{B}_M^2}{(1 - 2\tau^2)\gamma_2 - 2\gamma_3}} \tag{2.64}$$

where $r_i > 0$ $(i = 1, 2, 3)$ are design parameters. β, ρ, and η are defined as in (2.42), (2.43), and (2.44), respectively.

Noting that $\|\xi_a(k)\| \leq \sigma_{aM}\|\tilde{w}_a(k)\|$, $\|\xi_c(k)\| \leq \sigma_{cM}\|\tilde{w}_c(k)\|$, and by using (2.63) and (2.64), we can derive

$$\|\tilde{w}_a(k)\| > \frac{1}{\sigma_{aM}}\sqrt{\frac{\mathfrak{B}_M^2}{\gamma_3 - 2\gamma_1(\rho + \eta)}}$$

$$\|\tilde{w}_c(k)\| > \frac{1}{\sigma_{cM}}\sqrt{\frac{\mathfrak{B}_M^2}{(1 - 2\tau^2)\gamma_2 - 2\gamma_3}}.$$

By using the standard Lyapunov extension theorem [Lewis *et al.* (1999)], we can draw the conclusion that the tracking error vector $\boldsymbol{e}(k)$, the action NN weight estimation error $\tilde{w}_a(k)$ and the critic NN weight estimation error $\tilde{w}_c(k)$ are all UUB. Moreover, since the ideal weights of NNs are bounded, the uniform ultimate boundedness of $\|\tilde{w}_a(k)\|$ and $\|\tilde{w}_c(k)\|$ can guarantee that the estimated weights $\hat{w}_a(k)$ and $\hat{w}_c(k)$ are also UUB. □

2.5 Numerical Examples

In order to verify our theoretical results, two examples are employed for numerical experiments.

2.5.1 *Example 1*

Consider MIMO unknown nonaffine nonlinear DT systems described by

$$x_{11}(k+1) = x_{21}(k)$$

$$x_{12}(k+1) = x_{22}(k)$$

$$\begin{aligned}
x_{21}(k+1) &= 2x_{21}(k) - x_{11}(k) + (x_{21}(k) - x_{11}(k))^2 \\
&\quad + 0.04\Big[0.6u_1^3(k) + \big(2+\cos(x_{11}(k))\big)u_1(k) \\
&\quad - u_2(k) + x_{11}^2(k)\Big] + d_1(k) \\
x_{22}(k+1) &= 2x_{22}(k) - x_{12}(k) + \big(x_{22}(k) - x_{12}(k)\big)^2 \\
&\quad + 0.04\Big[-0.8u_1(k) + \big(1+x_{12}^2(k)\big)u_2^3(k) \\
&\quad + \big(2+\sin(x_{12}(k))\big)u_2(k) + x_{11}(k)\Big] \\
&\quad + \big(x_{21}(k) - x_{11}(k)\big)^2 + d_2(k) \\
y_1(k) &= x_{11}(k) \\
y_2(k) &= x_{12}(k)
\end{aligned} \tag{2.65}$$

where $x_1(k) = [x_{11}(k), x_{12}(k)]^\mathsf{T}$, $x_2(k) = [x_{21}(k), x_{22}(k)]^\mathsf{T}$ $u(k) = [u_1(k), u_2(k)]^\mathsf{T}$, and $y(k) = [y_1(k), y_2(k)]^\mathsf{T}$, $d(k) = [d_1(k), d_2(k)]^\mathsf{T}$, $d_i(k) = 0.01\sin(k)$, $i = 1, 2$.

The control objective is to control the system output $y(k)$ to track the desired trajectory $y_d(k) = [\sin(k), 0.6\sin(k)]^\mathsf{T}$. From (2.65), we can obtain

$$\begin{aligned}
&\det\left[\frac{\partial h\big(x(k), u(k)\big)}{\partial u(k)}\right] \\
&\quad = (0.04)^2 * \big[-0.8 + \big(1.8u_1^2(k) + 2 + \cos(x_{11}(k))\big) \\
&\qquad * \big(3(1+x_{12}^2(k))u_2^2(k) + 2 + \sin(x_{12}(k))\big)\big]
\end{aligned}$$

Obviously, $\det\big[\partial h\big(x(k), u(k)\big)/\partial u(k)\big] > 0$. Select $\Omega = [-1.2, 1.2] \times [-1, 1]$. Since $u(x(k))$ is a continuous function with respect to $x(k)$, we can conclude that $u(x(k))$ is bounded on Ω. Hence, $\partial h(x(k), u(k))/\partial u(k)$ is bounded on Ω.

The initial state is chosen to be $x_0 = [1, -0.5, 1, 0.4]^\mathsf{T}$. Let $\lambda_1 = 1$, $\lambda_2 = 0.25$ (i.e., $q^2 + \lambda_1 q + \lambda_2$ is stable). Meanwhile, we select $\alpha = 2$, $\beta = 1$, $\tau = 0.7$, and the prescribed threshold $\epsilon = 8 \times 10^{-3}$. The learning rates of the action NN and the critic NN are selected as $l_a = l_c = 0.01$.

Define

$$\Delta(k) = v_a(k) - f\big(x(k), v(k)\big) \tag{2.66}$$

where $f\big(x(k), v(k)\big)$ is defined as in (2.5). In fact, $\Delta(k)$ is the NN approximation error, which is utilized to show the performance of the approximation of the action NN canceling the nonlinearity of system (2.65).

Both the action NN and the critic NN are implemented by a single-hidden layer feedforward NN. Without loss of generality, the initial weights for the input layer to the hidden layer are selected randomly within an interval of [0, 1] and held constants. Meanwhile, the initial weights for the output layer are selected randomly within an interval of [−0.5, 0.5]. There are 12 nodes in the hidden-layer of the action NN, i.e., the structure of the action NN is 6–12–2. Meanwhile, the structure of the critic

NN is chosen to be 4–12–2. It is worth pointing out that selecting the structure of an NN is more of an art than science [Padhi *et al.* (2006)]. In this example, the number of neurons is derived by computer simulations, and we find that choosing 12 neurons for the hidden layer can lead to satisfactory simulation results.

The computer simulation results are shown by Figs. 2.3–2.7. Figs. 2.3 and 2.4 indicate the trajectories of $y_1(k)$ and $y_{1d}(k)$, and $y_2(k)$ and $y_{2d}(k)$, respectively. Fig. 2.5 presents tracking errors $e_0(k)$ and $e_1(k)$, which are the components of the tracking error vector $\boldsymbol{e}(k)$. Fig. 2.6 illustrates the control input $u_i(k)$ $(i = 1, 2)$,

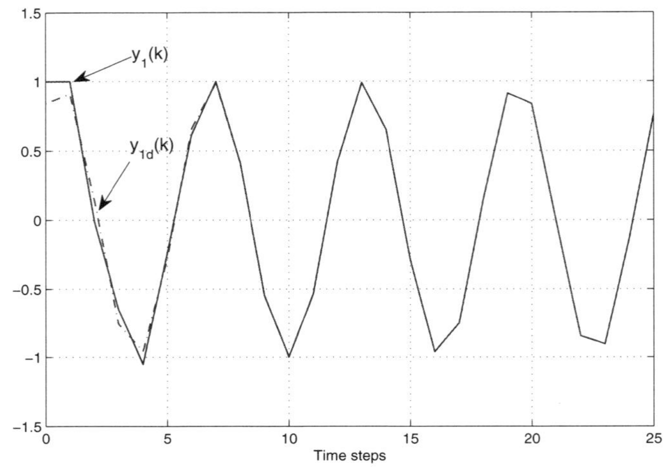

Fig. 2.3 Trajectories of $y_1(k)$ and $y_{1d}(k)$ in Example 1

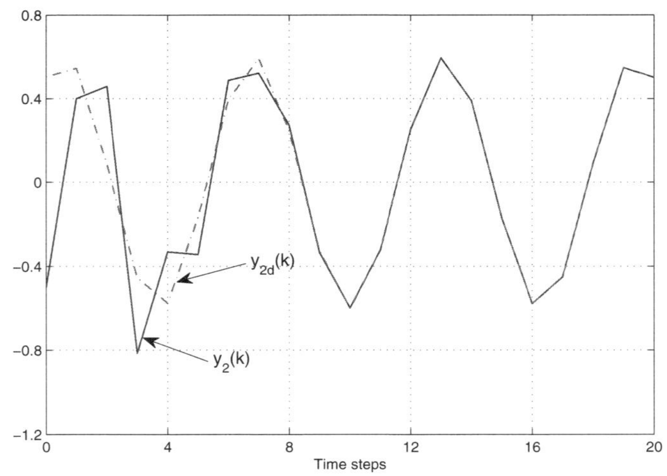

Fig. 2.4 Trajectories of $y_2(k)$ and $y_{2d}(k)$ in Example 1

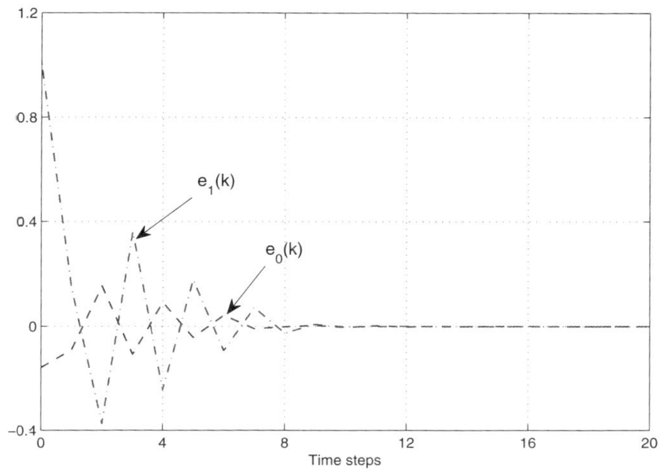

Fig. 2.5 Tracking errors $e_0(k)$ and $e_1(k)$ in Example 1

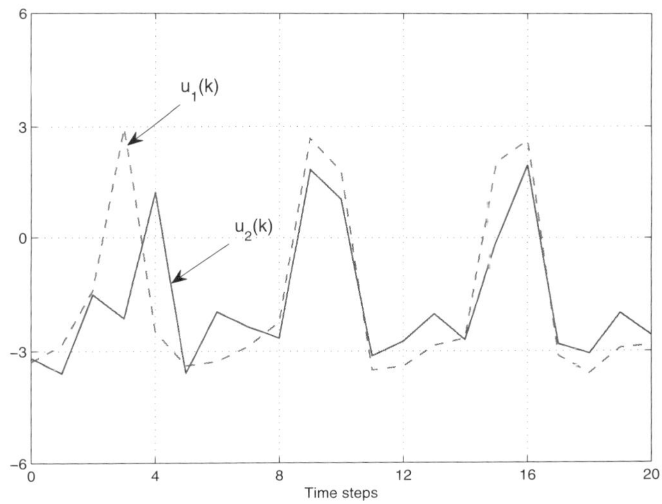

Fig. 2.6 Control input $u(k)$ in Example 1

which consist of the control vector $u(k)$. Fig. 2.7 describes the NN approximation error $\Delta_i(k)$ $(i = 1, 2)$, which consist of the NN approximation error vector $\Delta(k)$ defined as in (2.66). From Figs. 2.3–2.7, it is observed that the developed controller can make the system output $y(k)$ track the desired trajectory $y_d(k)$ very well. Meanwhile, the tracking error can converge to a small neighborhood of zero. In addition, we can notice that all signals involved in the closed-loop system are bounded.

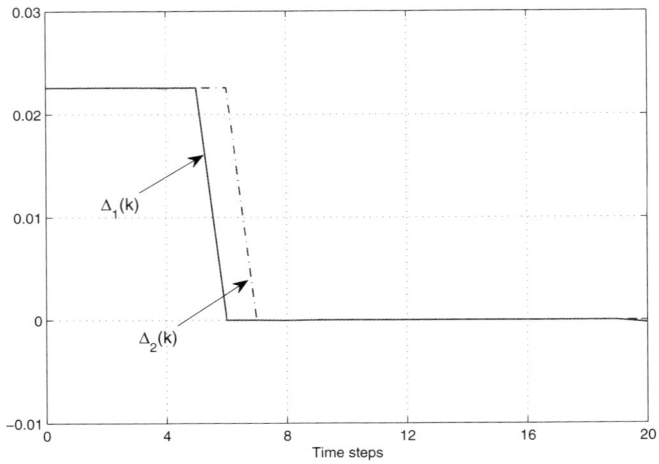

Fig. 2.7 NN approximation errors $\Delta_1(k)$ and $\Delta_2(k)$ in Example 1

2.5.2 *Example 2*

The purpose of this section is to further examine our method. Consider an MIMO nonaffine nonlinear DT system described by

$$\begin{aligned} x_{11}(k+1) &= x_{21}(k) \\ x_{12}(k+1) &= x_{22}(k) \\ x_{21}(k+1) &= 0.4x_{22}(k) - 0.3\cos(x_{21}(k)) \\ &\quad + 0.2u_1(k) - 0.1\tanh(u_2(k)) + d_1(k) \\ x_{22}(k+1) &= 0.1x_{11}(k) + 0.2u_2(k) \\ &\quad - 0.3\sin^2\big(x_{22}(k)\big)u_1(k) + d_2(k) \\ y_1(k) &= x_{11}(k) \\ y_2(k) &= x_{12}(k) \end{aligned} \tag{2.67}$$

where $x_1(k) = [x_{11}(k), x_{12}(k)]^\mathsf{T}$, $x_2(k) = [x_{21}(k), x_{22}(k)]^\mathsf{T}$, $u(k) = [u_1(k), u_2(k)]^\mathsf{T}$, and $y(k) = [y_1(k), y_2(k)]^\mathsf{T}$, $d(k) = [d_1(k), d_2(k)]^\mathsf{T}$, $d_i(k) = 0.01\sin(k)$, $i = 1, 2$.

The control objective is to control the system output $y(k)$ to track the desired trajectory $y_d(k)$ which is given by $y_d(k) = [2.6\sin(k\pi/200), 3\cos(k\pi/180)]^\mathsf{T}$.

From (2.67), we can derive

$$\partial h\big(x(k), u(k)\big)/\partial u(k) = \begin{bmatrix} 0.2 & -0.4/(e^x + e^{-x})^2 \\ -0.3\sin^2\big(x_{22}(k)\big) & 0.2 \end{bmatrix}.$$

Then, one can easily find that $\partial h\big(x(k), u(k)\big)/\partial u(k)$ is a positive definite matrix.

Meanwhile, one can also get

$$0.01I_2 \leq \partial h\big(x(k), u(k)\big)/\partial u(k) \leq 0.04I_2.$$

Choose $\Omega = [-3,3] \times [-3,3]$, and the initial state is selected to be $x_0 = [1, 0.31, 1, 0.31]^{\mathsf{T}}$. The design parameters are chosen as follows: $\lambda_1 = 1$, $\lambda_2 = 0.25$ (i.e., $z^2 + \lambda_1 z + \lambda_2$ is stable), $\alpha = 2$, $\beta = 2$, $\tau = 0.7$, and the prescribed threshold $\epsilon = 8 \times 10^{-3}$. The learning rates of the action NN and the critic NN are designed as $l_a = 0.01$ and $l_c = 0.001$, respectively. A single-hidden layer NN is applied to both the action NN and the critic NN, and the initial weights are selected in the same way as in Example 1. The structures of the action NN and the critic NN are designed as 6–30–2 and 4–24–2, respectively. In this example, the number of neurons in the hidden layers for both the action NN and the critic NN is obtained by computer simulations, and we find that it can lead to satisfactory simulation results.

The computer simulation results are presented in Figs. 2.8–2.12. Figs. 2.8 and 2.9 show the trajectories of $y_1(k)$ and $y_{1d}(k)$, and $y_2(k)$ and $y_{2d}(k)$, respectively. Fig. 2.10 describes tracking errors $e_0(k)$ and $e_1(k)$, which are the components of the tracking error vector $\boldsymbol{e}(k)$. Fig. 2.11 indicates the control input $u_i(k)$ $(i = 1, 2)$, which consist of the control vector $u(k)$. Fig. 2.12 describes the NN approximation error $\Delta_i(k)$ $(i = 1, 2)$, which consist of the NN approximation error vector $\Delta(k)$ defined as in (2.66). From the simulation results, it is observed that the system output $y(k)$ tracks the desired trajectory $y_d(k)$ very well, and the tracking errors converge to a small neighborhood of zero. It is also observed that the approximation of action NN can cancel the nonlinearity of system (2.67) rather well. Moreover, we can find that all signals involved in the closed-loop system are bounded.

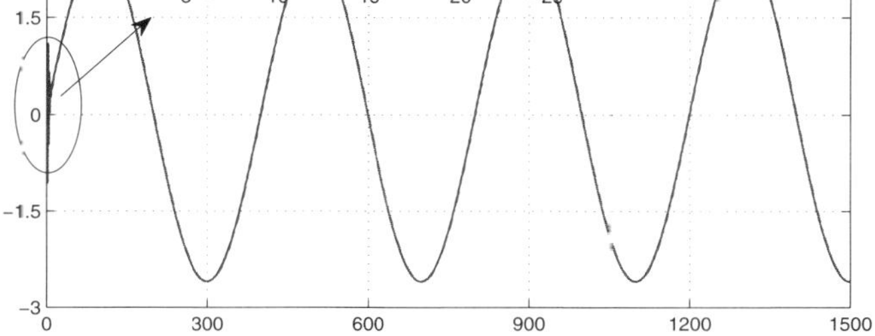

Fig. 2.8 Trajectories of $y_1(k)$ and $y_{1d}(k)$ in Example 2

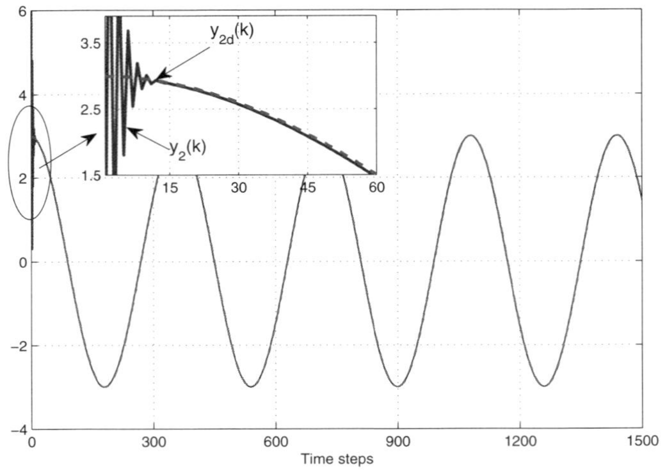

Fig. 2.9 Trajectories of $y_2(k)$ and $y_{2d}(k)$ in Example 2

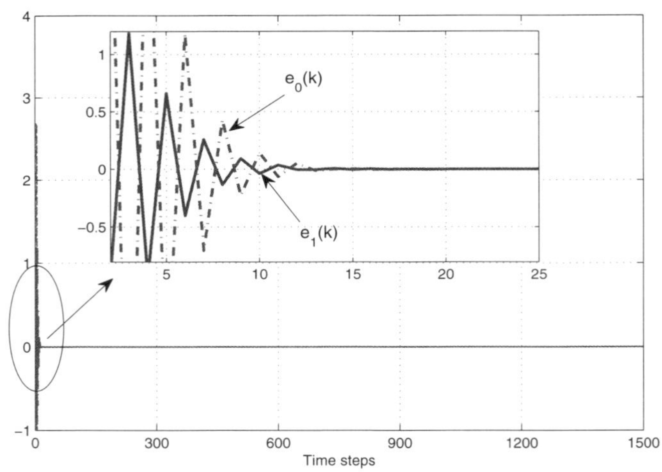

Fig. 2.10 Tracking errors $e_0(k)$ and $e_1(k)$ in Example 2

2.6 Conclusions

In this chapter, by using RL methods, an online direct adaptive control law is established to deliver a desired tracking performance for a class of unknown nonaffine nonlinear DT systems with unknown bounded disturbances. In order to utilize feedback linearization methods, the controller is divided into two parts: the first part of the controller is a feedback controller designed to stabilize linearized dynamics; the second part of the controller is a feedforward controller designed to cancel the non-

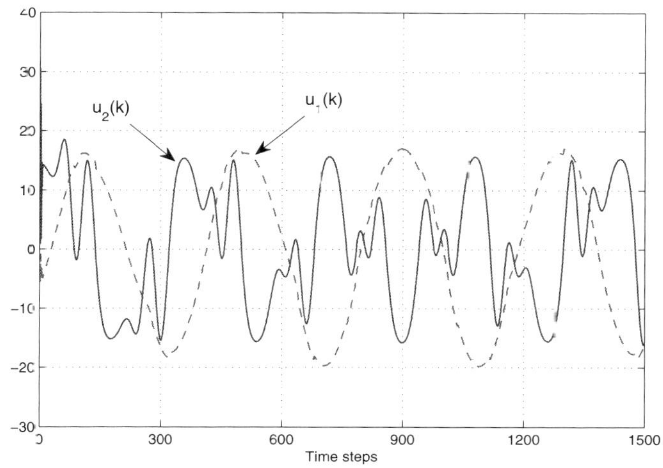

Fig. 2.11 Control input $u(k)$ in Example 2

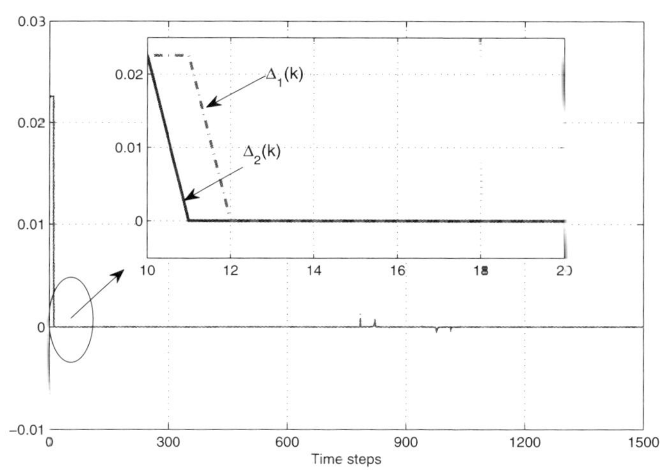

Fig. 2.12 NN approximation errors $\Delta_1(k)$ and $\Delta_2(k)$ in Example 2

linearity of nonaffine nonlinear DT systems. Two NNs are used in controller design. The action NN is applied to generate the control signal, and the critic NN is used to estimate the performance of the action NN. The weights of both the action NN and the critic NN are directly updated online. Moreover, both the tracking errors and the NN estimated weights are guaranteed to be UUB based on Lyapunov's direct method. The computer simulation results indicate that the developed online controller can perform control successfully and attain the desired performance. In our future work, we focus on designing an online direct adaptive control for MIMO unknown nonaffine nonlinear CT systems.

Acknowledgement

This work was supported in part by the National Natural Science Foundation of China under Grants 61034002, 61233001, 61273140, 61304086, and 61374105, and in part by Beijing Natural Science Foundation under Grant 4132078.

References

Al-Tamimi, A., Lewis, F. L. and Abu-Khalaf, M. (2008). Discrete-time nonlinear HJB solution using approximate dynamic programming: Convergence proof, *IEEE Transactions on Systems, Man, and Cybernetics, Part B: Cybernetics* **38**, 4, pp. 943–949.

Apostol, T. M. (1974). *Mathematical Analysis*, 2nd edn. (Pearson).

Bellman, R. E. (1957). *Dynamic Programming* (Princeton University Press).

Bertsekas, D. P. and Tsitsiklis, J. N. (1996). *Neuro-Dynamic Programming* (Athena Scientific).

Bhasin, S., Kamalapurkar, R., Johnson, M., Vamvoudakis, K. G., Lewis, F. L. and Dixon, W. E. (2013). A novel actor-critic-identifier architecture for approximate optimal control of uncertain nonlinear systems, *Automatica* **49**, 1, pp. 82–92.

Chemachema, M. (2012). Output feedback direct adaptive neural network control for uncertain SISO nonlinear systems using a fuzzy estimator of the control error, *Neural Networks* **36**, pp. 25–34.

Chen, F. C. and Khalil, H. K. (1995). Adaptive control of a class of nonlinear discrete-time systems using neural networks, *IEEE Transactions on Automatic Control* **40**, 5, pp. 791–801.

Deng, H., Li, H. X. and Wu, Y. H. (2008). Feedback-linearization-based neural adaptive control for unknown nonaffine nonlinear discrete-time systems, *IEEE Transactions on Neural Networks* **19**, 9, pp. 1615–1625.

Ge, S. S., Hang, C. C. and Zhang, T. (1999). Adaptive neural network control of nonlinear systems by state and output feedback, *IEEE Transactions on Systems, Man, and Cybernetics, Part B: Cybernetics* **29**, 6, pp. 818–828.

Haykin, S. (2008). *Neural Networks and Learning Machines*, 3rd edn. (Prentice Hall).

He, P. and Jagannathan, S. (2005). Reinforcement learning-based output feedback control of nonlinear systems with input constraints, *IEEE Transactions on Systems, Man, and Cybernetics, Part B: Cybernetics* **35**, 1, pp. 150–154.

Horn, R. A. and Johnson, C. R. (1985). *Matrix Analysis* (Cambridge University Press).

Hovakimyan, N., Nardi, F., Calise, A. and Kim, N. (2002). Adaptive output feedback control of uncertain nonlinear systems using single-hidden-layer neural networks, *IEEE Transactions on Neural Networks* **13**, 6, pp. 1420–1431.

Igelnik, B. and Pao, Y. H. (1995). Stochastic choice of basis functions in adaptive function approximation and the function-link net, *IEEE Transactions on Neural Networks* **6**, 6, pp. 1320–1329.

Lewis, F. L., Jagannathan, S. and Yesildirek, A. (1999). *Neural Network Control of Robot Manipulators and Nonlinear Systems* (Taylor and Francis).

Lewis, F. L. and Vamvoudakis, K. G. (2011). Reinforcement learning for partially observable dynamic processes: Adaptive dynamic programming using measured output data, *IEEE Transactions on Systems, Man, and Cybernetics, Part B: Cybernetics* **41**, 1, pp. 14–25.

Lewis, F. L. and Vrabie, D. (2009). Reinforcement learning and adaptive dynamic programming for feedback control, *IEEE Circuits and Systems Magazine* **9**, 3, pp. 32–50.

Lewis, F. L., Yesildirek, A. and Liu, K. (1996). Multilayer neural-net robot controller with guaranteed tracking performance, *IEEE Transactions on Neural Networks* **7**, 2, pp. 388–399.

Liu, D. (2005). Approximate dynamic programming for self-learning control, *Acta Automatica Sinica* **31**, 1, pp. 13–18.

Liu, D., Li, H. and Wang, D. (2013a). Neural-network-based zero-sum game for discrete-time nonlinear systems via iterative adaptive dynamic programming algorithm, *Neurocomputing* **110**, 13, pp. 92–100.

Liu, D., Wang, D. and Yang, X. (2013b). An iterative adaptive dynamic programming algorithm for optimal control of unknown discrete-time nonlinear systems with constrained inputs, *Information Sciences* **220**, pp. 331–342.

Murray, J. J., Cox, C. J., Lendaris, G. G. and Saeks, R. (2002). Adaptive dynamic programming, *IEEE Transactions on Systems, Man, and Cybernetics, Part C: Applications and Reviews* **32**, 2, pp. 140–153.

Nakanishi, J. and Schaal, S. (2004). Feedback error learning and nonlinear adaptive control, *Neural Networks* **17**, 10, pp. 1453–1465.

Narendra, K. S. and Mukhopadhyay, S. (1994). Adaptive control of nonlinear multivariable systems using neural networks, *Neural Networks* **7**, 5, pp. 737–752.

Noriega, J. R. and Wang, H. (1998). A direct adaptive neural-network control for unknown nonlinear systems and its applications, *IEEE Transactions on Neural Networks* **9**, 1, pp. 27–34.

Padhi, R., Unnikrishnan, N., Wang, X. and Balakrishnan, S. N. (2006). A single network adaptive critic (SNAC) architecture for optimal control synthesis for a class of nonlinear systems, *Neural Networks* **19**, 10, pp. 1648–1660.

Park, J. H., Huh, S. H., Kim, S. H., Seo, S. J. and Park, G. T. (2005). Direct adaptive controller for nonaffine nonlinear systems using self-structuring neural networks, *IEEE Transactions on Neural Networks* **16**, 2, pp. 414–422.

Prokhorov, D. V. and Wunsch, D. C. (1997). Adaptive critic designs, *IEEE Transactions on Neural Networks* **8**, 5, pp. 997–1007.

Rudin, W. (1991). *Functional Analysis*, 2nd edn. (McGraw-Hill).

Si, J. and Wang, Y. T. (2001). On-line learning control by association and reinforcement, *IEEE Transactions on Neural Networks* **12**, 2, pp. 264–276.

Sutton, R. S. and Barto, A. G. (1998). *Reinforcement Learning–An Introduction* (MIT Press).

Vamvoudakis, K. G. and Lewis, F. L. (2010). Online actor-critic algorithm to solve the continuous-time infinite horizon optimal control problem, *Automatica* **46**, 5, pp. 878–888.

Vamvoudakis, K. G. and Lewis, F. L. (2011). Multi-player non-zero-sum games: Online adaptive learning solution of coupled Hamilton-Jacobi equations, *Automatica* **47**, 8, pp. 1556–1569.

Wang, D., Liu, D. and Wei, Q. (2012a). Finite-horizon neuro-optimal tracking control for a class of discrete-time nonlinear systems using adaptive dynamic programming approach, *Neurocomputing* **78**, 1, pp. 14–22.

Wang, D., Liu, D., Wei, Q., Zhao, D. and Jin, N. (2012b). Optimal control of unknown nonaffine nonlinear discrete-time systems based on adaptive dynamic programming, *Automatica* **48**, 8, pp. 1825–1832.

Wang, F. Y., Jin, N., Liu, D. and Wei, Q. (2011). Adaptive dynamic programming for finite horizon optimal control of discrete-time nonlinear systems with ε-error bound, *IEEE Transactions on Neural Networks* **22**, 1, pp. 24–36.

Wang, F. Y., Zhang, H. and Liu, D. (2009). Adaptive dynamic programming: An introduction, *IEEE Computational Intelligence Magazine* **4**, 2, pp. 39–47.

Wei, Q. and Liu, D. (2012). An iterative ε-optimal control scheme for a class of discrete-time nonlinear systems with unfixed initial state, *Neural Networks* **32**, pp. 236–244.

Wei, Q., Zhang, H., Liu, D. and Zhao, Y. (2010). An optimal control scheme for a class of discrete-time nonlinear systems with time delays using adaptive dynamic programming, *Acta Automatica Sinica* **36**, 1, pp. 121–129.

Werbos, P. J. (1991). A menu of designs for reinforcement learning over time, in W. T. Miller, R. S. Sutton, and P. J. Werbos (eds.), *Neural Networks for Control* (MIT Press, Cambridge), pp. 67–95.

Werbos, P. J. (1992). Approximate dynamic programming for real-time control and neural modeling, in D. A. White and D. A. Sofge (eds.), *Handbook of Intelligent Control* (Van Nostrand Reinhold, New York), pp. 493–525.

Werbos, P. J. (2007). Using ADP to understand and replicate brain intelligence: The next level design, in L. I. Perlovsky and R. Kozma (eds.), *Neurodynamics of Cognition and Consciousness* (Springer, Berlin), pp. 109–123.

Werbos, P. J. (2008). ADP: The key direction for future research in intelligent control and understanding brain intelligence, *IEEE Transactions on Systems, Man, and Cybernetics, Part B: Cybernetics* **38**, 4, pp. 898–900.

Yang, L., Si, J., Tsakalis, K. S. and Rodriguez, A. A. (2009). Direct heuristic dynamic programming for nonlinear tracking control with filtered tracking error, *IEEE Transactions on Systems, Man, and Cybernetics, Part B: Cybernetics* **19**, 6, pp. 1617–1622.

Yang, Q., Vance, J. B. and Jagannathan, S. (2008). Control of nonaffine nonlinear discrete-time systems using reiforcement-learning-based linearly parameterized neural networks, *IEEE Transactions on Systems, Man, and Cybernetics, Part B: Cybernetics* **38**, 4, pp. 994–1001.

Yang, Q. and Jagannathan, S. (2012). Reinforcement learning controller design for affine nonlinear discrete-time systems using approximators, *IEEE Transactions on Systems, Man, and Cybernetics, Part B: Cybernetics* **42**, 2, pp. 377–390.

Yang, X., Liu, D. and Huang, Y. (2013). Neural-network-based online optimal control for uncertain nonlinear continuous-time systems with control constraints, *IET Control Theory and Applications* **7**, 17, pp. 2037–2047.

Yu, W. (ed.) (2009). *Recent Advances in Intelligent Control Systems* (Springer).

Zeidler, E. (1985). *Nonlinear Functional Analysis and its Applications: I: Fixed-Point Theorems* (Springer).

Zhang, H., Wei, Q. and Liu, D. (2011). An iterative adaptive dynamic programming method for solving a class of nonlinear zero-sum differential games, *Automatica* **47**, 1, pp. 207–214.

Chapter 3

Experimental Studies on Data-Driven Heuristic Dynamic Programming for POMDP

Zhen Ni, Haibo He and Xiangnan Zhong

Abstract Adaptive dynamic programming (ADP) has been a popular approach to seek the optimal control strategy in Markov decision process (MDP). Generally, this type of approach requires a complete set of system information/states to achieve the online optimal decision-making. However, the full system information/states are not usually available in practical situations. In many cases, the measured input/output data can only represent part of the system information and the system internal states are not available. In this chapter, we investigate a data-driven heuristic dynamic programming (HDP) architecture to tackle the partially observed Markov decision process (POMDP). In specific, we include a state estimator neural network to recover the full system information for the action network, so that the optimal control policy can still be achieved under the partially observed environment. We randomly initialize the weights in the state estimator network, and conduct online learning for the entire process. Both discrete-time and continuous-time system functions are tested. Simulation results and system trajectories justify the control performance of our proposed approach.

3.1 Introduction

During the recent years, adaptive dynamic programming (ADP) has become one of the most critical approaches and has been widely applied in both machine learning and adaptive/optimal control areas to seek the optimal solution for the decision-making agents [Werbos (2008, 2009, 2013)]. Markov decision process (MDP) can be generally formulated into a reinforcement learning (RL)/ADP task that satisfies the Markov property. RL/ADP is a agent-based method to optimize the control strategy based on the observed responses from environment [Bertsekas and Tsitsiklis (1996); Sutton and Barto (1998)].

In general, ADP can be categorized into three typical structures [Ferrari and Stengel (2004); Lewis and Vrabie (2009); Prokhorov and Wunsch (1997); Wang *et al.* (2009)]: *a*) heuristic dynamic programming (HDP), which was proposed to provide a control action sequence with the action network and use the critic network to critique the generated action by propagating a temporal difference (TD) between two consecutive estimates of the value function; *b*) dual heuristic dynamic programming (DHP), the key point of which is to employ the critic network to approximate the derivatives of value function with respect to the state vectors; *c*) globalized dual heuristic dynamic programming (GDHP), which takes the advantage of both HDP and DHP by employing the critic network to approximate both the value function and its derivatives with respect to the state vectors. Moreover, various versions have been developed based on these typical structures, such as the action dependent (AD) version by taking the control action as one of the inputs for the critic network [He *et al.* (2012, 2013); Ni *et al.* (2011); Si and Wang (2001)], and the model dependent version by taking the estimates of the model network as part of the inputs for the critic network [Venayagamoorthy *et al.* (2002); Zhang *et al.* (2008, 2011a)]. An iterative ADP algorithm was proposed and demonstrated its potential capabilities in various examples, such as zero-sum games [Liu *et al.* (2013a); Zhang *et al.* (2011b)], optimal control for nonlinear systems [Li and Liu (2012); Liu *et al.* (2013b); Wei and Liu (2013)]. Rigorous proof of convergence for this iterative ADP algorithm was also included in [Liu and Wei (2013); Wang and Liu (2013); Wang *et al.* (2011)]. In addition, the authors went one step further to prove the convergence of value function together with the partial derivatives of the value function in DHP/GDHP in [Liu *et al.* (2012); Wang *et al.* (2012)]. Many real applications on power system control, Helicopter trimming/tracking control and others have been demonstrated in [Enns and Si (2003); Liu *et al.* (2005, 2008); Tang *et al.* (2014)]. Recent research books also included the cutting-edge topics in the ADP/RL communities Lewis and Liu (2013); Si *et al.* (2004).

More recently, many papers presented the exploration of the internal reward (goal) and demonstrated its significance in the field. It has been proposed and demonstrated in He *et al.* (2012); He (2011) that a three-network architecture can achieve better control performance comparing with the traditional ADP design on several balancing benchmarks. In addition, hierarchical HDP design is presented with significant improvement with respect to the average successful trial number, comparing with both three-network design and the traditional ADP design in Ni *et al.* (2012); He *et al.* (2013). In addition, the stability analysis on a dual-critic design with a tracking controller has been provided and verified on numerical simulation benchmarks in Ni *et al.* (2013a,b). Maze navigation benchmark was also demonstrated with this novel ADP design and the performance is better than several traditional reinforcement learning approaches [Ni *et al.* (2013c); Ni and He (2013)].

The ADP approaches mentioned above generally require the full feedback of the entire system state. In real applications, however, full system state is usually not available and the internal system state is usually not measurable, either. In these cases, the MDP problem turns into the partially observable MDP (POMDP) problem and new techniques on the ADP method need to be developed to tackle this issue. Partially observable Markov decision process is a class of challenging problems in the areas of artificial intelligence and machine learning. Extensive work has been done to handle this problem in the past decades, yet many of them focus on the mathematics aspect to numerical calculate the belief state/distribution. The state-of-art solutions for POMDP are piecewise linear and convex (PWLC) for every horizon length and the heuristic search value iteration (HSVI) algorithm [POMDP; Smith and Simmons (2004)]. Yet both methods will cause intensive computational burden to obtain the belief state (probability distribution over the set of state S) if the dimension of the system state increases (i.e., curse of dimensionality). Reinforcement learning algorithm, especially Q-learning, have been applied on POMDP to derive new value function and iterative algorithm under the partially observable environment [Michael and Jordan (1995); Saad (2010)]. State estimation is also presented to be omitted under certain conditions in POMDP in Singh *et al.* (1994). Many other papers proposed alternative solutions for POMDP, however, they are all parameter/probability based approaches and require solid mathematic background to apply [Hauskrecht (2000); Kaelbling *et al.* (1998)]

Recently, there is a trend in the community that researchers proposed new ADP designs/algorithms to handle POMDP with only measured input/output data. For instance, in Vrabie *et al.* (2009), Vrabie *et al* presented a new scheme based on adaptive critic design (ACD) to find the optimal control solution for a linear continuous time system without knowing the internal dynamics of the system. In Lewis and Vamvoudakis (2011), Lewis *et al* proposed to use ADP for the partially observable dynamic process. They derived the kernel matrix for policy iteration (PI) and value iteration (VI), given that the LTI system is observable and previous measured data is available. In Lin and Mitchell (1993), the authors formulated the partially observable control problem into the reinforcement learning task (trial-and-error) with hidden states, and showed the preliminary control performance.

In many practical scenarios, system dynamics and internal system states may not be exactly observed/measured, and the measured output data can only partially represent system state/information. In these cases, it would be desirable to design an estimator using input/output data with very limited prior knowledge. Motivated by the POMDP agent that is decomposed into a state estimator (SE) and a policy generator (π) in Kaelbling *et al.* (1998) and the cascade action network HDP design in Enns and Si (2003), we integrate one state estimator network in HDP design to provide the "virtual state" for the action network. As we adopt neural network to recover the full information of the system state, we called this as a data-driven state estimator. The recovered virtual state has the same dimension as the system state, yet their values are not necessarily to be the same. The motivation is to represent the full system information (higher dimension state) with the observed data (lower dimension state) based on the proposed state estimator network. In this chapter, we apply multi-layer perceptron (MLP) neural network structure for the state estimator network. Comparing with the "old fashion" model network, the error function for the state estimator network is different. We do not set the error function for the state estimator network as the difference between the output and the true/physical values. Instead, we set the error function as the difference between the value function and the ultimate utility function (i.e., we adopt the same error function as that of the action network in literature). The reason is that we are using the recovered the virtual states to further minimize the total cost, which is the ultimate goal in the ADP design. On the other hand, the input/output of the state estimator network are also different with the traditional model network. We use the state estimator network to provide the full system information, which is not necessarily the same with the current physical states. While the traditional model network is adopted to output the system states for the next step (need to be as exactly as possible). The traditional model network is normally trained offline with input/output data and applied online with fixed weights. While for the state estimator network here, we random initialize the weights and let it learn to seek the optimal weights over time.

The major contribution of our proposed design can be summarized as: (1) there is no need to calculate or recover the exact physical system states and the computation burden can be reduced if the dimension of the state increases; (2) we conduct online learning (starts from scratch) for the data-driven state estimator and prior knowledge is not necessarily required; (3) the error function for the state estimator is defined as the total cost function J, which will be further minimized when we tune the weights in state estimator.

The rest of this chapter can be organized as follows: the background of MDP and POMDP are introduced and discussed in Section 3.2. The problem formulation with the proposed state estimator is introduced in Section 3.3. Section 3.4 provides the algorithm for the proposed data-driven HDP approach and Section 3.5 shows the simulation studies. Finally, the conclusion and future work are presented in Section 3.6.

3.2 Markov Decision Process and Partially Observable Markov Decision Process

3.2.1 *Markov decision process*

The MDP can be denoted as a quadruple $\{X, A, R, T\}$, where X refers to the state space, A refers to the action space, T refers to the state transition probability and R is the reward function [Bertsekas and Tsitsiklis (1996); Sutton and Barto (1998)]. The policy (π) of the MDP can be regarded as the mapping between the state space and the distribution over the action space. We define the policy as

$$a(k) = \pi(x(k)). \tag{3.1}$$

The objective for the decision maker is to seek the control policy π so that the discounted total future rewards can be maximized. Thus, the optimal policy π^* should satisfy that

$$J_\pi = \max_\pi E_\pi \left[\sum_{k=0}^{\infty} \gamma^k r(k) \right] \tag{3.2}$$

where γ is a discount factor ($0 < \gamma < 1$), and $r(k)$ is the reward function at time instance k. E_π refers to the expectation of the discounted total future reward with respect to the policy π. In ADP/RL field, J_π is generally called the value function based on the policy π.

The state-action value function (pair) $Q^\pi(x, a)$ is defined with the expected discounted total rewards when taking action a in state x as follows

$$Q^\pi(x, a) = E_\pi \left[\sum_{k=0}^{\infty} \gamma^k r(k) | x(0) = x, a(0) = a \right]. \tag{3.3}$$

The value function based on the policy π can be obtained from (3.3) as

$$V^\pi(x) = E_\pi \left[\sum_{k=0}^{\infty} \gamma^k r(k) | x(0) = x \right]. \tag{3.4}$$

According to the theory of dynamic programming, $Q^\pi(x, a)$ in (3.3) satisfies the Bellman's equation as

$$Q^\pi(x(k), a(k)) = E \left[r(x(k), a(k)) + \gamma \sum_{a \in A} Q^\pi(x(k+1), a(k+1)) \right] \tag{3.5}$$

and the value function can be written as

$$V^\pi(x(k)) = E \left[r(x(k)) + \gamma \sum V^\pi(x(k+1)) \right]. \tag{3.6}$$

The optimal state-action value function is then defined as

$$Q^*(x, a) = \max_\pi Q^\pi(x, a) \tag{3.7}$$

and the subsequently optimal value function can be obtained as

$$V^*(x) = \max_\pi V^\pi(x). \tag{3.8}$$

Therefore, the corresponding optimal policy that we need to seek is

$$\pi^*(x) = \arg\max_a Q^*(x, a). \tag{3.9}$$

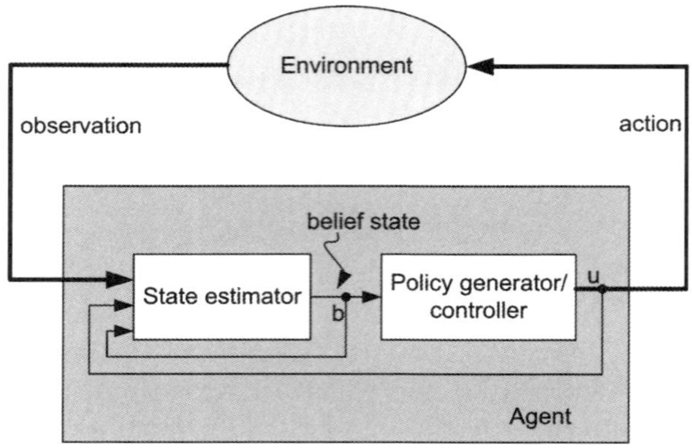

Fig. 3.1 Agent-environment interface under partially observable condition

3.2.2 *Partially observable Markov decision process*

The POMDP is defined with $M = \{X, A, T, R, Z, O\}$, where X refers to the state space, A refers to the action space, T refers to the state transition probability, R is the reward function, Z refers to the observations and O refers to the observation function (i.e., the quantity O is the probability of observation). In literature [Kaelbling *et al.* (1998); Saad (2010)], the state estimator is adopted to calculate the belief state or the distribution of the belief state, so that the control action can still be generated based on the belief state. The agent-environment interaction under the partially observable condition is presented in Fig. 3.1. When the agent is at belief state b, executing action a and observing o, it can calculate the new belief state $b' = \tau(b, a, o)$ as

$$\begin{aligned} b'(b,a,o) &= \Pr(x' \,|a,o,b) \\ &= \frac{\Pr(o\,|x',a,b)\Pr(x'\,|a,b)}{\Pr(o\,|a,b)} \\ &= \frac{O(a,x',o)\sum_{x\in X} T(x,a,x')b(x)}{\Pr(o\,|a,b)} \end{aligned} \tag{3.10}$$

where $\Pr(o\,|a,b)$ is a normalizing factor defined as

$$\Pr(o\,|b,a) = \sum_{x\in X} b(x) \sum_{x'\in X} T(x,a,x')O(a,x',o). \tag{3.11}$$

The traditional way to solve POMDP is through the belief space MDP, which is an MDP over the belief space of the POMDP as

$$R(b,a) = \sum_{x\in X} b(x)R(x,a). \tag{3.12}$$

$$T(b, a, b) = \sum_{o \in Z} \Pr(b' \,|b, a, o) \Pr(o \,|b, c). \tag{3.13}$$

$$\Pr(b' \,|b, a, o) = \begin{cases} 1, & \tau(b, a, o) = b' \\ 0, & \text{otherwise.} \end{cases} \tag{3.14}$$

$$\Pr(o \,|b, a) = \sum_{s \in S} b(s) \sum_{s' \in S} b(s)T(s, c, s')O(a, s', o). \tag{3.15}$$

The iteration for the value function V_{k+1} can thus be obtained as

$$V_{k+1}(b) = \max_{a \in A} \left(\sum_{s \in S} b(s)R(s, a) + \gamma \sum_{o \in Z} \Pr(o|b, a)V_k(\tau(b, a, o)) \right). \tag{3.16}$$

After the belief state is calculated, we can map the control policy to the belief state to achieve the final optimal policy. However, a common drawback for this state estimation is that it is always based on strong assumptions about the environment [Saad (2010)]. For instance, it is usually assumed that the number of the states is known in advance. Furthermore, the state estimation is computationally expensive and require a large amount of data, even if the true environment has finite number of states. If the number of state increases, it will also cause tremendous increasing of the computation (i.e., curse of dimensionality).

3.3 Problem Formulation with the State Estimator

We introduce the neural network structure of the state estimator network, and then define the input and output of the state estimator. The objective function of the state estimator network is also highlighted here. Let the system function be defined as

$$X(k+1) = AX(k) + Bu(k) \tag{3.17}$$

$$Y(k) = CX(k) \tag{3.18}$$

where X is the system state, which is also called the internal dynamics/states of the system here, and $X(k) = [x_1(k), x_2(k), \ldots, x_n(k)]$ $(n \in \Re)$. The output Y is the observed data (partially observed state vector), and $Y(k) = [y_1(k), y_2(k), \ldots, y_m(k)]$ $(m \in \Re, m \leq n)$. In our design, u and Y are available to measure, while X is the internal state that can not be measured. The parameters A, B and C are system dynamics that are assumed to be unknown here.

State estimation is commonly used in artificial intelligence and machine learning areas to find the belief state or the distribution of the belief state, in order to tackle POMDP problem. In literature, state estimation is normally applied in a mathematical/probability way to derive the belief state [Kaelbling *et al.* (1998); POMDP; Saad (2010); Michael and Jordan (1995)]. As presented in (3.10) and (3.11), the posterior probabilities need to be calculated for every possible case, which will cause the curse of dimensionality if the state number increases (i.e., the increasing of the state dimension will cost tremendous computation burden).

In this chapter, we are introducing a data-driven state estimator that is built with neural network. We only include the input/output data of the system as the input for the state estimator. Here the input/output data does not contain the full system states. Instead, we recover the full system information from input/output of the system using state estimator network. In order to recover the system information (virtual state) from the partially observed data, we keep the previous observed input/output data (i.e., $u(k-1)$, $u(k-2)$, $u(k-3)\ldots/Y(k-1)$, $Y(k-2)$, $Y(k-3)\ldots$). Therefore, we define the input of the state estimator as

$$\begin{aligned} X_{SE} = \big[Y(k),\ \ Y(k-1),\ \ldots,\ Y(k-k_1),\\ u(k),\ \ u(k-1),\ \ldots,\ u(k-k_2)\big] \end{aligned} \tag{3.19}$$

where k_1 and k_2 refer to the steps of time-delay for the observed output/input data. The selection of the values for both k_1 and k_2 will be discussed in the simulation case studies. Multilayer perceptron (MLP) neural network has been one of the most popular techniques to approximate the nonlinear function in the ADP community [Lewis and Liu (2013); Si *et al.* (2004)]. Therefore, we follow this trend and adopt MLP in our research for the state estimator as well. The neural network structure of the state estimator is presented in Fig. 3.2. We include the previous observed input/output data as the input vector for the state estimator. The output of the state estimator is a $n \times 1$ vector, which is the same dimension as the system state vector X.

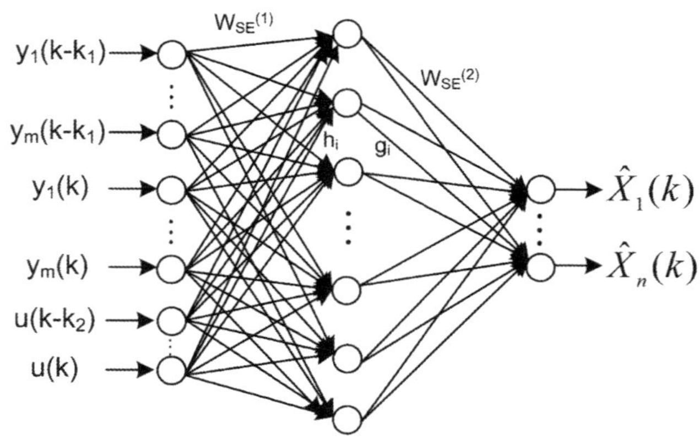

Fig. 3.2 The neural network structure of the state estimator

3.4 Data-Driven HDP Algorithm for POMDP

The ADP/HDP approaches generally require the feedback of the full system states so that the optimal control strategy can be achieved in response to the environment over time. In many of the real applications, however, it is normal to see that the full system states are not available and the internal system dynamics are unmeasurable. Therefore, it is desirable to design a state estimator to help to recover the full system information. Motivated by the cascade action network design in Enns and Si (2003), we include a state estimator network in our proposed data-driven HDP design. We provide the action network with full system information, which is named as "virtual" state here. The virtual states are not necessarily the same as those of the physical state, yet they can help minimizing the discounted total future cost (i.e., value function). The input of the state estimator is the measured input/output data (including the previous measured data) and the output is the recovered system information.

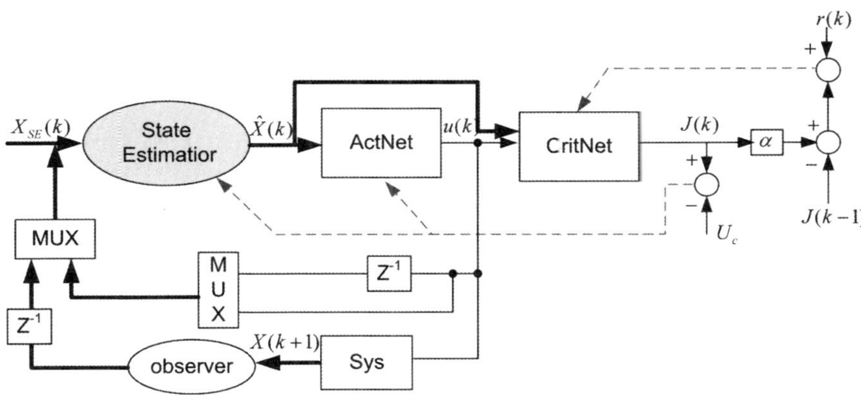

Fig. 3.3 Schematic diagram of HDP with state estimator

The schematic diagram of the proposed data-driven HDP approach is presented in Fig. 3.3. We adopt the state estimator to recover the full system information (i.e., virtual state) from the previous observed input/output data, so that the action network in our design can still have the same dimension of the system state vector (i.e., the virtual state has the same dimension as that of system state). The objective functions are provided with the dash arrows in the figure and the solid lines show the signal paths (bolded lines represent the vector).

The mechanism of the proposed data-driven HDP approach is similar as that in Enns and Si (2003); Si and Wang (2001). That is, the action network generates the control action and the critic network critique the control performance based on the corresponding control action so that the controller learns to seek the optimal policy over time. In order to provide the full system feedback for the action network under partially observable environment, we adopt the state estimator network to provide full system information based on the observed input/output data from the system. In addition, as our ultimate goal is to minimize the total cost function, we also set the objective function for the state estimator network the same as that for action network. In other words, we are actually further minimizing the total cost function with the state estimator network.

3.4.1 *Learning in the state estimator network*

Unlike previous designs for state estimator that the posterior probability of the belief states needs to be calculated and compared, we here introduce a different type of state estimator design. That is, we employ MLP neural network to build a data-driven state estimator, which requires no probability computation. Specifically, we build a mapping between the observed input/output data (previous observed input/output data are also used) and the full system information (i.e., virtual state). In this way, we save the computation when the dimension of the system state goes high. We provide an alternative choice to design the state estimator.

The forward paths of the state estimator network is presented as

$$\hat{X}(k) = \sum_{i=1}^{N_{SE}} \omega_{SE_i}^{(2)}(k) z_i(k), \tag{3.20}$$

$$z_i(k) = \phi(h_i(k)), \qquad i = 1, \ldots, N_{SE} \tag{3.21}$$

$$h_i(k) = \sum_{j=1}^{n} \omega_{SE_{i,j}}^{(1)}(k) X_{SE_j}(k) \tag{3.22}$$

where ϕ is the activation function defined as

$$\phi(x) = \frac{1 - e^{-x}}{1 + e^{-x}}. \tag{3.23}$$

As discussed above, the error function of the state estimator is defined as

$$e_{SE} = J(k) - U_c \tag{3.24}$$

and the objective function is

$$E_{SE}(k) = \frac{1}{2} e_{SE}^2(k). \tag{3.25}$$

The motivation for the error function (3.25) is that the learning of the state estimator network could further minimize the total cost, which is the ultimate objective of the ADP design. We adopt the gradient descent method for the online learning of the state estimator and derive the weights updating rule as follows

$$W_{SE}(k+1) = W_{SE}(k) + \Delta W_{SE}(k) \tag{3.26}$$

and

$$\Delta W_{SE}(k) = \eta_{SE}(k)\left[-\frac{\partial E_{SE}(k)}{\partial W_{SE}(k)}\right] \tag{3.27}$$

where η_{SE} is the learning rate for the state estimator network. In this design, we employ chain backpropagation algorithm to derive $\frac{\partial E_{SE}(k)}{\partial W_{SE}(k)}$ for the state estimator network as follows.

$$\Delta \hat{W}_{SE}(k) = \eta_{SE}(k)\left[-\frac{\partial E_{SE}(k)}{\partial \hat{W}_{SE}(k)}\right] \tag{3.28}$$

where

$$\frac{\partial E_{SE}(k)}{\partial \hat{W}_{SE}(k)} = \frac{\partial E_{SE}(k)}{\partial J(k)}\frac{\partial J(k)}{\partial u(k)}\frac{\partial u(k)}{\partial \hat{X}(k)}\frac{\partial \hat{X}(k)}{\partial \hat{W}_{SE}(k)}. \tag{3.29}$$

From (3.29), one can also see that the J function is actually minimized by state estimator network.

3.4.2 *Learning in the critic and the action network*

As discussed in the Section 3.2.1, the critic network is adopted to approximate the expected discounted total future cost. We define that the discounted total future cost as

$$R(k) = r(k+1) + \gamma r(k+2) + \gamma^2 r(k+3) + \cdots. \tag{3.30}$$

Thus, we have

$$R(k-1) = r(k) + \gamma r(k+1) + \gamma^2 r(k+2) + \cdots. \tag{3.31}$$

Subtracting (3.31) from (3.30), we can obtain that

$$R(k) = \frac{1}{\gamma}[R(k-1) - r(k)]. \tag{3.32}$$

In this design, we employ critic network to approximate the R function (i.e., the right-hand side of equation (3.32)). We use J to replace R in (3.32) and we can easily obtain the error function for critic network as

$$e_c(k) = \gamma J(k) - [J(k-1) - r(k)] \tag{3.33}$$

and

$$E_c(k) = \frac{1}{2}e_c^2(k). \tag{3.34}$$

The weights of the critic network is updated as follows

$$W_c(k+1) = W_c(k) + \Delta W_c(k), \tag{3.35}$$

and

$$\Delta W_c(k) = \eta_c(k)\Big[-\frac{\partial E_c(k)}{\partial W_c(k)}\Big] \tag{3.36}$$

where η_c is the learning rate for the critic network. Here we employ chain backpropagation algorithm to derive $\frac{\partial E_c(k)}{\partial W_c(k)}$ for the critic network as

$$\frac{\partial E_c(k)}{\partial W_c(k)} = \frac{\partial E_c(k)}{\partial J(k)}\frac{\partial J(k)}{\partial W_c(k)}. \tag{3.37}$$

We keep the similar action network design as that in literature and adopt the action network to provide optimal control strategy. The input of the action network is the virtual state (i.e., the recovered full system information) and the output is the control action. The error function of action network is defined as

$$e_a(k) = J(k) - U_c \tag{3.38}$$

and

$$E_a(k) = \frac{1}{2}e_a^2(k) \tag{3.39}$$

where U_c is the ultimate utility function. The value of U_c is critical in ADP design and it could be variant in different applications. Without loss of generality, we set U_c to be 0 here.

The weights of the action network is updated as follows

$$W_a(k+1) = W_a(k) + \Delta W_a(k) \tag{3.40}$$

and

$$\Delta W_a(k) = \eta_a(k)\Big[-\frac{\partial E_a(k)}{\partial W_a(k)}\Big] \tag{3.41}$$

where η_a is the learning rate for the action network. In this design, we employ chain backpropagation algorithm to derive $\frac{\partial E_a(k)}{\partial W_a(k)}$ for the action network as

$$\frac{\partial E_a(k)}{\partial W_a(k)} = \frac{\partial E_a(k)}{\partial J(k)}\frac{\partial J(k)}{\partial u(k)}\frac{\partial u(k)}{\partial W_a(k)}. \tag{3.42}$$

The learning process of the proposed data-driven HDP approach is regarded as completed when the critic network, action network and the state estimator network all finish its iterative weights updating. In this design, both action network and the state estimator network share the same error function and can further minimize the total cost every time step.

3.5 Simulation Study

We first conduct a discrete-time system with only measured input/output data. The typical trajectories of system state and the control action are presented to show the control performance. In addition, we also present the well-known cart-pole balancing example and the ball and beam balancing example with that only part of the system states are measurable.

3.5.1 *Case study one*

The discrete-time linear system function (modified system function from Lewis and Vrabie (2009)) is defined as follows

$$\begin{aligned} x(k+1) &= \begin{bmatrix} 1 & -0.5 \\ 1 & -0.1 \end{bmatrix} x(k) + \begin{bmatrix} 1 \\ 0 \end{bmatrix} u(k) \\ y(k) &= \begin{bmatrix} 1 & 0 \end{bmatrix} x(k) \end{aligned} \tag{3.43}$$

where y is a scalar here (i.e., $m = 1$), and we include 5 time-step delay of $y(k)$ for the input of the state estimator. Therefore, we define the input for the state estimator as

$$\begin{aligned} X_{SE} = [&y(k-5) \;\; y(k-4) \;\; y(k-3) \;\; y(k-2) \\ &y(k-1) \;\; y(k) \;\; u(k-1) \;\; u(k)]. \end{aligned} \tag{3.44}$$

In this case, X_{SE} is a 8×1 vector. The MLP structure for state estimator network, action network and critic network are 8-8-2 (i.e., there are 8 input nodes, 8 hidden nodes and 2 output nodes), 2-6-1, 3-6-1, respectively. The weights for all the neural networks are randomly initialized between $[-1, 1]$, and the system starts from $(1, 1)$. The reward function for the system is set as

$$r = x^{\mathsf{T}} Q x + u^{\mathsf{T}} R u \tag{3.45}$$

where Q and R are the identity matrices. The weights in the critic network, action network and state estimator network are tuned with their internal cycles N_c, N_a and N_{SE}, respectively. That is to say that three types of networks are updated at most N_c, N_a and N_{SE} in each time step, respectively, or stopped once the internal errors are within the threshold T_c, T_a and T_{SE}, respectively. We apply the adaptive learning rate (ALR) as follows: $\eta_c(0) = 0.1$, $\eta_a(0) = 0.1$ and $\eta_{SE}(0) = 0.1$. $\eta_c(0)$ was the initial learning rate for the critic network and would be later dropped to $\eta_c(f) = 0.005$, and we keep the same setting for the other two types of networks. The stopping criteria for the learning of the networks are: $N_c = 150$, $N_a = 200$, $N_{SE} = 220$, $T_c = 1e-3$, $T_a = 1e-3$, and $T_{SE} = 1e-4$.

The proposed data-driven HDP design shows the control performance using only the measured input/output data in Fig. 3.4. The trajectories of the system states are presented in Fig. 3.4, where one can see that system states can quickly converge to zero with our proposed controller.

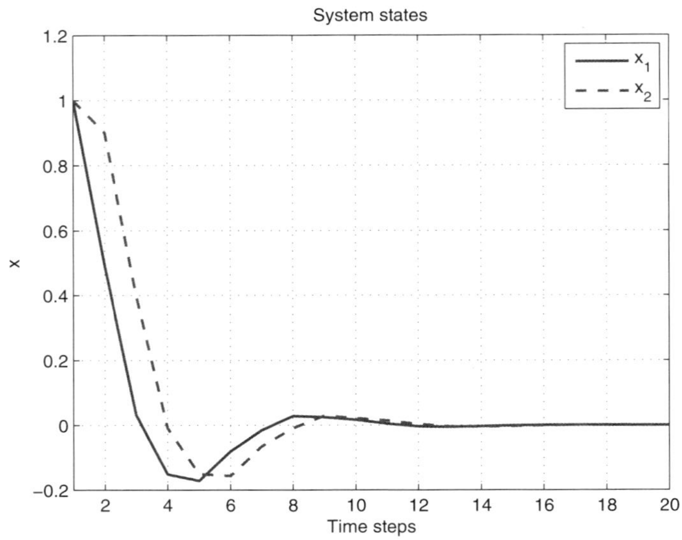

Fig. 3.4 Evolution of the system states using HDP with state estimator

3.5.2 *Case study two*

In this case study, we conduct the continuous nonlinear system model with cart-pole balancing system (same benchmark as that in Lin and Mitchell (1993); Si and Wang (2001)). The system model is presented as

$$\partial^2\theta/\partial t^2 = \frac{g\sin\theta + \frac{\cos\theta[-F - ml\dot{\theta}^2\sin\theta + \mu_c sgn(\dot{x})]}{m_c+m} - \frac{\mu_p\dot{\theta}}{ml}}{l(\frac{4}{3} - \frac{m\cos\theta^2}{m_c+m})} \tag{3.46}$$

$$\partial^2 x/\partial t^2 = \frac{F + ml[\dot{\theta}^2\sin\theta - \ddot{\theta}\cos\theta] - \mu_c sgn(\dot{x})}{m_c + m} \tag{3.47}$$

where the acceleration $g = 9.8\text{m/s}^2$, the mass of the cart $m_c = 1.0\text{kg}$, the mass of the pole $m = 0.1\text{kg}$, half-pole length $l = 0.5\text{m}$, the coefficient of friction of the cart $\mu = 0.0005$ and the coefficient of friction the pole $\mu_p = 0.000002$. The force F applied to the cart is continuous between 10 Newtons and -10 Newtons.

The state vector in this system model is

$$X = \begin{bmatrix} x_1 \; x_2 \; x_3 \; x_4 \end{bmatrix} \tag{3.48}$$

where x_1 is the angle θ of the pole with respect to the vertical axis, x_2 is the angle velocity $\dot{\theta}$, x_3 is the position (x) of the cart, and x_4 is the velocity of the cart. We keep the same partially observable settings as that in Lin and Mitchell (1993). That is to say that only x_1 and x_3 are observable. Here we also include one step backward of the observable data as the inputs for the state estimator network. Therefore, we define the input for the state estimator as

$$X_{SE} = [x_1(k-1)\ \ x_1(k)\ \ x_3(k-1)\ \ x_3(k)\ \ u(k-1)\ \ u(k)]. \tag{3.49}$$

The MLP structure for state estimator, action network, critic network are 6-8-4 (i.e., there are 6 input nodes, 8 hidden nodes and 4 output nodes), 4-6-1, 5-6-1, respectively. In this case study, we would apply the following parameter settings: $\eta_c(0) = 0.4$, $\eta_a(0) = 0.4$ and $\eta_{SE}(0) = 0.1$. $\eta_c(0)$ was the initial learning rate for the critic network and would be later dropped to $\eta_c(f) = 0.005$, and we keep the same setting for the other two type of networks. The stopping criteria for the learning of the networks are: $N_c = 150$, $N_a = 200$, $N_{SE} = 220$, $T_c = 1e-3$, $T_a = 1e-3$, and $T_{SE} = 1e-4$. Note we keep the other parameter settings the same as above.

In our current study, the objective is to keep balancing the pole on the cart for a certain period of time. Specifically, each run consists of a maximum of 3000 trials, and a trial will be considered successful if it can maintain the balance of the pole for $60,000$ time steps. That is to say a pole is considered fallen when the angular is outside the range of $[-12°, 12°]$ or the cart if beyond the range of $[-2.4\text{m}, 2.4\text{m}]$. Here we conduct 30 runs for the simulation, and the initial states for 30 runs are independent. We set the initial values for the angle of the pole and the position of the cart to be uniformly distributed in $[-0.1\text{rad}, 0.1\text{rad}]$ and $[-0.5\text{m}, 0.5\text{m}]$, respectively.

The proposed date-driven HDP approach can achieve 100% successful rate and the average trial number is 595.0. The trajectories of the system states (in the first 4000 steps) in a typical successful trial are presented in Fig. 3.5 and Fig. 3.6. The trajectories of the control action is presented in Fig. 3.7.

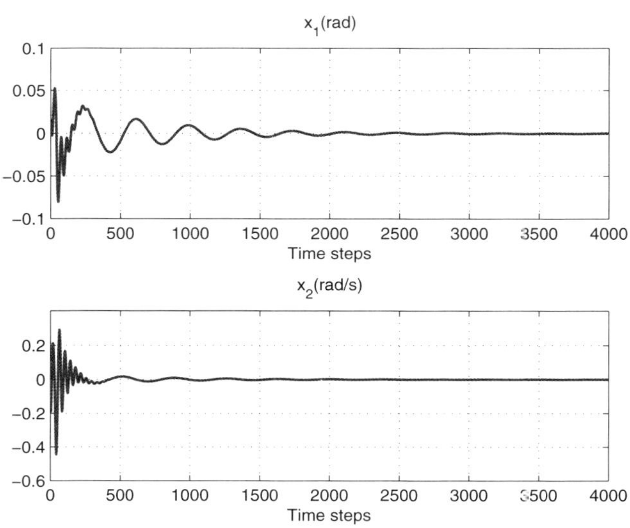

Fig. 3.5 Trajectories of x_1 and x_2 in the first 4000 steps in a typical successful run

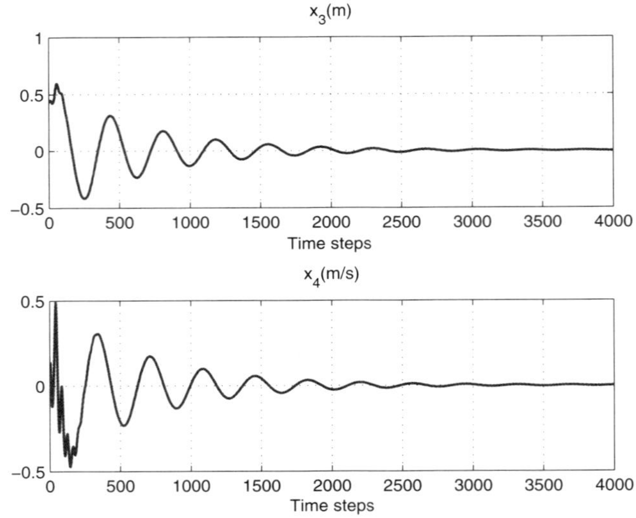

Fig. 3.6 Trajectories of x_3 and x_4 in the first 4000 steps in a typical successful run

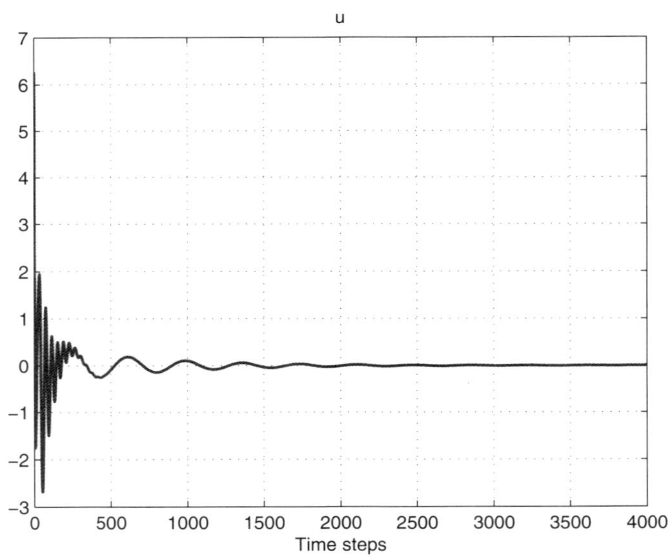

Fig. 3.7 Trajectory of the control action 4000 steps in a typical successful run

3.5.3 *Case study three*

In this case study, we test the proposed data-driven HDP approach on the ball and beam balancing system (same benchmark as that in He *et al.* (2013)). The schematic diagram for the system is provided in Fig. 3.8, and the system function

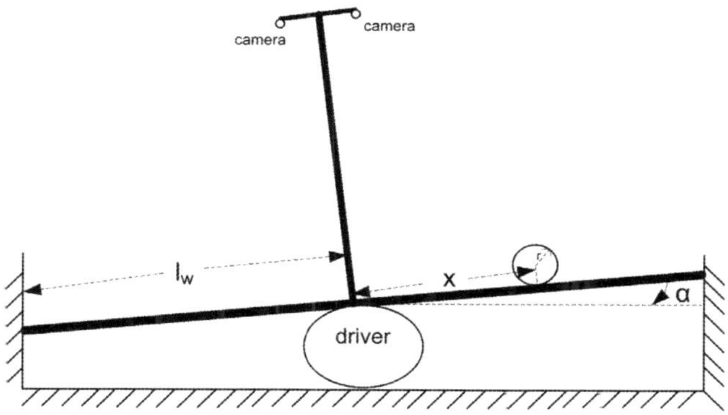

Fig. 3.8 The schematic diagram of the ball and beam balancing system

is presented as

$$\begin{bmatrix} A & B \\ C & D \end{bmatrix} \cdot \begin{bmatrix} \dot{x}_2 \\ \dot{x}_4 \end{bmatrix} = \begin{bmatrix} P_1 \\ P_2 \end{bmatrix} \tag{3.50}$$

where

$$P_1 = mx_1x_4^2 + mg(\sin x_3) \tag{3.51}$$

$$P_2 = (ul + mgx_1)\cos x_3 - (2mx_2x_1 + bl^2)x_4 - Kl^2x_3 \tag{3.52}$$

$$A = m + \frac{I_b}{r^2} \tag{3.53}$$

$$B = (mr^2 + I_b)\frac{1}{r} \tag{3.54}$$

$$C = (mr^2 + I_b)\frac{1}{r} \tag{3.55}$$

$$D = mx_1^2 + I_b + I_\omega. \tag{3.56}$$

The parameters for the system are listed as follows:

m: 0.0162kg, the mass of the ball;

r: 0.02m, the roll radius of the ball;

I_b: 4.32×10^{-5}kg·m^2, the inertia moment of the ball;

b: 1N/m, the friction coefficient of the drive mechanics;

l: 0.48m, the radius of force application;

l_ω: 0.5m, the radius of beam;

K: 0.001N/m, the stiffness of the drive mechanics;

g: 9.8N/kg, the gravity;

I_ω: 0.14025kg·m^2, the inertia moment of the beam;

u: the force of the drive mechanics.

Therefore, we can obtain the general form of this problem as described in (3.57)

$$\begin{bmatrix} \dot{x}_2 \\ \dot{x}_4 \end{bmatrix} = \begin{bmatrix} A & B \\ C & D \end{bmatrix}^{-1} \begin{bmatrix} P_1 \\ P_2 \end{bmatrix}, \tag{3.57}$$

and the other two terms in the state vector can be expressed as $\dot{x}_1 = x_2$ and $\dot{x}_3 = x_4$. Here the state vector is defined as $X = \begin{bmatrix} x_1 \; x_2 \; x_3 \; x_4 \end{bmatrix}$.

We keep the same environment setting as above and assume that only x_1 and x_3 are observable. Here we also include one step backward of the observable data as the inputs for the state estimator network. All simulation results presented here are based on 30 random runs with random initial weights for all the neural networks. Each run consists of a maximum of 3000 trials, and a run will be considered successful if the last trial (i.e., trial number less or equal than 3000) can maintain balance of the ball for 60, 000 time steps with a step size of 0.02s.

The proposed date-driven HDP approach can achieve 100% successful rate and the average trial number is 271.8. The trajectories of the system states (in the first 6000 steps) in a typical successful trial are presented in Fig. 3.9 and Fig. 3.10. The trajectories of the control action is presented in Fig. 3.11.

3.6 Conclusions and Discussion

In this chapter, we proposed a novel data-driven HDP design to handle the POMDP with only the measured input/output data. We include the state estimator network

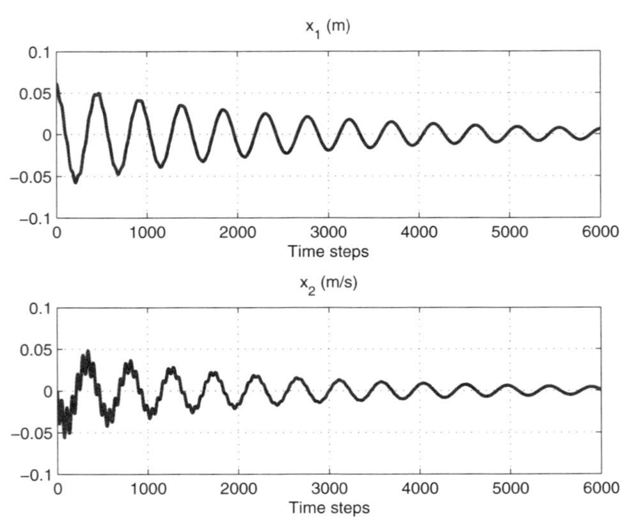

Fig. 3.9 Trajectories of x_1 and x_2 in a typical successful run

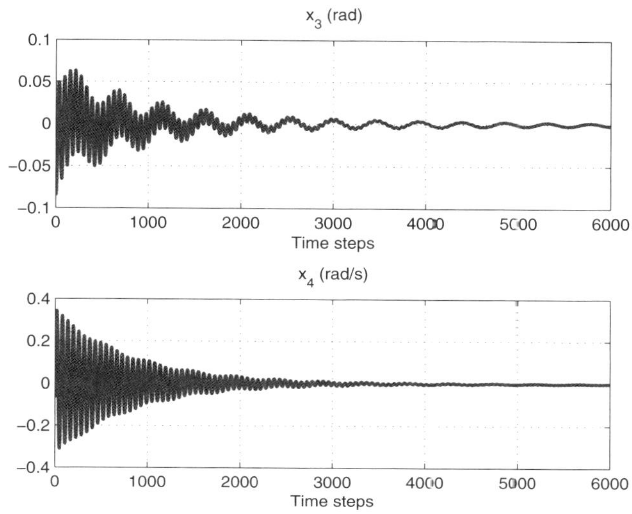

Fig. 3.10 Trajectories of x_3 and x_4 in a typical successful run

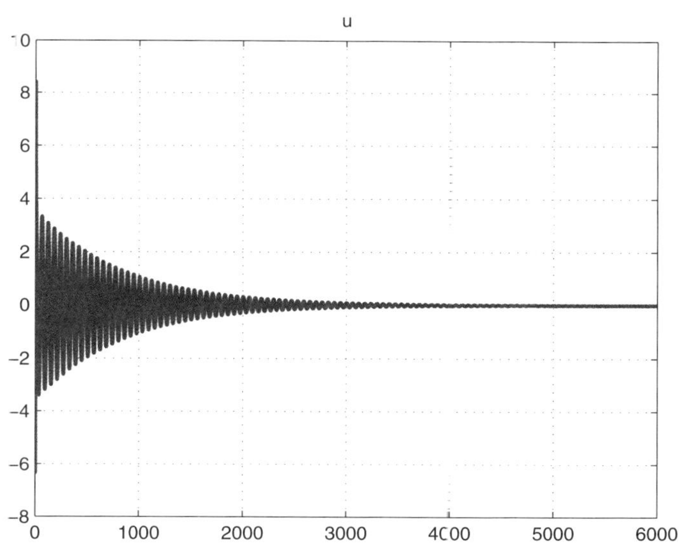

Fig. 3.11 Trajectory of the control action in a typical successful run

to provide the HDP controller with recovered system information. MLP neural network structure is used to build the state estimator network. It observes the input/output data (including the previous data) and provides the action network with full system information. The proposed data-driven HDP design on POMDP is presented with the gradient descent algorithm. We test it on both the discrete-time and continuous-time systems and simulation results justify the control performance.

Acknowledgement

This work was supported by the National Science Foundation (NSF) under grant CAREER ECCS 1053717, Army Research Office (ARO) under grant W911NF-12-1-0378, and NSF-DFG Collaborative Research on Autonomous Learning (a supplement grant to CNS 1117314). This work was also partially supported by the National Natural Science Foundation of China under grant 51228701.

References

Bertsekas, D. P. and Tsitsiklis, J. N. (1996). *Neuro-Dynamic Programming* (Athena Scientific).

Enns, R. and Si, J. (2003). Helicopter trimming and tracking control using direct neural dynamic programming, *IEEE Transactions on Neural Networks* **14**, 4, pp. 929–939.

Ferrari, S. and Stengel, R. F. (2004). Model-based adaptive critic designs, in J. Si, A. G. Barto, W. B. Powell, and D. Wunsch (eds.), *Handbook of Learning and Approximate Dynamic Programming* (Wiley-IEEE Press, New Jersey), pp. 65–95.

Hauskrecht, M. (2000). Value-function approximations for partially observable Markov decision processes, *Journal of Artificial Intelligence Research* **13**, 1, pp. 33–94.

He, H. (2011). *Self-Adaptive Systems for Machine Intelligence* (Wiley-IEEE Press).

He, H., Ni, Z. and Fu, J. (2012). A three-network architecture for on-line learning and optimization based on adaptive dynamic programming, *Neurocomputing* **78**, 1, pp. 3–13.

He, H., Ni, Z. and Zhao, D. (2013). Learning and optimization in hierarchical adaptive critic design, in F. L. Lewis and D. Liu (eds.), *Reinforcement Learning and Approximate Dynamic Programming for Feedback Control* (Wiley-IEEE Press, New Jersey), pp. 78–95.

Kaelbling, L. P., Littman, M. L. and Cassandra, A. R. (1998). Planning and acting in partially observable stochastic domains, *Artificial Intelligence* **101**, 1–2, pp. 99–134.

Lewis, F. L. and Liu, D. (2013). *Reinforcement Learning and Approximate Dynamic Programming for Feedback Control* (Wiley-IEEE Press).

Lewis, F. L. and Vamvoudakis, K. G. (2011). Reinforcement learning for partially observable dynamic processes: Adaptive dynamic programming using measured output data, *IEEE Transactions on Systems, Man, and Cybernetics, Part B: Cybernetics* **41**, 1, pp. 14–25.

Lewis, F. L. and Vrabie, D. (2009). Reinforcement learning and adaptive dynamic programming for feedback control, *IEEE Circuits and Systems Magazine* **9**, 3, pp. 32–50.

Li, H. and Liu, D. (2012). Optimal control for discrete-time affine nonlinear systems using general value iteration, *IET Control Theory and Applications* **6**, 18, pp. 2725–2736.

Lin, L. J. and Mitchell, T. M. (1993). Reinforcement learning with hidden states, in *Proceedings of the 2nd International Conference on the Simulation of Adaptive Behavior* (Cambridge), pp. 271–280.

Liu, D., Javaherian, H., Kovalenko, O. and Huang, T. (2008). Adaptive critic learning techniques for engine torque and air-fuel ratio control, *IEEE Transactions on Systems, Man, and Cybernetics, Part B: Cybernetics* **38**, 4, pp. 988–993.

Liu, D., Li, H. and Wang, D. (2013a). Neural-network-based zero-sum game for discrete-time nonlinear systems via iterative adaptive dynamic programming algorithm, *Neurocomputing* **110**, pp. 92–100.

Liu, D., Wang, D. and Yang, X. (2013b). An iterative adaptive dynamic programming algorithm for optimal control of unknown discrete-time nonlinear systems with constrained inputs, *Information Sciences* **220**, pp. 331–342.

Liu, D. and Wei, Q. (2013). Finite-approximation-error-based optimal control approach for discrete-time nonlinear systems, *IEEE Transactions on Cybernetics* **43**, 2, pp. 779–789.

Liu, D., Zhang, Y. and Zhang, H. (2005). A self-learning call admission control scheme for CDMA cellular networks, *IEEE Transactions on Neural Networks* **16**, 5, pp. 1219–1228.

Liu, D., Wang, D., Zhao, D., Wei, Q. and Jin, N. (2012). Neural-network-based optimal control for a class of unknown discrete-time nonlinear systems using globalized dual heuristic programming, *IEEE Transactions on Automation Science and Engineering* **9**, 3, pp. 628–634.

Michael, T. and Jordan, I. (1995). Reinforcement learning algorithm for partially observable Markov decision problems, in *Proceedings of the Advances in Neural Information Processing Systems* (Cambridge), pp. 345–352.

Ni, Z., Fang, X., He, H., Zhao, D. and Xu, X. (2013a). Real-time tracking control on adaptive critic design with uniformly ultimately bounded condition, in *IEEE Symposium on Adaptive Dynamic Programming and Reinforcement Learning* (Singapore), pp. 39–46.

Ni, Z. and He, H. (2013). Heuristic dynamic programming with internal goal representation, *Soft Computing* **17**, 11, pp. 2101–2108.

Ni, Z., He, H., Prokhorov, D. V. and Fu, J. (2011). An online actor-critic learning approach with Levenberg-Marquardt algorithm, in *International Joint Conference on Neural Networks* (San Jose, USA), pp. 2333–2340.

Ni, Z., He, H. and Wen, J. (2013b). Adaptive learning in tracking control based on the dual critic network design, *IEEE Transactions on Neural Networks and Learning Systems* **24**, 6, pp. 913–928.

Ni, Z., He, H., Wen, J. and Xu, X. (2013c). Goal representation heuristic dynamic programming on maze navigation, *IEEE Transactions on Neural Networks and Learning Systems* **24**, 12, pp. 2038–2050.

Ni, Z., He, H., Zhao, D. and Prokhorov, D. V. (2012). Reinforcement learning control based on multi-goal representation using hierarchical heuristic dynamic programming, in *International Joint Conference on Neural Networks* (Brisbane, Australia), pp. 1–8.

[Online]. `http://www.pomdp.org/pomdp/tutorial/index.shtml`.

Prokhorov, D. V. and Wunsch, D. C. (1997). Adaptive critic designs, *IEEE Transactions on Neural Networks* **8**, 5, pp. 997–1007.

Saad, E. (2010). Reinforcement learning in partially observable markov decision processes using hybrid probabilistic logic programs, *arXiv preprint arXiv: 1011.5951.*

Si, J., Barto, A. G., Powell, W. B. and Wunsch, D. C. (2004). *Handbook of Learning and Approximate Dynamic Programming* (Wiley-IEEE Press).

Si, J. and Wang, Y. T. (2001). On-line learning control by association and reinforcement, *IEEE Transactions on Neural Networks* **12**, 2, pp. 264–276.

Singh, S. P., Jaakkola, T. and Jordan, M. I. (1994). Learning without state-estimation in partially observable Markovian decision processes, in *International Conference on Machine Learning* (New Brunswick, USA), pp. 284–292.

Smith, T. and Simmons, R. (2004). Heuristic search value iteration for POMDPs, in *Proceedings of the 20th Conference on Uncertainty in Artificial Intelligence* (Arlington, USA), pp. 520–527.

Sutton, R. S. and Barto, A. G. (1998). *Reinforcement Learning–An Introduction* (MIT Press).

Tang, Y., He, H., Ni, Z., Wen, J. and Sui, X. (2014). Reactive power control of grid-connected wind farm based on adaptive dynamic programming, *Neurocomputing* **125**, pp. 125–133.

Venayagamoorthy, G. K., Harley, R. G. and Wunsch, D. C. (2002). Comparison of heuristic dynamic programming and dual heuristic programming adaptive critics for neurocontrol of a turbogenerator, *IEEE Transactions on Neural Networks* **13**, 3, pp. 764–773.

Vrabie, D., Pastravanu, O., Abu-Khalaf, M. and Lewis, F. L. (2009). Adaptive optimal control for continuous-time linear systems based on policy iteration, *Automatica* **45**, 2, pp. 477–484.

Wang, D. and Liu, D. (2013). Neuro-optimal control for a class of unknown nonlinear dynamic systems using SN-DHP technique, *Neurocomputing* **121**, pp. 218–225.

Wang, D., Liu, D., Wei, Q., Zhao, D. and Jin, N. (2012). Optimal control of unknown nonaffine nonlinear discrete-time systems based on adaptive dynamic programming, *Automatica* **48**, 8, pp. 1825–1832.

Wang, F. Y., Jin, N., Liu, D. and Wei, Q. (2011). Adaptive dynamic programming for finite horizon optimal control of discrete-time nonlinear systems with ε-error bound, *IEEE Transactions on Neural Networks* **22**, 1, pp. 24–36.

Wang, F. Y., Zhang, H. and Liu, D. (2009). Adaptive dynamic programming: An introduction, *IEEE Computational Intelligence Magazine* **4**, 2, pp. 39–47.

Wei, Q. and Liu, D. (2013). Numerical adaptive learning control scheme for discrete-time nonlinear systems, *IET Control Theory and Applications* **7**, 11, pp. 1472–1486.

Werbos, P. J. (2008). ADP: The key direction for future research in intelligent control and understanding brain intelligence, *IEEE Transactions on Systems, Man, and Cybernetics, Part B: Cybernetics* **38**, 4, pp. 898–900.

Werbos, P. J. (2009). Intelligence in the brain: A theory of how it works and how to build it, *Neural Networks* **22**, 3, pp. 200–212.

Werbos, P. J. (2013). Reinforcement learning and approximate dynamic programming (RLADP)–foundations, common misconceptions and challenges ahead, in F. L. Lewis and D. Liu (eds.), *Reinforcement Learning and Approximate Dynamic Programming for Feedback Control* (Wiley-IEEE Press, New Jersey), pp. 3–30.

Zhang, H., Song, R., Wei, Q. and Zhang, T. (2011a). Optimal tracking control for a class of nonlinear discrete-time systems with time delays based on heuristic dynamic programming, *IEEE Transactions Neural Networks* **22**, 12, pp. 1851–1862.

Zhang, H., Wei, Q. and Liu, D. (2011b). An iterative adaptive dynamic programming method for solving a class of nonlinear zero-sum differential games, *Automatica* **47**, 1, pp. 207–214.

Zhang, H., Wei, Q. and Luo, Y. (2008). A novel infinte-time optimal tracking control scheme for a class of discrete-time nonlinear systems via the greedy HDP iteration algorithm, *IEEE Transactions on Systems, Man, and Cybernetics, Part B: Cybernetics* **38**, 4, pp. 937–942.

Chapter 4

Online Reinforcement Learning for Continuous-State Systems

Yuanheng Zhu and Dongbin Zhao

Abstract In this part, we focus on the online reinforcement learning for continuous state-action systems. Compared to offline problem, online case lacks of system dynamics. And continuous state-space problems are more difficult than finite states problem, as the number of states is infinite. Two methods are involved here, respectively RLSPI (Recursive Least-Square Policy Iteration) and MSEC (Multi-Samples in Each Cell). Both of them have their advantages and drawbacks. RLSPI combines LSPI (Least-Squares Policy Iteration) with continuous policy parametrization and RLS (Recursive Least-Squares). So it can be applied to continuous-action systems. It has simplicity of implementation and low cost of computation. But RLSPI uses part of observed samples in the past, so its utilization of samples is still not sufficient. Besides, RLSPI adopts ε-greedy exploration technique, which has low efficiency and selects explorative actions randomly. MSEC overcomes the drawbacks of RLSPI. It utilizes samples efficiently so that the algorithm can be sure to achieve a policy that is optimal or near-optimal to the agent. And it adopts PAC (Probably Approximately Correct) principle to explore, so its time bound is polynomial in some parameters of the system. Examples of both two algorithms are simulated to reveal their performance.

4.1 Introduction

Reinforcement learning (RL) is a method that let agents interact with environment and observe reward from environment. By interaction, agents can tune its policy continually based on the reward and finally achieve an optimal or near-optimal policy which gains maximum calculated reward. Some famous algorithms of RL, such as Q-learning and SARSA, have been proposed for online learning and no relying on system dynamics. In most cases, RL considers Markov Decision Processes

(MDPs). More details about RL can be found in Busoniu *et al.* (2010a); Sutton and Barto (1998).

Some online algorithms were proposed by researchers to apply RL method to online implementation. Q-learning [Watkins (1989)], SARSA [Rummery and Niranjan (1994)] and TD(λ) [Sutton and Barto (1998)] belong to one kind of online algorithms which modify value function based on temporal difference (TD) error with gradient method. Si and Wang (2001) applied adaptive dynamic programming (ADP) for online learning of under-actuated control systems and presented great performance. However, the mentioned online algorithms update value function and policy based on TD errors at each step by gradient-based tuning. In algorithms, recent visited data have more impact on the tuning of Q function and policy than old data. So the convergence of these algorithms to the optimal solution relies on the learning rate and the sufficiency of visiting all state-action pairs [Tsitsiklis and Roy (1996, 1997)]. In this way, their efficiency of achieving optimal policy is low.

To solve those problems, Busoniu *et al.* (2010a,b) extended offline least-squares policy iteration (LSPI) [Lagoudakis and Parr (2003)] method to an online LSPI algorithm. LSPI is a method using PI and solves value function from a least-squares point, which benefits LSPI more efficient and accurate. The online LSPI algorithm of Busoniu *et al.* (2010a,b) employed Q function and the results revealed great performance for online learning. A batch least-squares method was used and only discrete-action policies were applied. In this way, the tuning of policy is based on a batch of samples, not just a single point in above TD methods. So online LSPI makes a higher use of online samples than the above mentioned algorithms and the efficiency is improved.

But online LSPI can only be applied to discrete-action systems, while considering continuous-action systems, action space must be discretized. To overcome the limit, we propose a RLSPI algorithm to apply online LSPI to continuous-action systems. A continuous policy approximation is used and during our algorithm, the continuous policy is tuned simultaneously. Moreover, a recursive least-squares (RLS) technique is adopted to facilitate the implementation of our algorithm.

Even though LSPI helps to improve the utilization of online samples, the results of those algorithms are still not satisfying. It is because it is not guaranteed that the final policy of online LSPI or RLSPI is optimal or near-optimal corresponding to agents. The drawbacks mainly focus on two parts. First, when tuning policies, algorithms only consider recent online samples and ignoring the past samples which may help to improve the policy. Besides, they all adopt ε-greedy exploration technique, which selects actions randomly. So their exploration is not efficient.

To solve this problem, a MSEC algorithm is proposed which is in the framework of PAC. PAC refers to a concept that if an algorithm can achieve optimal or near-optimal policy in a polynomial time, then it is called PAC. A lot of PAC algorithms have been proposed include E^3 [Kearns and Singh (2002)], RMAX [Brafman and Tennenholtz (2003)], MBIE [Strehl and Littman (2005)], Delayed Q-learning [Strehl *et al.* (2006)], ARL [Bernstein and Shimkin (2010)] and C-PACE [Pazis and Parr (2013)]. But they are either suitable for finite MDP, or relying on system dynamics. In our MSEC, we consider continuous state space and use state aggregation technique to partition the state space. More concretely, a grid is applied over state space. And useful online samples are saved in the cells of the grid. Then a near-upper Q function is calculated and the corresponding policy is derived. The policy either is prone to explore or reaches optimal with the systems. Our algorithm can successfully swing up an Inverted Pendulum from down position and balance it at up position.

The chapter is organized as follows. In Section 4.2, we introduce the background of reinforcement learning. And in Section 4.3, RLSPI is presented and its simulation is in Section 4.4. Section 4.5 and Section 4.6 are MSEC and its simulation respectively. In the end is our discussion and conclusion.

4.2 Background of Reinforcement Learning

The continuous MDPs with deterministic transition function can be represented by a 4-tuple (S, A, r, f). S denotes the continuous state space, A is action space, $r(s, a)$ is the reward function specifying the reward from environment at state $s \in S$ and action $a \in A$, and $f(s, a)$ is the deterministic transition function that indicates the next state at state-action pair (s, a). If A is finite, the system is called discrete-action system. Otherwise, it is continuous-action system. We assume that S is bounded, not infinitely extended. And reward function also has an interval, namely $r_{\min} \leq r(s, a) \leq r_{\max}$. Note that, in our algorithm, we assume r is known, but f is unknown.

To evaluate the performance of a policy, we adopt the *discounted return criterion*. Given a policy $\pi = \{\pi_t\}_{t=0}^{\infty}$ and initial state $s_0 = s$, the discounted return is defined as the sum of infinite horizon rewards, namely

$$J^{\pi}(s) \triangleq \sum_{t=0}^{\infty} \gamma^t r(s_t, a_t) \bigg| s_0 = s, a_t = \pi_t(s_t)$$

where γ is the *discounted factor* satisfying $0 < \gamma \leq 1$. The target of RL is maximizing the value of $J^{\pi}(s)$, namely finding the *optimal value function*, defined by $V(s) \triangleq \max_{\pi} J^{\pi}(s)$. And the corresponding policy is the *optimal policy*, $\pi^* \triangleq \arg\max_{\pi} J^{\pi}$. Sometimes, optimal policy is too difficult and strict to obtain, then a *near-optimal* policy with similar performance is acceptable. A policy π is called *ε-optimal* if $J^{\pi}(s) \geq V(s) - \varepsilon$ holds for all $s \in S$.

In some cases, another version of optimal value function-*optimal action-value function* or *optimal Q function* is preferred for the implementation of RL. Optimal Q function is defined as $Q(s, a) \triangleq r(s, a) + \gamma V(f(s, a))$, specifying the return of performing action a and executing the optimal policy at the following time. $Q(s, a)$

can be expressed in a Bellman principle

$$Q(s,a) = r(s,a) + \gamma \max_{a'} Q(f(s,a),a'))$$

and the optimal policy is generated by

$$\pi^*(s) = \arg\max_a Q(s,a).$$

In the following sections, our interest mainly focuses on Q function.

As we assume reward function is bounded in an interval $[r_{\min}, r_{\max}]$, it can be inferred that the discounted return of any policy has a low bound and an upper bound, respectively denoted by $V_{\min} = \frac{r_{\min}}{1-\gamma}$ and $V_{\max} = \frac{r_{\max}}{1-\gamma}$. And we define $V_b \triangleq \frac{1}{1-\gamma}(r_{\max} - r_{\min})$ as the maximum difference of returns between any two polices.

4.3 RLSPI Algorithm

In this part, we present RLSPI algorithm. Before that, we first introduce policy iteration (PI) method. For simplicity, we only consider undiscounted cases, namely $\gamma = 1$.

4.3.1 *Policy iteration*

In PI method, two-step iteration is included. First, policy evaluation of a given policy is implemented to calculate relevant value function. Next, a new improved policy is obtained based on the value function from the first step in policy improvement. Then, with the new policy, the two steps are iterated again until the policy converges to a fixed solution. As the policy is improved over and over again, the optimal policy can be finally obtained. Policy evaluation and policy improvement using Q function is presented as follows

PI method using Q function

Policy evaluation: given an admissible control policy $\pi^{(i)}$, calculate relevant Q function by

$$Q^{(i)}(s,a) = r(s,a) + Q^{(i)}(f(s,a), \pi^{(i)}(f(s,a))), \quad Q^{(i)}(0,0) = 0. \tag{4.1}$$

Policy improvement: generate a new improved policy $\pi^{(i+1)}$ using

$$\pi^{(i+1)}(s) = \arg\max_a Q^{(i)}(s,a). \tag{4.2}$$

It can be proved that with PI method, the sequence of Q functions is monotonically increasing and has an upper bound with optimal Q function. Namely,

Lemma 4.1. *Given $\pi^{(0)}$, use PI to obtain a Q function sequence $\{Q^{(0)}, Q^{(1)}, ...\}$ and a policy sequence $\{\pi^{(1)}, \pi^{(2)}, ...\}$. Then $\forall i \geq 0$, $\{Q^{(i)}\}$ is a monotonically increasing sequence with an upper bound Q.*

Then, the following theorem about PI is deduced. We omit the proofs here.

Theorem 4.1. *Consider a continuous MDP, if given $\pi^{(0)}$, a Q function sequence $\{Q^{(0)}, Q^{(1)}, ...\}$ generated by PI, and a policy sequence $\{\pi^{(1)}, \pi^{(2)}, ...\}$, then as $i \to \infty$, $Q^{(i)} \to Q$ and $\pi^{(i)} \to \pi^*$.*

Now, RLSPI is proposed to apply PI method to online learning.

4.3.2 RLSPI

(a) Q function and policy approximation

Here we consider continuous-state and continuous-action systems. In this way, the explicit presentation of Q function or policy is not possible. Then, the approximation structure is needed to approximate those functions. There may be kinds of approximation structures to be used, while the common and simple one is linear parametrization [Abu-Khalaf and Lewis (2005); Al-Tamimi *et al.* (2008); Vrabie and Lewis (2009); Vrabie *et al.* (2009)].

A linear parametrization of Q function can be expressed by

$$\hat{Q}(s, a) = \phi^{\mathsf{T}}(s, a)\theta \tag{4.3}$$

where $\phi(s, a) = [\phi_1(s, a), \ldots, \phi_N(s, a)]^{\mathsf{T}}$ is a vector of N basis functions (BFs), and $\theta \in R^N$ is a parameter vector. $\hat{Q}$ indicates the approximated Q . Like many others works [Abu-Khalaf and Lewis (2005); Xu *et al.* (2011)], polynomials are adopted here as BFs. Suppose $s = [s_1, \ldots, s_n]^{\mathsf{T}}$ and $a = [a_1, \ldots, a_m]^{\mathsf{T}}$ and polynomial BF ϕ_i has the following form

$$\phi_i(s, a) = s_1^{\alpha_1^i} s_2^{\alpha_2^i} \cdots s_n^{\alpha_n^i} a_1^{\beta_1^i} a_2^{\beta_2^i} \cdots a_m^{\beta_m^i} \tag{4.4}$$

where $0 \leq \alpha_1^i \leq c_1^i$, $\ldots$, $0 \leq \alpha_n^i \leq c_n^i$ and $0 \leq \beta_1^i \leq d_1^i$, $\ldots$, $0 \leq \beta_m^i \leq d_m^i$. $c_1^i, ..., c_n^i$ and $d_1^i, ..., d_m^i$ are the maximum power number of state and action with respect to Q function approximation.

Similarly, a linear parametrization

$$\hat{\pi}(s) = \omega^{\mathsf{T}}\varphi(s) \tag{4.5}$$

is adopted as the approximated policy $\hat{\pi}$, where $\varphi(s) = [\varphi_1(s), ..., \varphi_M(s)]^{\mathsf{T}}$ and $\omega = [\omega_1, ..., \omega_m] \in R^{M\times m}$, while each vector $\omega_j \in R^M$ is associated with action a_j. And the polynomial BF φ_j has the following form

$$\varphi_j(s) = s_1^{\gamma_1^j} s_2^{\gamma_2^j} \cdots s_n^{\gamma_n^j} \tag{4.6}$$

where $0 \leq \gamma_1^j \leq l_1^j, \ldots, 0 \leq \gamma_n^j \leq l_n^j$ and $l_1^j, ..., l_n^j$ are the maximum power number of state with respect to policy approximation.

(b) Policy evaluation with Q function approximation

With Q function approximation (4.3) and policy evaluation (4.1), the calculation of parameters vector θ can be implemented using online data. Suppose current policy is $\hat{\pi}^{(i)}$ and try to solve relevant Q function $\hat{Q}^{(i)}$ corresponding to parameters vector $\theta^{(i)}$. $\{(s_t, a_t, s_{t+1})\}$ denotes online data. For each sample $\{(s_t, a_t, s_{t+1})\}$, substitute (4.3) into (4.1) and we have

$$[\phi(s_t, a_t) - \phi(s_{t+1}, \hat{\pi}^{(i)}(s_{t+1}))]^\mathsf{T} \theta^{(i)} = r(s_t, a_t). \tag{4.7}$$

It is obvious that (4.7) is a linear equation to $\theta^{(i)}$. Besides, online samples are collected step by step. So RLS method is applied for learning $\theta^{(i)}$. The whole learning process is presented as follows

$$\begin{aligned}
z(t) &= r(s_t, a_t) \\
h(t) &= \phi(s_t, a_t) - \phi(s_{t+1}, \hat{\pi}^{(i)}(s_{t+1})) \\
q(t) &= P(t)h(t)\left[h(t)^\mathsf{T} P(t)h(t) + 1\right]^{-1} \\
P(t+1) &= \left[I - q(t)h(t)^\mathsf{T}\right] P(t) \\
\theta_{t+1}^{(i)} &= \theta_t^{(i)} + q(t)\left[z(t) - h(t)^\mathsf{T}\theta_t^{(i)}\right].
\end{aligned} \tag{4.8}$$

In order to ensure the implementation of RLS method at the beginning, matrix P is initialized to a large multiple of identity matrix. After enough samples, a sufficient accurate parameter $\theta^{(i)}$ is achieved. Here, a constant K_{update} denotes the number of steps for policy evaluation.

(c) Policy improvement with policy approximation

After $\theta^{(i)}$ is achieved, policy improvement continues to extract an improved policy $\hat{\pi}^{(i+1)}$. However, because of the linear parametrization (4.3) for Q function and polynomials (4.4) for BFs, it is difficult to solve policy improvement (4.2) directly and have an explicit solution of $\omega^{(i+1)}$ associated with $\hat{\pi}^{(i+1)}$. But, it is noted that both Q function and policy approximation is differentiable and the solution of (4.2) belongs to the points where the partial derivative of Q function to a is equal to zero. Therefore, a gradient-based method can be used for policy improvement, namely

$$\omega_j = \omega_j + \alpha \frac{\partial Q}{\partial a_j} \frac{\partial a_j}{\partial \omega_j} \tag{4.9}$$

where α is the learning rate.

To guarantee the accuracy of $\hat{\pi}^{(i+1)}$ or $\omega^{(i+1)}$ on the whole state space Ω, a training set which is evenly distributed over the state space is defined beforehand,

$\{S_x\}$, $x = 1, \ldots, N_x$. For any BF ϕ_i, its partial differential to action a_j has the following form

$$\frac{\partial \phi_i}{\partial u_j}(x,u) = \begin{cases} \beta_j^i \frac{\phi_i(x,u)}{u_j} & \text{if } u_j \neq 0 \\ 0 & \text{if } u_j = 0. \end{cases} \tag{4.10}$$

So partial differential of $\hat{Q}^{(i)}$ can be formulated by

$$\frac{\partial \hat{Q}^{(i)}}{\partial a_j}(s,a) = \phi_{a_j}^{\mathsf{T}}(s,a)\theta^{(i)} \tag{4.11}$$

where $\phi_{a_j} = \left[\frac{\partial \phi_1}{\partial a_j}, \ldots, \frac{\partial \phi_N}{\partial a_j}\right]^{\mathsf{T}}$. In this way, the gradient-based updating formula for $\omega_j^{(i+1)}$ on training set $\{S_x\}$ is obtained

$$\omega_{j,k}^{(i+1)} = \omega_{j,k-1}^{(i+1)} + \alpha\phi_{a_j}^{\mathsf{T}}(S_x, \varphi^{\mathsf{T}}(S_x)\omega_{j,k-1}^{(i+1)})\theta^{(i)}\varphi(S_x) \tag{4.12}$$

where $j = 1, \ldots, m$. It is noted that in order to generate an accurate parameter $\omega^{(i+1)}$, updating formula (4.12) on training set $\{S_x\}$ can be implemented for sufficient times.

(*d*) *Exploration*

For a good online algorithm, it can not only exploit current data obtained online and generate a good policy in a short time, but also achieve a good policy as optimal as possible instead of being stuck in a local optimum. The first part is called exploitation while the second part is exploration. And there is a confliction between these two parts. The above mentioned online RLSPI algorithm can be seen as an exploitation of online data, and next we discuss about the exploration for our algorithm.

Here, we introduce a common exploration technique–ε-greedy exploration and reset scheme in Busoniu *et al.* (2010a,b). At each step t, it has $1 - \varepsilon_t$ probability to apply the current policy directly and ε_t probability to add uniform random exploration noise n_t to the action associated with the current policy. It should be noted that the added exploration noise cannot be too large, otherwise the system may be deviated from the learning objective. Besides, at the beginning of the algorithm, the exploration probability is relatively large so it is benefit for policy improvement. Meanwhile, the exploration probability decreases gradually as the algorithm runs, so the proportion of the exploitation increases. In this way, a decay exploration is used here followed by

$$\varepsilon_t = \varepsilon_0 \varepsilon_d^t \tag{4.13}$$

where ε_0 is the initial value and ε_d is the decay factor with $0 < \varepsilon_d < 1$.

However, as the policy is admissible and the exploration noise is small, the system can still be stabilized after enough steps. At that time, the added exploration noise is not sufficient to drive the state away from the equilibrium and a new trial starting from non-equilibrium points is more benefit for the exploration. Hence, a reset scheme is included, namely after every T_{trial} steps, the state is reset away from equilibrium. The number of T_{trial} should be large enough to stabilize system.

Algorithm 4.1 presents our online RLSPI algorithm. It should be noted that during the implementation, no information of system dynamics is needed and the algorithm only relies on online data.

Algorithm 4.1 Online RLSPI algorithm

Input: BFs $\phi(s,a)$ and $\varphi(s)$
initial admissible policy π_0 and training set $\{S_x\}$
exploration coefficient $\{\varepsilon_t\}$ and learning rate α
policy evaluation length K_{update}
a large constant $a > 0$

initialize $\theta^{(0)} \leftarrow 0$, $\omega^{(0)} \leftarrow \pi_0$ and $P(0) = \lambda I_{N\times N}$
iteration number $i = 0$
measure initial state s_0
for every step $t = 0, 1, 2, \ldots$ **do**
  $$a_t = \begin{cases} \hat{\pi}^{(i)}(s_t) & \text{at } 1-\varepsilon_t \text{ probability} \\ \hat{\pi}^{(i)}(s_t) + n_t & \text{at } \varepsilon_t \text{ probability} \end{cases}$$
  apply a_t and measure next state s_{t+1} and reward r_t
  policy evaluation $\theta^{(i)}$ by (4.8)
  if $t = (i+1)K_{update}$ **then**
    policy improvement $\omega^{(i+1)}$ using (4.12) on $\{S_x\}$
    $i = i + 1$
  end if
end for

4.4 Examples of RLSPI

To test our online model-free RLSPI algorithm, two examples are simulated. The first one is a linear discrete-time system. Based on linear quadratic regulator (LQR) method, the explicit optimal solution of the linear discrete-time system with quadratic reward is available, which can be used to compare the results of our algorithm. The second one is a more complicated nonlinear discrete-time system.

4.4.1 *Linear discrete-time system*

The first linear system is a 4-order linearized power system used in Vrabie and Lewis (2009); Vrabie *et al.* (2009). The given system dynamics is continuous-time and is expressed by $\dot{s} = A_c s + B_c a$, where

$$A_c = \begin{bmatrix} -0.0665 & 11.5 & 0 & 0 \\ 0 & -2.5 & 2.5 & 0 \\ -9.5 & 0 & -13.736 & -13.736 \\ 0.6 & 0 & 0 & 0 \end{bmatrix}$$

$$B_c = \begin{bmatrix} 0 \; 0 \; 13.736 \; 0 \end{bmatrix}^{\mathsf{T}}.$$

To apply our discrete-time algorithm, zeros-order hold method [Lewis (1992)] is used to transform this system into a discrete-time form, namely $s_{t+1} = As_t + Ba_t$, where

$$A = \begin{bmatrix} 0.9920 & 0.5389 & 0.0277 & -0.0068 \\ -0.0232 & 0.8779 & 0.0843 & -0.0331 \\ -0.3507 & -0.1068 & 0.4992 & -0.4961 \\ 0.0299 & 0.0083 & 0.0003 & 0.9999 \end{bmatrix}$$

$$B = \begin{bmatrix} 0.0068 \; 0.0331 \; 0.4961 \; 0.00005 \end{bmatrix}^{\mathsf{T}}$$

And the reward function is a negative definite quadratic function with respect to state and action, $r(s_t, a_t) = -s_t^{\mathsf{T}} Q s_t - a_t^{\mathsf{T}} R a_t$, where $Q = I_{3\times3}$, $R = 1$. In this way, the solution of LQR for this linear discrete-time system is exactly the optimal policy and has a state-feedback form, $a_t = -Ks_t$, where $K = \begin{bmatrix} 0.4140 \; 2.1668 \; 0.5092 \; 0.1097 \end{bmatrix}^{\mathsf{T}}$.

For the RLSPI algorithm, we first define BFs, $\phi(s, a)$ and $\varphi(s)$. As we know, for the linear system, the optimal Q function has a quadratic form with respect to the state and action, and the optimal policy is a linear state-feedback controller. In this way, the following BFs are adopted

$$\phi(s, a) = \begin{bmatrix} s_1^2, s_2^2, s_3^2, s_4^2, a^2, \\ s_1 s_2, s_1 s_3, s_1 s_4, s_2 s_3, s_2 s_4, s_3 s_4, \\ s_1 a, s_2 a, s_3 a, s_4 a \end{bmatrix}^{\mathsf{T}}$$

$$\varphi(s) = [s_1, s_2, s_3, s_4]^{\mathsf{T}}.$$

So the numbers of θ and ω elements are 15 and 4 respectively.

Initial state of each trial is set to $[0, 0.1, 0, 0]^{\mathsf{T}}$. And each trial has a length of 300 steps which is sufficient to stabilize the system and the total length of implementation is 6000 steps, including 20 trials altogether. The policy evaluation length K_{update} is set to 50. Besides, uniform random exploration noise is limited between $[-1, 1]$. Initial exploration value ε_0 is equal to 1, and decay factor ε_d is set to 0.99962. And $\{-0.2, -0.1, 0, 0.1, 0.2\}^4$ is chosen for the policy improvement training set $\{S_x\}$. As the system is self-stable, policy equal to 0 is adopted as the initial admissible policy. And discounted factor γ is set to 1.

After 6000 steps of the RLSPI algorithm, the results are presented in Fig. 4.1, where (a) shows 10 representative trials during implementation, respectively at 0th, 600th, 1200th, ..., 5400th step. It is noted that with the added exploration

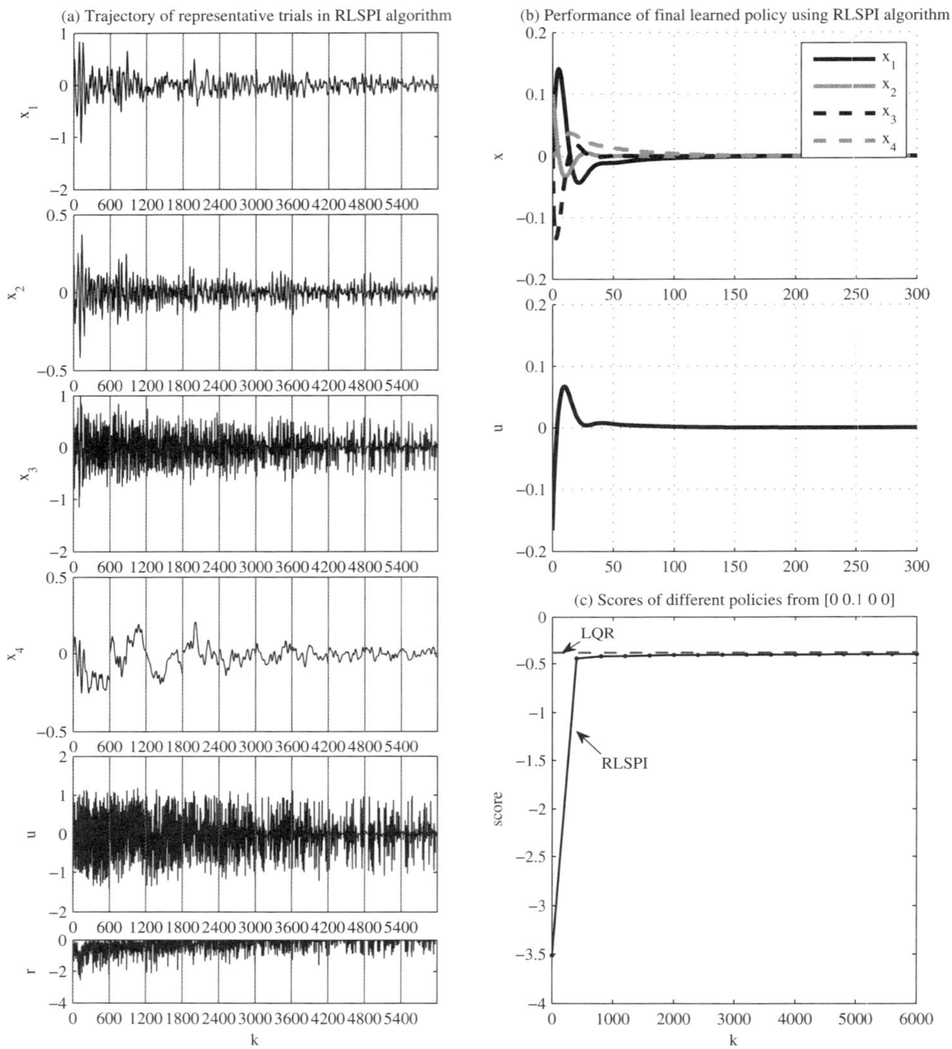

Fig. 4.1 Implementation of RLSPI algorithm for linear discrete-time system (a) Trajectory of representative trials in RLSPI algorithm (b) Performance of finial learned policy using RLSPI algorithm (c) Scores of different policies from $[0, 0.1, 0, 0]^{\mathsf{T}}$

noise, the state, action and reward oscillate during the implementation. And the oscillation weakens gradually as the step moving. It is not only because of the decay exploration, but also the policy improvement during the algorithm, associated with better and better performance. (b) reveals the performance of the final learned policy of the RLSPI algorithm starting from $[0, 0.1, 0, 0]^{\mathsf{T}}$. And (c) presents scores of different policies starting from $[0, 0.1, 0, 0]^{\mathsf{T}}$. Here, the score of a certain policy is defined as the sum of the reward along trajectory indicated by the policy starting from an initial state. The dark solid curve plots scores of different policies in RLSPI at the end of different trials with respect to their step index, while the light dashed line represents the score of optimal policy or the LQR solution. It is obvious that in the RLSPI algorithm, policies are improved very fast, e.g., within 2000 steps or 6 trials, the learned policies achieve almost the same performance of the optimal one.

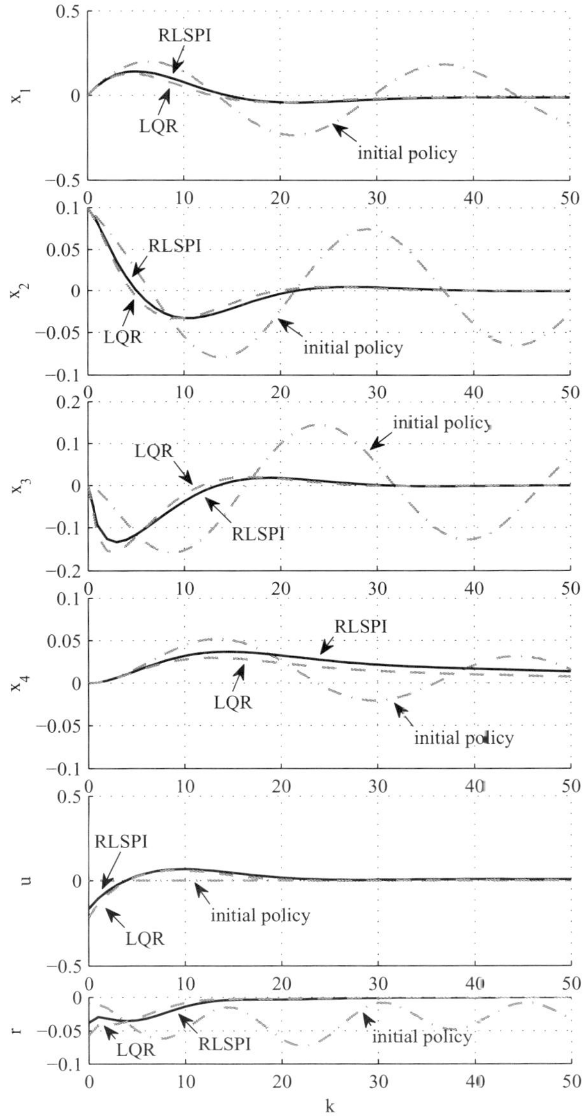

Fig. 4.2 Trajectories of three different policies, initial policy, RLSPI and LQR starting from $[0, 0.1, 0, 0]^T$

Fig. 4.2 presents trajectories of the state, action and reward by initial 0-value policy, the final learned policy using RLSPI and LQR starting from $[0, 0.1, 0, 0]^\mathsf{T}$. The initial policy has the worst performance with much smaller reward and undamped oscillation. Compared RLSPI with LQR, trajectories of RLSPI are all close to LQR with respect to the state, action and reward, but LQR performs a little better with shorter stabilization time. The reason for the difference between RLSPI and LQR can be explained that even with ε-greedy exploration, some state-action pairs are still not visited during the online RLSPI algorithm, which makes the final policy is not global optimal, or only near-optimal compared to LQR.

4.4.2 *Nonlinear discrete-time system*

The next nonlinear discrete-time system is a mass-spring system, presented in Zhang *et al.* (2009) with the following system dynamics

$$s_{t+1} = \begin{bmatrix} s_t(1) + 0.05 s_t(2) \\ -0.0005 s_t(1) - 0.0335 s_t^3(1) + s_t(2) \end{bmatrix} + \begin{bmatrix} 0 \\ 0.05 \end{bmatrix} a_t$$

Reward function adopted here has the same quadratic form as above, but with $Q = 0.5 I_{3\times 3}, R = 1$.

As the system is nonlinear, the associated Q function or policy is more complicated than that of the linear system, and more BFs are required to achieve higher precision. In this way, up to 6-order $\phi(s, a)$ and up to 5-order $\varphi(s)$ are used

$$\varphi(s,a) = \begin{bmatrix} s_1^2, s_2^2, a^2, s_1 s_2, s_1 a, s_2 a, s_1^4, s_2^4, a^4, s_1^3 s_2, s_1^2 s_2^2, s_1 s_2^3, \\ a[s_1^3, s_1^2 s_2, s_1 s_2^2, s_2^3], a^2[s_1^2, s_1 s_2, s_2^2], a^3[s_1, s_2], \\ s_1^6, s_2^6, a^6, s_1^5 s_2, s_1^4 s_2^2, s_1^3 s_2^3, s_1^2 s_2^4, s_1 s_2^5, \\ a[s_1^5, s_1^4 s_2, s_1^3 s_2^2, s_1^2 s_2^3, s_1 s_2^4, s_2^5], a^2[s_1^4, s_1^3 s_2, s_1^2 s_2^2, s_1 s_2^3, s_2^4], \\ a^3[s_1^3, s_1^2 s_2, s_1 s_2^2, s_2^3], a^4[s_1^2, s_1 s_2, s_2^2], a^5[s_1, s_2] \end{bmatrix}^\mathsf{T}$$

$$\varphi(s) = \left[s_1, s_2, s_1^3, s_1^2 s_2, s_1 s_2^2, s_2^3, s_1^5, s_1^4 s_2, s_1^3 s_2^2, s_1^2 s_2^3, s_1 s_2^4, s_2^5\right]^\mathsf{T}.$$

So the numbers of θ and ω elements increase to 49 and 12 respectively.

For this system, the initial state of each trial is set to $[-0.2, 0.2]^\mathsf{T}$. And the length of each trial T_{trial} is extended to 1000 steps, so that the system can be stabilized within that time. As the nonlinear system is more difficult to learn, RLSPI algorithm is set to run 10^5 steps, namely 100 trials are included. Besides, the policy evaluation length K_{update} increases to 200 and the training set $\{S_x\}$ selects more elements at each dimension, namely $\{-0.1, -0.08, ..., 0.08, 0.1\}^2$. The uniform random exploration noise is limited between $[-0.1, 0.1]$. The initial exploration value ε_0 is still 1, but the decay factor ε_d is changed to 0.999977. As the system is self-stable, policy equal to 0 is adopted as the initial admissible policy. And discounted factor γ is still set to 1.

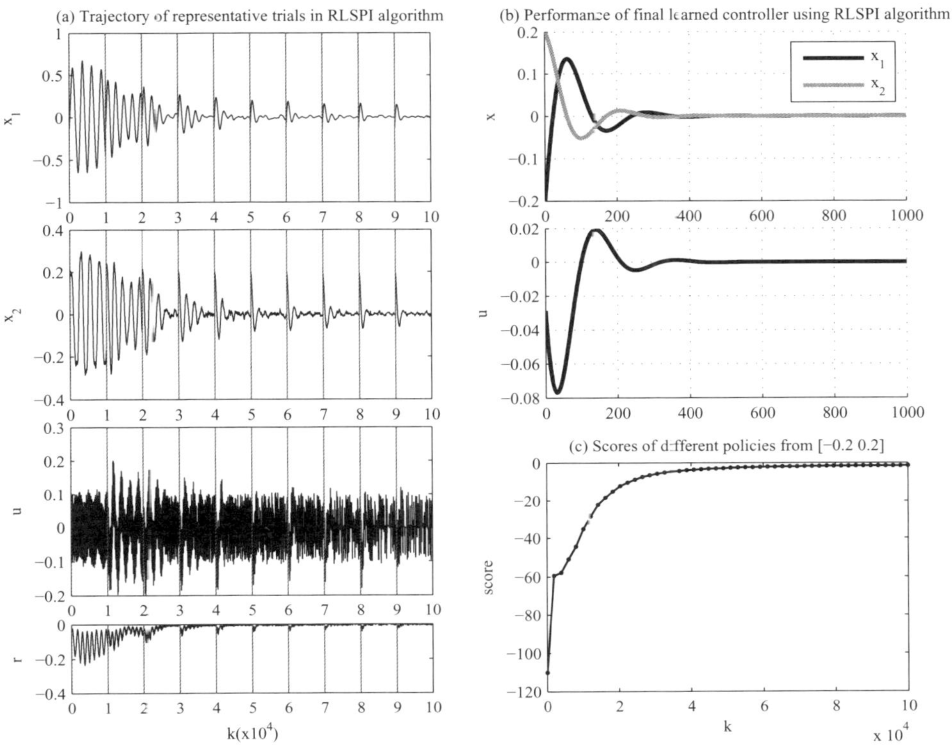

Fig. 4.3 Implementation of RLSPI algorithm for nonlinear discrete-time system (a) Trajectory of representative trials in RLSPI algorithm (b) Performance of finial learned policy using RLSPI algorithm (c) Scores of different policies from $[-0.2, 0.2]^{\mathsf{T}}$

The results of the RLSPI algorithm is presented in Fig. 4.3, where (a) reveals 10 representative trials during implementation, respectively at 0th, 10000th, 20000th, ..., 90000th step. It is obvious that in the first two representative trials, the state cannot be stabilized to zeros within the length of a trial. The main reason is that initial policies are not improved a lot. Then, as the algorithm keeps running, the successive trials have less and less oscillation and the stabilization time is reduced more and more. Meantime, the reward in each trial is also improved in latter trials. All those indicate that policies are improved during the RLSPI algorithm. The performance of the finial learned policy using RLSPI is presented in (b) starting from $[-0.2, 0.2]^{\mathsf{T}}$. And (c) reveals scores of policies at the end of different trials in RLSPI with respect to step index, starting from $[-0.2, 0.2]^{\mathsf{T}}$. At the beginning, policies perform worse with lower scores. As step moves, policies are improved gradually and relevant scores increase. At the end, they converge to an optimal or near-optimal one. Compared to the results of the linear system in example 1, the improvement of policies is slower for nonlinear implementation, as the nonlinear system is more complicated and difficult to learn with more BFs.

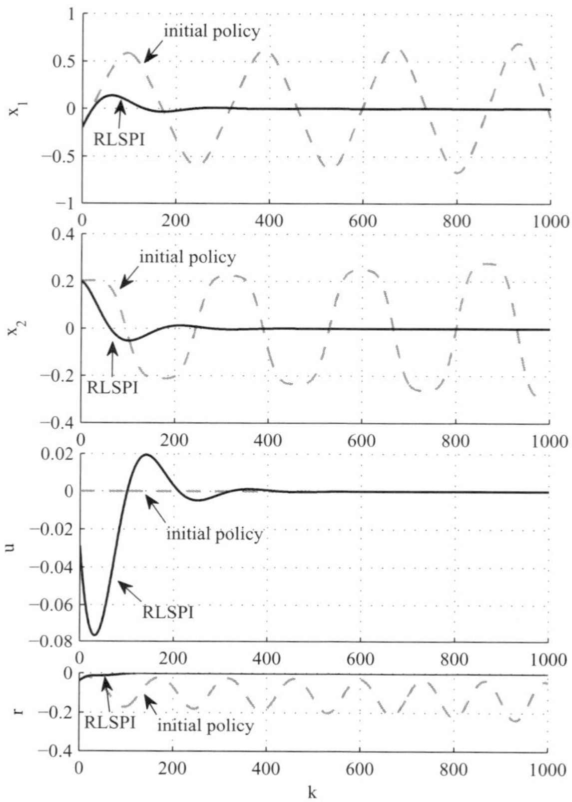

Fig. 4.4 Trajectories of two different policies, initial policy and RLSPI starting from $[-0.2, 0.2]^\mathsf{T}$

Fig. 4.4 shows two trajectories of the initial 0-value policy and the final learned policy in RLSPI. Even though the initial policy oscillates the system, the final policy can stabilize within 400 steps.

However, even RLSPI can sometimes find a good policy to the agents, but it is not guaranteed that it can always achieve an optimal or near-optimal policy every time. It is because RLSPI does not utilize samples efficiently and adopts random exploration technique. To solve this problem, next section proposes an algorithm in PAC framework.

4.5 MSEC Algorithm

PAC is defined that agents can learn a near-optimal policy within a time or error bound that is polynomial in the environment size and approximation parameters, with high probability. There are many metrics to evaluate if an algorithm is PAC, and here we adopt Policy-Mistake Count (PMC). PMC specifies the total number of steps when an algorithm performs non ε-optimal policies. PMC as a metric for the evaluation of PAC algorithms has been used a lot [Bernstein and Shimkin (2010); Strehl and Littman (2005); Strehl *et al.* (2006)].

Definition 4.1. In an algorithm, $h_t = \{s_0, a_0, s_1, a_1, \ldots, s_{t-1}, a_{t-1}, s_t\}$ is its history of states and actions and $A_t = \{\pi_k\}_{k=t}^{\infty}$ is the policy that the algorithm implements at time t. The discounted return of A_t from time t is denoted by $J^{A_t}(s_t) \triangleq \sum_{k=t}^{\infty} \gamma^{(k-t)} r(s_k, a_k)\Big| a_k = \pi_k(s_k)$. So the policy-mistake count is defined as

$$PMC(\varepsilon) \triangleq \sum_{t=0}^{\infty} \mathbb{I}\Big\{J^{A_t}(s_t) < V(s_t) - \varepsilon\Big\}$$

where $\mathbb{I}\{\cdot\}$ is a signal function. If the event in brace occurs, it outputs 1, otherwise 0.

After PMC steps, we know that the algorithm achieves near-optimal policy. If the count is polynomial in the parameters of the system, we state the algorithm is PAC. Next, we introduce state aggregation technique which is used in MSEC.

4.5.1 *State aggregation*

State aggregation is a technique that partitions a continuous state space into disjoint cells. Each cell contains a subset of the state space and is non-overlapping to each other. Here, we apply a grid over the whole state space and each cell is a rectangle or hyperrectangle. Suppose there are N_{grid} cells in the grid and each cell is denoted by C_i $(1 \leq i \leq N_{grid})$. Furthermore, we use $\Omega(C_i)$ to represent the area of state space in C_i.

4.5.2 MSEC

Now, we give a detailed presentation of MSEC algorithm. In this algorithm, only continuous-state and discrete-action systems are considered. The principle of MSEC is selectively collecting online observed samples into a data set, and adjusting the online implemented policy based on the non-stationary data set. Generally, MSEC can be divided into 3 parts.

(i) Data set

At the current time t, suppose that we have the data set $D_t = \{(\hat{s}_k, a_k, r(\hat{s}_k, a_k), f(\hat{s}_k, a_k))\}_{0 \leq k \leq t-1}$, in which are selective samples of the past time. Here, we use the notation $\hat{s}_k$ to indicate that it is an observed sample with a_k in the past and collected in D_t. So the corresponding $r(\hat{s}_k, a_k)$ and $f(\hat{s}_k, a_k)$ are known to the agent.

For any $(\hat{s}_k, a_k) \in D_t$, $\hat{s}_k$ must belong to one cell in the grid. Suppose C_i is the cell that contains $\hat{s}_k$. And in D_t, there may exist more than one samples (multi-samples) that belong to C_i. For convenience, we use notation $D_t(C_i, a)$ to specify the set of samples in D_t at a that belong to C_i. Moreover, if there exist no samples locating in C_i at a, we denote that by $D_t(C_i, a) = \emptyset$.

Given a cell C_i, if $D_t(C_i, a) \neq \emptyset$, it means that there exists at least one sample $(\hat{s}, a) \in D_t(C_i, a)$ satisfying $\hat{s} \in \Omega(C_i)$. However, even though samples in $D_t(C_i, a)$ are close to each other with a maximum distance, their next states may be far away and fall into different cells at a same action. Fig. 4.5 is an example indicating that there are three samples $\hat{s}_1, \hat{s}_2, \hat{s}_3$ in a same cell, but their next states at the same action fall into difference cells. But the distance between those next states is bounded.

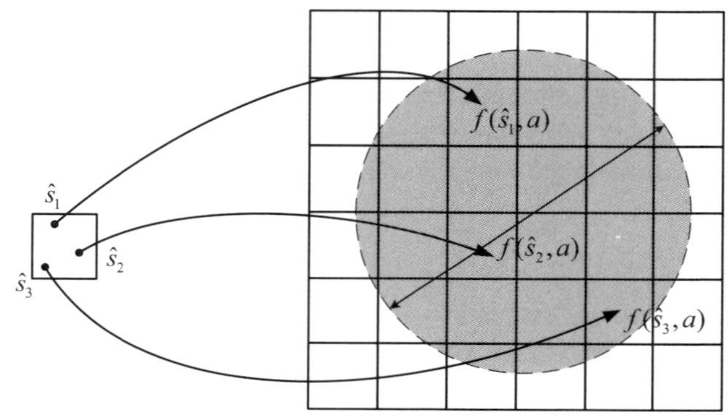

Fig. 4.5 An example of samples in a same cell falling into different cells at next states

(ii) Near-upper Q iteration

Now, a definition of *near-upper Q iteration* (NUQI) operator mapping function to function is given as follows.

Definition 4.2. Given a function $g : S \times A \to \mathbb{R}$ and arbitrary s satisfying $s \in \Omega(C_i)$, NUVI operator $\overline{\mathcal{T}}$ is defined as

$$\overline{\mathcal{T}}g(s,a) = \begin{cases} \min\limits_{\hat{s} \in D_t(C_i,a)} \left[r(\hat{s},a) + \gamma \max\limits_{a'} g(f(\hat{s},a),a') \right] & \text{if } D_t(C_i,a) \neq \emptyset \\ V_{\max} & \text{else} \end{cases} \tag{4.14}$$

NUQI operator means for a pair (s, a), if its corresponding C_i has no samples in D_t, we assign the value of this pair with the upper bound of value function, $V_{\max}$. Otherwise, we use the samples in $D_t(C_i, a)$ to approach its value, more concretely, the minimum one corresponding to the right side of the equation in (4.14).

We can prove that $\overline{\mathcal{T}}$ is a contraction, so it has a fixed solution. And as $\overline{\mathcal{T}}$ is defined based on D_t at t, a *near-upper Q function* (NUQF) $\bar{Q}_t$ is used to denote the fixed solution of $\overline{\mathcal{T}}$ with respect to D_t, namely $\bar{Q}_t = \overline{\mathcal{T}}\bar{Q}_t$.

As $\overline{\mathcal{T}}$ is a contraction, value iteration (VI) method can be used for solving $\bar{Q}_t$. Note that for all states in a same cell C_i, they have an equal value of $\bar{Q}_t$ because they share the same samples in $D_t(C_i, a)$ when substituting into the right side of (4.3). For simplicity, let $\bar{Q}_t(C_i, a)$ specify the value of any $s \in \Omega(C_i)$ at a. In this way, in VI there is no need to calculate $\bar{Q}_t$ for all states but just the value of each cell. More details about VI is available in Busoniu *et al.* (2010a); Li and Liu (2012).

Then the greedy policy is produced from $\bar{Q}_t$ and applied to the agent

$$\pi_t(s) = \arg\max_a \bar{Q}_t(s, a). \tag{4.15}$$

(iii) Known and unknown

With the greedy policy π_t applied to the agent at time t, a new observation $(s_t, a_t, r(s_t, a_t), f(s_t, a_t))$ is obtained. To decide whether this observation is added into D_t or just ignored, a new concept of known is given as follows.

Definition 4.3. Given an observation $(s, a, r(s, a), f(s, a))$, suppose $s \in \Omega(C_i)$. The pair (s, a) is called *known* in D_t only if $D_t(C_i, a) \neq \emptyset$ and there exists a sample $(\hat{s}, a, r(\hat{s}, a), f(\hat{s}, a)) \in D_t(C_i, a)$ such that $f(s, a)$ and $f(\hat{s}, a)$ locate in a same cell. Otherwise, we say (s, a) is *unknown*.

It is inferred that an observation is known because not only its current state is close to some samples in data set, but also the next state of this observation is also close to the next state of one of the samples. Examples of known and unknown are revealed in Fig. 4.6.

If $(s_t, a_t, r(s_t, a_t), f(s_t, a_t))$ at time t is known in D_t, this observation is ignored and the algorithm continues. Otherwise, $(s_t, a_t, r(s_t, a_t), f(s_t, a_t))$ is added into D_t and D_t becomes D_{t+1}. Then, $\bar{Q}_{t+1}$ is recalculated based on D_{t+1} and a new π_{t+1} is obtained.

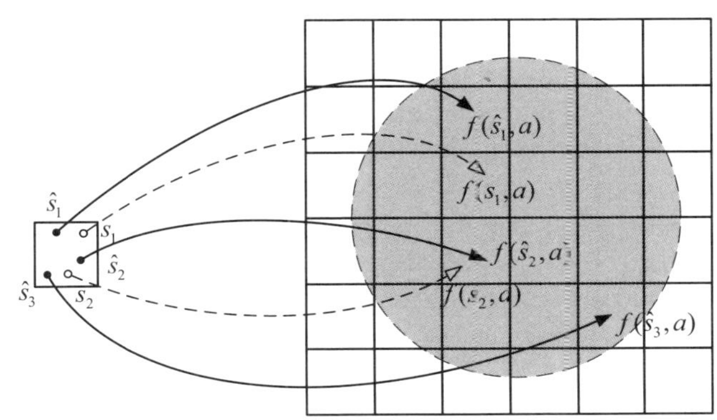

Fig. 4.6 Continue to use Fig. 4.5 and suppose all samples in the cell are presented, denoted by $\hat{s}_1, \hat{s}_2, \hat{s}_3$. By definition, (s_2, a) is known as $s_2, \hat{s}_2$ and $f(s_2, a), f(\hat{s}_2, a)$ are all in the same cells, while (s_1, a) is unknown as no next states of samples locate in the cell of $f(\hat{s}_1, a)$

An outline of the whole MSEC algorithm is presented in Algorithm 4.2. The multi-samples in each cell are formed as different samples are saved in each cell and near-upper Q function utilizes those multi-samples to produce policy. Note that no parameters about the system are involved in the algorithm.

It can be proved that the maximum running time of our algorithm to achieve near-optimal policy is a polynomial in some parameters of the system and grid. So MSEC is in the framework of PAC. And the near optimality of the learned policy is related to the resolution of grid. The finer grid is applied to state space, the more optimal is the final policy. As the limit of space, the theorem analysis is omitted here.

Algorithm 4.2 MSEC algorithm

Input: Value function upper bound $V_{\max}$
A grid over state space $\{C_i\}$

Initialize $D_0 \leftarrow \emptyset$, $\bar{Q}_0 \leftarrow V_{\max}$ and $\pi_0(s) = \arg\max_a \bar{Q}_0(s,a)$
For $t = 0, 1, 2, ...$ **do**
  Observe $(s_t, a_t, r(s_t, a_t), f(s_t, a_t))$
  If (s_t, a_t) is unknown in D_t **then**
    $(s_t, a_t, r(s_t, a_t), f(s_t, a_t))$ is added into D_t
    Update $\bar{Q}_t$ according to (4.14)
    Produce π_t according to (4.15)
  End if
  Execute π_t on agent
End for no escape event happens

4.6 Examples of MSEC

In this section, we apply our proposed MSEC algorithm to the swing-up problem of inverted pendulum. Before the implementation, some modification of MSEC is made to deal with swing-up or other similar problems with better performance. First, we note that the dynamics of inverted pendulum system is symmetric. If an observation $(s_t, a_t, r(s_t, a_t), f(s_t, a_t))$ is obtained, we know that $(-s_t, -a_t, r(-s_t, -a_t), -f(s_t, a_t))$ is also valid with respect to the system. So both two observations can be utilized by the algorithm. Second, as the target is to find a near-optimal policy that swings the agent from a starting point up to a balancing point, so a trial time T_{trial} is defined in the algorithm. When MSEC starts running, for every T_{trial} time, the system is reset to starting state. And if during a whole process of a trial, no escape event happens, that means MSEC finishes its running and outputs the policy. The selection of T_{trial} should be long enough for swinging up. Last, to speed up our algorithm, the calculation of $\bar{Q}_t$ is executed only at the end of each trial. This helps to avoid repeatedly calculating $\bar{Q}_t$ whenever an unknown observation is added into D_t.

4.6.1 *Inverted pendulum*

Inverted pendulum is a device that rotates a mass in a vertical plan and is driven by a DC motor. A schematic is presented in Fig. 4.7 and its dynamics can be expressed by

$$\ddot{\alpha} = \frac{1}{J}\left(mgl\sin(\alpha) - b\dot{\alpha} - \frac{K^2}{R}\dot{\alpha} + \frac{K}{R}u\right)$$

where α and $\dot{\alpha}$ are the angle and angular velocity of the pendulum, respectively satisfying the bound $[-\pi, \pi)$rad and $[-15\pi, 15\pi]$rad/s. u is the control action applied to DC motor and constrained to $[-3, 3]$V. Note that the control action is not powerful enough to swing the pendulum directly in a single rotation, so pendulum has to swing back and forth to gather energy before being pushed up.

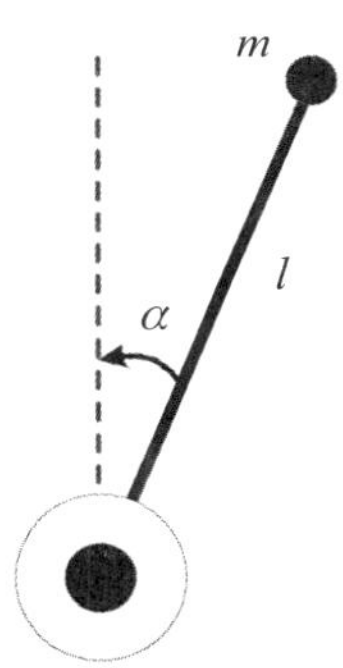

Fig. 4.7 A schematic of inverted pendulum

The goal of our algorithm is swinging up the pendulum from a starting position (pointing down) and balancing it at the top position (pointing up). So the state input is $s = [\alpha, \dot{\alpha}]^\mathsf{T}$ and control action is discretized into 3 discrete values, $A = \{-3, 0, 3\}$. And to consider both swinging up and balancing, we design the reward as $r(s, a) = -s^\mathsf{T}Qs$, where $Q = \text{diag}(5, 0.1)$. The discounted factor is set $\gamma = 0.98$. For simulation, the parameters of dynamics are adopted the same as Busoniu *et al.* (2010a), given in Table 4.1 and the sample time is set to 0.01s. The trial length T_{trial} is 6s, long enough for swinging up and each trial starts from $[\pi, 0]^\mathsf{T}$. The error bound of value iteration is chosen to 0.1. Note that in the implementation, the dynamics is unknown to the algorithm, including any information about the system. So it is totally model-free.

Table 4.1 Parameters of inverted pendulum

Symbol	Value	Units	Meaning
m	0.055	kg	mass
g	9.81	m/s^2	gravitational acceleration
l	0.042	m	distance from center of disk to mass
J	1.91e-4	kg · m^2	moment of inertia
b	3e-6	Nms/rad	viscous damping
K	0.0536	Nm/A	torque constant
R	9.5	Ω	rotor resistance

4.6.2 *Results of MSEC*

First, we apply a 100×100 grid over the state space to run MSEC. The algorithm stops at the 75th trial. And a total of 30420 samples are saved in data set. The learned policy is presented in Fig. 4.8 and Fig. 4.9 reveals the final near-upper Q functions with different actions and corresponding near-upper value function, $\max_a \bar{Q}(\alpha, \dot{\alpha}, a)$. Different colors in Fig. 4.8 indicate different actions, respectively black to -3, gray to 0 and white to 3. It is clear to see that in Fig. 4.9, some regions of state space reach 0, namely the upper bound $V_{\max}$ of value function. That is because in the implementation, no samples in those cells are collected. So they are assigned to $V_{\max}$ by the definition of operator $\overline{\mathcal{T}}$. Executing the learned policy from starting state $[\pi, 0]^{\mathsf{T}}$, we have the swinging-up trajectory of inverted pendulum in Fig. 4.10. The learned policy can swing up the pendulum, but it is revealed in Fig. 4.10 that this process requires two rounds of back and forth to push

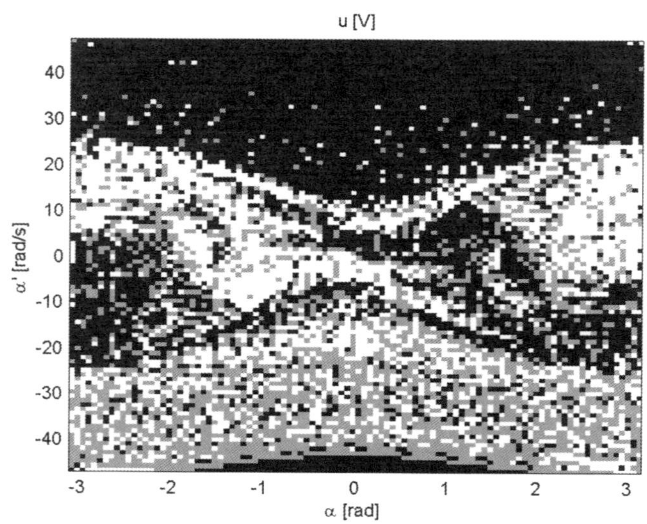

Fig. 4.8 The learned policy of MSEC with 100×100 grid

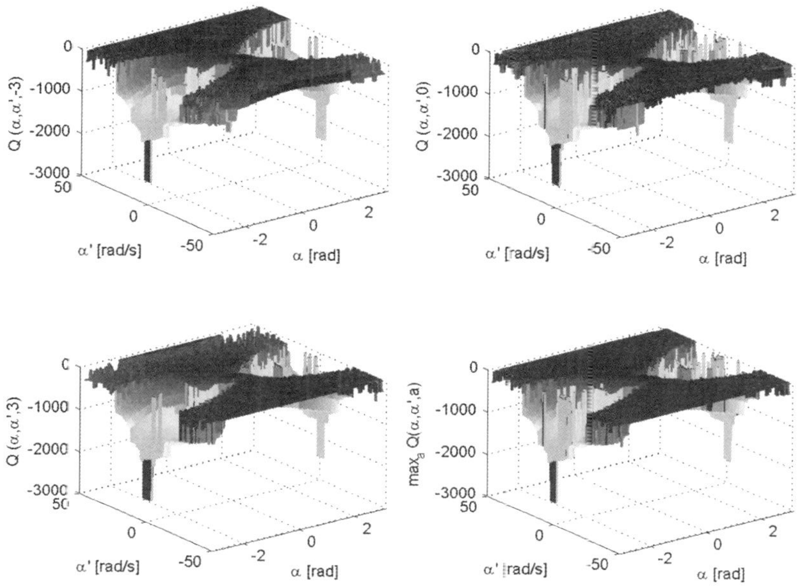

Fig. 4.9 The finial near-upper Q functions of 100 × 100 grid

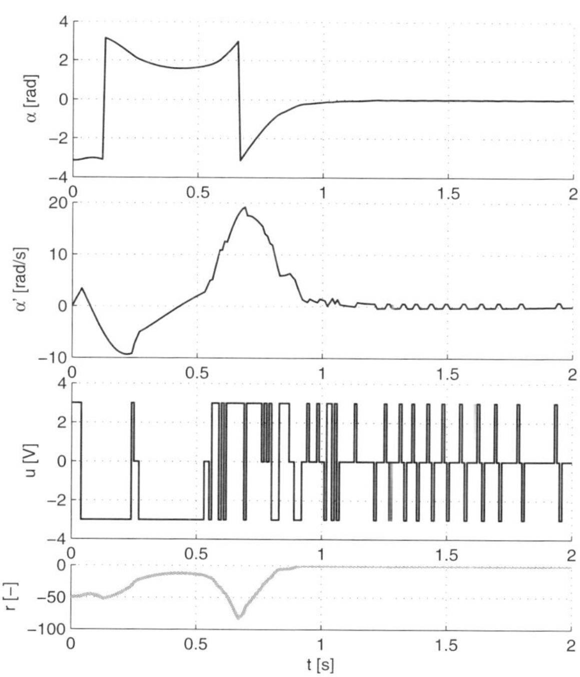

Fig. 4.10 Swinging up trajectory following the learned policy of 100 × 100 grid

pendulum up. The reason is explained that as the resolution of cells in 100×100 grid is not small enough, so the performance of learned policy is far near-optimal, corresponding to the fact that it takes long time to swing up.

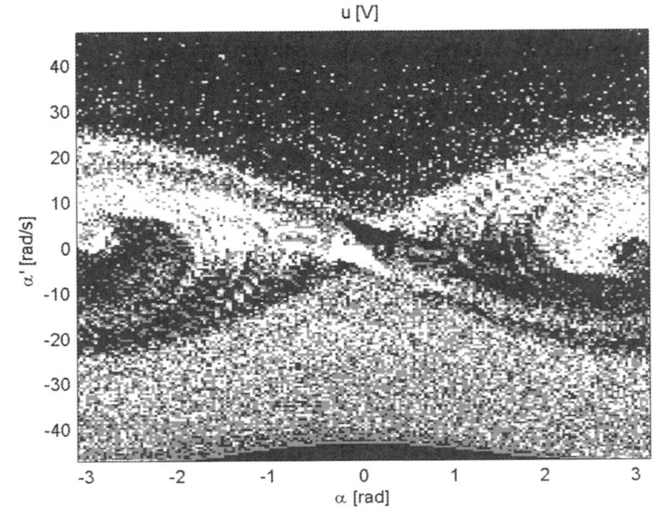

Fig. 4.11 The learned policy of MSEC with 200×200 grid

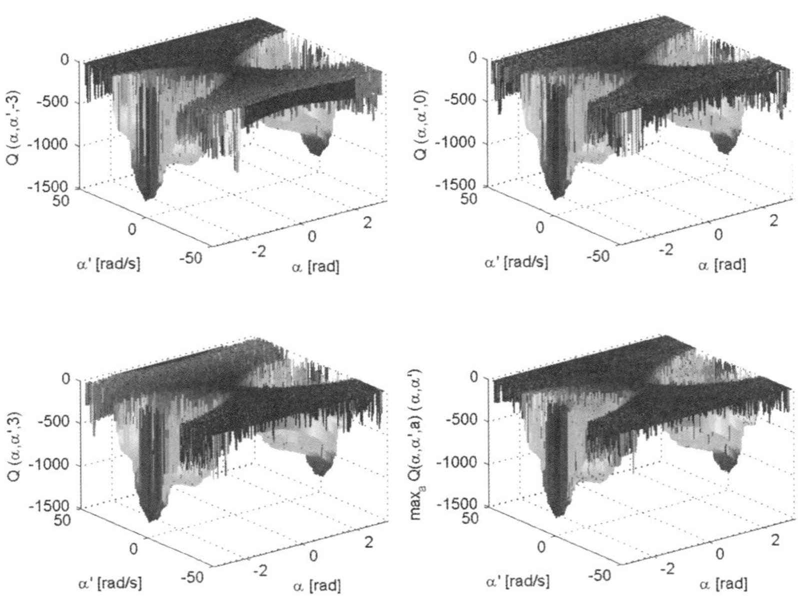

Fig. 4.12 The finial near-upper Q functions of 200×200 grid

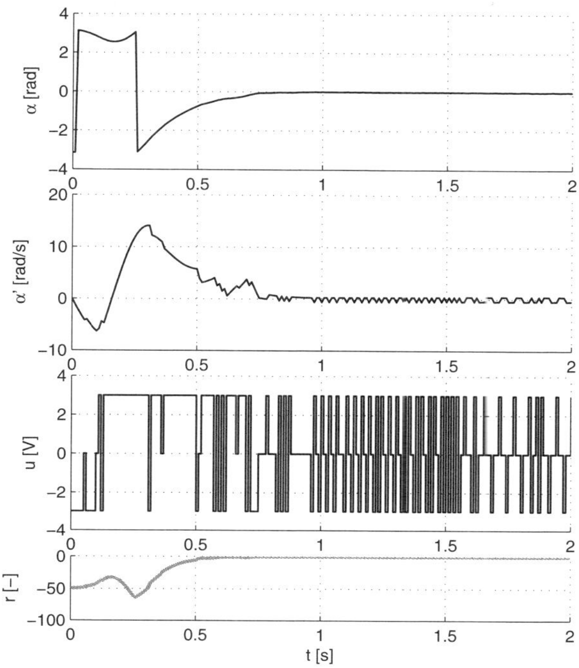

Fig. 4.13 Swinging up trajectory following the learned policy of 200 × 200 grid

To improve the performance, we increase the number of cells and a 200 × 200 grid is used. In this case, the difficulty of learning is increased associated with the fact that algorithm stops at the 150th trial and collects 100997 samples in data set. The results are presented in Fig. 4.11, Fig. 4.12 and Fig. 4.13. Compared to the former simulation with 100 × 100 grid, policy and value functions have much better resolution in Fig. 4.11 and Fig. 4.12. Besides, Fig. 4.13 reveals that the policy is more efficient with only one single round of back and forth to swing up the pendulum. This result is consistent with our theorem that the optimality is related to the resolution of grid.

4.7 Conclusions

In this chapter, we study the application of RL on online learning. Two algorithms are proposed, respectively RLSPI and MSEC. RLSPI adopts LSPI principle and uses linear parametrization to approach value function and continuous policy. It is easy to implement and costs few computation. RLSPI can find a good policy for agents but it has its own drawbacks. Only recent observed samples are utilized, which makes the algorithm lost lots of information about the agents. Besides, the exploration technique used in RLSPI is ε-greedy exploration, so its explorative actions are selected randomly. The algorithm is not guaranteed to achieve optimal policy every time and the efficiency of RLSPI is not high. On the contrary, MSEC adopts PAC principle and saves observed samples selectively to store the information of agents. And it selects explorative actions purposefully, which helps to improve its efficiency. So it is guaranteed that MSEC achieves a near-optimal policy in polynomial time. However, MSEC still faces some challenges. In order to achieve a highly near-optimal policy, the grid must be designed into small resolution. The solving problems difficulty and computational memory space rise exponentially as state dimension increases. Besides, the calculation of near-upper Q function is another main part that occupies a lot computation cost because of the slow convergence speed of value iteration. Those issues still need our further research.

Acknowledgement

This work was supported in part by National Natural Science Foundation of China (61273136 and 61034002), and Beijing Natural Science Foundation (4122083).

References

Abu-Khalaf, M. and Lewis, F. L. (2005). Nearly optimal control laws for nonlinear systems with saturating actuators using a neural network HJB approach, *Automatica* **41**, 5, pp. 779–791.

Al-Tamimi, A., Lewis, F. L. and Abu-Khalaf, M. (2008). Discrete-time nonlinear HJB solution using approximate dynamic programming: Convergence proof, *IEEE Transactions on Systems, Man, and Cybernetics, Part B: Cybernetics* **38**, 4, pp. 943–949.

Bernstein, A. and Shimkin, N. (2010). Adaptive-resolution reinforcement learning with polynomial exploration in deterministic domains, *Machine Learning* **81**, 3, pp. 359–397.

Brafman, R. I. and Tennenholtz, M. (2003). R-max–a general polynomial time algorithm for near-optimal reinforcement learning, *Journal of Machine Learning Research* **3**, pp. 213–231.

Busoniu, L., Babuska, R., De Schutter, B. and Ernst, D. (2010a). *Reinforcement Learning and Dynamic Programming Using Function Approximators* (CRC Press).

Busoniu, L., Ernst, D., De Schutter, B. and Babuska, R. (2010b). Online least-squares policy iteration for reinforcement learning control, in *American Control Conference* (Baltimore, USA), pp. 486–491.

Kearns, M. and Singh, S. (2002). Near-optimal reinforcement learning in polynomial time, *Machine Learning* **49**, 2-3, pp. 209–232.

Lagoudakis, M. G. and Parr, R. (2003). Least-squares policy iteration, *Journal of Machine Learning Research.* **4**, pp. 1107–1149.

Lewis, F. L. (1992). *Applied Optimal Control and Estimation* (Prentice Hall).

Li, H. and Liu, D. (2012). Optimal control for discrete-time affine nonlinear systems using general value iteration, *IET Control Theory and Applications* **6**, 18, pp. 2725–2736

Pazis, J. and Parr, R. (2013). PAC optimal exploration in continuous space markov decision processes, in *Proceedings of 27th AAAI Conference on Artificial Intelligence* (Bellevue, Washington), pp. 774–781.

Rummery, G. A. and Niranjan M. (1994). On-line Q-learning using connectionist systems, Technical Report CUED/F-INFENG/TR 166, Department of Engineering, University of Cambridge, UK.

Si, J. and Wang, Y. (2001). Online learning control by association and reinforcement, *IEEE Transactions on Neural Networks* **12**, 2, pp. 264–276.

Strehl, A. L. and Littman, M. L. (2005). A theoretical analysis of model-based interval estimation, in *Proceedings of the 22nd International Conference on Machine Learning* (New York, USA), pp. 856–863.

Strehl, A. L., Li, L., Wiewiora, E., Langford, J. and Littman, M. L. (2006). PAC model-free reinforcement learning, in *Proceedings of the 23rd International Conference on Machine Learning* (New York, USA), pp. 881–888.

Sutton, R.S. and Barto, A.G. (1998). *Reinforcement Learning–An Introduction* (MIT Press).

Tsitsiklis, J. N. and Van Roy, B. (1996). Feature-based methods for large scale dynamic programming, *Machine Learning* **22**, 1–3, pp. 59–94.

Tsitsiklis, J. N. and Van Roy, B. (1997). An analysis of temporal difference learning with function approximation, *IEEE Transactions on Automatic Control* **42**, 5, pp. 674–690.

Vrabie, D. and Lewis, F. (2009). Neural network approach to continuous-time direct adaptive optimal control for partially unknown nonlinear systems, *Neural Networks* **22**, 3, pp. 237–246.

Vrabie, D., Pastravanu, O., Abu-Khalaf, M. and Lewis, F. L. (2009). Adaptive optimal control for continuous-time linear systems based on policy iteration, *Automatica* **45**, 2, pp. 477–484.

Watkins, C. (1989). *Learning From Delayed Rewards*, Ph.D. Dissertation, University of Cambridge, Cambridge, UK.

Xu, X., Liu, C. and Hu, D. (2011). Continuous-action reinforcement learning with fast policy search and adaptive basis function selection, *Soft Computing* **15**, 6, pp. 1055–1070.

Zhang, H., Luo, Y. and Liu, D. (2009). Neural-network-based near-optimal control for a class of discrete-time affine nonlinear systems with control constraints, *IEEE Transactions on Neural Networks* **20**, 9, pp. 1490–1503.

Chapter 5

Adaptive Iterative Learning Control of Robot Manipulators in the Presence of Environmental Constraints

Xiongxiong He, Zhenhua Qin and Xianqing Wu

Abstract In this chapter, a novel adaptive iterative learning algorithm is proposed for a class of constraint robotic manipulators with uncertainties and external disturbances. The controller design is based on the reduced form of the robot model. The uncertain parameters are estimated in the time domain whereas the repetitive disturbances is compensated in the iteration domain. With the adoption of saturated learning, all the signals in the closed loop are guaranteed to be bounded. By constructing a Lyapunov-Krasovskii-like composite energy function, the proposed controller achieves both control objectives in the presence of dynamic parameter uncertainty. We prove the states of the closed system asymptotically converge to the desired trajectory while ensuring the constrained force remains bounded. Simulation results show the effectiveness of the proposed algorithm.

5.1 Introduction

From the control point of view, the tasks of industrial robots may be divided into two categories. The first is the so-called free motion task, such as pick-and-placing, spray painting, and welding. The second category, however, involves interactions between the robot end-effector and the environment.Typical tasks of this kind include grinding, polishing, inserting, fastening, deburring, scribing, etc. [Lewis *et al.* (1999)]. Many robot applications in manufacturing involve some kind of contact between the end-effector and the environment. Constrained robots have become a useful mathematical method to model the physical and dynamic effects of a robot when it is engaged in one of the contact tasks [McClamroch (1986)]. Unlike free motion control, where the only control objective is trajectory tracking or set-point regulation, the control of a constrained robot has an additional and more difficult objectivethe control of constrained force. If the exact robot dynamic model is available to the controller, both trajectory tracking and force control are manageable with a constrained robot [Chang (2008); Krishnan (1990); Raibert and Craig (1981)].

If the dynamics are exactly known for the control of a robot in constrained manipulation, the methods as mentioned above can be used for designing the effective controllers for the constrained motion of a robot. However, from a practical viewpoint, exact knowledge about the complex robot dynamics is not precisely known in advance. If there exist uncertainties, such as unknown disturbances, the controller so designed may give a degraded performance. To the best of our knowledge, the sliding mode control [Lian and Lin (1998); Su *et al.* (1992)], sliding-adaptive control [Su *et al.* (1995)], and the others [Kwan (1995)]. Asymptotic convergence to zero tracking error is able to be achieved as time approaches infinity, but control chattering occurs due to the use of discontinuous control laws. Iterative learning control (ILC) has found wide application in the robotics community [Hoelzle *et al.* (2011); Barton and Alleyne (2011)] as an intelligent teaching mechanism. The ILC always use the previous control error to improve the present control signal, which requires less a priori knowledge about the system dynamics and also less computational effort than many other kinds of control. The learning control is attributed to a class of self-tuning process with the system performance of a specified task being improved based on the previous performance of identical tasks. In many applications, the repeatability of operation is one of main technical features of present industrial robots. A number of learning control methods for unconstrained robots have been reported in the literatures, such as Cheah and Wang (1994); Nakada and Naniwa (2006); Park *et al.* (1997). To enhance the dynamic performance of a robotic system in the presence of uncertainties, we present a learning control strategy in this chapter. The main objective of the learning control strategy is to enhance the tracking accuracy from operation to operation for systems executing repetitive tasks. The challenge is that how to design the controller to ensure the desired trajectory and constraint force tracking under the parametric uncertainties and external disturbances. First, we use a nonlinear transformation that was introduced in Sun *et al.* (2006) to simplify the dynamic model. Then, based on the reduced dynamic model and the defined references signals, taking advantage of the repetition, the adaptive iterative learning scheme is attributed. At each trial, The uncertain parameters are estimated in the time domain whereas the repetitive disturbances is compensated in the iteration domain, which are used for the main part of the controller, and additional part based sliding mode technique is used to compensate the force tracking error and achieve robustness of the control system.

The chapter is organized as follows. Section 5.2 provides the problem statement and preliminaries. Section 5.3 develops an adaptive iterative learning control approaches. Section 5.4 shows the stability analysis and the performance of the closed-loop systems. Simulation results are presented in Section 5.5 to verify the effectiveness of the established theory. Finally, several conclusions are derived in Section 5.6.

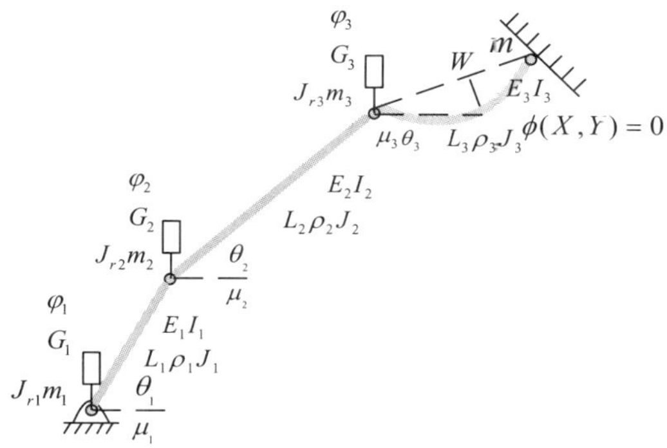

Fig. 5.1 Model of rigid-flexible robot

5.2 Problem Statement and Preliminaries

5.2.1 *Dynamics of constrained robot systems*

In order to study the controller design problems for constrained robot, phenomenon, a physical model of rigid-flexible constrained robot with three links is presented. Fig. 5.1 presents the structure of the flexible robot considered in this study. The flexible joints are dynamically considered as linear damping μ_1, μ_2, μ_3. The inertias of joints are J_{r1}, J_{r2}, J_{r3}. The masses of three joints are m_1, m_2 and m_3, and the payload mass is m. Joints connect with motors whose inertias are J_{m1}, J_{m2}, J_{m3}. The rotation angles of motor are φ_1, φ_2, φ_3. Gear ratios are G_1, G_2 and G_3. Considering θ_1 be the angle of a line pointing from the first joint to second joint, θ_2 is the angle of a line pointing from the second joint to the third joint, θ_3 is the angle of a line pointing from the third joint to the tip mass. The links are assumed as an Euler-Bernoulli beam with the lengths L_1, L_2 and L_3. Mass and elastic properties are assumed to be distributed uniformly along the elastic arm. Their flexural rigidities are E_1I_1, E_2I_2, E_3I_3; mass densities are ρ_1, ρ_2, ρ_3. The inertias of links are J_1, J_2, J_3. x_i $(i = 1, 2, 3)$ denotes the vector point on the i-th link. $W(x_3, t)$ shows the transversal deformation of the flexible rod on the x_i point. According to the vibration theory, $W(x_3, t)$ can be described as follows equation:

$$W(x_3, t) = \sum_{j=1}^{n} \nu_j(t)\gamma(x_3) \tag{5.1}$$

where $\nu_j(t)(j = 1, 2, \ldots, n)$ express time dependent generalized coordinates; $\gamma(x_3)$ denotes the mode shape function.

Let (X, Y) designate an inertial Cartesian coordinate variables and let the constraint surface be described as

$$\Phi(X, Y) = 0. \tag{5.2}$$

The constraint surface will produce a reaction force:

$$F_n = \lambda \frac{\partial \Phi}{\partial R_3} \tag{5.3}$$

where λ denotes the Lagrange operator.

The kinetic energies of the system are expressed as

$$\begin{aligned}
T_1 &= \frac{1}{2} J_{r1}\dot{\theta}_1^2 + \frac{1}{2} J_{r2}\dot{\theta}_2^2 + \frac{1}{2} J_{r3}\dot{\theta}_3^2 \\
T_2 &= \frac{1}{2} m_2 \dot{R}_1^{\mathsf{T}} \dot{R}_1 + \frac{1}{2} m_3 \dot{R}_2^{\mathsf{T}} \dot{R}_2 + \frac{1}{2} m \dot{R}_3^{\mathsf{T}} \dot{R}_3 \\
T_3 &= \frac{1}{2} J_1 \dot{\theta}_1^2 + \frac{1}{2} J_2 \dot{\theta}_2^2 + \frac{1}{2} J_3 \dot{\theta}_3^2 + \frac{1}{2} \int_0^{L_1} \rho_1 \dot{R}_1^{\mathsf{T}} \dot{R}_1 \mathrm{d}x_1 \\
&\quad + \frac{1}{2} \int_0^{L_2} \rho_2 \dot{R}_2^{\mathsf{T}} \dot{R}_2 \mathrm{d}x_2 + \frac{1}{2} \int_0^{L_3} \rho_3 \dot{R}_3^{\mathsf{T}} \dot{R}_3 \mathrm{d}x_3 \\
T_4 &= \frac{1}{2} J_{m1} G_1^2 \dot{\varphi}_1^2 = \frac{1}{2} J_{m2} G_2^2 \dot{\varphi}_2^2 + \frac{1}{2} J_{m3} G_3^2 \dot{\varphi}_3^2.
\end{aligned} \tag{5.4}$$

The total kinetic energy of the system can be written as

$$T = T_1 + T_2 + T_3 + T_4. \tag{5.5}$$

The potential energy of the flexible robot consists of four parts. They are corresponding to elastic deformation of flexible link, deformation of motors, and gravitational energy of links, joints and tip mass.

$$\begin{aligned}
V_1 &= \frac{1}{2} \int_0^{L_3} E_3 I_3 (W'')^2 \mathrm{d}x_3 \\
V_2 &= \int_0^{L_1} \rho_1 g Y_1 \mathrm{d}x_1 + \int_0^{L_2} \rho_2 g Y_2 \mathrm{d}x_2 + \int_0^{L_3} \rho_3 g Y_3 \mathrm{d}x_3 \\
V_3 &= m_2 g L_1 \sin\theta_1 + m_3 g (L_1 \sin\theta_1 + L_2 \sin\theta_2) \\
&\quad + m g (L_1 \sin\theta_1 + L_2 \sin\theta_2 + L_3 \sin\theta_3) \\
V_4 &= \frac{1}{2} K_1 (\varphi_1 - \theta_1)^2 + \frac{1}{2} K_2 (\varphi_2 - \theta_2)^2 + \frac{1}{2} K_3 (\varphi_3 - \theta_3)^2
\end{aligned} \tag{5.6}$$

where K_i denotes the constants of the equivalent springs, $i = 1, 2, 3$.

From the above equations, we can obtain that the potential energy for flexible deformation have no relation with the whole rigid movement, the coupling between the rigid movement and the elastic deformation can not be found. Therefore, the total potential energy is obtained as

$$V = V_1 + V_2 + V_3 + V_4. \tag{5.7}$$

Under the affection of the joint damping, flexible links will produce energy dissipation in the process of rotation; the dissipation function can be expressed as

$$Q = \frac{1}{2}\mu_1\dot{\theta}_1^2 + \frac{1}{2}\mu_2\dot{\theta}_2^2 + \frac{1}{2}\mu_3\dot{\theta}_3^2. \tag{5.8}$$

We assume $\Phi = [\varphi_1 \quad \varphi_2 \quad \varphi_3 \quad \theta_1 \quad \theta_2 \quad \theta_3 \quad \nu_1 \cdots \nu_n]^\mathsf{T}$, which is the generalized coordinate vector, consists of motor angle, link angle and modal variables.

$F = [\tau_1 \quad \tau_2 \quad \tau_3 \quad F_1 \quad F_2 \quad F_3 \quad 0 \cdots 0]^\mathsf{T}$ corresponding to the generalized force vector, F_i are reaction forces. τ_i are torques. According to the Lagrangian method, we can obtain the dynamic equations of flexible robot system as

$$\begin{aligned}
&\begin{bmatrix} J_{m1} & & 0 \\ & J_{m2} & \\ 0 & & J_{m3} \end{bmatrix}\begin{bmatrix} \ddot{\varphi}_1 \\ \ddot{\varphi}_2 \\ \ddot{\varphi}_3 \end{bmatrix} + \begin{bmatrix} K_1 & & 0 \\ & K_2 & \\ 0 & & K_3 \end{bmatrix}\begin{bmatrix} \varphi_1 - \theta_1 \\ \varphi_2 - \theta_2 \\ \varphi_3 - \theta_3 \end{bmatrix} = \begin{bmatrix} \tau_1 \\ \tau_2 \\ \tau_3 \end{bmatrix} \\
&\begin{bmatrix} M_{11} & M_{12} & M_{13} & M_{1k} \\ M_{21} & M_{22} & M_{23} & M_{2k} \\ M_{31} & M_{32} & M_{33} & M_{3k} \\ M_{k1} & M_{k2} & M_{k3} & M_{kk} \end{bmatrix}\begin{bmatrix} \ddot{\theta}_1 \\ \ddot{\theta}_2 \\ \ddot{\theta}_3 \\ \ddot{\nu}_j \end{bmatrix} + \begin{bmatrix} D_{11} & D_{12} & D_{13} & D_{1k} \\ D_{21} & D_{22} & D_{23} & D_{2k} \\ D_{31} & D_{32} & D_{33} & D_{3k} \\ D_{k1} & D_{k2} & D_{k3} & D_{kk} \end{bmatrix}\begin{bmatrix} \dot{\theta}_1 \\ \dot{\theta}_2 \\ \dot{\theta}_3 \\ \dot{\nu}_j \end{bmatrix} + \\
&\begin{bmatrix} G_1 \\ G_2 \\ G_3 \\ G_k \end{bmatrix} + \begin{bmatrix} K_1 & & & 0 \\ & K_2 & & \\ & & K_3 & \\ 0 & & & E_3 I_3 k_j^3(\frac{1}{4}\sin 2k_j L_3 - \frac{1}{2}k_j L_3) \end{bmatrix}\begin{bmatrix} \theta_1 \\ \theta_2 \\ \theta_3 \\ \nu_j \end{bmatrix} + \\
&\begin{bmatrix} \mu_1 & & & 0 \\ & \mu_2 & & \\ & & \mu_3 & \\ 0 & & & 0 \end{bmatrix}\begin{bmatrix} \dot{\theta}_1 \\ \dot{\theta}_2 \\ \dot{\theta}_3 \\ \dot{\nu}_j \end{bmatrix} = \begin{bmatrix} F_1 \\ F_2 \\ F_3 \\ 0 \end{bmatrix}.
\end{aligned} \tag{5.9}$$

$[M]$ represents the inertia matrix; $[D]$ represents Coriolis and centrifugal forces matrix. $[K]$ represents the stiffness matrix, and $[\mu]$ represents the damping matrix.

$$M_{11} = J_{r1} + J_1 + \left(m_2 + m_3 + m + \frac{\rho_1 L_1}{3} + \rho_2 L_2 + \rho_3 L_3\right) L_1^2$$

$$M_{12} = M_{21} = \left(m_3 + m + \frac{\rho_2 L_2}{2} + \rho_3 L_3\right) L_1 L_2 \cos(\theta_1 - \theta_2)$$

$$\begin{aligned}
M_{13} = M_{31} &= m L_1 L_3 \cos(\theta_1 - \theta_3) - m L_1 \sin(\theta_1 - \theta_3) \sum_{j=1}^{n} \nu_j \sin k_j L_3 \\
&+ \rho_3 L_1 \sin(\theta_1 - \theta_3) \sum_{j=1}^{n} \frac{1}{k_j}\nu_j(\cos k_j L_3 - 1) + \frac{1}{2}\rho_3 L_1 L_3^2 \cos(\theta_1 - \theta_3)
\end{aligned}$$

$$M_{1k} = \rho_3 L_1 \cos(\theta_1 - \theta_3)\frac{1}{k_j}(\cos k_j L_3 - 1) - m L_1 \cos(\theta_1 - \theta_3) \sin k_j L_3$$

$$M_{22} = J_{r2} + J_2 + \left(m_3 + m + \frac{\rho_2 L_2}{3} + \rho_3 L_3\right) L_2^2$$

$$M_{23} = M_{32} = mL_2L_3\cos(\theta_2 - \theta_3) - mL_2\sin(\theta_2 - \theta_3)\sum_{j=1}^{n}\nu_j\sin k_jL_3$$

$$+ \rho_3 L_2\sin(\theta_2 - \theta_3)\sum_{j=1}^{n}\frac{1}{k_j}\nu_j(\cos k_jL_3 - 1) + \frac{1}{2}\rho_3L_2L_3^2\cos(\theta_2 - \theta_3)$$

$$M_{2k} = \rho_3L_2\cos(\theta_2 - \theta_3)\frac{1}{k_j}(\cos k_jL_3 - 1) - mL_2\cos(\theta_2 - \theta_3)\sin k_jL_3$$

$$M_{33} = J_{r3} + J_3 + \left(m + \frac{\rho_3L_3}{3}\right)L_3^2 + m\sum_{j=1}^{n}\nu_j^2\sin k_jL_3$$

$$+ \rho_3\sum_{j=1}^{n}\frac{1}{2k_j}\nu_j^2(k_jL_3 - \sin k_jL_3)$$

$$M_{3k} = \rho_3\frac{1}{k_j^2}\left(k_jL_3\cos k_jL_3 - \sin k_jL_3\right) - mL_3\sin k_jL_3$$

$$M_{ki} = M_{ik}$$

$$M_{kk} = m\sin^2 k_jL_3 + \frac{1}{2}\rho_3\left(k_j^2L_3 - \frac{1}{2}\sin 2k_jL_3\right)$$

$$D_{12} = \left(m_3 + m + \frac{\rho_2L_2}{2} + \rho_3L_3\right)L_1L_2\sin(\theta_1 - \theta_2)\dot{\theta}_2$$

$$D_{13} = \rho_3L_1\left[\frac{1}{2}L_3^2\sin(\theta_1 - \theta_3) - \cos(\theta_1 - \theta_3)\sum_{j=1}^{n}\frac{1}{k_j}\nu_j(\cos k_jL_3 - 1)\right]\dot{\theta}_3$$

$$+ mL_1\left[\cos(\theta_1 - \theta_3)\sum_{j=1}^{n}\nu_j\sin k_jL_3 + L_3\sin(\theta_1 - \theta_3)\right]\dot{\theta}_3$$

$$D_{1k} = \left[2\rho_3L_1\sin(\theta_1 - \theta_3)\frac{1}{k_j}(\cos k_jL_3 - 1) - mL_1\sin(\theta_1 - \theta_3)\sin k_jL_3\right]\dot{\theta}_3$$

$$D_{21} = -\left(m_3 + m + \frac{\rho_2L_2}{2} + \rho_3L_3\right)L_1L_2\sin(\theta_1 - \theta_2)\dot{\theta}_1$$

$$D_{23} = \rho_3L_2\left[\frac{1}{2}L_3^2\sin(\theta_2 - \theta_3) - \cos(\theta_2 - \theta_3)\sum_{j=1}^{n}\frac{1}{k_j}\nu_j(\cos k_jL_3 - 1)\right]\dot{\theta}_3$$

$$+ mL_2\left[\cos(\theta_2 - \theta_3)\sum_{j=1}^{n}\nu_j\sin k_jL_3 + L_3\sin(\theta_2 - \theta_3)\right]\dot{\theta}_3$$

$$D_{2k} = \left[2\rho_3L_2\sin(\theta_2 - \theta_3)\frac{1}{k_j}(\cos k_jL_3 - 1) - 2mL_2\sin(\theta_2 - \theta_3)\sin k_jL_3\right]\dot{\theta}_3$$

$$D_{31} = -\rho_3 L_1 \left[\frac{1}{2}L_3^2 \sin(\theta_1 - \theta_3) - \cos(\theta_1 - \theta_3)\sum_{j=1}^{n}\frac{1}{k_j}\nu_j(\cos k_j L_3 - 1)\right]\dot{\theta}_1$$
$$- mL_1\left[\cos(\theta_1 - \theta_3)\sum_{j=1}^{n}\nu_j \sin k_j L_3 + L_3 \sin(\theta_1 - \theta_3)\right]\dot{\theta}_1$$

$$D_{32} = -\rho_3 L_2 \left[\frac{1}{2}L_3^2 \sin(\theta_2 - \theta_3) - \cos(\theta_2 - \theta_3)\sum_{j=1}^{n}\frac{1}{k_j}\nu_j(\cos k_j L_3 - 1)\right]\dot{\theta}_2$$
$$- mL_1\left[\cos(\theta_1 - \theta_3)\sum_{j=1}^{n}\nu_j \sin k_j L_3 + L_3 \sin(\theta_1 - \theta_3)\right]\dot{\theta}_2$$

$$D_{3k} = \left[\rho_3 \frac{1}{k_j}\nu_j(k_j L_3 - \frac{1}{2}\sin 2k_j L_3) + 2m\nu_j \sin^2 k_j L_3\right]\dot{\theta}_3$$

$$D_{k1} = \left[mL_1 \sin(\theta_1 - \theta_3)\sin k_j L_3 - \rho_3 L_1 \sin(\theta_1 - \theta_3)\frac{1}{k_j}(\cos k_j L_3 - 1)\right]\dot{\theta}_1$$

$$D_{k2} = \left[mL_2 \sin(\theta_2 - \theta_3)\sin k_j L_3 - \rho_3 L_2 \sin(\theta_2 - \theta_3)\frac{1}{k_j}(\cos k_j L_3 - 1)\right]\dot{\theta}_2$$

$$D_{k3} = \left[m\nu_j \sin^2 k_j L_3 - \rho_3 \frac{1}{2k_j}\nu_j(k_j L_3 - \frac{1}{2}\sin 2k_j L_3)\right]\dot{\theta}_3$$

$$G_1 = \left[\frac{1}{2}\rho_1 L_1^2 + \rho_2 L_1 L_2 + \rho_3 L_1 L_3 + (m + m_2 + m_3)L_1\right] g\cos\theta_1$$

$$G_2 = \left[\frac{1}{2}\rho_2 L_2^2 + \rho_3 L_2 L_3 + (m + m_3)L_2\right] g\cos\theta_2$$

$$G_3 = \left[\frac{1}{2}\rho_3 L_3^2 + mL_3 + \rho_3 \sum_{j-1}^{n}\frac{1}{k_j}\nu_j(1 - \cos k_j L_3)\right] g\cos\theta_3$$

$$G_k = \rho_3 g\cos\theta_3 \frac{1}{k_j}(\cos k_j L_3 - 1)$$

$$F_i = f_i + \tau_i - K_i\varphi_i.$$

Others are zero.

From the above expression, it is concluded that this is a high nonlinear coupled mechanical system, including two types of nonlinearity; one is a coupling of rigid movements, for the performance of the equation in the angular velocity of the secondary and the trigonometric function in coefficient factor, which is the main factor to determine the overall movement. Another nonlinearity movement is the coupling between rigidity and the overall elastic deformation, for the performance of rigid movement variables and elastic deformation variables of the cross-term, which will lead to the high-frequency disturbances of the flexible link and affect the precision of the end position. The complexity resulted from the nonlinearity of the system makes it hard to control. Some authors have studied these problems by the fussy control method and neural network methods to predict the links deformation, and then to determine the position of the end.

5.3 Adaptive Iterative Learning Control Design

The purpose of this section is to develop an adaptive iterative learning controller by using Lyapunov methods. Two subsections are included in this section. First, a transformation Model is developed about constrained model. Then, the design of the iterative learning controller is proposed.

5.3.1 *The dynamic model conversion and assumption*

From last section we can obtain the dynamics of a constrained robot about n rigid bodies under external disturbance, which is described as

$$M(\boldsymbol{q})\ddot{\boldsymbol{q}} + C(\boldsymbol{q},\dot{\boldsymbol{q}})\dot{\boldsymbol{q}} + g(\boldsymbol{q}) + \boldsymbol{\tau}_d(t) = \boldsymbol{\tau} + \boldsymbol{f} \tag{5.10}$$

where $\boldsymbol{q} \in \mathbb{R}^n$ is the joint position, $\boldsymbol{\tau} \in \mathbb{R}^n$ is the generalized torque, $M(\boldsymbol{q}) \in \mathbb{R}^{n\times n}$ is the symmetric and positive definite inertia matrix, $C(\boldsymbol{q},\dot{\boldsymbol{q}}) \in \mathbb{R}^{n\times n}$ is the centripetal-Coriolis matrix, $\boldsymbol{q} \in \mathbb{R}^n$ is the gravity vector, and $\boldsymbol{\tau}_d(t) \in \mathbb{R}^n$ is the unknown bounded disturbance vector that is repetitive for each iteration, $\boldsymbol{f} \in \mathbb{R}^n$ is the interaction force due to contact with the environment. Suppose the constraint equations is described by

$$\boldsymbol{\Phi}(\boldsymbol{q}) = 0 \in \mathbb{R}^m, \quad m \le n. \tag{5.11}$$

So we can get

$$\frac{\partial \Phi}{\partial \boldsymbol{q}}\dot{\boldsymbol{q}} = J(\boldsymbol{q})\dot{\boldsymbol{q}} = 0 \tag{5.12}$$

where $J(\boldsymbol{q})$ is an $m \times n$ matrix. The constrained force can be expressed as $f = J^{\mathsf{T}}(q)\boldsymbol{\lambda}$, $\boldsymbol{\lambda} \in \mathbb{R}^m$ is the generalized contact force vector associated with the constraints. When the desired trajectory $\boldsymbol{q}_d$ and a desired constraint force $\boldsymbol{f}_d$, or a desired constraint multiplier $\boldsymbol{\lambda}_d(t)$ is known as a reference input for the system of equation (5.10), the fundamental control problem is to find a control input $\boldsymbol{\tau}(t)$ with which the system output $\boldsymbol{q}(t)$ follows $\boldsymbol{q}_d(t)$ and $\boldsymbol{\lambda}(t)$ follows $\boldsymbol{\lambda}_d(t)$ for all $t \in [0, \mathrm{T}]$ as closely as possible. In the framework of learning control this objective can be stated as follows:

Given the desired trajectory $\boldsymbol{q}_d(t) \in \mathbf{C^2}(t \in [0, \mathrm{T}])$ and a desired constraint multiplier $\boldsymbol{\lambda}_d(t) \in \mathbb{R}^m(t \in [0, \mathrm{T}])$ is known as a reference input for the system of (5.1), we will propose a sequence of piecewise continuous control input $\boldsymbol{\tau}^j(t) \in \mathbb{R}^n(t \in [0, \mathrm{T}])$ which guarantees

1) The perfect motion tracking, i.e., $\lim_{j\to\infty} \boldsymbol{q}^j(t) = \boldsymbol{q}^d(t)$.
2) The force tracking error is bounded and adjustable for $t \in [0, \mathrm{T}]$.

For learning controller design, we assume the system has the following properties.

Assumption 5.1. Each output trajectory can be measured and hence the error signal: $\boldsymbol{e}^j = \boldsymbol{q}^j - \boldsymbol{q}_d$ can be utilized.

Assumption 5.2. The resetting condition is satisfied, i.e., $\boldsymbol{q}^j(0) = \boldsymbol{q}_d(0)$, $\dot{\boldsymbol{q}}^j(0) = \dot{\boldsymbol{q}}_d(0)$, $\forall j \in \mathbf{Z}_+$.

One can partition the joint position vector $\boldsymbol{q}$ as $\boldsymbol{q} = [\boldsymbol{q}_1^\mathsf{T}, \boldsymbol{q}_2^\mathsf{T}]^\mathsf{T}$, for $\boldsymbol{q}_1^\mathsf{T} \in \mathbb{R}^{n-m}, \boldsymbol{q}_2^\mathsf{T} \in \mathbb{R}^m$ and $J(\boldsymbol{q}) = [J_1(\boldsymbol{q}), J_2(\boldsymbol{q})] = [\partial\Phi/\partial\boldsymbol{q}_1, \partial\Phi/\partial\boldsymbol{q}_2]$. Then, it follows $J(\boldsymbol{q})\dot{\boldsymbol{q}} = J(\boldsymbol{q})_1\dot{\boldsymbol{q}}_1 + J(\boldsymbol{q})_2\dot{\boldsymbol{q}}_2 = 0$. There exists a proper partition such that $\det[J(\boldsymbol{q})_2] \neq 0$, for all $\boldsymbol{q}$. Then, we can get $\dot{\boldsymbol{q}}_2 = -J_2^{-1}(\boldsymbol{q})J_1(\boldsymbol{q})\dot{\boldsymbol{q}}_1$. We can conclude

$$\dot{\boldsymbol{q}} = L(\boldsymbol{q}_1)\dot{\boldsymbol{q}}_1 \tag{5.13}$$

$$\ddot{\boldsymbol{q}} = L(\boldsymbol{q}_1)\ddot{\boldsymbol{q}}_1 + \dot{L}(\boldsymbol{q}_1)\dot{\boldsymbol{q}}_1 \tag{5.14}$$

$$L(\boldsymbol{q}_1) = \begin{pmatrix} I_{n-m} \\ \frac{\partial\Omega^\mathsf{T}(\boldsymbol{q}_1)}{\partial\boldsymbol{q}_1} \end{pmatrix}. \tag{5.15}$$

So we can conclude

$$L^\mathsf{T}(\boldsymbol{q})J^\mathsf{T}(\boldsymbol{q}) = J_1^\mathsf{T}(\boldsymbol{q}) - J_1^\mathsf{T}(\boldsymbol{q}) = 0. \tag{5.16}$$

Then, the constrained system given by (5.10) and (5.11) can be transformed by substituting (5.13), (5.14) into (5.10). The result is given by

$$\begin{aligned} M(\boldsymbol{q})L(\boldsymbol{q}_1)\ddot{\boldsymbol{q}}_1 + M(\boldsymbol{q})\dot{L}(\boldsymbol{q}_1)\dot{\boldsymbol{q}}_1 + C(\boldsymbol{q}, \dot{\boldsymbol{q}})L(\boldsymbol{q}_1)\dot{\boldsymbol{q}}_1 \\ + g(\boldsymbol{q}) + \boldsymbol{\tau}_d(t) = \boldsymbol{\tau} + J^\mathsf{T}\boldsymbol{\lambda}. \end{aligned} \tag{5.17}$$

A multiplication of $L^\mathsf{T}(\boldsymbol{q}_1)$ to the above equation leads to

$$\bar{M}(\boldsymbol{q})\ddot{\boldsymbol{q}}_1 + \bar{C}(\boldsymbol{q}, \dot{\boldsymbol{q}})\dot{\boldsymbol{q}}_1 + \bar{g}(\boldsymbol{q}) + \bar{\boldsymbol{\tau}}_d(t) = L^\mathsf{T}\boldsymbol{\tau} \tag{5.18}$$

where $\bar{M} = L^\mathsf{T}ML$, $\bar{C} = L^\mathsf{T}(M\dot{L} + CL)$, $\bar{g} = L^\mathsf{T}g$, $\bar{\tau}_d = L^\mathsf{T}\boldsymbol{\tau}_d$. We now state some fundamental properties of the dynamic motion equation (5.18) as follows:

(P1) Matrix $\bar{M}$ is uniformly bounded and symmetric positive definite.

(P2) Matrix $\dot{\bar{M}} - 2\bar{C}$ is skew-symmetric, i.e., $\boldsymbol{y}^\mathsf{T}(\dot{\bar{M}} - 2\bar{C})\boldsymbol{y} = 0$, $\forall \boldsymbol{y} \in \mathbb{R}^n$. $\bar{M}$ and $\bar{C}$ in (5.18) satisfy $\dot{\bar{M}} - 2\bar{C} = L^\mathsf{T}(\dot{M} - 2C)L$, so $\boldsymbol{\xi}^\mathsf{T}(\dot{\bar{M}} - 2\bar{C})\boldsymbol{\xi} = 0$, $\forall \boldsymbol{\xi} \in \mathbb{R}^n$.

(P3) Linear parameterization with a suitable selected vector of robot and load parameters

$$M(\boldsymbol{q})\dot{\boldsymbol{\xi}} + C(\boldsymbol{q}, \dot{\boldsymbol{q}})\boldsymbol{\xi} + g(\boldsymbol{q}) = Y(\boldsymbol{q}, \dot{\boldsymbol{q}}, \boldsymbol{\xi}, \dot{\boldsymbol{\xi}})\boldsymbol{\theta}$$

where $\boldsymbol{\xi}$ is an intermediate variable, and $Y(\cdot)$ is a known regression matrix, $\boldsymbol{\theta}$ is the unknown parameters of robot.

5.3.2 *Adaptive iterative learning control design*

We assume the desired signals $\boldsymbol{q}_1^d$, $\dot{\boldsymbol{q}}_1^d$, $\ddot{\boldsymbol{q}}_1^d$, λ_d and $\dot{\lambda}_d$ and are bounded for all $t \in [0, \mathrm{T}]$, while $\boldsymbol{q}_1^d$ satisfies the constraint. For controller design, we define the motion error, force error, and auxiliary signal vector: $\boldsymbol{e}_{\mathrm{m}}^j = \boldsymbol{q}_1^d - \boldsymbol{q}_1^j$, $\boldsymbol{e}_{\mathrm{m}}^j \in \mathbb{R}^{n-m}$, $\boldsymbol{e}_\lambda^j = \int(\boldsymbol{\lambda}^j - \boldsymbol{\lambda}_d)\mathrm{d}t$, $\boldsymbol{e}_\lambda^j \in \mathbb{R}^m$, $\dot{\boldsymbol{q}}_{1r}^j = \dot{\boldsymbol{q}}_1^d + \Lambda_{\mathrm{m}}\boldsymbol{e}_{\mathrm{m}}^j$, and the filtered tracking error at the j-th cycle as

$$\boldsymbol{s}^j = \Lambda_{\mathrm{m}}\boldsymbol{e}_{\mathrm{m}}^j + \dot{\boldsymbol{e}}_{\mathrm{m}}^j, \quad \boldsymbol{s}^j \in \mathbb{R}^{n-m} \tag{5.19}$$

where Λ_{m} is the symmetric positive-definite matrix.

Consider the dynamics (5.18) and error dynamics in (5.19), we get

$$\begin{aligned}\bar{M}\dot{\boldsymbol{s}}^j &= L^\top\big(ML\ddot{\boldsymbol{q}}_{1\mathrm{r}}^j + M\dot{L}\dot{\boldsymbol{q}}_{1\mathrm{r}}^j + CL\dot{\boldsymbol{q}}_{1\mathrm{r}}^j + g\big)\\ &\quad - \bar{C}\boldsymbol{s}^j + \bar{\boldsymbol{\tau}}_d(t) - L^\top\boldsymbol{\tau}^j.\end{aligned} \tag{5.20}$$

The definitions of these variables pave the way for the proposed controller, which is chosen in the form:

$$\boldsymbol{\tau}^j = Y\hat{\boldsymbol{\theta}}^j + \hat{\boldsymbol{\tau}}_d^j(t) + KL\boldsymbol{s}^j - J^\top\boldsymbol{u}_{\mathrm{a}}^j \tag{5.21}$$

$$\boldsymbol{u}_{\mathrm{a}}^j = \boldsymbol{\lambda}_d - K_{\mathrm{P}}\dot{\boldsymbol{e}}_\lambda^j - K_{\mathrm{S}}\boldsymbol{e}_\lambda^j \tag{5.22}$$

where K, K_P, K_S are positive definite.

Consider the positive definite function

$$V^j(t) = \frac{1}{2}\big(\boldsymbol{s}^{j\top}\bar{M}\boldsymbol{s}^j + \tilde{\boldsymbol{\theta}}^{j\top}\Gamma_1^{-1}\tilde{\boldsymbol{\theta}}^j\big). \tag{5.23}$$

Taking the time derivative of (5.23) along (5.20), we obtain

$$\begin{aligned}\dot{V}^j(t) &= \boldsymbol{s}^{j\top}\bar{M}\dot{\boldsymbol{s}}^j + \frac{1}{2}\boldsymbol{s}^{j\top}\dot{\bar{M}}\boldsymbol{s}^j + \tilde{\boldsymbol{\theta}}^{j\top}\Gamma_1^{-1}\dot{\tilde{\boldsymbol{\theta}}}^j\\ &= \boldsymbol{s}^{j\top}\bar{M}\dot{\boldsymbol{s}}^j + \boldsymbol{s}^{j\top}\bar{C}\boldsymbol{s}^j + \tilde{\boldsymbol{\theta}}^{j\top}\Gamma_1^{-1}\dot{\tilde{\boldsymbol{\theta}}}^j\\ &= -\boldsymbol{s}^{j\top}L^\top KL\boldsymbol{s}^j + \boldsymbol{s}^{j\top}L^\top Y\tilde{\boldsymbol{\theta}}^j - \boldsymbol{s}^{j\top}L^\top\tilde{\boldsymbol{\tau}}_d^j(t) + \tilde{\boldsymbol{\theta}}^{j\top}\Gamma_1^{-1}\dot{\tilde{\boldsymbol{\theta}}}^j.\end{aligned} \tag{5.24}$$

So in order to offset the item $\boldsymbol{s}^{j\top}L^\top Y\tilde{\boldsymbol{\theta}}^j$, the parameter adaptation is chosen as

$$\dot{\hat{\boldsymbol{\theta}}}^j(t) = \Gamma_1 Y^\top(t)L\boldsymbol{s}^j(t) \tag{5.25}$$

$$\hat{\boldsymbol{\theta}}^j(0) = \hat{\boldsymbol{\theta}}^{j-1}(\mathrm{T}) \tag{5.26}$$

$$\hat{\boldsymbol{\theta}}^{-1}(\mathrm{T}) = 0. \tag{5.27}$$

Substituting (5.25) back into (5.24) yields

$$\dot{V}^j(t) = -\boldsymbol{s}^{j\mathsf{T}} L^\mathsf{T} K \boldsymbol{s}^j L - \boldsymbol{s}^{j\mathsf{T}} L^\mathsf{T} \tilde{\boldsymbol{\tau}}_d^j(t) \tag{5.28}$$

where $\tilde{\boldsymbol{\tau}}_d^j(t) = \boldsymbol{\tau}_d(t) - \hat{\boldsymbol{\tau}}_d^j(t)$, the learning law is designed as

$$\hat{\boldsymbol{\tau}}_d^j(t) = \text{sat}\big(\hat{\boldsymbol{\tau}}_d^{j-1}(t)\big) + \Gamma_2 L \boldsymbol{s}^j(t), \forall t \in [0, \mathrm{T}] \tag{5.29}$$

where $\Gamma_2 > 0$ is learning gain. Initial value $\hat{\boldsymbol{\tau}}_d^{-1}(t) = 0$.

5.4 Stability and Convergence Analysis of the Closed-Loop System

The convergence property of the proposed adaptive iterative learning control law is summarized in the following theorem.

Theorem 5.1. *Consider the constrained robot system* (5.10), (5.11) *with the external disturbances, using the control law* (5.21) *with adaptation law* (5.25)–(5.27) *and learning update laws* (5.29), *then the following holds:*
(a) All the signals $\boldsymbol{s}^j$, $\dot{\boldsymbol{s}}^j$, $\boldsymbol{q}_1^j$, $\dot{\boldsymbol{q}}_1^j$ *and* $\hat{\boldsymbol{\theta}}^j$ *in the closed-loop are bounded.*
(b) $\lim_{j\to\infty} \boldsymbol{q}^j(t) = \boldsymbol{q}^d(t)$, $\lim_{\mathrm{j}\to\infty} \dot{\boldsymbol{q}}^j(t) = \dot{\boldsymbol{q}}^d(t)$, *for all* $t \in [0, \mathrm{T}]$.
(c) $\boldsymbol{e}_\lambda^j$ *is bounded and adjustable for all* $t \in [0, \mathrm{T}]$.

Proof. Define a Lyapunov-Krasovskii-like CEF at the j-th iteration as

$$E^j(t) = V^j(t) + \frac{1}{2}\int_0^t \tilde{\boldsymbol{\tau}}_d^{j\mathsf{T}} \Gamma_2^{-1} \tilde{\boldsymbol{\tau}}_d^j \mathrm{d}s. \tag{5.30}$$

(a) First, considering $t = T$, we compute the difference of $E^j(T)$ at the j-th iteration, which is

$$\begin{aligned}\Delta E^j(T) &= E^j(T) - E^{j-1}(T) \\ &= \frac{1}{2}\int_0^T \left[\tilde{\boldsymbol{\tau}}_d^{j\mathsf{T}} \Gamma_2^{-1} \tilde{\boldsymbol{\tau}}_d^j - (\tilde{\boldsymbol{\tau}}_d^{j-1})^\mathsf{T} \Gamma_2^{-1} \tilde{\boldsymbol{\tau}}_d^{j-1}\right] \mathrm{d}s \\ &\quad + V^j(0) - V^{j-1}(T) + \int_0^T \dot{V}^j(s)\mathrm{d}s. \end{aligned} \tag{5.31}$$

By simple algebraic manipulation, we have the following inequality:

$$\begin{aligned}\frac{1}{2}\int_0^T \left[\tilde{\boldsymbol{\tau}}_d^{j\mathsf{T}} \Gamma_2^{-1} \tilde{\boldsymbol{\tau}}_d^j - (\tilde{\boldsymbol{\tau}}_d^{j-1})^\mathsf{T} \Gamma_2^{-1} \tilde{\boldsymbol{\tau}}_d^{j-1}\right] \mathrm{d}s &\le -\int_0^T \left[\hat{\boldsymbol{\tau}}_d^j(s) - \hat{\boldsymbol{\tau}}_d^{j-1}(s)\right]^\mathsf{T} \Gamma_2^{-1} \tilde{\boldsymbol{\tau}}_d^j \mathrm{d}s \\ &\le -\int_0^T \boldsymbol{s}^{j\mathsf{T}} L^\mathsf{T} \tilde{\boldsymbol{\tau}}_d^j \mathrm{d}s. \end{aligned} \tag{5.32}$$

So we get

$$\Delta E^j(T) \leq V^j(0) - V^{j-1}(T) - \int_0^T \boldsymbol{s}^{j\mathsf{T}} L^\mathsf{T} K L \boldsymbol{s}^j \mathrm{d}s \tag{5.33}$$

which implies

$$E^j(T) - E^{j-1}(T) \leq V^j(0) - V^{j-1}(T) - \int_0^T s^{j\mathsf{T}} L^\mathsf{T} K L \boldsymbol{s}^j \mathrm{d}s. \tag{5.34}$$

Since Assumption 5.2 ensures that $s^j(0) = 0$, and from (5.25)–(5.27) we know $\tilde{\boldsymbol{\theta}}^j(0) = \tilde{\boldsymbol{\theta}}^{j-1}(T)$, which results in

$$V^j(0) - V^{j-1}(T) = -\frac{1}{2}\boldsymbol{s}^{j\mathsf{T}}(T)\bar{M}\boldsymbol{s}^j(T) \leq 0. \tag{5.35}$$

Thus, we obtain

$$E^j(T) - E^{j-1}(T) \leq -\int_0^T \boldsymbol{s}^{j\mathsf{T}} L^\mathsf{T} K L \boldsymbol{s}^j \mathrm{d}s. \tag{5.36}$$

It can be seen that the boundedness of $E^j(T)$ is ensured provided $E^0(T)$ is uniformly finite for $t \in [0, T]$. So next, we prove the uniform finiteness of $E^0(T)$.

$$\begin{aligned}
\dot{E}^0(t) &= \dot{V}^0(t) + \frac{1}{2}\tilde{\boldsymbol{\tau}}_d^{0\mathsf{T}}\Gamma_2^{-1}\tilde{\boldsymbol{\tau}}_d^0 \\
&= -\boldsymbol{s}^{0T} L^\mathsf{T} K L \boldsymbol{s}^0 - \boldsymbol{s}^{0T} L^T \tilde{\boldsymbol{\tau}}_d^0(t) + \frac{1}{2}\tilde{\boldsymbol{\tau}}_d^{0\mathsf{T}}\Gamma_2^{-1}\tilde{\boldsymbol{\tau}}_d^0 \\
&\leq -\boldsymbol{s}^{0\mathsf{T}} L^\mathsf{T} \tilde{\boldsymbol{\tau}}_d^0(t) + \frac{1}{2}\tilde{\boldsymbol{\tau}}_d^{0\mathsf{T}}\Gamma_2^{-1}\tilde{\boldsymbol{\tau}}_d^0 \\
&\leq \hat{\boldsymbol{\tau}}_d^{0\mathsf{T}}(t)\Gamma_2^{-1}\tilde{\boldsymbol{\tau}}_d^0(t) + \frac{1}{2}\tilde{\boldsymbol{\tau}}_d^{0\mathsf{T}}\Gamma_2^{-1}\tilde{\boldsymbol{\tau}}_d^0 \\
&\leq \boldsymbol{\tau}_d^\mathsf{T}(t)\Gamma_2^{-1}\tilde{\boldsymbol{\tau}}_d^0(t) - \frac{1}{2}\tilde{\boldsymbol{\tau}}_d^{0\mathsf{T}}\Gamma_2^{-1}\tilde{\boldsymbol{\tau}}_d^0 \\
&\leq \frac{1}{2}\boldsymbol{\tau}_d^\mathsf{T}\Gamma_2^{-1}\boldsymbol{\tau}_d.
\end{aligned} \tag{5.37}$$

Since $\boldsymbol{\tau}_d$ is bounded, we get the boundedness of $\dot{E}^0(t)$. Noting $E^0(0) = 0$, which implies that $E^0(t)$ is uniformly continuous and bounded over $[0, T]$, thus $E^0(T)$ is bounded, so we can conclude that $E^j(T)$ is bounded. From (5.33), we obtain

$$\begin{aligned}
E^j(t) &\leq E^{j-1}(t) + V^j(0) - V^{j-1}(t) - \int_0^t \boldsymbol{s}^{j\mathsf{T}} L^\mathsf{T} K L \boldsymbol{s}^j \mathrm{d}s \\
&= \frac{1}{2}\int_0^t \tilde{\boldsymbol{\tau}}_d^{j\mathsf{T}}\Gamma_2^{-1}\tilde{\boldsymbol{\tau}}_d^j \mathrm{d}s - \int_0^t \boldsymbol{s}^{j\mathsf{T}} L^\mathsf{T} K L \boldsymbol{s}^j \mathrm{d}s + V^j(0).
\end{aligned} \tag{5.38}$$

(b) Summing $E^j(t) - E^{j-1}(t)$ from $j = 1$ to k gives

$$\begin{aligned} E^k(t) &= E^0(t) + \sum_{j=1}^{k} \Delta E_t^j \\ &\leq E^0(t) - \sum_{j=1}^{k} \int_0^t \boldsymbol{s}^{j\mathsf{T}} L^{\mathsf{T}} K L \boldsymbol{s}^j \mathrm{d}s \end{aligned} \tag{5.39}$$

which implies that $\sum\limits_{j=1}^{k} \int_0^t \boldsymbol{s}^{j\mathsf{T}} L^{\mathsf{T}} K L \boldsymbol{s}^j \mathrm{d}s \leq E^0(t)$.

Since $E^0(t)$ is finite and $E^k(t)$ is positive, $\sum\limits_{j=1}^{\infty} \int_0^t \boldsymbol{s}^{j\mathsf{T}} L^{\mathsf{T}} K L \boldsymbol{s}^j \mathrm{d}s$ converges. According to the convergence theorem of the sum of series, $\lim\limits_{j\to\infty} \int_0^t \boldsymbol{s}^{j\mathsf{T}} \boldsymbol{s}^j \mathrm{d}s = 0$. In addition, from (5.18) and (5.19), $\dot{\boldsymbol{s}}^j$ is bounded on $[0, T]$. Thus, $\boldsymbol{s}^j$ converge to zero on $[0, T]$, as $j \to \infty$. From (5.19), we can conclude that $\lim\limits_{j\to\infty} \boldsymbol{e}_{\mathrm{m}}^j = \lim\limits_{j\to\infty} \dot{\boldsymbol{e}}_{\mathrm{m}}^j = 0$.

(c) Substituting the control law (5.21) into the reduced order dynamic system model (5.17) yields

$$\begin{aligned} J^{\mathsf{T}}\left[(I + K_{\mathrm{P}})\dot{\boldsymbol{e}}_\lambda^j + K_{\mathrm{S}}\boldsymbol{e}_\lambda^j\right] &= M(\boldsymbol{q}^j)L(\boldsymbol{q}_1^j)\ddot{\boldsymbol{q}}_1^j + M(\boldsymbol{q}^j)\dot{L}(\boldsymbol{q}_1^j)\dot{\boldsymbol{q}}_1^j + g(\boldsymbol{q}^j) + \boldsymbol{\tau}_d(t) \\ &\quad + C(\boldsymbol{q}^j, \dot{\boldsymbol{q}}^j)L(\boldsymbol{q}_1^j)\dot{\boldsymbol{q}}_1^j - Y\hat{\boldsymbol{\theta}}^j - \hat{\boldsymbol{\tau}}_d^j(t) - KL\boldsymbol{s}^j \\ &= \boldsymbol{\xi}(\boldsymbol{q}_1, \dot{\boldsymbol{q}}_1, \boldsymbol{q}_{1\mathrm{r}}, \dot{\boldsymbol{q}}_{1\mathrm{r}}). \end{aligned} \tag{5.40}$$

Since $\boldsymbol{s}^j$, $\dot{\boldsymbol{s}}^j$, $\boldsymbol{q}_1^j$, $\dot{\boldsymbol{q}}_1^j$ are all bounded, so $\boldsymbol{\xi}$ is bounded and we get $J^{\mathsf{T}}[(I + K_{\mathrm{P}})\dot{\boldsymbol{e}}_\lambda^j + K_{\mathrm{S}}\boldsymbol{e}_\lambda^j]$ is bounded. If we appropriately choose $K_{\mathrm{P}} = \mathrm{diag}[k_{\mathrm{p},i}]$, $k_{\mathrm{p},i} > -1$ and $K_{\mathrm{S}} = \mathrm{diag}[k_{\mathrm{s},i}]$, $k_{\mathrm{s},i} > 0$ to make $G_i(p) = \frac{1}{(k_{\mathrm{p},i}+1)p + k_{\mathrm{s},i}}$, $p = \mathrm{d}/\mathrm{d}t$, a strictly proper exponential stable transfer function, then it can be concluded that $\boldsymbol{e}_\lambda^j \in L_\infty$, $\dot{\boldsymbol{e}}_\lambda^j \in L_\infty$. The size of $\boldsymbol{e}_\lambda^j \in L_\infty$ can be adjusted by choosing the proper gain matrices K_{P} and K_{S}. □

Since all three terms on the right hand side of (5.36) are bounded, so $E^j(t)$ is bounded on $[0, T]$. From the definition of $E^j(t)$, we know that $\boldsymbol{s}^j$ and $\hat{\boldsymbol{\theta}}^j$ are all bounded on $[0, T]$ for all j. From the fact that $\boldsymbol{q}_1^d$ and $\dot{\boldsymbol{q}}_1^d$ are all bounded, we can conclude that both $\boldsymbol{q}_1^j$ and $\dot{\boldsymbol{q}}_1^j$ are bounded on $[0, T]$ for all j.

5.5 Numerical Examples

A two-link robot with a circular path constraint is used to verify the validity of the learning controller presented in this chapter. The matrices of the original model which is in the form of (5.10) can be written as

$$M(\boldsymbol{q}) = \begin{pmatrix} \theta_1 + \theta_2 + 2\theta_3\cos(q_2) & \theta_2 + \theta_3\cos(q_2) \\ \theta_2 + \theta_3\cos(q_2) & \theta_1 \end{pmatrix}$$

$$C(\boldsymbol{q},\dot{\boldsymbol{q}})=\begin{pmatrix}-\theta_3\dot{q}_2\sin(q_2) & -\theta_3(\dot{q}_1+\dot{q}_2)\sin(q_2)\\ \theta_3\dot{q}_1\sin(q_2) & 0\end{pmatrix}$$

$$g(\boldsymbol{q})=\frac{9.81}{l_1}\begin{pmatrix}\theta_1\cos(q_2)+\theta_3\cos(q_1+q_2)\\ \theta_3\cos(q_1+q_2)\end{pmatrix}.$$

The constraint surface is expressed as

$$\Phi(\boldsymbol{q})=\sqrt{l_1^2+2l_1l_2\cos(q_2)}-a \tag{5.41}$$

where a is a constraint and l_1 and l_2 are the lengths of the two robot links. It then follows that $J_{\mathrm{c}}(\boldsymbol{q})=[0,-2l_1l_2\sin(q_2)]$, $L=[1,0]^{\mathsf{T}}$, i.e., $\dot{q}_2=0$. The desired trajectory and the desired constraint force are respectively assumed to be $q_{1d}=\cos(3t)$, $q_{2d}=\frac{\pi}{4}$ $\lambda_d=20$ for $t\in[0,20]$. The disturbance is chosen as $\tau_d=[0.5+0.5\cos(10t),1+0.5\sin(10t)]^{\mathsf{T}}$. The three unknown parameters are: $\theta_1=(m_1+m_2)l_1^2$, $\theta_2=m_2l_2^2$, $\theta_3=m_2l_1l_2$. The robot parameters are: $T=20\mathrm{s}$, $l_1=l_2=0.5$, $m_1=10$, $m_2=5$, $\Gamma_1=\mathrm{diag}(0.015,0.015,0.015)$, $\Gamma_2=\mathrm{diag}(0.05,0.05)$, $K=\mathrm{diag}(60,60)$, $\boldsymbol{\theta}=[\theta_1,\theta_2,\theta_3]^{\mathsf{T}}$, $K_{\mathrm{P}}=30$, $K_{\mathrm{S}}=50$.

The simulation results are shown in Fig. 5.2–Fig. 5.6. As shown in Fig. 5.2 and Fig. 5.3, the controller we designed for the constrained robot cannot achieve good performance mainly due to the disturbances and parameter uncertainties. However, we increase the number of iterations into 20 times, the system trajectory can achieve perfect tracking and the force error remains bounded as small as possible as the

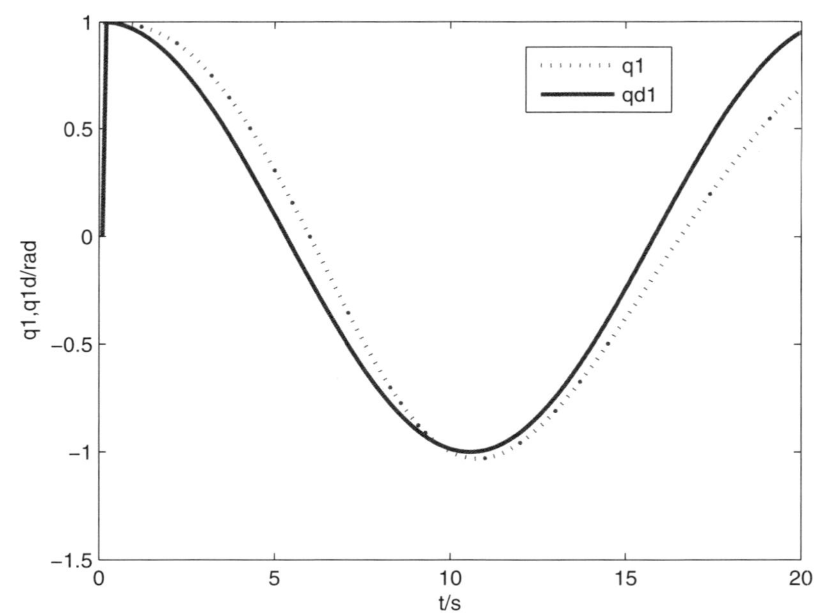

Fig. 5.2 Simulated responses of q_1, q_{1d} at $j=1$

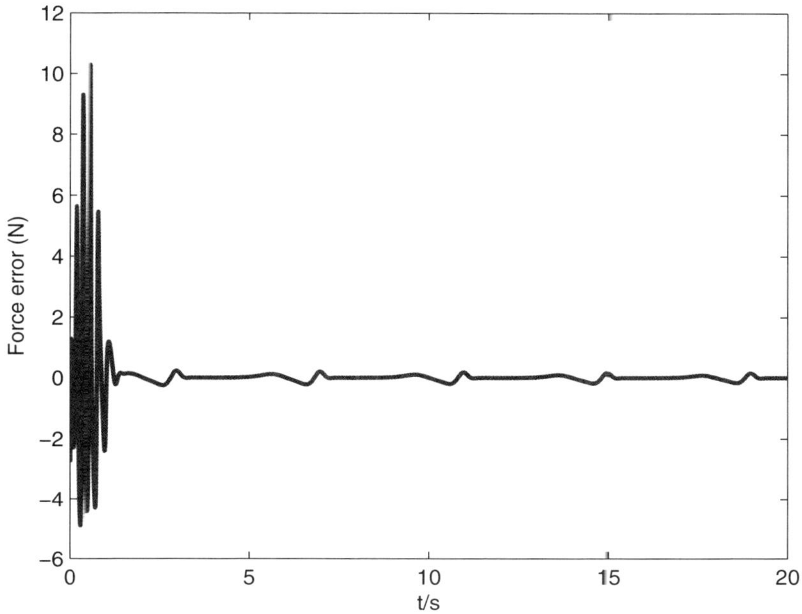

Fig. 5.3 Simulated responses of force tracking error at $j = 1$

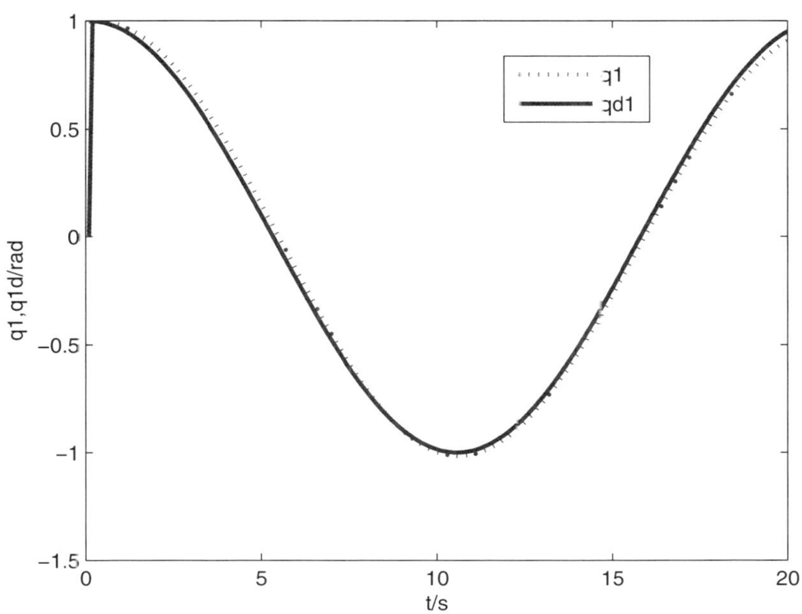

Fig. 5.4 Simulated responses of q_1, q_{1d} at $j = 20$

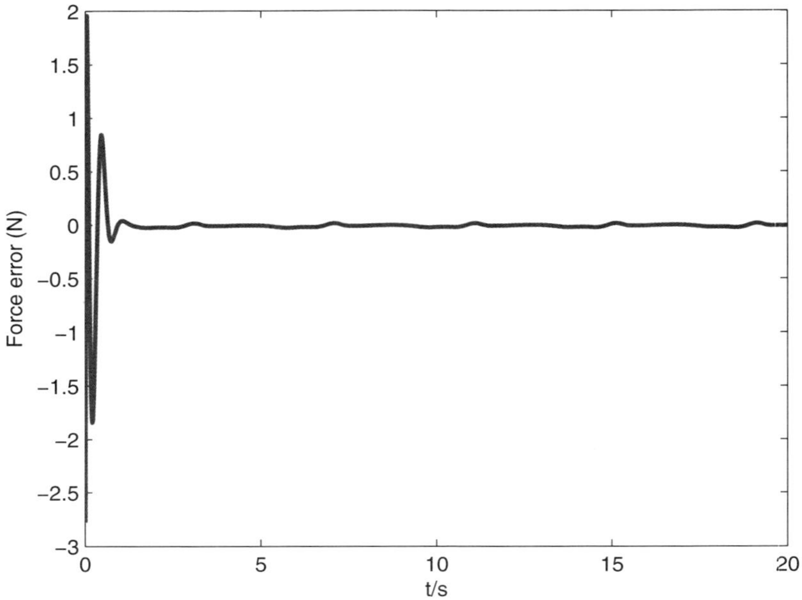

Fig. 5.5 Simulated responses of force tracking error at $j = 20$

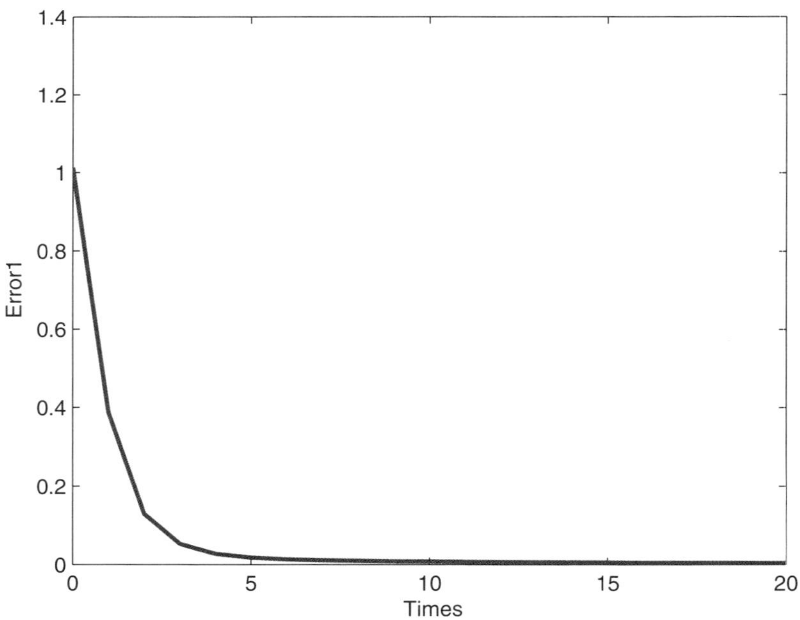

Fig. 5.6 The maximum tracking error1 with iteration numbers

results shown in Fig. 5.4 and Fig. 5.5. From Fig. 5.6 we can easily get that with the iteration number increasing, the maximum tracking error of link 1 gradually tends to zero. The example validates the effectiveness of the control law in Theorem 5.1.

5.6 Conclusions

In this chapter, an adaptive iterative learning control law is established to deliver a desired tracking performance for a class of constrained robot systems with uncertainties and external disturbances. First, a physical model of rigid-flexible constrained robot with three links is presented according to the Lagrangian method. To simplify the controller design, we reduce the dynamic model, based on the reduced dynamic model, the uncertain parameters are estimated in the time domain while repetitive disturbances are compensated in the iteration domain. Based on the Lyapunov method, the convergence and stability analysis of the overall system is also given. Both theoretical analysis and simulation results demonstrate that the proposed approach can achieve the perfect tracking at the same time the force error is bounded and adjustable, and rejects repetitive disturbances. The computer simulation results indicate that the developed adaptive iterative learning controller can perform successful control and attain the desired performance.

References

Barton, K. L. and Alleyne, A. G. (2011). A norm optimal approach to time-varying ILC with application to a multi-axis robotic testbed, *IEEE Transactions on Control Systems Technology* **19**, 1, pp. 166–180.

Chang, C. T. and Huang, H. P. (2008). Robust joint model-based control for constrained robots, in *IEEE Workshop on on Advanced Robotics and Its Social Impacts* (Taipei, Taiwan), pp. 1–6.

Cheah, C. C. and Wang, D. (1994). Learning control for a class of nonlinear differential-algebraic systems with application to constrained robots in *American Contorl Conference* (Baltimore, Maryland), pp. 1734–1741.

Hoelzle, D. J., Alleyne, A. G. and Johnson, A. J. W. (2011). Basis task approach to iterative learning control with applications to micro-robotic deposition, *IEEE Transactions on Control Systems Technology* **19**, 5, pp. 1138–1148.

Krishnan, H. and McClamroch, N. H. (1990). A new approach to position and contact force regulation in constrained robot systems, in *Proceedings of the IEEE International Conference on Robotics and Automation* (Cincinnati, USA), pp. 1344–1349.

Kwan, C. M. (1995). Hybrid force/position control for manipulators with motor dynamics using a sliding-adaptive approach, *IEEE Transactions on Automatic Control* **40**, 5, pp. 963–968.

Lewis, F. L., Jagannathan. S. and Yesildirek, A. (1999). *Neural Network Control of Robot Manipulators and Nonlinear Systems* (Taylor and Francis).

Lian, K. Y. and Lin, C. R. (1998). Sliding-mode motion/force control of constrained robots, *IEEE Transactions on Automatic Control* **43**, 8, pp. 1101–1103.

Luo, J., Li, Z., Ming, A. and Ge, S. S. (2006). Robust motion/force control of holonomic constrainted nonholonomic mobile manipulators using hybrid joints, in *The Sixth World Congress on Intelligent Control and Automation* (Dalian, China), pp. 408–412.

McClamroch, N. H. (1986). Singular systems of differential equations and dynamic models for constrained robot systems, in *Proceedings of the IEEE International Conference on Robotics and Automation* (San Francisco, USA), pp. 21–28.

Mnif, F. A. (2005). A robust optimal control for constrained robot manipulators, *International Journal of Computational Cognition* **3**, 1, pp. 35–43.

Nakada, S. and Naniwa, T. (2006). A hybrid controller of adaptive and learning control for geometrically constrained robot manipulators, in *IEEE/RSJ International Conference on Intelligent Robots and Systems* (Beijing, China), pp. 4515–4520.

Park, B. H., Lee, J. S. and Park, S. Y. (1997). An adaptive learning control method for constrained motion of uncertain robotic systems, in *Proceedings of the IEEE International Conference on Robotics and Automation* (Albuquerque, New Mexico), pp. 531–536.

Raibert, M. H. and Craig, J. J. (1981). Hybrid position/force control of manipulators, *Journal of Dynamic Systems, Measurement, and Control* **103**, 2, pp. 126–133.

Su, C. Y., Lenug, T. P. and Zhou, Q. J. (1992). Force/motion control of constrained robots using sliding mode, *IEEE Transactions on Automatic Control* **37**, 5, pp. 668–672.

Su, C. Y., Stepanenko, Y. and Leung, T. P. (1995). Combined adaptive and variable structure control for constrained robots, *Automatica* **31**, 3, pp. 483–488.

Sun, M., Ge, S. S. and Mareels, I. (2006). Adaptive repetitive learning control of robotic manipulators without the requirement for initial repositioning, *IEEE Transactions on Robotics* **22**, 3, pp. 563–568.

Chapter 6

Neural Network Control of Nonlinear Discrete-Time Systems in Affine Form in the Presence of Communication Network

Hao Xu, Avimanyu Sahoo and Sarangapani Jagannathan

Abstract Stochastic adaptive control of a nonlinear system enclosed by a communication network or referred to as a nonlinear networked control system (NNCS) is a challenging problem due to the presence of unknown network imperfections such as network-induced delays and packet losses. Moreover, the known system dynamics of the original nonlinear system become uncertain and stochastic after the incorporation of the network imperfections due to the communication network within the feedback loop. Therefore, first, a novel NNCS representation incorporating the system uncertainties and network imperfections are derived in this chapter. Subsequently, an online neural network (NN) identifier is developed to identify the control coefficient matrix of the stochastic nonlinear discrete-time system for the purpose of the controller design. Further, critic and action NNs are proposed along with identified system dynamics to determine time-based stochastic optimal adaptive control of NNCS in a forward-in-time manner. Lyapunov stability theory is utilized to demonstrate that all the closed-loop signals are uniformly ultimately bounded (UUB) in the mean. Eventually, to reduce the network traffic between the nonlinear system and the controller, an emerging event-triggered control scheme is developed for the NNCS. The performance of both the controllers is contrasted via simulation.

6.1 Introduction

Nonlinear Networked Control Systems (NNCS) [Tipsuwan and Chow (2003)], which utilizes a communication network to connect the nonlinear plant with a remote controller, has been considered as next-generation control system due to enormous benefits such as high flexibility, efficiency, low installation cost and so on. However, while data is exchanged among devices via the communication network, network imperfections such as random delays and packet losses will result and can cause several challenging issues thus degrading the control system performance even causing instability [Tipsuwan and Chow (2003)].

Therefore, a stability region of linear networked control system (LNCS) has been developed in Zhang *et al.* (2009). Subsequently, authors in Walsh *et al.* (2001) analyzed asymptotic behavior of NNCS in the presence of network-induced delays. In van de Wouw *et al.* (2012), a discrete-time framework is introduced to analyze NNCS stability with both delays and packet losses. The above schemes maintain the stability of NNCS while assuming that the system dynamics and network imperfections are known beforehand. However, due to time-varying network imperfections which are traditionally are uncertain; the NNCS dynamics are not known. In addition, optimality, which includes stability, is generally preferred over stability alone [Hu and Zhu (2003); Tabbara (2008)].

On the other hand, neuro dynamic programming (NDP) technique, proposed by Bertsekas and Tsitsiklis (1996); Werbos (1990), intends to obtain the optimal control of uncertain nonlinear systems in a forward-in-time manner by using value and/or policy iterations in contrast to traditional backward-in-time optimal control techniques which normally requires complete knowledge of system dynamics. Moreover, reinforcement learning can be incorporated with NDP to obtain optimal control [Lewis and Vrabie (2009)]. Traditionally, the cost function is minimized over an infinite horizon in early works [Xu and Jagannathan (2013b)].

In the recent literature, authors in Wang *et al.* (2011); Zhang *et al.* (2009) proposed policy or value iteration-based NDP schemes to attain finite horizon optimal control of nonlinear system with unknown internal system dynamics. However, to compute an optimal solution, these iteration-based NDP methods require significant number of iterations within a fixed sampling interval which is a bottleneck for real-time implementation [Dierks and Jagannathan (2012)].

Therefore, Dierks and Jagannathan (2012) developed a time-based NDP scheme to obtain infinite horizon optimal control of affine nonlinear discrete-time system in the presence of unknown internal dynamics. In Dierks and Jagannathan (2012), historical state information and cost function approximation are utilized instead of iteration-based approach. Nevertheless, these existing NDP schemes are unsuitable for finite horizon optimal control of NNCS since *a*) these schemes are only developed to solve infinite horizon optimal control [Lewis and Syrmos (1995)] partial system dynamics in the form of control coefficient matrix are required, and *b*) network imperfections resulting from the communication network are ignored.

Further, authors in Xu *et al.* (2012) derived a model-free infinite horizon optimal control of LNCS in the presence of uncertain dynamics and network imperfections by using approximate dynamic programming (ADP) based technique. However, finite horizon optimal control is more difficult to solve for NNCS due to the presence of terminal constraints whereas the terminal constraint is ignored in infinite horizon based schemes [Hu and Zhu (2003); Tabbara (2008)]. To the best knowledge of the authors, this chapter for the first time considers the optimal control design over finite horizon for NNCS by incorporating terminal constraints.

In this chapter, an optimal adaptive control scheme using time-based NDP is undertaken to obtain finite horizon stochastic optimal regulations of NNCS in the presence of uncertain system dynamics due to network imperfections. Subsequently, the finite horizon is extended to the infinite horizon. First, to relax the NNCS dynamics, a novel neural network (NN) identifier is proposed to learn online the control coefficient matrix. Then, by using an initial admissible control, a critic NN [Jagannathan (2006)] is introduced and tuned in a forward-in-time manner to approximate the stochastic value function by using Hamilton-Jacobi-Bellman (HJB) equation [Lewis and Syrmos (1995)], given the terminal constraints. Eventually, an action NN is introduced to generate optimal control input by minimizing the estimated stochastic value function.

Compared with the traditional stochastic optimal controller design [Chen *et al.* (2007)] which requires full knowledge of system dynamics. our proposed stochastic optimal controller design for NNCS can relax the requirement on the system dynamics due to network imperfection as well as value or policy iterations by using novel time-based NDP technique. Moreover, the available control techniques for time-delay systems with known deterministic delays [Luck and Ray (1990); Mahmoud and Ismail (2005)] are unsuitable here since network imperfection from NNCS result in delays and packet losses which cannot be handled by the time-delay control techniques.

Finally, to reduce the network traffic [Anta and Tabuada (2010)] further, a novel NN based event-triggered state feedback linearization controller [Sahoo *et al.* (2013)] has been developed for NNCS with packet losses in this chapter. The uncertain dynamics of the system is relaxed by designing a NN based feedback linearization in the context of aperiodic transmission. The event-based transmission and the packet losses preclude the possibility of utilizing traditional periodic weight update law for the NN. Hence, a novel nonperiodic NN weight tuning law for the controller is derived in order to approximate feedback linearization control input. Eventually, a novel event condition is designed to decide transmission of feedback measurements and maintain system stability.

Though the event triggered feedback linearization controller is not optimal, the future work will involve obtaining an event-triggered optimal adaptive control.

The main contribution of this chapter includes:

1. Development of time-based finite and infinite horizon NDP-based stochastic optimal control schemes for uncertain NNCS.
2. The design of a novel online identifier to obtain the NNCS dynamics.
3. Development of a novel NN based event triggered feedback linearization controller for NNCS.
4. Demonstration of the closed-loop stability via a combination of Lyapunov and geometric sequence analysis.

This chapter is organized as follows. First, Section 6.2 presents the background of NNCS and traditional optimal control. Subsequently, Section 6.3 develops the finite horizon stochastic optimal control scheme for NNCS in the presence of network imperfections. Moreover, Section 6.4 extends results of Section 6.3 to infinite horizon case. Finally, an emerging event-triggered control scheme is developed for the NNCS in Section 6.5.

6.2 NNCS Background and Traditional Optimal Control

6.2.1 *Nonlinear networked control systems representation*

The block diagram representation of general NNCS is shown in Fig. 6.1 where control loop is closed by using a communication network. Fig. 6.1 shows a typical NNCS that is utilized in many real time applications. In addition, since the communication network is shared [Goldsmith (2005); Stallings (2004)], the NNCS in this chapter considers the network imperfections including:

1. $\tau_{ca}(t)$: sensor-to-controller delay.
2. $\tau_{ca}(t)$: controller-to-actuator delay.
3. $\gamma(t)$: indicator of network-induced packet losses.

In the recent NCS and communication network protocol development literature [Goldsmith (2005); Stallings (2004)], the following assumptions [Hu and Zhu (2003)] are asserted for the controller design.

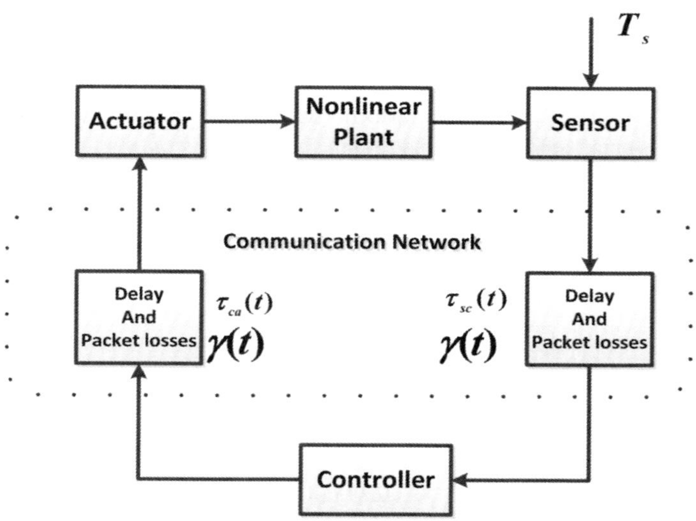

Fig. 6.1 General NNCS block diagram

Assumption 6.1.

(a) For the wide area network considered as the communication network, two types of network-induced delays are considered independent, ergodic and unknown whereas their probability distribution functions are considered known. The sensor-to-controller delay is assumed to be less than a sampling interval.
(b) The sum of two delays is considered to be bounded while initial state of system is assumed to be deterministic [Hu and Zhu (2003)].

By incorporating network-induced delays and packet losses, the nonlinear system in affine form can be represented as

$$\dot{x} = f(x(t)) + \gamma(t)g(x(t))u(t - \tau(t)) \tag{6.1}$$

where

$$\gamma(t) = \begin{cases} \mathbf{I}^{n\times n} & \text{if control input is received by the actuator at time } t, \\ \mathbf{0}^{n\times n} & \text{if control input is lost at time } t, \end{cases}$$

$\mathbf{I}^{n\times n}$ is $n \times n$ identity matrix, $x(t) \in \mathbb{R}^n$, $u(t) \in \mathbb{R}^m$, $f(x) \in \mathbb{R}^n$ and $g(x) \in \mathbb{R}^{n\times m}$ represents system state, control inputs, nonlinear internal dynamics and control coefficient matrix respectively.

By integrating (6.1) over a sampling interval similar to Xu *et al.* (2012), with network-induced delays and packet losses, the NNCS can be represented as

$$x_{k+1} = X_{\tau,\gamma}(x_k, u_{k-1}, \ldots, u_{k-\bar{d}}) + P_{\tau,\gamma}(x_k, u_{k-1}, \ldots, u_{k-\bar{d}})u_k \tag{6.2}$$

where $\bar{d}T_s$ is the upper bound on network-induced delay, T_s is the sampling interval, $x(kT_s) = x_k$, and $u((k-i)T_s) = u_{k-i}$, $\forall i = 0, 1, \ldots, \bar{d}$ and are discretized system state and historical control inputs, and $X_{\tau,\gamma}(\cdot)$, $P_{\tau,\gamma}(\cdot)$ are defined similar to Xu *et al.* (2012).

Next, define a new augment state variable $z_k = \begin{bmatrix} x_k^{\mathsf{T}} & u_{k-1}^{\mathsf{T}} & \cdots & u_{k-\bar{d}}^{\mathsf{T}} \end{bmatrix}^{\mathsf{T}} \in \mathbb{R}^{n+\bar{d}m}$. (6.2) can be expressed equivalently as

$$z_{k+1} = F(z_k) + G(z_k)u_k \tag{6.3}$$

where $F(\cdot)$ denotes NNCS internal dynamics and $G(\cdot)$ is control coefficient matrix which are derived similar to Xu *et al.* (2012) with $\|G(z_k)\| \le G_M$ where $\|\cdot\|_F$ denotes the Frobenius norm [Jagannathan (2006)] and G_M is a positive constant. Due to the presence of network-induced delays and packet losses, (6.3) becomes uncertain and stochastic thus needing adaptive control methods.

6.2.2 *Traditional stochastic optimal control*

Consider a nonlinear discrete-time system described by

$$x_{d,k+1} = f_d(x_{d,k}) + g_d(x_{d,k})u_{d,k} \tag{6.4}$$

where $x_{d,k}$, $u_{d,k}$ represent the system state and input while $f_d(x_{d,k})$ and $g_d(x_{d,k})$ denote the internal dynamics and control coefficient matrix respectively. According to Lewis and Syrmos (1995), optimal control input is derived by minimizing value function expressed as

$$\begin{cases} V_k(x_{d,k},k) = E\big[\phi_N(x_{d,N}) + \sum_{l=k}^{N-1} r(x_{d,l},u_{d,l},l)\big] \\ \qquad\qquad\quad = E\big[\phi_N(x_{d,N}) + \sum_{l=k}^{N-1} (Q_d(x_{d,l},l) + u_{d,l}^{\mathsf{T}} R_d u_{d,l})\big] \\ V_k(x_{d,N},N) = E\big[\phi_N(x_{d,N})\big] \end{cases} \tag{6.5}$$

where the cost-to-go is denoted as $r(x_{d,k},u_{d,k},k) = Q_d(x_{d,k},k) + u_{d,k}^{\mathsf{T}} R_d u_{d,k}$, $\forall k = 0,1,\ldots,N-1$, NT_s is the final time instant, $Q_d(x,k) \geq 0$, $\phi_N(x) \geq 0$ and R_d being symmetric positive definite matrix, $E(\cdot)$ and is the expectation operator(the mean value). Here $\phi_N(x)$ being the terminal constraint which needs to be satisfied in the finite horizon optimal control design. (6.5) can also be rewritten as

$$\begin{aligned} V_k(x_{d,k},k) &= E\big[r(x_{d,k},u_{d,k},k) + \phi_N(x_{d,N}) + \sum_{l=k}^{N-1} r(x_{d,l},u_{d,l},l)\big] \\ &= E\big[(Q_d(x_{d,k},k) + u_{d,k}^{\mathsf{T}} R_d u_{d,k}) + V_{k+1}(x_{d,k+1})\big], \\ & k = 0,\ldots,N-1. \end{aligned} \tag{6.6}$$

According to the observability condition [Chen and Guo (1991)], when $x = 0$, $V_k(x,k) = 0$, the value function $V_k(x,k)$ serves as a Lyapunov function [Jagannathan (2006)]. Based on Bellman principle of optimality [Lewis and Syrmos (1995)], the optimal value function also satisfies the discrete-time Hamilton-Jacob-Bellman (HJB) equation which is given by

$$\begin{cases} V^*(x_{d,k},k) = \min\limits_{u_{d,k}}(V_k(x_{d,k},k)) \\ \qquad\qquad\quad = \min\limits_{u_{d,k}} \big\{E\big[(Q_d(x_{d,k},k) + u_{d,k}^{\mathsf{T}} R_d u_{d,k}) + V_{k+1}(x_{d,k+1})\big]\big\} \\ V^*(x_{d,N},N) = E\big[\phi_N(x_{d,N})\big]. \end{cases} \tag{6.7}$$

Differentiating (6.7), the optimal control $u_{d,k}^*$ is obtained as

$$E\left[\frac{\partial(Q_d(x_{d,k},k) + u_{d,k}^{\mathsf{T}} R_d u_{d,k})}{\partial u_{d,k}} + \frac{\partial x_{d,k+1}^{\mathsf{T}}}{\partial u_{d,k}} \frac{\partial V^*(x_{d,k+1},k+1)}{\partial x_{d,k+1}}\right] = 0. \tag{6.8}$$

In other words,

$$\begin{cases} u^*(x_{d,k}) = -\dfrac{1}{2} E\left[R_d^{-1} g_d^{\mathsf{T}}(x_{d,k}) \dfrac{\partial V^*(x_{d,k+1},k+1)}{\partial x_{d,k+1}}\right], \ k = 0,\ldots,N-1 \\ u^*(x_{d,N-1}) = -\dfrac{1}{2} E\left[R_d^{-1} g_d^{\mathsf{T}}(x_{d,N}) \dfrac{\partial \phi_N(x_{d,N},N)}{\partial x_{d,N}}\right]. \end{cases} \tag{6.9}$$

Substituting (6.9) into (6.7), the discrete-time HJB equation (6.7) becomes

$$V^*(x_{d,k},k) = E\Big[Q_d(x_{d,k},k) + \frac{1}{4}\frac{\partial V^{*\mathsf{T}}(x_{d,k+1},k+1)}{\partial x_{d,k+1}} g_d(x_{d,k+1}) R_d^{-1} g_d(x_{d,k+1}) \times \frac{\partial V^{*\mathsf{T}}(x_{d,k+1})}{\partial x_{d,k+1}} + V^{*\mathsf{T}}(x_{d,k+1},k+1)\Big],\ \forall k = 0,\ldots,N-1$$

$$V^*(x_{d,N-1}) = E\Big[Q_d(x_{d,N-1},N-1) + \frac{1}{4}\frac{\partial \phi_N^{\mathsf{T}}(x_{d,N})}{\partial x_{d,N}} g_d(x_{d,N-1}) R_d^{-1} g_d(x_{d,N-1}) \times \frac{\partial \phi_N(x_{d,N})}{\partial x_{d,N}} + \phi_N(x_{d,N})\Big]. \tag{6.10}$$

It is worthwhile to observe that obtaining a closed-form solution to the discrete-time HJB is difficult since future system state x_{k+1} and system dynamics are needed at KT_s. To circumvent this issue, normally value and policy iteration-based schemes are utilized [Wang *et al.* (2011); Zhang *et al.* (2009)] to approximate the value function as a solution to the HJB equation. However, iteration-based methods are unsuitable for real-time control due to large number of iterations needed within a sampling interval. Inadequate number of iterations will lead to instability [Dierks and Jagannathan (2012)]. Therefore, time-based NDP finite horizon optimal controller design is presented next for the NNCS.

6.3 Stochastic Optimal Controller Design for NNCS

In this section, novel time-based NDP technique is derived to obtain finite horizon stochastic optimal regulation of NNCS with uncertain system dynamics due to network imperfections. First, an online identifier is introduced to obtain the control coefficient matrix. Then, the critic NN is proposed to approximate the stochastic value function within the finite horizon. Eventually, by using action NN, identified NNCS dynamics and estimated stochastic value function, the finite horizon stochastic optimal control of NNCS is derived. The details are given next.

6.3.1 *Online NN-identifier design*

The control coefficient matrix is normally required for the optimal control of affine nonlinear system [Wang *et al.* (2011)]. However, the control coefficient matrix is not normally known in advance. To overcome this deficiency, a novel NN-based identifier is proposed to estimate the control coefficient matrix denoted as $G(z)$.

Based on Dankert *et al.* (2006) and universal function approximation property, the NNCS internal dynamics and control coefficient matrix can be represented on a compact set Ω as

$$F(z_k) = \underset{\tau,\gamma}{E}\left[W_F^{\mathsf{T}} \upsilon_F(z_k) + \epsilon_{F,k}\right], \ \forall k = 0, 1, \ldots, N$$
$$G(z_k) = \underset{\tau,\gamma}{E}\left[W_G^{\mathsf{T}} \upsilon_G(z_k) + \epsilon_{G,k}\right], \ \forall k = 0, 1, \ldots, N \tag{6.11}$$

where W_F, W_G denote the target NN weights, $\upsilon_F(\cdot)$, $\upsilon(\cdot)$ are activation functions and $\epsilon_{F,k}$, $\epsilon_{G,k}$ represent reconstruction errors respectively. Substituting (6.11) into (6.3) to get

$$\begin{aligned} z_k &= F(z_{k-1}) + G(z_{k-1})u_{k-1} \\ &= \underset{\tau,\gamma}{E}\left[\left[W_F^{\mathsf{T}} \ W_G^{\mathsf{T}}\right]\begin{bmatrix}\upsilon_F(z_{k-1}) & 0 \\ 0 & \upsilon_G(z_{k-1})\end{bmatrix}\begin{bmatrix} I_n \\ u_{k-1}\end{bmatrix} + \epsilon_{I,k-1}\right] \\ &= \underset{\tau,\gamma}{E}\left[W_I^{\mathsf{T}} \upsilon_I(z_{k-1})U_{k-1} + \epsilon_{I,k-1}\right], \quad \forall k = 0, 1, \ldots, N \end{aligned} \tag{6.12}$$

where $W_I = \left[W_F^{\mathsf{T}} \ W_G^{\mathsf{T}}\right]^{\mathsf{T}}$, $\upsilon_I(z_{k-1}) = \mathrm{diag}\{\left[\upsilon_F(z_{k-1}) \ \upsilon_G(z_{k-1})\right]\}$ are NN identifier target weight and activation function respectively, augment control input $U_{k-1} \left[I_n^{\mathsf{T}} \ u_{k-1}^{\mathsf{T}}\right]^{\mathsf{T}}$ includes historical input values u_{k-1} and $I_n = \left[1 \ 1 \ \cdots \ 1\right]^{\mathsf{T}} \in \mathbb{R}^n$, $\epsilon_{I,k-1} = \epsilon_{F,k-1} + \epsilon_{G,k-1}$ represent the NN identifier reconstruction error, and $E_{\tau,\gamma}(\cdot)$ is the expectation operator in terms of network-induced delays and packet losses. Since the NN activation function and augmented control input from previous time instants are considered bounded with an initial bounded input, the term $\left\|\underset{\tau,\gamma}{E}\left[\upsilon_I(z_{k-1}U_k - 1)\right]\right\| \leq \zeta_M$, where ζ_M is a positive constant. Moreover, the NN identifier reconstruction error is considered to be bounded, i.e., $\left\|\underset{\tau,\gamma}{E}\left[\epsilon_{I,k-1}\right]\right\| \leq \epsilon_{I,M}$, where denotes a positive constant. Therefore, given the bounded NN activation functions $\upsilon_F(\cdot)$, $\upsilon_G(\cdot)$, $\upsilon_I(\cdot)$ the NNCS control coefficient matrix $G(z)$ can be identified once the NN identifier weight matrix, W_I, is obtained.

Next, the update law for the NN identifier will be introduced.

The NNCS system state z_k can be approximated by using a NN identifier as

$$\hat{z}_k = \underset{\tau,\gamma}{E}\left[\hat{W}_{I,k}^{\mathsf{T}} \upsilon_I(z_{k-1})U_{k-1}\right], \ \forall k = 0, 1, \ldots, N \tag{6.13}$$

where $\hat{W}_{I,k}$ is actual identifier NN weight matrix at time instant kT_s, and $\underset{\tau,\gamma}{E}\left[\overline{\upsilon}_I(z_k)U_k\right]$ is the basis function of NN identifier. Based on (6.12) and (6.13), the identification error can be expressed as

$$\underset{\tau,\gamma}{E}(e_{I,k}) = \underset{\tau,\gamma}{E}(z_k - \hat{z}_k) = \underset{\tau,\gamma}{E}(z_k) - \underset{\tau,\gamma}{E}\left[\hat{W}_{I,k}^{\mathsf{T}} \upsilon_I(z_k - 1)U_{k-1}\right]. \tag{6.14}$$

Moreover, identification error dynamics can be derived as

$$\underset{\tau,\gamma}{E}(e_{I,k+1}) = \underset{\tau,\gamma}{E}(z_{k+1}) - \underset{\tau,\gamma}{E}\left[\hat{W}_{I,k+1}^{\mathsf{T}} \upsilon_I(z_k)U_k\right], \ \forall k = 0, 1, \ldots, N. \tag{6.15}$$

According to Xu and Jagannathan (2013a) and using the history from NNCS, an auxiliary identification error vector can be defined as

$$\underset{\tau,\gamma}{E}(\Xi_{I,k}) = \underset{\tau,\gamma}{E}(Z_k - \hat{Z}_k) = \underset{\tau,\gamma}{E}(Z_k) - \underset{\tau,\gamma}{E}\left[\hat{W}_{I,k}^{\mathsf{T}}\bar{\upsilon}_I(z_{k-1})\overline{U}_{k-1}\right] \tag{6.16}$$

where $Z_k = [z_k \; z_{k-1} \cdots z_{k+1-l}]$, $\bar{\upsilon}_I(z_{k-1}) = [\upsilon_I(z_{k-1}) \; \upsilon_I(z_{k-2}) \; \cdots \; \upsilon_I(z_{k-l})]$, $\bar{U}_{k-1} = \text{diag}\{U_{k-1}^{\mathsf{T}} \; \cdots \; U_{k-l}^{\mathsf{T}}\}^{\mathsf{T}}$ and $\bar{\varepsilon}_{I,k-1} = [\varepsilon_{I,k-1} \; \cdots \; \varepsilon_{I,k-l}]$ with $0 < l < k-1$. The l previous identification errors (6.17) are recalculated by using the most recent estimated NN identifier weight matrix (i.e., $\underset{\tau,\gamma}{E}(\hat{W}_{I,k})$).

Then, dynamics of auxiliary identification error can be represented as

$$\underset{\tau,\gamma}{E}(\Xi_{I,k+1}) = \underset{\tau,\gamma}{E}(Z_{k+1}) - \underset{\tau,\gamma}{E}[\hat{W}_{I,k+1}^{\mathsf{T}}\bar{\upsilon}_I(z_k)\bar{U}_k], \; \forall k = 0, 1, \ldots, N-1. \tag{6.17}$$

To force the estimated NN identifier weight matrix close to its target within fixed for $\underset{\tau,\gamma}{E}(\hat{W}_{I,k})$ can be expressed as

$$\underset{\tau,\gamma}{E}(\hat{W}_{I,k+1}) = \underset{\tau,\gamma}{E}\left[\bar{U}_k\bar{\upsilon}_I(z_k)(\bar{\upsilon}_I^{\mathsf{T}}(z_k)\bar{U}_k^{\mathsf{T}}\bar{U}_k\bar{\upsilon}_I(z_k))^{-1}(Z_k - \alpha_I\Xi_{I,k})^{\mathsf{T}}\right],$$
$$\forall k = 0, \ldots, N-1 \tag{6.18}$$

where α_I is the tuning parameter satisfying $0 < \alpha_I < 1$. Substituting update law (6.18) into auxiliary error dynamics (6.17), the error dynamics $\underset{\tau,\gamma}{E}(\Xi_{I,k+1})$ can be represented as

$$\underset{\tau,\gamma}{E}(\Xi_{I,k+1}) = \alpha_I \underset{\tau,\gamma}{E}(\Xi_{I,k}), \quad \forall k = 0, 1, \ldots, N-1. \tag{6.19}$$

In order to learn the NNCS control coefficient matrix $G(z)$, $\underset{\tau,\gamma}{E}[\bar{\upsilon}_I(z_k)U_k]$ has to be persistently existing (PE) [Jagannathan (2006)] long enough. Namely, there exists a positive constant $\zeta_{\min}$ such that $0 < \zeta_{\min} \leq \left\|\underset{\tau,\gamma}{E}[\bar{\upsilon}_I(z_k)U_k]\right\|$ is satisfied for $k = 0, 1, \ldots, N$.

Recalling (6.12), the identification error dynamics (6.15) can be represented as

$$\underset{\tau,\gamma}{E}(e_{I,k+1}) = \underset{\tau,\gamma}{E}[\tilde{W}_{I,k+1}^{\mathsf{T}}\upsilon_I(z_k)U_k + \varepsilon_{I,k}], \; \forall k = 0, 1, \ldots, N-1 \tag{6.20}$$

where $\tilde{W}_{I,k} = W_I - \hat{W}_{I,k}$ is NN identifier weight estimation error at time kT_s. Using the NN identifier update law and (6.20), the weight estimation error dynamics of the NN identifier can be derived as

$$\underset{\tau,\gamma}{E}\left(\tilde{W}_{I,k+1}^{\mathsf{T}}\upsilon_I(z_k)U_k\right) = \underset{\tau,\gamma}{E}\left(\alpha_I(\tilde{W}_{I,k}^{\mathsf{T}}\upsilon_I(z_k)U_{k-1}) + \alpha_I\varepsilon_{I,k-1} - \varepsilon_{I,k}\right). \tag{6.21}$$

Lemma 6.1. *Given the initial NN identifier weight matrix $W_{I,0}$ which resides in a compact set Ω, let the proposed NN identifier be defined as (6.13), its update law be given by (6.18). Assuming that $\underset{\tau,\gamma}{E}[\bar{\upsilon}_I(z_k)U_k]$ satisfies the PE condition within fixed final time (i.e., $t \in [0, NT_s]$), there exists a positive tuning parameter α_I*

where $0 < \alpha_I < \frac{\zeta_{\min}}{\sqrt{2}\zeta_M}$ *such that identification error (6.14) and NN identifier weight estimation error* $\underset{\tau,\gamma}{E}(\tilde{W}_{I,k})$ *are all uniformly ultimately bounded (UUB) in the mean within fixed final time.*

6.3.2 *Stochastic value function setup and critic NN design*

According to the value function defined in Wang *et al.* (2011); Zhang *et al.* (2009) and given NNCS dynamics (6.3), the stochastic value function can be expressed in terms of augment state z_k as

$$\begin{cases} V(z_k,k) = \underset{\tau,\gamma}{E}\left[\phi_N(z_N) + \sum_{l=k}^{N-1}(Q_z(z_l,l) + u_l^{\mathsf{T}}R_z u_l)\right], \forall k = 0,\ldots,N-1 \\ V(z_N,N) = \underset{\tau,\gamma}{E}\left[\phi_N(z_N)\right] \end{cases} \tag{6.22}$$

where $Q_z(z_k,k) \geq 0$ and $R_z = \frac{1}{d}R_d$. Compared with stochastic value function under infinite horizon case, a terminal constraint (i.e., $V_N(z_N,N) = \underset{\tau,\gamma}{E}[\phi_N(z_N)]$) needs to be considered while developing stochastic optimal controller. Next, according to the universal approximation property of NN [Jagannathan (2006)], stochastic value function (6.22) can be represented by using a critic NN as

$$V(z_k,k) = \underset{\tau,\gamma}{E}\left(W_V^{\mathsf{T}}\psi(z_k,N-k) + \varepsilon_{V,k}\right),\ \forall k = 0,1,\ldots,N \tag{6.23}$$

where W_V, $\varepsilon_{V,k}$ denote critic NN target weight matrix and NN reconstruction error respectively, and $\psi(z_k,N-k)$ represents the time-dependent critic NN activation function. Since the activation function explicitly depends upon time, the finite horizon design is different and difficult when compared to the infinite horizon case [Dierks and Jagannathan (2012)]. Consistent with the NN literature, the target weight matrix of the critic NN is considered bounded in the mean as $\left\|\underset{\tau,\gamma}{E}(W_V)\right\| \leq W_{VM}$ with W_{VM} being a positive constant, and the reconstruction error is also considered bounded in the mean such that $\left\|\underset{\tau,\gamma}{E}(\varepsilon_{V,k})\right\| \leq \varepsilon_{VM}$ with ε_{VM} being a positive constant. In addition, the gradient of the NN reconstruction error is assumed to be bounded in the mean as $\left\|\underset{\tau,\gamma}{E}(\partial\varepsilon_{V,k}/\partial z_k)\right\| \leq \varepsilon'_{VM}$ with ε'_{VM} being a positive constant [Dierks and Jagannathan (2012)].

Next, approximated stochastic value function (6.23) can be represented as

$$\hat{V}(z_k,k) = \underset{\tau,\gamma}{E}\left(\hat{W}_{V,k}^{\mathsf{T}}\psi(z_k,N-k)\right),\ \forall k = 0,1,\ldots,N \tag{6.24}$$

where $\underset{\tau,\gamma}{E}(\hat{W}_{V,k})$ is the estimated critic NN weight matrix and $\psi(z_k,N-k)$ represents the time dependent activation function selected from a basis function set whose elements in the set are linearly independent [Dierks and Jagannathan (2012)]. Also, since the activation function is continuous and smooth, two time independent functions $\psi_{\min}(z_k)$, $\psi_{\max}(z_k)$ can be found such that $\|\psi_{\min}(z_k)\| \leq \|\psi(z_k, N-k)\| \leq \|\psi_{\max}(z_k)\|$, $\forall k = 0, \dots, N$. Then, the target stochastic value function is considered bounded as $\|W_V^{\mathsf{T}}\psi_{\min}(z_k)\| \leq \|V(z_k, k)\| \leq \|W_V^{\mathsf{T}}\psi_{\max}(z_k)\|$ with stochastic value function estimation error satisfying $\|\tilde{W}_{V,k}^{\mathsf{T}}\psi_{\min}(z_k)\| \leq \|\tilde{V}(z_k, k)\| \leq \|\tilde{W}_{V,k}^{\mathsf{T}}\psi_{\max}(z_k)\|$.

Recall HJB equation (6.7) and substitute (6.22) into (6.7) to get

$$\begin{aligned} \underset{\tau,\gamma}{E}(\varepsilon_{V,k} - \varepsilon_{V,k+1}) = & \underset{\tau,\gamma}{E}\left[W_V^{\mathsf{T}}(\psi(z_{k+1}, N-k-1) - \psi(z_k, N-k))\right] \\ & + \underset{\tau,\gamma}{E}\left[Q_z(z_k, k) + u_k^{\mathsf{T}} R_z u_k\right], \ \forall k = 0, 1, \dots, N-1. \end{aligned} \tag{6.25}$$

Namely,

$$\underset{\tau,\gamma}{E}\left(W_V^{\mathsf{T}}\Delta\psi(z_k, N-k)\right) + r(z_k, u_k, k) = \Delta\varepsilon_{V,k}, \ \forall k = 0, 1, \dots, N-1 \tag{6.26}$$

where $\Delta\psi(z_k, N-k) = \psi(z_{k+1}, N-k-1) - \psi(z_k, N-k)$, $r(z_k, u_k, k) = \underset{\tau,\gamma}{E}\left[Q_z(z_k, k) + u_k^{\mathsf{T}} R_z u_k\right]$ and $\Delta\varepsilon_{V,k} = \underset{\tau,\gamma}{E}(\varepsilon_{V,k} - \varepsilon_{V,k+1})$ with $\|\Delta\varepsilon_{V,k}\| = \Delta\varepsilon_{VM}$, $\forall k = 0, \dots, N-1$. However, (6.26) cannot be held while utilizing the approximated critic NN, $\hat{V}(z_k, k)$, instead of $V(z_k, k)$. Similar to Dierks and Jagannathan (2012); Xu *et al.* (2012), using delayed values for convenience, the residual error dynamics associated with (6.26) are derived as

$$\begin{aligned} \underset{\tau,\gamma}{E}(e_{\mathrm{HJB},k}) &= \underset{\tau,\gamma}{E}\left[Q_z(z_k, k) + u_{k-1}^{\mathsf{T}} R_z u_{k-1}\right] + \hat{V}(z_{k+1}, k+1) - \hat{V}(z_k, k) \\ &= \underset{\tau,\gamma}{E}\left(\hat{W}_{V,k}^{\mathsf{T}}\Delta\psi(z_k, N-k)\right) + r(z_k, u_k, k), \ \forall k = 0, \dots, N-1 \end{aligned} \tag{6.27}$$

with $\underset{\tau,\gamma}{E}(e_{\mathrm{HJB},k})$ being the residual error of the HJB equation for the finite horizon scenario. Moreover, since $r(z_k, u_k, k) = \Delta\varepsilon_{V,k} - \underset{\tau,\gamma}{E}\left(W_V^{\mathsf{T}}\Delta\psi(z_k, N-k)\right)$, $\forall k = 0, 1, \dots, N$, the residual error dynamics are represented as

$$\begin{aligned} \underset{\tau,\gamma}{E}(e_{\mathrm{HJB},k}) &= \underset{\tau,\gamma}{E}\left(\hat{W}_{V,k}^{\mathsf{T}}\Delta\psi(z_k, N-k)\right) - \underset{\tau,\gamma}{E}\left(W_V^{\mathsf{T}}\Delta\psi(z_k, N-k)\right) + \Delta\varepsilon_{V,k} \\ &= -\underset{\tau,\gamma}{E}\left(\tilde{W}_{V,k}^{\mathsf{T}}\Delta\psi(z_k, N-k)\right) + \Delta\varepsilon_{V,k}, \ \forall k = 0, \dots, N-1 \end{aligned} \tag{6.28}$$

where $\underset{\tau,\gamma}{E}(\tilde{W}_{V,k}) = \underset{\tau,\gamma}{E}(W_V) - \underset{\tau,\gamma}{E}(\hat{W}_{V,k})$ denotes the critic NN weight estimation error.

Next, in order to take into account the terminal constraint, the estimation error $\underset{\tau,\gamma}{E}(e_{\mathrm{FC},k})$ is defined as

$$\underset{\tau,\gamma}{E}(e_{\mathrm{FC},k}) = \underset{\tau,\gamma}{E}\left[\phi_N(z_N)\right] - \underset{\tau,\gamma}{E}\left(\hat{W}_{V,k}^{\mathsf{T}}\psi(\hat{z}_{N,k}, 0)\right), \ \forall k = 0, 1, \dots, N \tag{6.29}$$

where $\hat{z}_{N,k}$ is the estimated final NNCS system state at time kT_s by using NN identifier (i.e., $\hat{F}(\cdot), \hat{G}(\cdot)$). Recalling (6.23), (6.29) can be represented in terms of critic NN weight estimation error as

$$\begin{aligned}\underset{\tau,\gamma}{E}\left(e_{\mathrm{FC},k}\right) &= \underset{\tau,\gamma}{E}\left(W_V^{\mathsf{T}}\psi(z_N,0)+\varepsilon_{V,0}\right) - \underset{\tau,\gamma}{E}\left(\hat{W}_{V,k}^{\mathsf{T}}\psi(\hat{z}_{N,k},0)\right),\ \forall k=0,1,\ldots,N \\ &= \underset{\tau,\gamma}{E}\left(\tilde{W}_{V,k}^{\mathsf{T}}\psi(\hat{z}_{N,k},0)\right) + \underset{\tau,\gamma}{E}\left(W_V^{\mathsf{T}}\tilde{\psi}(z_N,\hat{z}_{N,k},0)\right) + \underset{\tau,\gamma}{E}\left(\varepsilon_{V,0}\right) \end{aligned} \tag{6.30}$$

with $\tilde{\psi}(z_N,\hat{z}_{N,k},0) = \psi(z_N,0) - \psi(\hat{z}_{N,k},0)$. Since the critic NN activation function $\psi(\cdot)$ is bounded, i.e., $\|\psi(\cdot)\| \leq \psi_M$ [Dierks and Jagannathan (2012); Xu and Jagannathan (2013b)], we have $\left\|\tilde{\psi}(z_N,\hat{z}_{N,k},0)\right\| \leq 2\psi_M$ with ψ_M being a positive constant.

Combining both the HJB residual and terminal constraint estimation errors and using the gradient descent scheme, the update law of critic NN weight can be given by

$$\begin{aligned}\underset{\tau,\gamma}{E}\,(\hat{W}_{V,k+1}) &= \underset{\tau,\gamma}{E}\,(\hat{W}_{V,k}) + \alpha_V \underset{\tau,\gamma}{E}\left(\frac{\psi(\hat{z}_{N,k})e_{\mathrm{FC},k}^{\mathsf{T}}}{\psi^{\mathsf{T}}(\hat{z}_{N,k})\psi(\hat{z}_{N,k})+1}\right) \\ &\quad - \alpha_V \underset{\tau,\gamma}{E}\left(\frac{\Delta\psi(z_k,N-k)e_{\mathrm{HJB},k}^{\mathsf{T}}}{\Delta\psi^{\mathsf{T}}(z_k,N-k)\Delta\psi(z_k,N-k)+1}\right),\ \forall k=0,\ldots,N-1.\end{aligned} \tag{6.31}$$

Remark 6.1. When NNCS system state becomes zero (i.e., $z_k = 0$, $\forall k = 0,1,\ldots,N$), both the stochastic value function (6.23) and critic NN approximation (6.24) become zero. Therefore, the critic NN will stop updating once the system state converges to zero. According to Dierks and Jagannathan (2012); Xu *et al.* (2012), this can be considered as a persistently existing requirement for the input to the critic NN. In other words, the system state has to persistently existing long enough in order for the critic NN to learn the stochastic value function within the finite time (i.e., $t \in [0, NT_s]$. Similar to many recent NN works [Dierks and Jagannathan (2012); Xu *et al.* (2012)], persistently existing requirement can be satisfied by introducing exploration noise here such that $0 < \psi_{\min} < \|\psi_{\min}(z)\| \leq \|\psi(z,k)\|$ and $0 < \Delta\psi_{\min} < \|\Delta\psi_{\min}(z_k)\| \leq \|\Delta\psi(z_k,N-k)\|$ with $\psi_{\min}$ and $\Delta\psi_{\min}$ being positive constants.

Recalling the definition of critic NN weight estimation error $\underset{\tau,\gamma}{E}\,(\tilde{W}_{V,k})$, the dynamics of $\underset{\tau,\gamma}{E}\,(\tilde{W}_{V,k})$ can be expressed as

$$\begin{aligned}\underset{\tau,\gamma}{E}\,(\tilde{W}_{V,k+1}) &= \underset{\tau,\gamma}{E}\,(\tilde{W}_{V,k}) - \alpha_V \underset{\tau,\gamma}{E}\left(\frac{\psi(\hat{z}_{N,k},0)\psi^{\mathsf{T}}(\hat{z}_{N,k},0)\tilde{W}_{V,k}}{\psi^{\mathsf{T}}(\hat{z}_{N,k},0)\psi(\hat{z}_{N,k},0)+1}\right) \\ &\quad -\alpha_V \underset{\tau,\gamma}{E}\left(\frac{\psi(\hat{z}_{N,k},0)\tilde{\psi}^{\mathsf{T}}(z_N,\hat{z}_{N,k},0)W_V}{\psi^{\mathsf{T}}(\hat{z}_{N,k},0)\psi(\hat{z}_{N,k},0)+1}\right) \\ &\quad -\alpha_V \underset{\tau,\gamma}{E}\left(\frac{\psi(\hat{z}_{N,k},0)\varepsilon_{V,0}}{\psi^{\mathsf{T}}(\hat{z}_{N,k},0)\psi(\hat{z}_{N,k},0)+1}\right)\end{aligned}$$

$$-\alpha_V \mathop{E}_{\tau,\gamma}\left(\frac{\Delta\psi(z_k,N-k)\Delta\psi^{\mathsf{T}}(z_k,N-k)\tilde{W}_{V,k}}{\Delta\psi^{\mathsf{T}}(z_k,N-k)\Delta\psi(z_k,N-k)+1}\right)$$
$$-\alpha_V \mathop{E}_{\tau,\gamma}\left(\frac{\Delta\psi(z_k,N-k)\Delta\varepsilon_{V,k}}{\Delta\psi^{\mathsf{T}}(z_k,N-k)\Delta\psi(z_k,N-k)+1}\right),\ \forall k=0,\ldots,N-1. \tag{6.32}$$

Theorem 6.1. *Given an initial NNCS admissible control policy $u_0(z_k)$, and the initial critic NN weight $WV,0$ which resides in a compact set Ω. Let critic NN weight update law be (6.31). Then, there exists a positive constant α_V satisfying $0<\alpha_V<\frac{2-\chi}{\chi+5}$ where $0<\chi=\frac{\psi_{\min}^2+\Delta\psi_{\min}^2+2}{(\psi_{\min}^2+1)(\Delta\psi_{\min}^2+1)}<2$ such that critic NN weight estimation error (6.32) is UUB in the mean within fixed final time. Furthermore, the ultimate bound depends upon final time, NT_s, and initial bounded critic NN weight estimation error $B_{WV,0}$.*

6.3.3 *Action NN estimation of optimal control input*

According to the universal approximation property of NN, the ideal finite horizon stochastic optimal control input can be expressed by using action NN as

$$u^*(z_k)=\mathop{E}_{\tau,\gamma}\left[W_u^{\mathsf{T}}\vartheta(z_k,k)+\varepsilon_{u,k}\right],\ \forall k=0,1,\ldots,N \tag{6.33}$$

where $\mathop{E}_{\tau,\gamma}(W_u)$ and $\mathop{E}_{\tau,\gamma}(\varepsilon_{u,k})$ denote the target weight and reconstruction error of the action NN respectively, and $\vartheta(z_k,k)$ represents the smooth time-varying action NN activation function. Moreover, two time independent function $\vartheta_{\min}(z_k)$, $\vartheta_{\max}(z_k)$ can be found such that $\|\vartheta_{\min}(z_k)\|\le\|\vartheta(z_k,k)\|\le\|\vartheta_{\max}(z_k)\|$, $\forall k=0,\ldots,N$. Also, the ideal action NN weight matrix, activation function and reconstruction are all considered to be bounded such that $\left\|\mathop{E}_{\tau,\gamma}(W_u)\right\|\le W_{uM}$, $\left\|\mathop{E}_{\tau,\gamma}(\vartheta(z_k,k))\right\|\le\vartheta_M$ and $\left\|\mathop{E}_{\tau,\gamma}(\varepsilon_{u,k})\right\|\le\varepsilon_{uM}$ with W_{uM}, ϑ_M and ε_{uM} being positive constants.

Next, similar to Dierks and Jagannathan (2012); Xu *et al.* (2012), the action NN estimation of (6.33) can be represented as

$$\hat{u}(z_k)=\mathop{E}_{\tau,\gamma}\left[\hat{W}_{u,k}^{\mathsf{T}}\vartheta(z_k,k)\right],\ \forall k=0,1,\ldots,N \tag{6.34}$$

where $\mathop{E}_{\tau,\gamma}(\hat{W}_{u,k})$ represents the estimated weights for action NN. Moreover, the estimation error of the action NN can be defined as the difference between the actual control inputs (6.34) applied to the NNCS and control policy which minimizes the estimated stochastic value function (6.24) with identified control coefficient matrix $\hat{G}(z_k)$ during the interval $t\in[0,kT_s]$ as

$$\mathop{E}_{\tau,\gamma}(e_{u,k})=\mathop{E}_{\tau,\gamma}\left[\hat{W}_{u,k}^{\mathsf{T}}\vartheta(z_k,k)+\frac{1}{2}R_z^{-1}\hat{G}^{\mathsf{T}}(z_k)\frac{\partial\psi^{\mathsf{T}}(z_{k+1},N-k-1)}{\partial z_{k+1}}\hat{W}_{V,k}\right]. \tag{6.35}$$

Select the update law for the actual action NN weight matrix as

$$\underset{\tau,\gamma}{E}\left(\hat{W}_{u,k+1}\right) = \underset{\tau,\gamma}{E}\left(\hat{W}_{u,k}\right) - \alpha_u \underset{\tau,\gamma}{E}\left[\frac{\vartheta(z_k,k)}{\vartheta^{\mathsf{T}}(z_k,k)\vartheta(z_k,k)+1}e_{u,k}^{\mathsf{T}}\right], \forall k = 0,\ldots,N-1 \tag{6.36}$$

with the tuning parameter α_u satisfying $0 < \alpha_u < 1$. Further, the ideal action NN output (6.33) should be equal to the control policy that minimizes the ideal stochastic value function (6.23) which is given by

$$\underset{\tau,\gamma}{E}\left[W_u^{\mathsf{T}}\vartheta(z_k,k) + \varepsilon_{u,k}\right] = -\frac{1}{2}\underset{\tau,\gamma}{E}\left[R_z^{-1}G^{\mathsf{T}}(z_k)\frac{\partial(\psi^{\mathsf{T}}(z_{k+1},N-k-1)W_V + \varepsilon_{V,k}^{\mathsf{T}})}{\partial z_{k+1}}\right]. \tag{6.37a}$$

In other words,

$$\underset{\tau,\gamma}{E}\left[W_u^{\mathsf{T}}\vartheta(z_k) + \varepsilon_{u,k} + \frac{1}{2}R_z^{-1}G^{\mathsf{T}}(z_k)\left(\frac{\partial\vartheta^{\mathsf{T}}(z_{k+1})W_V}{\partial z_{k+1}} + \frac{\partial\varepsilon_{V,k}^{\mathsf{T}}}{\partial z_{k+1}}\right)\right] = 0. \tag{6.37b}$$

Substitute (6.37b) into (6.35), the action NN estimation error can be represented equivalently as

$$\begin{aligned}\underset{\tau,\gamma}{E}(e_{u,k}) =& -\underset{\tau,\gamma}{E}\left[\tilde{W}_{u,k}^{\mathsf{T}}\vartheta(z_k,k)\right] - \frac{1}{2}\underset{\tau,\gamma}{E}\left[R_z^{-1}\hat{G}^{\mathsf{T}}(z_k)\frac{\partial\psi^{\mathsf{T}}(z_{k+1},N-k)}{\partial z_{k+1}}\tilde{W}_{V,k}\right]\\ &-\frac{1}{2}\underset{\tau,\gamma}{E}\left[R_z^{-1}\tilde{G}^{\mathsf{T}}(z_k)\frac{\partial\psi^{\mathsf{T}}(z_{k+1},N-k)}{\partial z_{k+1}}W_V\right]\\ &-\frac{1}{2}\underset{\tau,\gamma}{E}\left[R_z^{-1}G^{\mathsf{T}}(z_k)\frac{\partial\varepsilon_{V,k}^{\mathsf{T}}}{\partial z_{k+1}} - \varepsilon_{u,k}\right], \quad \forall k = 0,\ldots,N\end{aligned} \tag{6.38}$$

where $E(\tilde{W}_{u,k}) = E(W_u) - E(\hat{W}_{u,k})$ is the action NN weight estimation error with $\tilde{G}(z_k) = G(z_k) - \hat{G}(z_k)$. Next, substitute (6.38) into (6.36), the action NN weight estimation error dynamics can be expressed as

$$\underset{\tau,\gamma}{E}\left(\tilde{W}_{u,k+1}\right) = \underset{\tau,\gamma}{E}\left(\tilde{W}_{u,k}\right) + \alpha_u \underset{\tau,\gamma}{E}\left[\frac{\vartheta(z_k,k)}{\vartheta^{\mathsf{T}}(z_k,k)\vartheta(z_k,k)+1}e_{u,k}^{\mathsf{T}}\right]. \tag{6.39}$$

In this approach, due to novel NN identifier, need for NNCS control coefficient matrix, $G(z_k)$, is relaxed which itself is a contribution when compared to Dierks and Jagannathan (2012); Wang *et al.* (2011); Zhang *et al.* (2009).

6.3.4 *Closed-loop stability*

In this section, we will prove that closed-loop NNCS system is UUB in the mean within fixed final time with the ultimate bounds are dependent upon initial conditions and final time. Moreover, when the final time instant goes to infinity, $k \to \infty$, the estimated control input approaches the infinite horizon optimal control input. Before demonstrating the main theorem on closed-loop stability, the flowchart of the proposed novel time-based NDP finite horizon optimal control design is presented in Fig. 6.2. In that figure, first the identifier weights are updated, and then the critic NN and action NN are updated. At the same time the value function and control input are obtained based on the identifier and value function estimation. The update continues until the final time.

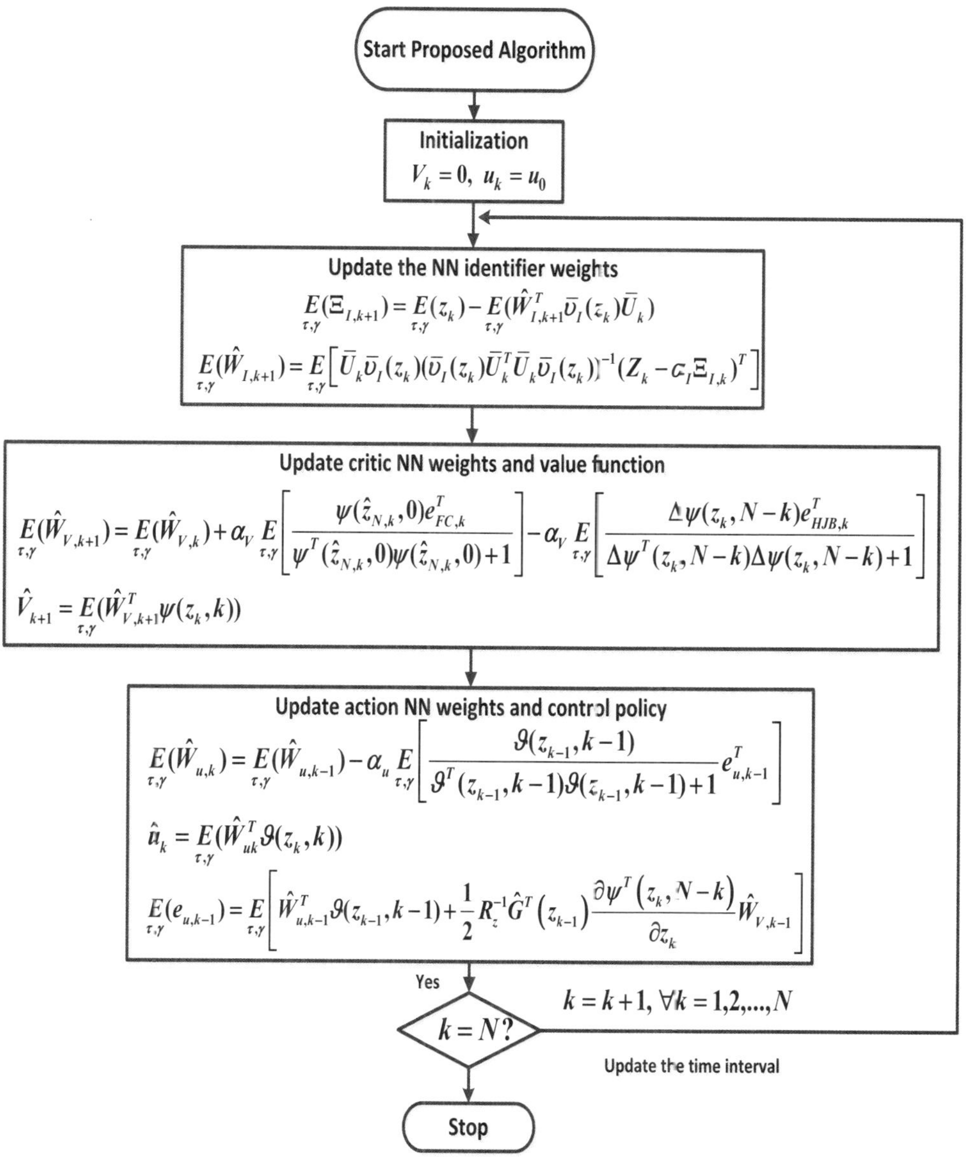

Fig. 6.2 Flowchart of proposed finite horizon stochastic optimal control

Similar to Dierks and Jagannathan (2012), the initial NNCS system state is considered to reside in a compact set Ω due to the initial admissible control input $u_0(z_k)$. Further, the action NN activation function, the critic NN activation function and its gradient are all considered bounded in Ω as $\left\| \underset{\tau,\gamma}{E} (\psi(z_k, k)) \right\| \leq \psi_M$, $\left\| \underset{\tau,\gamma}{E} \left[\frac{\partial \psi(z_k)}{\partial z_k} \right] \right\| \leq \psi_M^{'}$ and $\left\| \underset{\tau,\gamma}{E} (\vartheta(z_k, k)) \right\| \leq \vartheta_M$. In addition, the PE condition will be satisfied by introducing exploration noise [Dierks and Jagannathan (2012)]. The three NN tuning parameters α_I, α_V and α_u will be derived to guarantee that all future system state vector remains in Ω. In order to proceed, the following lemma is needed before introducing the theorem.

Lemma 6.2. *Let an optimal control policy be utilized for the controllable NNCS (6.3) such that (6.3) is asymptotically stable in the mean [Xu and Jagannathan (2013a)]. Then, the closed-loop NNCS dynamics,* $\underset{\tau,\gamma}{E} [F(z_k) + G(z_k)u^*(z_k)]$, *satisfy*

$$\left\| \underset{\tau,\gamma}{E} [F(z_k) + G(z_k)u^*(z_k)] \right\| \leq l_o \left\| \underset{\tau,\gamma}{E} (z_k) \right\|, \; \forall k = 0, 1, \ldots, N \tag{6.40}$$

where $u^*(z_k)$ *is the optimal control input where* $0 < l_o < 1$ *is a positive constant.*

Theorem 6.2. *Let* $u_0(z_k)$ *be any initial admissible control policy for the NNCS (6.3) such that (6.40) holds with* $0 < l_o < \frac{1}{2}$, *the initial NN identifier weight matrix* $WI, 0$, *critic and action NNs weights* $WV, 0$, $Wu, 0$ *all reside in a compact set* Ω. *Given the NN weight update laws for identifier, critic and action NN as (6.18), (6.31) and (6.36) respectively, there exists three positive tuning parameters* α_I, α_V, α_u *satisfying* $0 < \alpha_I < \min\left\{ \frac{1}{2\varsigma_M}, \frac{\varsigma_{\min}}{\sqrt{2}\varsigma_M} \right\}$, $0 < \alpha_V < \frac{2-\chi}{\chi+5}$, $0 < \alpha_u < 1$ *with* $0 < \chi = \frac{\psi_{\min}^2 + \Delta\psi_{\min}^2 + 2}{(\psi_{\min}^2+1)(\Delta\psi_{\min}^2+1)} < 2$ *defined in Theorem 6.1 such that NNCS system state* $\underset{\tau,\gamma}{E} (z_k)$, *identification error* $\underset{\tau,\gamma}{E} (e_{I,k})$, *NN identifier weight estimation error* $\underset{\tau,\gamma}{E} (\tilde{W}_{I,k})$, *and critic and action NN weight estimation errors* $\underset{\tau,\gamma}{E} (\tilde{W}_{V,k})$, $\underset{\tau,\gamma}{E} (\tilde{W}_{u,k})$ *are all UUB in the mean within fixed final time* $t \in [0, NT_s]$. *In addition, the ultimate bounds are dependent upon the final time instant,* NT_s, *bounded initial state* $B_{z,0}$, *identification error* $B_{eI,0}$ *and weight estimation error for NN identifier, critic and action NNs* $B_{WI,0}$, $B_{WV,0}$, $B_{Wu,0}$.

6.3.5 *Simulation results*

The performance of proposed finite horizon stochastic optimal regulation control of NNCS in presence of unknown system dynamics and network imperfections has been evaluated. Continuous-time version of original two-link robot system given in Lewis *et al.* (1999) is expressed as

$$\dot{x} = f(x) + g(x)u \tag{6.41}$$

where the internal dynamics $f(x)$ and control coefficient matrix $g(x)$ are given as

$$f(x) = \begin{bmatrix} x_3 \\ x_4 \\ \dfrac{\begin{pmatrix} -(2x_3x_4 + x_4^2 - x_3^2 - x_3^2\cos x_2)\sin x_2 \\ +20\cos x_1 - 10\cos(x_1 + x_2)\cos x_2 \end{pmatrix}}{\cos^2 x_2 - 2} \\ \dfrac{\begin{pmatrix} (2x_3x_4 + x_4^2 + 2x_3x_4\cos x_2 + x_4^2\cos x_2 + 3x_3^2 \\ +2x_3^2\cos x_2 + 20(\cos(x_1 + x_2) - \cos x_1)\times \\ (1 + \cos x_2) - 10\cos x_2\cos(x_1 + x_2) \end{pmatrix}}{\cos^2 x_2 - 2} \end{bmatrix}, \tag{6.42}$$

$$g(x) = \begin{bmatrix} 0 & 0 \\ 0 & 0 \\ \frac{1}{2-\cos^2 x_2} & \frac{-1-\cos x_2}{2-\cos^2 x_2} \\ \frac{-1-\cos x_2}{2-\cos^2 x_2} & \frac{3+2\cos x_2}{2-\cos^2 x_2} \end{bmatrix}.$$

The network parameters are selected as follows [Xu and Jagannathan (2013b)]: 1) The sampling time: $T_s = 10$ms; 2) The upper bound of network-induced delay is given as two, i.e., $\bar{d} = 2$; 3) The network-induced delays: $E(\tau_{sc}) = 8$ms and $E(\tau) = E(\tau) = 15$ms; 4) Network-induced packet losses follow Bernoulli distribution with $\bar{\gamma} = 0.3$ and 5) The final time is set as $t_f = NT_s = 20$s with simulation time steps $N = 2000$. The distribution of network-induced delays and packet losses is shown in Fig. 6.3 and Fig. 6.4.

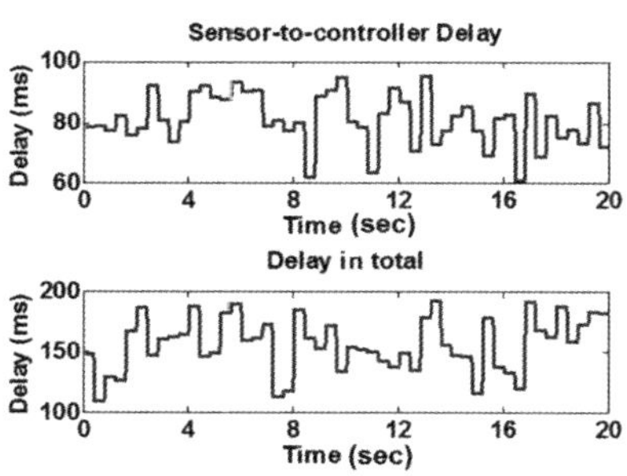

Fig. 6.3 The distribution of network-induced delay in NNCS

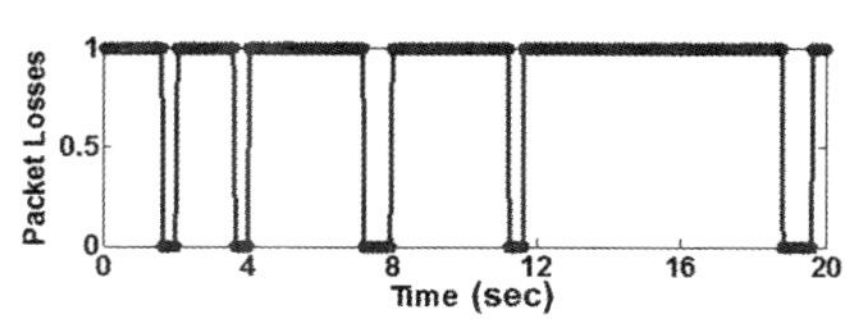

Fig. 6.4 The distribution of network-induced packet losses

(a) State regulation errors and controller performance

First, the state regulation error and stochastic optimal control inputs are studied. Note that the problem attempted in this chapter is optimal regulation which implies that NNCS states should converge to the origin in an optimal manner. After incorporating the network imperfections into NNCS, the augment state is presented as $z_k = [x_k^{\mathsf{T}}\ u_{k-1}^{\mathsf{T}}\ u_{k-2}^{\mathsf{T}}]^{\mathsf{T}} \in \mathbb{R}^{8\times 1}$ and admissible control and the initial state are selected as $u_o(z_k) = \begin{bmatrix} -100 & 0 & -100 & 0 & 0\,0\,0\,0 \\ 0 & -100 & 0 & -100 & 0\,0\,0\,0 \end{bmatrix} z_k$ and

$x_0 = \left[-\frac{\pi}{6}\ \frac{\pi}{6}\ 0\ 0\right]^{\mathrm{T}}$ respectively. Moreover, the NN identifier activation function is given as $\tanh\{(z_{k,1})^2, z_{k,1}z_{k,2}, \ldots, (z_{k,8})^2, \ldots, (z_{k,1})^6, (z_{k,1})^5(z_{k,2}), \ldots, (z_{k,8})^6\}$, state dependent part of activation function for the critic NN is defined as sigmoid of sixth order polynomial (i.e., $\mathrm{sigmoid}\{(z_{k,1})^2, z_{k,1}z_{k,2}, .., (z_{k,8})^2, \ldots, (z_{k,1})^6, (z_{k,1})^5(z_{k,2}), \ldots, (z_{k,8})^6\}$) and time dependent part of critic NN activation function is selected as saturation polynomial time function (i.e., $\mathrm{sat}\{(N-k)^{31}, (N-k)^{30}, \ldots, 1; \ldots; 1, (N-k)^{31}, \ldots, N-k\}$), and activation function of action NN are selected as the gradient of critic NN activation function. The saturation operator for time function is added in order to ensure the magnitude of time function stays within a reasonable range such that the NN weights are computable.

The tuning parameters of NN identifier, critic NN and action NN are defined as $\alpha_I = 0.03$, $\alpha_V = 0.01$ and $\alpha_u = 0.5$ with the initial weights of NN identifier and critic NN are selected as zeros whereas action NN weight matrix is set to reflect the initial admissible control at the beginning of simulation. The results are shown from Fig. 6.5-6.9. As shown in Figs. 6.5 and 6.6, the proposed stochastic optimal controller can force the NNCS state regulation errors converge close to zero within fixed final time even in the presence of uncertain NNCS dynamics and network imperfections. Moreover, the stochastic control signal is also bounded in the mean. The initial admission control selection affects the transient performance similar to the NN tuning parameters and hence the initial admissible controller has to be carefully selected for suitable transient performance.

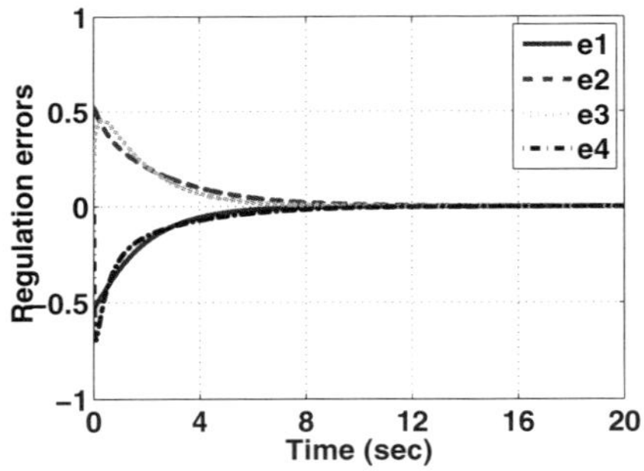

Fig. 6.5 State regulation errors with the proposed controller

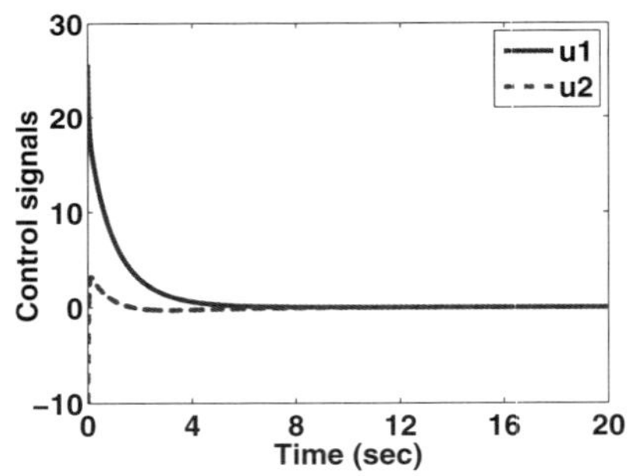

Fig. 6.6 Proposed stochastic optimal control inputs

Next, the effect of network imperfections is evaluated. According to Jagannathan (2006), an NN-based feedback linearization control can maintain the stability of the two-link robot system (6.41). However, after introducing the network imperfections such as network-induced delays and packet losses shown in Fig. 6.3 and 6.4, this feedback linearization controller cannot retain the stability of NNCS as shown in Fig. 6.7. This in turn confirms that a controller should be carefully designed after incorporating the effects of network imperfections for NNCS.

Now, the evolution of the NN weights is studied. In Fig. 6.8, the actual weights of critic and action NN are shown. Within fixed final time (i.e., $t \in [0, 20\text{s}]$), the actual weights of critic and action NN converge and remain UUB in the mean consistent with Theorem 6.2. Further, as shown in Fig. 6.9, the identification error converges close to zero which indicates that the NN identifier learns the system dynamics properly.

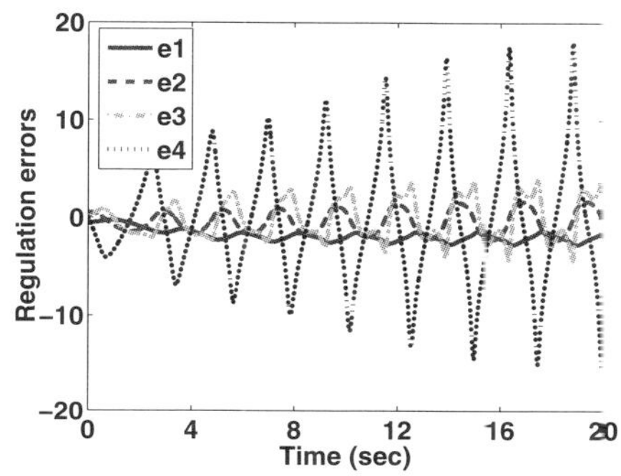

Fig. 6.7 State regulation errors of NN-based feedback linearization controller

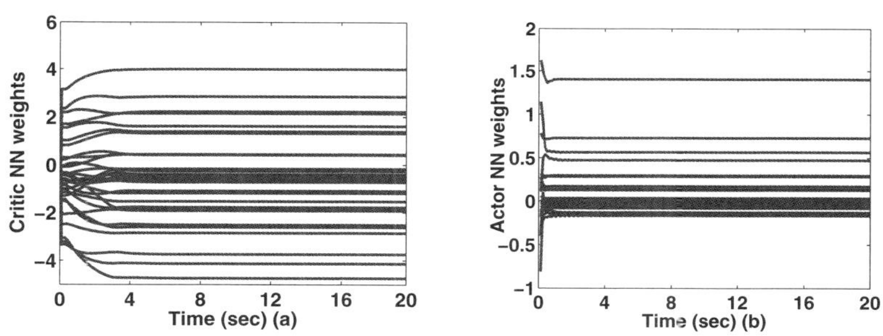

Fig. 6.8 Estimated NN weights of (a) critic NN (b) actor NN

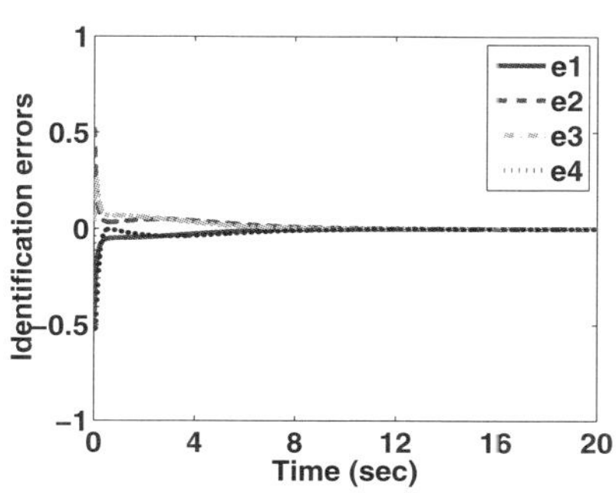

Fig. 6.9 Identification errors

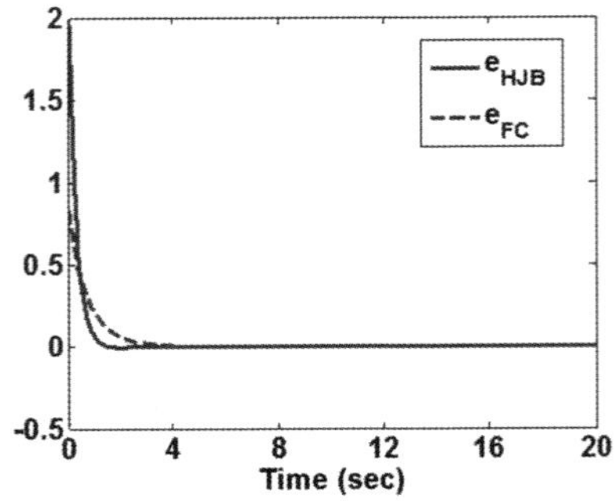

Fig. 6.10 HJB equation and terminal constraint estimation errors

(b) HJB equation and terminal constraint estimation errors

In this part, the HJB equation and terminal constraint estimation errors have been analyzed. It is well known that proposed control input will approach finite horizon optimal control [Chen *et al.* (2007); Lewis and Syrmos (1995)] input only when the control input satisfies both HJB and terminal constraint errors. If the HJB equation error is near zero, then the solution of the HJB equation is optimal and the control input that uses the value function becomes optimal. In Fig. 6.10, within fixed final time (i.e., $t \in [0, 20\text{s}]$), not only the HJB equation error but also terminal constraint estimation errors converge close to zero. This indicates that the proposed stochastic optimal control inputs approach the finite horizon optimal control inputs.

(c) Cost function comparison

Subsequently, the cost function of the proposed finite horizon stochastic optimal controller is studied. For comparison, with known system dynamics and network imperfections, both a conventional NN-based feedback linearization control [Jagannathan (2006)] and an ideal offline finite horizon optimal control [Chen *et al.*

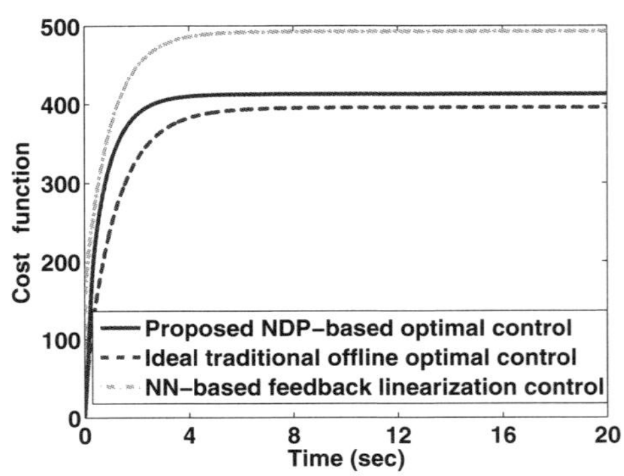

Fig. 6.11 Comparison of cost functions

(2007)] for NNCS have been included. In Fig. 6.11, the cost function comparison result is shown for the three controllers. Compared with conventional NN-based NNCS feedback linearization control, proposed optimal control design can deliver a much better performance since optimality is neglected in feedback linearization controller. Moreover, in contrast to traditional offline NNCS finite horizon optimal control [Chen *et al.* (2007)], cost function of proposed scheme is slight higher due to transient performance and system uncertainties whereas proposed design is more practical since the prior knowledge on network imperfections and system dynamics are not needed unlike in the case of traditional offline optimal controller design.

However, computational complexity of an optimal controller is higher than a traditional controller. Despite the increase in computational cost for the optimal controller, these advanced controllers can still be realized in practice cheaply due to a drastic increase in processor speed. Therefore, the advanced controllers, such as the one proposed, can be utilized on NNCS for generating an improvement in performance over traditional controllers.

In summary, the simulation results given in Figs. 6.3-6.11 confirm that the proposed time-based NDP scheme renders acceptable performance in the presence of uncertain NNCS dynamics due to network imperfections. Next, the extension of the finite horizon optimal adaptive controller to infinite horizon is introduced.

6.4 Extension to Infinite Horizon

It is important to note that infinite horizon optimal control design can be derived when the terminal constraint is ignored and time goes to infinity i.e., $N \to \infty$. The details of infinite horizon optimal adaptive design are given from Xu and Jagannathan (2013b) as follows.

6.4.1 *Approximation of the optimal stochastic value function and control policy using NN*

In Jagannathan (2006), by using universal approximation property of NN, the stochastic value function (6.7) and control policy (6.9) can be represented with critic and action NN as

$$V^*(z_k) = \underset{\tau,\gamma}{E}\left(W_V^{\mathsf{T}}\vartheta(z_k) + \varepsilon_{Vk}\right) \tag{6.43}$$

$$u^*(z_k) = \underset{\tau,\gamma}{E}\left(W_u^{\mathsf{T}}\varphi(z_k) + \varepsilon_{uk}\right) \tag{6.44}$$

respectively, where $\underset{\tau,\gamma}{E}(W_V)$ and $\underset{\tau,\gamma}{E}(W_u)$ represent the constant target NN weights, ε_{Vk}, ε_{uk} are the reconstruction errors for critic and action NN, respectively, and $\vartheta(\cdot)$ and $\varphi(\cdot)$ are the vector activation functions for two NNs, respectively. The upper bounds in the mean for the two target NN weights are defined as $\left\|\underset{\tau,\gamma}{E}(W_V)\right\| \le W_{VM}$

and $\left\| \underset{\tau,\gamma}{E}(W_u) \right\| \leq W_{uM}$ where W_{VM}, W_{uM} are positive constants [Jagannathan (2006)], and the approximation errors are also considered bounded in the mean as $\left\| \underset{\tau,\gamma}{E}(\varepsilon_{Vk}) \right\| \leq \varepsilon_{VM}$ and $\left\| \underset{\tau,\gamma}{E}(\varepsilon_{uk}) \right\| \leq \varepsilon_{uM}$ where ε_{VM}, ε_{uM} are also positive constants [Jagannathan (2006)], respectively. Additionally, the gradient of approximation error is assumed to be bounded in the mean as $\left\| \underset{\tau,\gamma}{E}(\partial\varepsilon_{Vk}/\partial z_{k+1}) \right\| \leq \varepsilon'_{VM}$ with ε'_{VM} being a positive constant [Jagannathan (2006)].

The critic and action NN approximations of (6.43) and (6.44) can be expressed as

$$\hat{V}(z_k) = \underset{\tau,\gamma}{E}\left[\hat{W}_{Vk}^{\mathsf{T}}\vartheta(z_k)\right] \tag{6.45}$$

$$\hat{u}(z_k) = \underset{\tau,\gamma}{E}\left[\hat{W}_{uk}^{\mathsf{T}}\varphi(z_k)\right] \tag{6.46}$$

where $\underset{\tau,\gamma}{E}(\hat{W}_{Vk})$ and $\underset{\tau,\gamma}{E}(\hat{W}_{uk})$ represent the estimated values of target weights W_V and W_u, respectively. In this work, the activation functions $\vartheta(\cdot)$, $\varphi(\cdot)$ are selected to be a basis function set and linearly independent [Jagannathan (2006)]. Since it is required that $V\left(\underset{\tau,\gamma}{E}(z_k) = 0\right) = 0$ and $u\left(\underset{\tau,\gamma}{E}(z_k) = 0\right) = 0$, the basis functions $\vartheta(\cdot)$, $\varphi(\cdot)$ are chosen such that $\vartheta\left(\underset{\tau,\gamma}{E}(z_k) = 0\right) = 0$, $\varphi\left(\underset{\tau,\gamma}{E}(z_k) = 0\right) = 0$, respectively.

Substituting (6.46) into (6.7), it can be rewritten as

$$\underset{\tau,\gamma}{E}\left[W_V^{\mathsf{T}}(\vartheta(z_{k+1}) - \vartheta(z_k))\right] + \underset{\tau,\gamma}{E}\left[z_k^{\mathsf{T}}Q_z z_k + u_k^{\mathsf{T}}R_z u_k\right] = \underset{\tau,\gamma}{E}(\varepsilon_{Vk} - \varepsilon_{Vk+1}).$$

In other words,

$$\underset{\tau,\gamma}{E}\left[W_V^{\mathsf{T}}\Delta\vartheta(z_{k+1})\right] + r(z_k, u_k) = \Delta\varepsilon_{Vk} \tag{6.47}$$

where $r(z_k, u_k) = \underset{\tau,\gamma}{E}(z_k^{\mathsf{T}}Q_z z_k + u_k^{\mathsf{T}}R_z u_k)$, $\Delta\vartheta(z_k) = \underset{\tau,\gamma}{E}[\vartheta(z_{k+1}) - \vartheta(z_k)]$ and $\Delta\varepsilon_{vk} = \underset{\tau,\gamma}{E}(\varepsilon_{vk} - \varepsilon_{vk+1})$.

However, when implementing the estimated value function (6.45), (6.47) does not hold. Therefore, using delayed values for convenience, the residual error or cost-to-go error where (6.47) can be expressed as

$$\begin{aligned}\underset{\tau,\gamma}{E}(e_{Vk}) &= \underset{\tau,\gamma}{E}\left(z_k^{\mathsf{T}}Q_z z_k + u_k^{\mathsf{T}}R_y u_k\right) + \hat{V}(z_{k+1}) - \hat{V}(z_k) \\ &= r(z_k, u_k) + \underset{\tau,\gamma}{E}\left[\hat{W}_{Vk}^{\mathsf{T}}\Delta\vartheta(z_k)\right].\end{aligned} \tag{6.48}$$

Based on gradient descent algorithm, the update law of critic NN weights is given by

$$\begin{aligned}\underset{\tau,\gamma}{E}\left(\hat{W}_{Vk+1}\right) &= \underset{\tau,\gamma}{E}\left(\hat{W}_{Vk}\right) - \alpha_v \underset{\tau,\gamma}{E}\left(\frac{\Delta\vartheta\left(z_k\right)}{\Delta\vartheta^{\mathsf{T}}\left(z_k\right)\Delta\vartheta\left(z_k\right)+1}e_{Vk}^{\mathsf{T}}\right)\\ &= \underset{\tau,\gamma}{E}\left(\hat{W}_{Vk}\right) - \alpha_v \underset{\tau,\gamma}{E}\left[\frac{\Delta\vartheta\left(z_k\right)}{\Delta\vartheta^{\mathsf{T}}\left(z_k\right)\Delta\vartheta\left(z_k\right)+1}\right.\\ &\quad \left.\times\left[r^{\mathsf{T}}\left(z_k,u_k\right)+\Delta\vartheta^{\mathsf{T}}\left(z_k\right)\hat{W}_{Vk}\right]\right]. \end{aligned} \tag{6.49}$$

Remark 6.2. It is important to note that the stochastic value function (6.7) and critic NN approximation (6.45) all become zero only when $z_k = 0$. Therefore, once the system states have converged to zero, the value function approximation is no longer updated. This can be also viewed as a PE requirement for the inputs to the critic NN where the system outputs must be persistently exiting long enough for the approximation so that critic NN learns the optimal stochastic value function. In this chapter, the PE condition is met by introducing exploration noise.

As a final step in the critic NN design, define the weight estimation error as $\underset{\tau,\gamma}{E}(\tilde{W}_{Vk}) = \underset{\tau,\gamma}{E}(W_V) - \underset{\tau,\gamma}{E}(\hat{W}_{Vk})$. Since $r^{\mathsf{T}}(z_k,u_k) = -\underset{\tau,\gamma}{E}\left[\Delta\vartheta^{\mathsf{T}}(z_{k+1})W_V\right] + \Delta\varepsilon_{Vk}^{\mathsf{T}}$ in (6.47), the dynamics of the critic NN weights estimation error can be rewritten as

$$\begin{aligned}\underset{\tau,\gamma}{E}\left(\tilde{W}_{Vk+1}\right) &= \underset{\tau,\gamma}{E}\left(\tilde{W}_{Vk}\right) - \alpha_V \underset{\tau,\gamma}{E}\left(\frac{\Delta\vartheta\left(z_k\right)\Delta\vartheta^{\mathsf{T}}\left(z_k\right)}{\Delta\vartheta^{\mathsf{T}}\left(z_k\right)\Delta\vartheta\left(z_k\right)+1}\tilde{W}_{Vk}\right)\\ &\quad +\alpha_V \underset{\tau,\gamma}{E}\left(\frac{\Delta\vartheta\left(z_k\right)\Delta\varepsilon_{Vk}}{\Delta\vartheta^{\mathsf{T}}\left(z_k\right)\Delta\vartheta\left(z_k\right)+1}\right). \end{aligned} \tag{6.50}$$

Next, the boundedness in the mean of the critic NN estimation error dynamics $\underset{\tau,\gamma}{E}\left(\tilde{W}_{Vk}\right)$ given by (6.50) is demonstrated in the following theorem.

Theorem 6.3. *Let $u_0(z_k)$ be any admissible control policy for nonlinear NCS (6.3), the initial critic NN weights $WV,0$, which reside in a compact set Ω, and let the critic NN weights update law be given by (6.49). Then, there exists a positive constant α_V satisfying $0 < \alpha_V < \frac{1}{2}$ and computable positive constant B_{Wv}, such that the critic NN weights estimation error (6.50) is UUB in the mean with ultimate bounds in the mean given by $\left\|\underset{\tau,\gamma}{E}\left(\tilde{W}_{Vk}\right)\right\| \leq B_{Wv}$.*

Proof. Refer to Section 6.7.1. □

Now we need to find the control policy via action NN (6.46) which minimizes the approximated value function (6.45). First, the action NN estimation errors are defined to be the difference between the actual optimal control input (6.46) that is being applied to NNCS (6.3) and the control input that minimizes the estimated value function (6.45) with identified NNCS dynamics $\hat{G}(z_k)$. This error can be expressed as

$$\underset{\tau,\gamma}{E}\left(e_{uk}\right)=\underset{\tau,\gamma}{E}\left[\hat{W}_{uk}^{\mathsf{T}}\varphi\left(z_k\right)+\frac{1}{2}R_z^{-1}\hat{G}^{\mathsf{T}}\left(z_k\right)\frac{\partial\vartheta^{\mathsf{T}}\left(z_{k+1}\right)}{\partial z_{k+1}}\hat{W}_{Vk}\right]. \tag{6.51}$$

The update law for action NN weights is defined as

$$\underset{\tau,\gamma}{E}\left(\hat{W}_{uk+1}\right)=\underset{\tau,\gamma}{E}\left(\hat{W}_{uk}\right)-\alpha_u\underset{\tau,\gamma}{E}\left[\frac{\varphi\left(z_k\right)}{\varphi^{\mathsf{T}}\left(z_k\right)\varphi\left(z_k\right)+1}e_{uk}^{\mathsf{T}}\right] \tag{6.52}$$

where $0<\alpha_u<1$ is a positive constant. By selecting the control policy u_k to minimize the desired value function (6.43), it follows that

$$\underset{\tau,\gamma}{E}\left[W_u^{\mathsf{T}}\varphi\left(z_k\right)+\varepsilon_{uk}\right]=-\frac{1}{2}\underset{\tau,\gamma}{E}\left[R_z^{-1}\hat{G}^{\mathsf{T}}\left(z_k\right)\left(\frac{\partial\vartheta^{\mathsf{T}}\left(z_{k+1}\right)}{\partial z_{k+1}}W_V+\frac{\partial\varepsilon_{Vk}^{\mathsf{T}}}{\partial z_{k+1}}\right)\right].$$

In other words,

$$\underset{\tau,\gamma}{E}\left[W_u^{\mathsf{T}}\varphi\left(z_k\right)+\varepsilon_{uk}+\frac{1}{2}R_z^{-1}\hat{G}^{\mathsf{T}}\left(z_k\right)\left(\frac{\partial\vartheta^{\mathsf{T}}\left(z_{k+1}\right)}{\partial z_{k+1}}W_V+\frac{\partial\varepsilon_{Vk}^{\mathsf{T}}}{\partial z_{k+1}}\right)\right]=0. \tag{6.53}$$

Substituting (6.53) into (6.51), the action NN estimation error dynamics can be rewritten as

$$\begin{aligned}\underset{\tau,\gamma}{E}\left(e_{uk}\right) &= -\underset{\tau,\gamma}{E}\left[\tilde{W}_{uk}^{\mathsf{T}}\varphi\left(z_k\right)\right]-\frac{1}{2}\underset{\tau,\gamma}{E}\left[R_z^{-1}\hat{G}^{\mathsf{T}}\left(z_k\right)\frac{\partial\vartheta^{\mathsf{T}}\left(z_{k+1}\right)}{\partial z_{k+1}}\tilde{W}_{Vk}\right]\\ &\quad+\frac{1}{2}\underset{\tau,\gamma}{E}\left[R_y^{-1}\tilde{G}^{\mathsf{T}}\left(z_k\right)\frac{\partial\varepsilon_{Vk}^{\mathsf{T}}}{\partial z_{k+1}}-\varepsilon_{ek}\right]\end{aligned} \tag{6.54}$$

where $\underset{\tau,\gamma}{E}\left(\tilde{W}_{uk}\right)=\underset{\tau,\gamma}{E}\left(W_u\right)-\underset{\tau,\gamma}{E}\left(\hat{W}_{uk}\right)$, $\tilde{G}\left(z_k\right)=G\left(z_k\right)-\hat{G}\left(z_k\right)$, $\varepsilon_{ek}=\underset{\tau,\gamma}{E}\left[\varepsilon_{uk}+\frac{1}{2}R_z^{-1}G^{\mathsf{T}}\left(z_k\right)\frac{\partial\varepsilon_{Vk}^{\mathsf{T}}}{\partial z_{k+1}}\right]$ satisfying $\left\|\underset{\tau,\gamma}{E}\left(\varepsilon_{ek}\right)\right\|\le\varepsilon_{eM}$ with ε_{eM} being a positive constant, and $\left\|\underset{\tau,\gamma}{E}\left[\frac{\partial\varepsilon_{Vk}^{\mathsf{T}}}{\partial z_{k+1}}\right]\right\|\le\varepsilon_{VM}'$.

The action NN weight estimation error dynamics can be represented as

$$\begin{aligned}\underset{\tau,\gamma}{E}\left(\tilde{W}_{uk+1}\right) &= \underset{\tau,\gamma}{E}\left(\tilde{W}_{uk}\right)+\alpha_u\underset{\tau,\gamma}{E}\left[\frac{\varphi\left(z_k\right)}{\varphi^{\mathsf{T}}\left(z_k\right)\varphi\left(z_k\right)+1}e_{uk}^{\mathsf{T}}\right]\\ &= \underset{\tau,\gamma}{E}\left(\tilde{W}_{uk}\right)-\alpha_u\underset{\tau,\gamma}{E}\left[\frac{\varphi\left(z_k\right)}{\varphi^{\mathsf{T}}\left(z_k\right)\varphi\left(z_k\right)+1}\right.\\ &\quad\times\left[\tilde{W}_{uk}^{\mathsf{T}}\varphi\left(z_k\right)+\frac{1}{2}R_z^{-1}\hat{G}^{\mathsf{T}}\left(z_k\right)\frac{\partial\vartheta^{\mathsf{T}}\left(z_{k+1}\right)}{\partial z_{k+1}}\tilde{W}_{Vk}\right.\\ &\quad\left.\left.-\frac{1}{2}R_z^{-1}\tilde{G}^{\mathsf{T}}\left(z_k\right)\frac{\partial\varepsilon_{Vk}^{\mathsf{T}}}{\partial z_{k+1}}+\varepsilon_{ek}\right]\right].\end{aligned} \tag{6.55}$$

Remark 6.3. In this work, the proposed NN-based identifier relaxes the need for partial NNCS dynamics $G\left(z_k\right)$. Compared to Dierks and Jagannathan (2012), the knowledge of the input transformation matrix $G(z_k)$ and internal dynamics $F(z_k)$ are considered unknown here.

Theorem 6.4. *Let $u_0(z_k)$ be any initial stabilizing control policy for the NNCS in (6.3) and $0 < k^* < 1/2$, the initial NN identifier weight matrix $WI, 0$, critic and action NNs weights $WV, 0$, $Wu, 0$ which reside in a compact set Ω. Let the NN weight tuning for the identifier, critic and the action NN be provided by (6.18), (6.49) and (6.52), respectively. Then, there exists several positive constants α_C, α_u, α_V satisfying $0 < \alpha_C < \min\left\{1, \frac{\Psi_{\min}}{2\sqrt{2}\Psi_M}\right\}$, $\frac{1}{4} < \alpha_V < \frac{3+\sqrt{3}}{12}$ and $\frac{1}{6} < \alpha_u < \frac{1}{3}$ respectively, and positive constants b_z, b_V, b_{WC}, b_{ez} and b_u such that the system output vector $\underset{\tau,\gamma}{E}(z_k)$, NN identification error $\underset{\tau,\gamma}{E}(e_{zk})$, weight estimation errors $\underset{\tau,\gamma}{E}(\tilde{W}_{Ck})$, critic and action NN weight estimation errors $\underset{\tau,\gamma}{E}(\tilde{W}_{Vk})$ and $\underset{\tau,\gamma}{E}(\tilde{W}_{uk})$, respectively, are all UUB in the mean for all $k \geq k_0 + T$ with ultimate bounds in the mean given by $\left\|\underset{\tau,\gamma}{E}(z_k)\right\| \leq b_z$, $\left\|\underset{\tau,\gamma}{E}(e_{zk})\right\| \leq b_{ez}$, $\left\|\underset{\tau,\gamma}{E}(\tilde{W}_{Ck})\right\| \leq b_{WC}$, $\left\|\underset{\tau,\gamma}{E}(\tilde{W}_{Vk})\right\| \leq b_V$ and $\left\|\underset{\tau,\gamma}{E}(\tilde{W}_{uk})\right\| \leq b_u$. Further, $\left\|\underset{\tau,\gamma}{E}\left[\hat{u}(z_k) - u^*(z_k)\right]\right\| \leq \delta_u$ for a small positive constant δ_u.*

Proof. Refer to the Section 6.7.2. □

Remark 6.4. It is important to note that 6.2 and 6.3 demonstrated UUB in the mean of NN identifier and Critic NN estimation errors, respectively. In 6.4, boundedness in the mean of NN identifier, Critic NN and estimated stochastic optimal control from action NN are all considered together.

6.4.2 *Simulation results*

In this section, stochastic optimal control of NNCS with uncertain dynamics in the presence of unknown random delays and packet losses is evaluated. The continuous-time version of original nonlinear affine system is given by

$$\dot{x} = f(x) + g(x)u \tag{6.56}$$

where $f(x) = \begin{bmatrix} -x_1 + x_2 \\ -0.5x_1 - 0.5x_2(1 - (\cos(2x_1) + 2)^2) \end{bmatrix}$, $g(x) = \begin{bmatrix} 0 \\ \cos(2x_1) + 2 \end{bmatrix}$. The network parameters of NNCS are selected as Hu and Zhu (2003): (1) The sampling time: $T_s = 100$ms; (2) The bound of delay is set as two, i.e., $\bar{d} = 2$; (3) The mean random delay values are given by $E(\tau_{sc}) = 80$ms, $E(\tau) = 150$ms; (4) Packet losses follow Bernoulli distribution with $p = 0.3$.

First, the effect of random delays and packet losses for NNCS is studied. The initial state is taken as $x_0 = \begin{bmatrix} 5 & -3 \end{bmatrix}^T$. The initial static control $u_k = \begin{bmatrix} -2 & -5 \end{bmatrix} x_k$, which maintains the original nonlinear affine system (6.37) stable, is shown in Fig. 6.12(a). By contrast, this controller cannot maintain system stable for NNCS in presence of random delays and packet losses as Fig. 6.12(b).

Next, the proposed stochastic optimal control is implemented for the NNCS with unknown system dynamics in presence of random delays and packet losses.

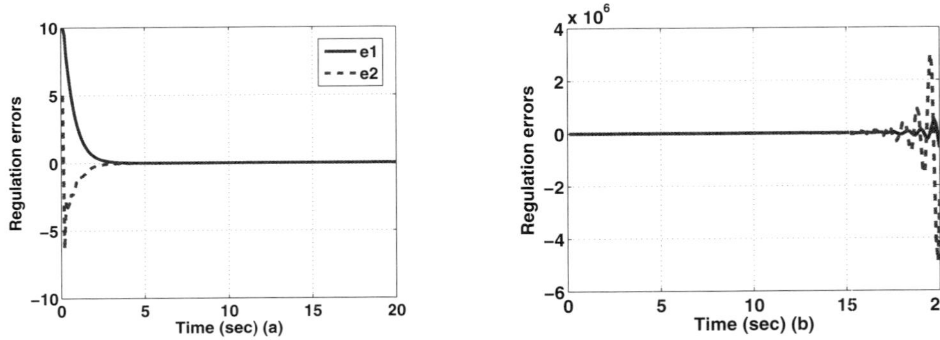

Fig. 6.12 Performance of a static feedback controller: (a) random delays and packet losses are not present (b) with random delays and packet losses

The augment state z_k is generated as $\mathbb{R}^{41}$, and the initial stabilizing policy for proposed algorithm was selected as $u_o(z_k) = [-2\ -5\ -1\ -1]\ z_k$ generated by using standard pole placement method, while the activation functions for NN-based identifier were generated as $\tanh\left\{(z_1)^2, z_1z_2, \ldots, (z_4)^2, (z_1)^4, (z_1)^3 z_2, \ldots, (z_4)^6\right\}$, critic NN activation function were selected as sigmoid of the sixth order polynomial $\left\{(z_1)^2, z_1z_2, \ldots, (z_4)^2, (z_1)^4, (z_1)^3 z_2, \ldots, (z_4)^6\right\}$ and action NN activation function were generated from the gradient of critic NN activation function.

The design parameters for NN-based identifier, critic NN and action NN were selected as $\alpha_C = 0.002$, $\alpha_V = 10^{-4}$ and $\alpha_u = 0.005$ while the NN-based identifier and critic NN weights are set to zero at the beginning of the simulation. The initial weights of the action NN are chosen to reflect the initial stabilizing control. The simulation was run for 20 seconds (200 time steps), for the first 10 seconds (100 time steps), exploration noise with mean zero and variance 0.06 was added to the system in order to ensure the persistency of excitation (PE) condition (See Remark 6.2).

The performance of proposed stochastic optimal controller is evaluated from several aspects: 1) as shown in Fig. 6.13(a), the proposed stochastic optimal controller can make the NNCS state regulation errors converge to zero even when the NNCS dynamics are uncertain which implies that the proposed controller can make the NNCS system stable in the mean; 2) the proposed critic NN and action NN parameters converge to constant values and remain bounded in the mean consistent with Theorem 6.3 as shown in Fig. 6.13(b); 3) The optimal control input for NNCS with uncertain dynamics is shown in Fig. 6.13 (c) which is bounded in the mean. For comparison, the HDP value iteration (VI) [Al-Tamimi *et al.* (2008)] based scheme is also implemented for NNCS with known dynamics $G(\cdot)$ by incorporating the $g(\cdot)$ (6.56) and information of delays and packet losses which are normally not known beforehand. The initial admissible control, critic NN and action NN activation function are same as proposed time-based stochastic optimal control.

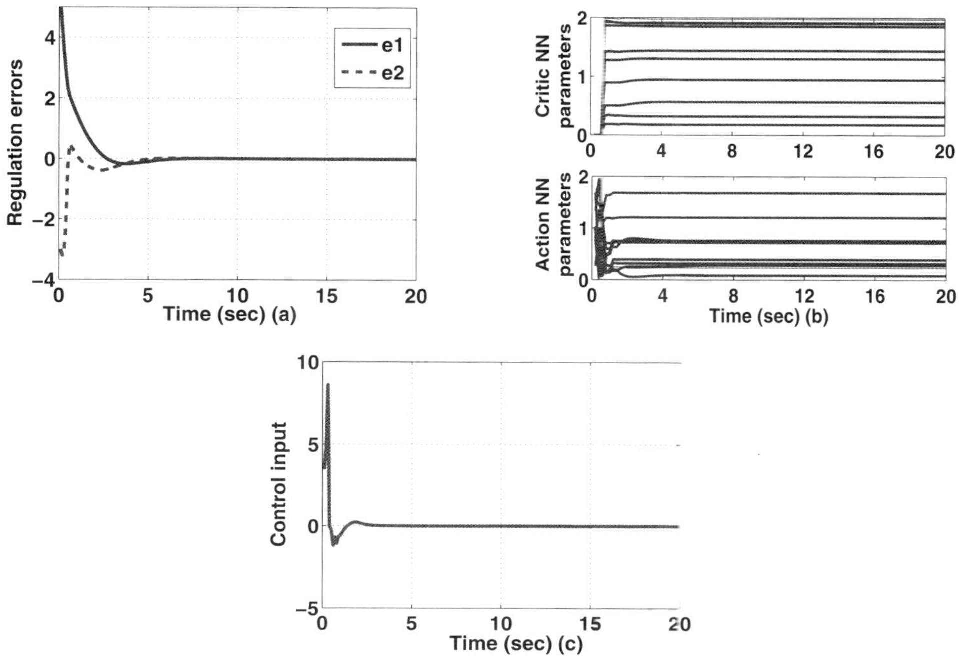

Fig. 6.13 Performance of stochastic optimal controller for NNCS: (a) State regulation errors (b) Critic NN and Action NN parameters (c) Control input

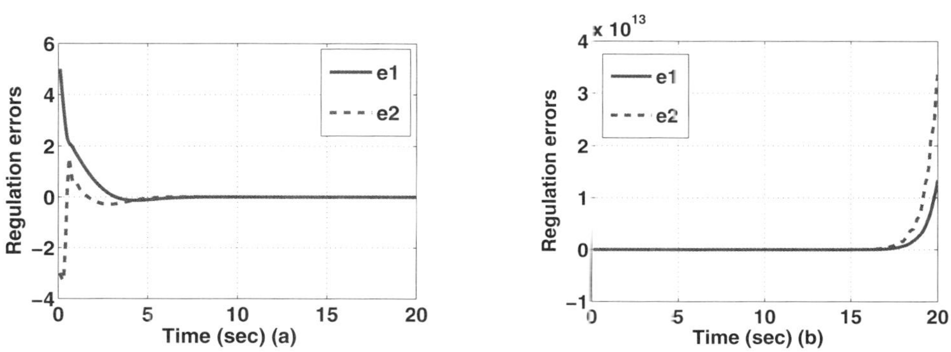

Fig. 6.14 Performance of HDP value iteration for NNCS: (a) Iterations = 100 times/sample (b) Iterations = 10 times/sample

As shown in Fig. 6.14(a), the HDP VI method can make the NNCS state regulation errors converge to zero when the number of iterations is 100 times/sample. By contrast, HDP VI cannot maintain NNCS stable in the mean when iterations become 10 times/sample as shown in Fig. 6.14(b). It implies that HDP VI scheme not only needs partial knowledge of original nonlinear affine system dynamics, $g(\cdot)$, but also information on delays and packet losses. The number of iterations required for a given nonlinear system is unknown. Due to these drawbacks, the HDP VI is not preferred for NNCS implementation in real-time.

Based on the results presented in Fig. 6.12 through Fig. 6.14, the proposed stochastic optimal control scheme with uncertain NNCS dynamics and unknown network imperfections can overcome the drawbacks of HDP-based value iteration method and will render nearly the same performance as that of an optimal controller for NNCS when the system dynamics, random delays and packet losses are known.

6.5 Event-Triggered Control of NNCS

In the previous section, an optimal controller design fo the networked control system is discussed. The proposed scheme not only relaxed the uncertain dynamics of the system but also rendered the performance of the feedback linearization controller in the presence of the network artifacts, such as, delays and packet dropout. On the other hand, it uses the traditional periodic transmission scheme to transfer the feedback data through the communication channel. This requires higher bandwidth of the communication channel which leads to higher traffic and congestion. In one of the pioneer works by Astrom and Bernhardsson (2002), it is being shown that aperiodic sampling is advantageous than the traditional periodic sampling especially for the networked control systems. The decision of the aperiodic sampling instants, for a feedback control system, plays a key role in maintaining system stability and desired performance. Event-triggered control, which is emerged in the recent time, provides a solution to the problem with additional benefits.

6.5.1 *Event-triggered control design of uncertain continuous time nonlinear system*

The structure of the event-triggered networked control system (NCS) for a continuous time uncertain nonlinear system is given in the Fig. 6.15. Consider the plant is in an affine form given as

$$\begin{aligned} \dot{x} &= f(x) + g(x)\,u \\ y &= h(x) \end{aligned} \tag{6.57}$$

where $x \in \mathbb{R}^n$, $u \in \mathbb{R}^m$ and $y \in \mathbb{R}^p$ are the system state, input and output, respectively. $f\colon \mathbb{R}^n \to \mathbb{R}^n$ and $g\colon \mathbb{R}^n \to \mathbb{R}^{n\times m}$ are the uncertain nonlinear functions. Due to the digital communication network, the controller is implemented in a sampled data approach. In a periodic sampled data framework the sampling time $t_{k+1} - t_k = T_s$ is a fixed time interval. We assume that the system is stabilizable and there exist an ideal feedback linearization controller as

$$u^* = \xi(x) \tag{6.58}$$

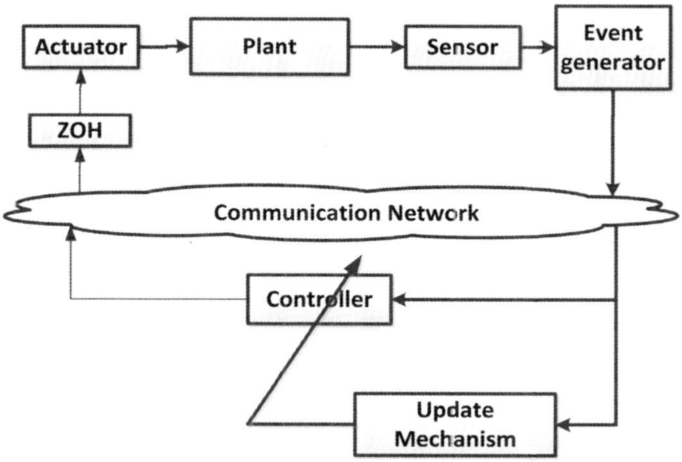

Fig. 6.15 Event-triggered control of nonlinear networked control system with pack dropouts

which renders the closed loop discrete time system asymptotically stable. This feedback linearization controller design requires the knowledge of the system dynamics which are considered unknown. Next, the assumption regarding the boundedness of the control coefficient matrix is presented.

Assumption 6.2. The nonlinear function, or control coefficient matrix $g(x)$ is bounded above, i.e., $0 < \|g(x)\| \leq g_{\max}$.

This assumption is standard for practical systems, for example, robot manipulators where $g(x)$ denotes the inverse of the mass matrix and the upper limit is known. In the proposed scheme, as shown in Fig. 6.15, the system state vector and the controller input vector are not transmitted periodically to the controller and the actuator, respectively. The sensor samples the system state vector periodically and feeds it to the event generator. The transmission instants are decided by the event-generator by evaluating the event condition and initiating the transmission. Upon receiving the transmitted state vector, the controller updates and transmits the control input vector to the actuator. The actuator holds the last transmitted control input until the next update is received. This reduces the transmission through the communication network, thus reducing the congestion which in turn delays due to congestion.

Traditionally, the event condition consists of the system error and a state dependent threshold which is explicitly computed using the known system dynamics. In contrast, the design of the event condition with the uncertain system dynamics complicates the design approach and leads to a trade-off in computation. Furthermore, in presence of the network artifacts, such as packet loss and delay the problem becomes more challenging. Towards the first step in designing the event-triggered control for an uncertain nonlinear system we assume there are no delays, which is considered reasonable, because of the reduced communication but random packet losses occur during transmission. We further assume that the distribution of the packet losses is known. In the presence of the packet loss the system dynamics can be expressed as

$$\begin{aligned}\dot{x} &= f(x) + \gamma(t)\, g(x)\, u \\ y &= h(x)\end{aligned} \tag{6.59}$$

where $\gamma(t)$ is an indicator function for the packet dropout and defined as

$$\gamma(t) = \begin{cases} I^{n\times n}, & \text{if control input is received by the actuator at time } t, \\ 0^{n\times n}, & \text{if control input is lost at time } t, \end{cases}$$

where $I^{n\times n}$ is an identity matrix. The inclusion of random packet drop out in the system dynamics leads to a stochastic uncertain nonlinear system.

Since the controller is implemented in sampled data format the dynamics of the system can be discretized by integrating (6.59) over the interval $(kT_s, (k+1)T_s]$ for all $k \in \mathbb{N}$. This yields

$$\int_{kT_s}^{(k+1)T_s} \dot{x}\mathrm{d}t = \int_{kT_s}^{(k+1)T} f(x)\mathrm{d}t + \gamma(kT_s) \int_{kT_s}^{(k+1)T} g(x)\, u(kT_s)\, \mathrm{d}t \tag{6.60}$$

which can be rewritten as

$$x_{k+1} = F(x_k) + G(x_k)\, u_k \tag{6.61}$$

where $F(x_k) = \int_{kT_s}^{(k+1)T} f(x)\mathrm{d}t$, $G(x_k) = \gamma(kT_s)\int_{kT_s}^{(k+1)T} g(x)\mathrm{d}t$, $x(kT_s) = x_k$, $u(kT_s) = u_k$. We assume the stochastic nonlinear discrete time system (6.60) is controllable and there exists an ideal feedback linearization controller for the stochastic system represented as

$$u_k^* = \mathop{E}_{\gamma}(\xi(x_k)) \tag{6.62}$$

which guarantees asymptotic stability of the system. For instance, the ideal controller for a system with known system dynamics can be designed by using feedback linearization. This will lead to a linearized stable closed loop dynamics in the form

$$x_{k+1} = Kx_k \tag{6.63}$$

where K is a stable control gain matrix designed using pole placement. Furthermore, considering a Lyapunov function

$$V = \mathop{E}_{\gamma}\left(x_k^{\top} P x_k\right) \tag{6.64}$$

where $E_\gamma(*)$ denotes the expectation operator (the mean value) in terms of network-induced packet losses, and P is a symmetric positive definite matrix, it is clear that the first difference

$$\Delta V \leq -\sigma_{\min}(Q)\|x_k\|^2 \tag{6.65}$$

which is less than zero where $\sigma_{\min}(Q)$ is the minimum singular value of Q. The positive definite matrices P and Q satisfy the Lyapunov equation

$$K^{\mathsf{T}}PK - P + Q = 0. \tag{6.66}$$

Remark 6.5. As K is Hurwitz, the Lyapunov equation (6.66) has a positive definite solution P for a given positive definite Q. Hence, the closed-loop system is asymptotically stable.

In addition, the assumption for asymptotic stability of the system is necessary for the event-triggered design and a weaker assumption than the input to state stability of the system which is used for designing traditional event triggered control [Tabuada (2007)].

In the event-triggered paradigm the ideal control input (6.62) cannot be implemented because of the unavailability of the system state at the controller at all sampling instants. Hence, the control input is updated at every trigger instant as mentioned earlier in the proposed scheme. This introduces a state transmission error known as event-trigger state error similar to that of a measurement error in a perturb system. Let $\{k_i\}$ for $i = 1, 2, \ldots$ be a monotonically increasing sequence of time instants at which the event generator generates events. Define the difference between the last transmitted state and the current state is the event-trigger state error represented as

$$\underset{\gamma}{E}\left(e_k^s\right) = \underset{\gamma}{E}\left(x_k - x_{k_i}\right) \tag{6.67}$$

where x_{k_i} is the last transmitted system state. It is important to mention here that the packet transmitted from the sensor to the controller at the transmission instants may reach the controller or may not due to the packet drop in the communication network. In addition, the event-trigger state error becomes zero at the transmission instants as the new system state is transmitted. Hence, the event-trigger state error is a discontinuous function and discontinuous at every k_i for $i = 1, 2, \ldots$. The control input in the event-triggered frame work is modified as

$$u_k = \underset{\gamma}{E}\left(\xi\left(x_{k_i}\right)\right) \quad \text{for } k_i \leq k < k_{i+1} \tag{6.68}$$

where x_{k_i} is the last transmitted system state and a piece wise continuous signal. Due to the uncertain system dynamics, the above event-based feedback linearization controller is approximated using a linearly parameterized NN. Based on the universal approximation property, the control input can be represented in a compact set $\Omega \subset \mathbb{R}^n$ as

$$u_k = \underset{\gamma}{E}\left(W_u^{\mathsf{T}}\upsilon\left(x_{k_i}\right) + \varepsilon_{u,k}\right) \quad \text{for } k_i \leq k < k_{i+1},\ k \in \mathbb{N} \tag{6.69}$$

where $W_u \in \mathbb{R}^{l\times b}$ is the unknown target NN weight matrix and $\upsilon(x_{k_i}) \in \mathbb{R}^l$ is the stochastic activation function with l neurons in the hidden layer and b outputs. $\varepsilon_{u,k}$ is the NN reconstruction error. Next, the standard assumption for NN weights, activation function and reconstruction error is presented.

Assumption 6.3. The NN weights, activation function and the reconstruction error are bounded in a compact set $\Omega \subset \mathbb{R}^n$. i.e., $\left\| \underset{\gamma}{E}\left(W_u^{\mathsf{T}}\right) \right\| \le W_{\max}$, $\left\| \underset{\gamma}{E}\left(\upsilon(x_k)\right) \right\| \le \upsilon_{\max}$ and $\left\| \underset{\gamma}{E}\left(\varepsilon_{u,k}\right) \right\| \le \varepsilon_{\max}$. We further assume that the activation function is also Lipschitz continuous in the compact set $\Omega \subset \mathbb{R}^n$. This implies there exists a constant L_υ such that $\|\upsilon(x_k) - \upsilon(x_{k_i})\| \le L_\upsilon \|x_k - x_{k_i}\|$.

The control input can be estimated by using

$$\hat{u}_k = \underset{\gamma}{E}\left(\hat{W}_{u,k}^{\mathsf{T}}\upsilon(x_{k_i})\right) \tag{6.70}$$

where $\hat{W}_{u,k}^{\mathsf{T}}$ is the estimated NN weight matrix updated by following update law

$$\underset{\gamma}{E}\left(\hat{W}_{u,k+1}\right) = \underset{\gamma}{E}\left(\hat{W}_{u,k}\right) - \delta_k\alpha \underset{\gamma}{E}\left(\upsilon(x_k)\, e_{k+1}^{s^{\mathsf{T}}}\right) N - \kappa \underset{\gamma}{E}\left(\hat{W}_{u,k}\right) \tag{6.71}$$

where $\alpha > 0$ is the learning gain and δ_k is the indicator function for indicating the transmission instants and defined as

$$\delta_k = \begin{cases} 1, & \text{transmission is initiated,} \\ 0, & \text{transmission is not initiated.} \end{cases}$$

Moreover, $\kappa > 0$ is a constant and similar to the σ-modification in the continuous time counterpart [Narendra and Annaswamy (1989)]. As the transmission instants decided by the event generator and aperiodic in nature, the indicator function δ_k enables the NN weight update law only at the trigger instants, hence, becomes aperiodic. During the inter event time the NN weights are not updated and held at the previous values. Next, using the above approximation based control design and NN update law, an event condition along with the system stability is presented in the next section.

6.5.2 *Closed-loop system stability and event condition design*

In this section, we formulate the closed-loop dynamics and designed an event condition. The UUB of the system state and the NN estimation error are proved using the Lyapunov technique.

The NN weight estimation error dynamics can be computed using (6.71)

$$\underset{\gamma}{E}\left(\tilde{W}_{u,k+1}\right) = \underset{\gamma}{E}\left(\tilde{W}_{u,k}\right) - \delta_k\alpha \underset{\gamma}{E}\left(\upsilon(x_k)\, x_{k+1}^{\mathsf{T}}\right) N - \kappa \underset{\gamma}{E}\left(\hat{W}_{u,k}\right) \tag{6.72}$$

where $\tilde{W}_u = W_u - \hat{W}_u$ is the NN weight estimation error.

Similarly, the closed-loop system dynamics can be formulated using (6.61) and (6.70) and written as

$$\begin{aligned} x_{k+1} &= F(x_k) + G(x_k) \underset{\gamma}{E}\left(\xi\left(x_{k_i}\right)\right) \\ &= F(x_k) + G(x_k) \underset{\gamma}{E}\left(\xi\left(x_k + e_k^s\right)\right). \end{aligned} \tag{6.73}$$

Adding and subtracting the ideal control input leads to

$$\begin{aligned} x_{k+1} &= F(x_k) + G(x_k)\Big(\underset{\gamma}{E}\left(\xi\left(x_k + e_k^s\right)\right) + \underset{\gamma}{E}\left(\xi\left(x_k\right)\right) - \underset{\gamma}{E}\left(\xi\left(x_k\right)\right)\Big) \\ &= F(x_k) + G(x_k)\underset{\gamma}{E}\left(\xi\left(x_k\right)\right) \\ &\quad + G(x_k)\Big(\underset{\gamma}{E}\left(\xi\left(x_k + e_k^s\right)\right) - \underset{\gamma}{E}\left(\xi\left(x_k\right)\right)\Big). \end{aligned} \tag{6.74}$$

Using the approximated control input (6.71), the closed-loop dynamics becomes

$$\begin{aligned} x_{k+1} &= F(x_k) + G(x_k) \underset{\gamma}{E}\left(\xi\left(x_k\right)\right) \\ &\quad + G(x_k)\Big(\underset{\gamma}{E}\big(\hat{W}_{u,k}^{\top} \upsilon\left(x_k + e_k^s\right)\big) - \underset{\gamma}{E}\big(\xi(x_k)\big)\Big). \end{aligned} \tag{6.75}$$

Eventually, the event-trigger state error dynamics $e_{k+1}^s = x_{k+1} - x_{k_i}$.

The event condition which consists of the event-trigger state error and a threshold based on system state and NN weight estimates can be analytically computed by using Lyapunov direct approach for system stability. The basic idea behind computing the threshold is to allow the event-trigger state error e_k^s to increase to its limiting value where the control input need to be updated to avoid instability. Now, defining the following criteria as event condition

$$\underset{\gamma}{E}\left(\|e_k^s\|\right) \leq \frac{\sigma_{\min}(Q)}{2g_{\max}^2 L_\upsilon^2 \|P\|} \underset{\gamma}{E}\left(\frac{\|x_k\|}{\|\hat{W}_u\|^2}\right) \tag{6.76}$$

where K is the control gain matrix, $g_{\max}$ is the bound for the control input function L_υ is the Lipschitz constant for activation function and $\hat{W}_u$ is the NN estimated weight matrix. $\sigma_{\min}(Q)$ is the minimum singular value of the Q. P and Q are symmetric positive definite matrices satisfies the Lyapunov equation $K^{\top}PK - P + Q = 0$. The feedback data need to be transmitted and the control should be updated at the violation of the above event condition (6.76).

Theorem 6.5. *Consider the continuous time uncertain nonlinear system (6.57) represented as a stochastic nonlinear discrete time system (6.59) along with the approximation based control inputs (6.75). Let the NN weights are updated as (6.71). Then, let Assumption 6.3 hold, the closed-loop event-triggered stochastic nonlinear discrete time system UUB if the events are triggered and control is updated due to the violation of event condition (6.76).*

6.5.3 *Simulation results*

In this section, we have considered the two link robot manipulator to demonstrate the effectiveness of the event-triggered control in reducing the traffic. The nonlinear continuous time dynamics of the two link robot manipulator is given in (6.41).

Moreover, the controller u_k is designed as (6.75) with 12 neurons in the hidden layer. The input weight vectors are selected at random from a uniform distribution in the interval 0 to 0.9 and the output layer weight vectors W_u is chosen at random in the interval 0 to 0.05. Sigmoid activation functions are used, (i.e., $\tanh(\cdot)$). The Lipschitz constant for the activation function computed to be 5.8. Learning gains α is selected as 0.97 and the sigma modification parameter κ is 0.001. Initial system states are taken to be $\left[\pi/3 \ -\pi/10 \ 0 \ -0\right]^{\mathsf{T}}$, the control gains $K_1 = \operatorname{diag}\left(0.6, 0.5, 0.3, 0.7\right)$, $\mu = 0.99$, $Q = I$, and $g_{\max} = 3$. The event conditions are obtained by using (6.76). The system is simulated for 30 Sec with a sampling time 10ms. Network-induced packet losses follow Bernoulli distribution with $\gamma = 0.3$.

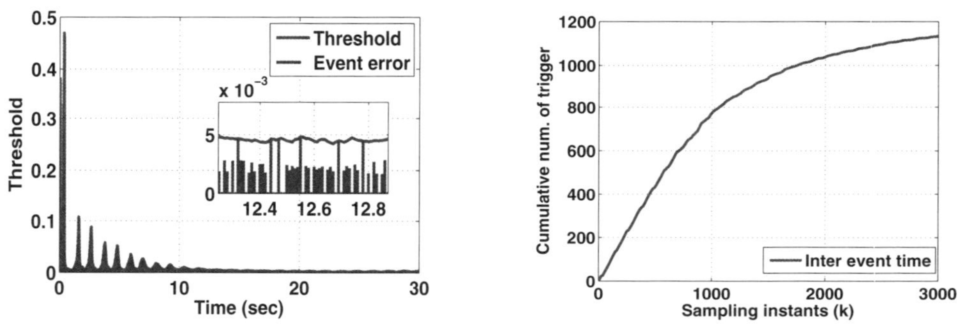

Fig. 6.16 (Left) Evolution of the threshold, (right) cumulative number triggers instants indicating reduction in transmission.

Fig. 6.16 (left) shows the evolution threshold for the event condition. The threshold value changes along with the system state and NN weight estimates and finally converge near zero along with the system state. The vertical lines indicate the instants of triggering and the transmissions are initiated. The figure is zoomed to indicate the aperiodical transmission instants. The staircases plot in Fig. 6.16 (right) gives the counts for the trigger instants during the simulation time of 30 seconds. With a sampling period of 10ms (i.e., 3000 sample instants), it is clearly evident from the right figure that the events are triggered almost at every sampling instant during the initial phase of simulation because of the approximation based feedback linearization controller and the NN is in learning phase. Once the control inputs are approximated by the NN-based feedback linearization controllers, i.e., the NN weight vectors converged to the bound, the triggering is reduced. And the stairs in the plot elongated and the plot flattens showing the reduction in trigger instants. The horizontal step size in the plot is the inter-event time and it increases with the approximation of the control input as the time evolves. This shows a reduction of the trigger instants, alternatively the traffic through the communication network in comparison to the traditional periodic sampled data approach.

Fig. 6.17 shows the convergence of the system states and the approximated control inputs near zero. This indicates the approximation based control input in the event based transmission regulates the system state and achieves the desired performance along with the reduction in the traffic through the communication network.

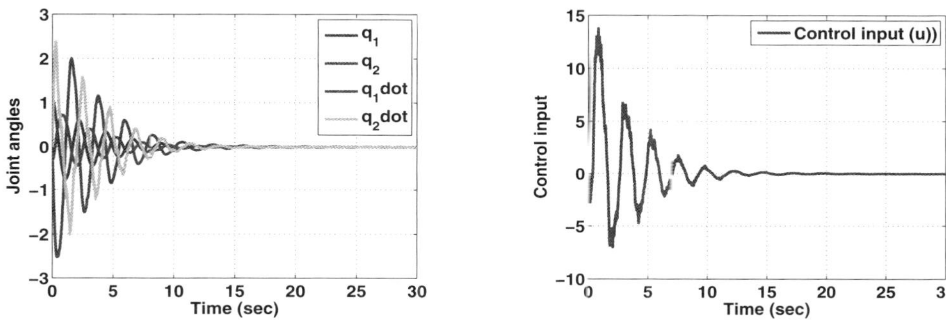

Fig. 6.17 (Left) Convergence of system states, (right) approximated control input for the nonlinear event-triggered control.

6.6 Conclusions

In this chapter, a novel time-based finite horizon NDP scheme was proposed for NNCS by using NN identifier, critic and action NNs to obtain stochastic optimal control policy in the presence of uncertain system dynamics due to network imperfections. By using historical inputs and NN identifier the requirement on both internal dynamics and control coefficient matrix was relaxed. Further, critic NN was derived to estimate the HJB equation solution online while satisfying the terminal constraint. An initial admissible control ensures that NNCS is stable when NN identifier, critic and action NN were being tuned. Using Lyapunov and geometric sequence analysis, the NNCS system state, identification error, weight estimation errors of NN-identifier, critic and action NNs have been proven to be UUB in the mean with ultimate bounds dependent upon initial condition and final time instant. While the final time instant increases, all the ultimate bounds would decrease and converge to bounds derived for the infinite horizon case. Moreover, a novel NN based event-triggered control is proposed for NNCS to reduce the communication traffic. This event triggered controller uses adaptive feedback linearization whereas it is not optimal. Future work includes the development of an optimal event triggered controller.

6.7 Appendix

6.7.1 *The proof of Theorem 6.3*

Consider the Lyapunov function candidate

$$L_V = \operatorname{tr}\left\{ \underset{\tau,\gamma}{E}\left(\tilde{W}_{Vk}^{\mathsf{T}} \tilde{W}_{Vk} \right) \right\}. \tag{6.77}$$

The first difference of (6.77) is given by

$$\Delta L_V = \operatorname{tr}\left\{ \underset{\tau,\gamma}{E}\left(\tilde{W}_{Vk+1}^{\mathsf{T}} \tilde{W}_{Vk+1} \right) \right\} - \operatorname{tr}\left\{ \underset{\tau,\gamma}{E}\left(\tilde{W}_{Vk}^{\mathsf{T}} \tilde{W}_{Vk} \right) \right\},$$

and using (6.50) yields

$$\begin{aligned}
\Delta L_V = {} & \operatorname{tr}\left\{ \underset{\tau,\gamma}{E}\left(\tilde{W}_{Vk}^{\mathsf{T}} \tilde{W}_{Vk} \right) \right\} - \underset{\tau,\gamma}{E}\left[\frac{2\alpha_V}{\Delta\vartheta^{\mathsf{T}}(z_k)\,\Delta\vartheta(z_k) + 1} \right. \\
& \left. \times \operatorname{tr}\left\{ \tilde{W}_{Vk}^{\mathsf{T}} \Delta\vartheta(z_k)\,\Delta\vartheta^{\mathsf{T}}(z_k)\,\tilde{W}_{Vk} \right\} \right] \\
& + \underset{\tau,\gamma}{E}\left[\frac{2\alpha_V}{\Delta\vartheta^{\mathsf{T}}(z_k)\,\Delta\vartheta(z_k) + 1} \operatorname{tr}\left\{ \Delta\varepsilon_{Vk}^{\mathsf{T}} \Delta\vartheta^{\mathsf{T}}(z_k)\,\tilde{W}_{Vk} \right\} \right] \\
& + \underset{\tau,\gamma}{E}\left[\frac{\alpha_V^2 \operatorname{tr}\left\{ \tilde{W}_{Vk}^{\mathsf{T}} \Delta\vartheta(z_k)\,\Delta\vartheta^{\mathsf{T}}(z_k)\,\Delta\vartheta(z_k)\,\Delta\vartheta^{\mathsf{T}}(z_k)\,\tilde{W}_{Vk} \right\}}{\left(\Delta\vartheta^{\mathsf{T}}(z_k)\,\Delta\vartheta(z_k) + 1 \right)^2} \right] \\
& + \underset{\tau,\gamma}{E}\left[\frac{2\alpha_V^2 \operatorname{tr}\left\{ \Delta\varepsilon_{Vk}^{\mathsf{T}} \Delta\vartheta^{\mathsf{T}}(z_k)\,\Delta\vartheta(z_k)\,\Delta\vartheta^{\mathsf{T}}(z_k)\,\tilde{W}_{Vk} \right\}}{\left(\Delta\vartheta^{\mathsf{T}}(z_k)\,\Delta\vartheta(z_k) + 1 \right)^2} \right] \\
& + \underset{\tau,\gamma}{E}\left[\frac{\alpha_V^2 \operatorname{tr}\left\{ \Delta\varepsilon_{Vk}^{\mathsf{T}} \Delta\vartheta^{\mathsf{T}}(z_k)\,\Delta\vartheta(z_k)\,\Delta\varepsilon_{Vk} \right\}}{\left(\Delta\vartheta^{\mathsf{T}}(z_k)\,\Delta\vartheta(z_k) + 1 \right)^2} \right] \\
& - \operatorname{tr}\left\{ \underset{\tau,\gamma}{E}\left(\tilde{W}_{Vk}^{\mathsf{T}} \tilde{W}_{Vk} \right) \right\} \\
\le {} & -\frac{\alpha_V (1 - 2a_V)\,\Delta\vartheta_{\min}^2 \left\| \underset{\tau,\gamma}{E}(\tilde{W}_{Vk}) \right\|^2}{(2 - a_V)\left(\Delta\vartheta_M^2 + 1 \right)} + \frac{\alpha_V (2a_V + 1)\,\Delta\varepsilon_{VM}^2}{(2 - a_V)\left(\Delta\vartheta_M^2 + 1 \right)}
\end{aligned} \tag{6.78}$$

where $0 < \Delta\vartheta_{\min} < \left\| \underset{\tau,\gamma}{E}\left(\Delta\vartheta(z_k) \right) \right\|$ is ensured by the PE condition described in Remark 6.2 and $\|\Delta\varepsilon_{Vk}\| \le \Delta\varepsilon_{VM}$ for a constant $\Delta\varepsilon_{VM}$ is ensured by the boundness of ε_{Vk}. Therefore, $\Delta L_V < 0$ if

$$\left\| \underset{\tau,\gamma}{E}\left(\tilde{W}_{Vk} \right) \right\| > \sqrt{\frac{2a_V + 1}{(1 - 2a_V)\,\Delta\vartheta_{\min}^2} \Delta\varepsilon_{VM}^2} \equiv B_{Wv}. \tag{6.79}$$

Using standard Lyapunov theory [Jagannathan (2006)], it can be concluded that ΔL_V is less than zero outside of a compact set rendering the critic NN weights estimation errors to be UUB in the mean.

6.7.2 *The proof of Theorem 6.4*

Consider the Lyapunov function candidate

$$L = L_{DN} + L_{uN} + L_{VN} + L_{CN} + L_{AN} + L_{BN} \tag{6.80}$$

where $L_{DN} = \underset{\tau,\gamma}{E}\left[z_k^{\mathsf{T}} z_k\right]$, L_{uN}, L_{VN}, L_{CN}, L_{AN} and L_{BN} are defined as

$$\begin{aligned}
&L_{uN} = \operatorname{tr}\left\{\underset{\tau,\gamma}{E}\left(\tilde{W}_{uk}^{\mathsf{T}}\Omega\tilde{W}_{uk}\right)\right\}; \quad L_{VN} = \operatorname{tr}\left\{\underset{\tau,\gamma}{E}\left(\tilde{W}_{Vk}^{\mathsf{T}}\Lambda\tilde{W}_{Vk}\right)\right\}\\
&L_{AN} = \left(\operatorname{tr}\left\{\underset{\tau,\gamma}{E}\left(\tilde{W}_{Vk}^{\mathsf{T}}\Gamma\tilde{W}_{Vk}\right)\right\}\right)^2; \; L_{CN} = \operatorname{tr}\left\{\underset{\tau,\gamma}{E}\left(e_{yk}^{\mathsf{T}}e_{yk}\right)\right\} + \operatorname{tr}\left\{\underset{\tau,\gamma}{E}\left(\tilde{W}_{Ck}^{\mathsf{T}}\mathrm{O}\tilde{W}_{Ck}\right)\right\}\\
&L_{BN} = \left(\operatorname{tr}\left\{\underset{\tau,\gamma}{E}\left(\tilde{W}_{Ck}^{\mathsf{T}}\Theta\tilde{W}_{Ck}\right)\right\}\right)^2
\end{aligned} \tag{6.81}$$

where $\Omega = \frac{24G_M^2\varphi_M^2\left(\varphi_M^2+1\right)}{\varphi_{\min}^2}\mathbf{I}$, $\Lambda = \frac{288\varphi_M^2\Xi^2\left(\Delta\vartheta_M^2+1\right)}{\varphi_{\min}^2\Delta\vartheta_{\min}^2} \times \left(G_M^2 + 12\frac{\psi_M^2\Delta\varepsilon_{VM}^2}{\Delta\vartheta_{\min}^2}\right)\mathbf{I}$, $\Theta = \left(\sqrt{\frac{24(\Xi\psi_M\varphi_M)^2}{\varphi_{\min}^2}}\right)\mathbf{I}$, $\mathrm{O} = 2\left[\Psi_M^2 + \frac{9\left(\varepsilon_{VM}'\psi_M\right)^2\varphi_M^2\Xi^2}{2\varphi_{\min}^2} + 6\left(\Xi\psi_M\right)^2\varphi_M^2\frac{\varphi_M^2\Delta\bar{\varepsilon}_{CM}^2}{\varphi_{\min}^2\Psi_{\min}^2}\right]\mathbf{I}$ and $\Gamma = \left(85\sqrt{\frac{(\Xi\psi_M\varphi_M)^2\left(\Delta\vartheta_M^2+1\right)^2}{\varphi_{\min}^2\Delta\vartheta_{\min}^4}}\right)\mathbf{I}$ are positive definite matrices, $\mathbf{I}$ is identity matrix, Ξ is defined as $\lambda_{\max}\left(R^{-1}\right)\vartheta_M' G_M$, and $\lambda_{\max}\left(R^{-1}\right)$ is the maximum singular value of R. The first difference of (6.80) is given by $\Delta L = \Delta L_{DN} + \Delta L_{uN} + \Delta L_{VN} + \Delta L_{CN} + \Delta L_{AN} + \Delta L_{BN}$.

Considering first difference $\Delta L_{DN} = \underset{\tau,\gamma}{E}\left[\left(z_{k+1}\right)^{\mathsf{T}} z_{k+1}\right] - \underset{\tau,\gamma}{E}\left[\left(z_k\right)^{\mathsf{T}} z_k\right]$, using the NNCS dynamics (6.3), and applying the Cauchy-Schwartz inequality reveals that the first difference becomes

$$\begin{aligned}
\Delta L_{DN} \le &-\left(1-2k^*\right)\left\|\underset{\tau,\gamma}{E}\left(y_k^o\right)\right\|^2 + 4\Psi_M^2\left\|\underset{\tau,\gamma}{E}\left(\tilde{W}_{Ck}^{\mathsf{T}}\right)\right\|^2\\
&+ 8G_M^2\varphi_M^2\left\|\underset{\tau,\gamma}{E}\left(\tilde{W}_{uk}\right)\right\|^2 + 8G_M^2\varepsilon_{uM}^2.
\end{aligned} \tag{6.82}$$

Next, first different L_u can be expressed as

$$\begin{aligned}
\Delta L_{uN} \le &-\frac{\left(3\alpha_u - 6\alpha_u^2\right)\varphi_{\min}^2\|\Omega\|}{\varphi_M^2+1}\left\|\underset{\tau,\gamma}{E}\left(\tilde{W}_{uk}\right)\right\|^2 + \frac{\left(\alpha_u^2+\alpha_u\right)\Xi^2\|\Omega\|}{2\left(\varphi^{\mathsf{T}}\left(z_k\right)\varphi\left(z_k\right)+1\right)}\left\|\underset{\tau,\gamma}{E}\left(\tilde{W}_{Vk}\right)\right\|^2\\
&+\frac{\left(2\alpha_u^2+\alpha_u\right)\left(\Xi\psi_M\right)^2\|\Omega\|}{4\left(\varphi_M^2+1\right)G_M^2}\left\|\underset{\tau,\gamma}{E}\left(\tilde{W}_{Ck}\right)\right\|^4\\
&+\frac{\left(2\alpha_u^2+\alpha_u\right)\left(\Xi\psi_M\right)^2\|\Omega\|}{4\left(\varphi_M^2+1\right)G_M^2}\left\|\underset{\tau,\gamma}{E}\left(\tilde{W}_{Vk}\right)\right\|^4\\
&+\frac{\left(2\alpha_u^2+\alpha_u\right)\left(\Xi\varepsilon_{VM}'\psi_M\right)^2\|\Omega\|}{2\left(\varphi_M^2+1\right)G_M^2}\left\|\underset{\tau,\gamma}{E}\left(\tilde{W}_{Ck}\right)\right\|^2 + \|\Omega\|\Delta\varepsilon_{eM}^2
\end{aligned} \tag{6.83}$$

where $0 < \varphi_{\min} < \left\| \underset{\tau,\gamma}{E} \left(\varphi\left(z_k\right)\right) \right\|$, is ensured by PE condition described in Remark 6.2, $\left\| \underset{\tau,\gamma}{E} \left(\Delta\varepsilon_{euk}\right) \right\|^2 = \frac{\alpha_u \left\| \underset{\tau,\gamma}{E} \left(\varepsilon_{ek}\right) \right\|^2}{\left(\varphi^{\mathrm{T}}(z_k)\varphi(z_k)+1\right)} \le \Delta\varepsilon_{eM}^2$, which is a bounded positive constant.

$$\Delta L_{AN} \le -\frac{\left(\alpha_V - 2a_V^2 - \frac{1}{12}\right)\Delta\vartheta_{\min}^2 \|\Gamma\|^2 \left\| \underset{\tau,\gamma}{E} \left(\tilde{W}_{Vk}\right) \right\|^4}{\left(\Delta\vartheta_M^2 + 1\right)} + \|\Gamma\|^2 \frac{4\Delta\varepsilon_{VM}^2 \left\| \underset{\tau,\gamma}{E} \left(\tilde{W}_{Vk}\right) \right\|^2}{3\left(\Delta\vartheta_M^2 + 1\right)} + \frac{4\|\Gamma\|^2 \Delta\varepsilon_{VM}^4}{9\left(\Delta\vartheta_M^2 + 1\right)^2}. \tag{6.84}$$

Next, the first difference of L_{BN} can be expressed as

$$\Delta L_{BN} \le -\|\Theta\|^2 \left(1 - 4\alpha_C^4 \frac{\Psi_M^4}{\Psi_{\min}^4}\right) \left\| \underset{\tau,\gamma}{E} \left(\tilde{W}_{Ck}\right) \right\|^4 + \frac{8\alpha_C^2 \Psi_M^2 \|\Theta\|^2}{\Psi_{\min}^2} \left\| \underset{\tau,\gamma}{E} \left(\tilde{W}\right)_{Ck} \right\|^2 \Delta\bar{\varepsilon}_{CM}^2 + \frac{4\|\Theta\|^2}{\Psi_{\min}^4} \Delta\bar{\varepsilon}_{CM}^4. \tag{6.85}$$

Next, using (6.78), (6.82), (6.83), (6.84) and (6.85) to form as:

$$\begin{aligned}
\Delta L \le & -(1-2k^*) \left\| \underset{\tau,\gamma}{E} (z_k) \right\|^2 - 288A\left(\alpha_V - 2a_V^2 - \frac{1}{12}\right) \left\| \underset{\tau,\gamma}{E} \left(\tilde{W}_{Vk}\right) \right\|^2 \\
& - \left(1-\alpha_C^2\right) \left\| \underset{\tau,\gamma}{E} (e_{zk}) \right\|^2 - 24G_M^2 \varphi_M^2 \left(3\alpha_u - 6\alpha_u^2 - \frac{1}{3}\right) \left\| \underset{\tau,\gamma}{E} \left(\tilde{W}_{uk}\right) \right\|^2 \\
& - 4M\left(1 - 4\alpha_C^2 \frac{\Psi_M^2}{\Psi_{\min}^2}\right) \left\| \underset{\tau,\gamma}{E} \left(\tilde{W}_{Ck}\right) \right\|^2 - 4\alpha_C^2 \frac{\Psi_M^2}{\Psi_{\min}^2} \left\| \underset{\tau,\gamma}{E} \left(\tilde{W}_{Ck}\right) \right\|^2 \\
& - \frac{216\left(\alpha_V - 2a_V^2 - \frac{1}{9}\right)\left(\Xi\psi_M\varphi_M\right)^2}{\varphi_{\min}^2} \left\| \underset{\tau,\gamma}{E} \left(\tilde{W}_{Vk}\right) \right\|^4 \\
& - \left(1 - 8\alpha_C^4 \frac{\Psi_M^4}{\Psi_{\min}^4}\right) \frac{6\left(\Xi\psi_M\varphi_M\right)^2}{\varphi_{\min}^2} \left\| \underset{\tau,\gamma}{E} \left(\tilde{W}_{Ck}\right) \right\|^4 + \varepsilon_{TM}
\end{aligned} \tag{6.86}$$

where $\Xi = \left(\lambda_{\max}\left(R_y^{-1}\right)\vartheta'_M G_M\right)$, $\eta = \left[\Psi_M^2 + \frac{9\left(\varepsilon'_{VM}\psi_M\right)^2 \varphi_M^2 \Xi^2}{2\varphi_{\min}^2} + \frac{6(\Xi\psi_M)^2 \varphi_M^2 \Delta\bar{\varepsilon}_{CM}^2}{\varphi_{\min}^2 \Psi_{\min}^2}\right]$ and $\rho = \frac{\varphi_M^2 \Xi^2}{\varphi_{\min}^2}\left(G_M^2 + 12\frac{\psi_M^2 \Delta\varepsilon_{VM}^2}{\Delta\vartheta_{\min}^2}\right)$ are positive constant and ε_{TM} is $\varepsilon_{TM} = 8G_M^2 \varepsilon_{uM}^2 + 16\frac{\Psi_M^2}{\Psi_{\min}^2}\Delta\bar{\varepsilon}_{CM}^2 + \frac{24\left(\varphi_M^2+1\right)G_M^2\varphi_M^2}{\varphi_{\min}^2}\Delta\varepsilon_{eM}^2 + \frac{\left(23G_M\varphi_M\Xi\right)^2}{\varphi_{\min}^2\Delta\vartheta_{\min}^2}\Delta\varepsilon_{VM}^2 + \frac{72\left(\varepsilon'_{VM}\psi_M\right)^2\varphi_M^2\Xi^2}{\varphi_{\min}^2}\Delta\bar{\varepsilon}_{CM}^2 + \frac{48\left(\Xi\psi_M\varphi_M\right)^2}{\varphi_{\min}^2\Psi_{\min}^4}\Delta\bar{\varepsilon}_{CM}^4 + \frac{96\left(\Xi\psi_M\varphi_M\right)^2\Delta\varepsilon_{VM}^4}{\varphi_{\min}^2\Delta\vartheta_{\min}^2\left(\Delta\vartheta_M^2+1\right)} + \frac{\left(68\Xi\psi_M\varphi_M\right)^2\Delta\varepsilon_{VM}^4}{\varphi_{\min}^2\Delta\vartheta_{\min}^4} + \frac{96\left(\Xi\psi_M\right)^2\varphi_M^2\Delta\bar{\varepsilon}_{CM}^4}{\varphi_{\min}^2\Psi_{\min}^4}$.

Therefore, ΔL is less than zero when the following inequalities hold

$$\left\| \underset{\tau,\gamma}{E}\left(e_{zk}\right) \right\| > \sqrt{\frac{\varepsilon_{TM}}{\left(1-\alpha_C^2\right)}} \equiv b_{ez},$$

OR

$$\left\| \underset{\tau,\gamma}{E}\left(\tilde{W}_{Ck}\right) \right\| > \max\left\{ \sqrt{\frac{\varepsilon_{TM}\Psi_{\min}^2}{4\left(\Psi_{\min}^2 - 4\alpha_C^2\Psi_M^2\right)M}}, \sqrt[4]{\frac{\varepsilon_{TM}\Psi_{\min}^4\varphi_{\min}^2}{6\left(\Psi_{\min}^4 - 8\alpha_C^4\Psi_M^4\right)\left(\Xi\psi_M\varphi_M\right)^2}} \right\}$$

$$\equiv b_{WC},$$

OR

$$\left\| \underset{\tau,\gamma}{E}\left(\tilde{W}_{Vk}\right) \right\| > \max\left\{ \sqrt{\frac{\varepsilon_{TM}}{288A\left(\alpha_V - 2a_V^2 - \frac{1}{12}\right)}}, \sqrt[4]{\frac{\varphi_{\min}^2}{216\left(\alpha_V - 2a_V^2 - \frac{1}{9}\right)\left(\Xi\psi_M\varphi_M\right)^2}} \right\}$$

$$\equiv b_{WV},$$

OR

$$\left\| \underset{\tau,\gamma}{E}\left(\tilde{W}_{uk}\right) \right\| > \sqrt{\frac{\varepsilon_{TM}}{8G_M^2\varphi_M^2\left(9\alpha_u - 18\alpha_u^2 - 1\right)}} \equiv b_{Wu},$$

OR

$$\left\| \underset{\tau,\gamma}{E}\left(z_k\right) \right\| > \sqrt{\frac{\varepsilon_{TM}}{\left(1-2k^*\right)}} \equiv b_z$$

provided the tuning gains are selected according to (6.18) (6.49), and (6.52) for the NNCS (6.3). Using the standard Lyapunov extension [Jagannathan (2006)], the system outputs, NN identifier and weight estimation errors, critic and action NN estimation errors are UUB in the mean while the system outputs never leave the compact set.

Next using (6.44) and (6.46), we have

$$\underset{\tau,\gamma}{E}\left[\hat{u}\left(z_k\right) - u^*\left(z_k\right)\right] = -\underset{\tau,\gamma}{E}\left[\tilde{W}_u^{\mathsf{T}}\varphi\left(z_k\right) + \varepsilon_{uk}\right].$$

When $k \to \infty$, the upper bound of $\underset{\tau,\gamma}{E}\left[\hat{u}\left(z_k\right) - u^*(z_k)\right]$ can be represented as

$$\begin{aligned} \left\| \underset{\tau,\gamma}{E}\left[\hat{u}\left(z_k\right)\right] - \underset{\tau,\gamma}{E}\left[u^*\left(z_k\right)\right] \right\| &\le \left\| \underset{\tau,\gamma}{E}\left[\tilde{W}_u^{\mathsf{T}}\varphi\left(z_k\right)\right] \right\| + \left\| \underset{\tau,\gamma}{E}\left(\varepsilon_{uk}\right) \right\| \\ &\le \left\| \underset{\tau,\gamma}{E}\left[\tilde{W}_u^{\mathsf{T}}\varphi\left(z_k\right)\right] \right\| + \varepsilon_{uM} \\ &\le b_{Wu} + \varepsilon_{uM} \equiv \varepsilon_{bu}. \end{aligned} \tag{6.87}$$

Now, if the NN identifier, critic and action NN approximation errors are neglected as in Enns and Si (2003) and when $k \to \infty$, ε_{TM} in (6.87) and ε_{bu} in (6.87) will become zero in the mean. In this case, it can be shown that the NN-based identification, action NN and critic NN estimation errors converge to zero asymptotically in the mean, i.e., $\underset{\tau,\gamma}{E}\left[\hat{u}\left(z_k\right)\right] \to \underset{\tau,\gamma}{E}\left[u^*\left(z_k\right)\right]$.

Acknowledgement

The authors acknowledge partial support from the National Science Foundation (NSF) under grant ECCS-1128281 and Intelligent Systems Center (ISC) at the Missouri University of Science and Technology. The authors thank the ICICIP Book Editors for inviting the authors and providing constructive comments.

References

Al-Tamimi, A., Lewis, F. L. and Abu-Khalaf, M. (2008). Discrete-time nonlinear HJB solution using approximate dynamic programming: Convergence proof, *IEEE Transactions on Systems, Man, and Cybernetics, Part B: Cybernetics* **38**, 4, pp. 943–949.

Anta, A. and Tabuada, P. (2010). To sample or not to sample: Self-triggered control for nonlinear systems, *IEEE Transactions on Automatic Control* **55**, 9, pp. 2030–2042.

Astrom, K. J. and Bernhardsson, B. M. (2002). Comparison of Riemann and Lebesque sampling for first order stochastic systems, in *Proceeding of the 41st IEEE Conference on Decision and Control* (Las Vegas, USA), pp. 2011–2016.

Bertsekas, D. P. and Tsitsiklis J. (1996). *Neuro-Dynamic Programming* (Athena Scientfic).

Chen, H. F. and Guo, L. (1991). *Identification and Stochastic Adaptive Control* (Birkhäuser).

Chen, T., Lewis, F. L. and Abu-khalaf, M. (2007). Fixed-final-time-constrained optimal control of nonlinear system using neural network HJB approach, *IEEE Transactions on Neural Networks* **18**, 6, pp. 1725–1737.

Dankert, J., Yang, L. and Si, J. (2006). A performance gradient perspective on approximate dynamic programming and its application to partially observable markov decision process, in *IEEE International Symposium on Intelligent Control* (Munich, Germany), pp. 458–463.

Dierks, T. and Jagannathan, S. (2012). Online optimal control of affine nonlinear discrete-time systems with unknown internal dynamics by using time-based policy update, *IEEE Transactions on Neural Networks and Learning Systems* **23**, 7, pp. 1118–1129.

Enns, R. and Si, J. (2003). Helicopter trimming and tracking control using direct neural dynamic programming, *IEEE Transactions on Neural Networks* **14**, 4, pp. 929–939.

Goldsmith, A. (2005). *Wireless Communication* (Cambridge University Press).

Hu, S. and Zhu, Q. (2003). Stochastic optimal control and analysis of stability of networked control system with long delay, *Automatica* **39**, 11, pp. 1877–1884.

Jagannathan, S. (2006). *Neural Network Control of Nonlinear Discrete-Time Systems* (CRC Press).

Lewis, F. L., Jagannathan, S. and Yesilderik, A. (1999). *Neural Network Control of Robot Manipulators and Nonlinear Systems* (Taylor and Francis).

Lewis, F. L. and Syrmos, V. L. (1995). *Optimal Control*, 2nd edn. (Wiley).

Lewis, F. L. and Vrabie, D. (2009). Reinforcement learning and adaptive dynamic programming for feedback control, *IEEE Circuits and Systems Magazine* **9**, 3, pp. 32–50.

Luck, R. and Ray, A. (1990). An observer-based compensator for distributed delays, *Automatica* **26**, 5, pp. 903–908.

Mahmoud, M. S. and Ismail, A. (2005). New results on delay-dependent control of time-delay systems, *IEEE Transactions on Automatic Control* **50**, 1, pp. 95–100.

Narendra, K. S. and Annaswamy, A. M. (1989). *Stable Adaptive Control* (Prentice Hall).

Sahoo, A., Xu, H. and Jagannathan, S. (2013). Neural network-based adaptive event-triggered control of affine nonlinear discrete time systems with unknown internal dynamics, in *American Control Conference* (Washington, USA), pp. 6418–6423.

Stallings, W. (2004). *Wireless Communications and Networks*, 2nd edn. (Prentice Hall).

Tabbara, M., Nesic, D. and Martins, N. C. (2008). A linear quadratic Gaussian framework for optimal networked control system design, in *American Control Conference* (Washington, USA), pp. 3804–3809.

Tabuada, P. (2007). Event-triggered real-time scheduling of stabilizing control tasks, *IEEE Transactions on Automatic Control* **52**, 9, pp. 1680–1685.

Tipsuwan, Y. and Chow, M. Y. (2003). Control methodologies in networked control systems, *Control Engineering Practice* **11**, 10, pp. 1099–1111.

van de Wouw, N., Nesic, D. and Heemels, W. (2012). A discrete-time framework for stability analysis of nonlinear networked control systems, *Automatica* **48**, 6, pp. 1144–1153.

Walsh, G. C., Beldiman, O. and Bushnell, L. G. (2001). Asymptotic behavior of nonlinear networked control systems, *IEEE Transactions on Automatic Control* **46**, 7, pp. 1093–1097.

Wang, F. Y., Jin, N., Liu, D. and Wei, Q. L. (2011). Adaptive dynamic programming for finite horizon optimal control of discrete-time nonlinear systems with ε-error bound, *IEEE Transactions on Neural Networks* **22**, 1, pp. 24–36.

Werbos, P. J. (1990). A menu of designs for reinforcement learning over time, in W. T. Miller, P. J. Werbos, and R. S. Sutton (eds.), *Neural Networks for Control* (MIT press, Cambridge), pp. 67–95.

Xu, H. and Jagannathan, S. (2013a). Neural network based finite horizon stochastic optimal controller design for nonlinear networked control systems, in *International Joint Conference on Neural Networks* (Dallas, USA), pp. 1–7.

Xu, H. and Jagannathan, S. (2013b). Stochastic optimal controller design for uncertain nonlinear networked control system via neuro dynamic programming, *IEEE Transactions on Neural Networks and Learning Systems* **24**, 3, pp. 471–484.

Xu, H., Jagannathan, S. and Lewis, F. L. (2012). Stochastic optimal control of unknown linear networked control system in presence of random delays and packet losses, *Automatica* **48**, 6, pp. 1017–1030.

Zhang, H., Luo, Y. and Liu, D. (2009). Neural-network-based near-optimal control for a class of discrete-time affine nonlinear systems with control constraints, *IEEE Transactions on Neural Networks* **20**, 9, pp. 1490–1503.

Chapter 7

Nonlinear and Robust Model Predictive Control of Systems with Unmodeled Dynamics Based on Supervised Learning and Neurodynamic Optimization

Zheng Yan and Jun Wang

Abstract Model predictive control is an optimization-based control strategy which has achieved enormous successes in numerous real-world applications. MPC generates control signals by means of real-time optimization of a performance index over a finite moving horizon of predicted future, subject to system constraints. A major challenge of the MPC research and development lies in the realization of nonlinear and robust MPC approaches, especially to cases where unmodeled dynamics exist. This chapter presents novel MPC approaches to nonlinear systems with undoeled dynamics based on neural networks. At first, a nonlinear system with unmodeled dynamics is decomposed by means of Jacobian linearization to an affine part and a higher-order unknown term. The linearization residues, together with the unmodeled dynamics, are then modeled by using a feedforward neural network called the Extreme Learning Machine via supervised learning. When the controlled system is affected by bounded additive disturbances, the minimax methodology is exploited to achieve robustness. The nonlinear and robust MPC problems are formulated as constrained convex optimization problems and iteratively solved by applying neurodynamic optimization approaches. The applied neurodynamic optimization approaches can compute the optimal control signals in real-time, which shed a light for real-time implementability of MPC technology. Simulation results are provided to substantiate the effectiveness and characteristics of the proposed approaches.

7.1 Introduction

Model predictive control (MPC), also known as receding horizon control, is a model-based optimal control technique. MPC generates control signals by solving a finite-horizon constrained optimization problem in real time, using the current state as an initial state. The optimization problem is iteratively formulated as solved in receding horizon framework. Compared with other control techniques, MPC has distinctive advantages; e.g., it handles multivariable control problems naturally, it takes account of input and output constraints, and it adapts structural changes [Mayne *et al.* (2000)].

One important issue for MPC synthesis is the computational efficiency of real-time optimization, which determines whether an MPC approach can be effectively implemented. For nonlinear control problems, a nonlinear system model generally results in nonconvex optimization problems. However, tackling nonconvex optimization in real time is computational demanding. Reducing the computational burden of nonlinear MPC has attracted increasing interest in the control community. Recent developments in neural computation have brought new opportunities in this direction. Several approaches have been proposed in recent years. The first method is MPC using neural models with online linearization [Colin *et al.* (2007); Lawrynczuk and Tatjewski (2010); Mu *et al.* (2005); Tatjewski (2003)]. Linearization enables MPC to be synthesized by quadratic programming, which can be solved very efficiently. The second method is to use neural networks to approximate the time-varying feedback control law. The control signal is calculated explicitly without any on-line optimization [Lawrynczuk (2010)]. The third method is to use neural networks to directly compute control signals without any optimization [Akesson and Toivonen (2006); Cavagnari *et al.* (1999)]. One common limitation of most existing methods is that resulting control signals are inherent suboptimal. There is always an incentive to further improve the optimality to seek for better control performances.

In the past two decades, neurodynamic optimization based on recurrent neural networks (RNNs) emerged as promising computational tools for real time optimization. The essence of neurodynamic optimization lies in its inherent nature of parallel and distributed information processing and the availability of hardware implementation. Based on the duality and projection methods, various neurodynamic optimization models have been developed for convex optimization [Hu and Wang (2007, 2008); Liu and Wang (2006, 2008); Xia and Wang (2004a); Xia *et al.* (2005, 2008)], pseudoconvex optimization [Guo *et al.* (2011); Liu *et al.* (2012)], and nonsmooth optimization [Cheng *et al.* (2011); Forti *et al.* (2004); Liu and Wang (2011, 2013)]. These neural networks have shown good performance in terms of guaranteed optimality, extended applicability, improved convergence properties, and reduced model complexity. The recent advance of neurodynamic optimization offers a new paradigm for synthesizing MPC.

Werbos *et al.* (1992) firstly proposed to incorporate neural networks into MPC design. Their work inspired many researchers to develop MPC approaches based on neural networks in the last two decades. Generally speaking, the applications of neural networks in MPC can be classified into three categories: first, using neural networks for system modeling and identification [Draeger *et al.* (1995); Huang and Lewis (2003); Liu and Kadirkamanathan (1998); Piche *et al.* (2000); Song *et al.* (2007)]; second, using neural networks for real time optimization problems [Cheng *et al.* (2007); Pan and Wang (2012); Wang and Wan (2001); Yan and Wang (2012a,b)]; third, using neural networks to approximate MPC laws off-line [Akesson and Toivonen (2006)]. The use of neural networks manifested important progress in MPC research.

As MPC is a model-based control technique, the control performance is largely determined by the prediction model. Two main modeling approaches are commonly used for nonlinear MPC design. One is to use a full nonlinear model for prediction without simplifications. The full model provides better accuracy, but it generally results in a nonconvex optimization. An alternative is to find linear approximations to original plants. A benefit of linearization is that the MPC optimization task becomes a quadratic programming. As linearizaiton approximates a nonlinear model locally, it would result in large prediction errors, particularly in the presence of unmodeled dynamics and model uncertainties.

Robustness against uncertainties is another important but challenging issue in MPC. Two types of uncertainties commonly present, one is parametric uncertainties and the other one is external disturbances. There are three typical approaches for designing MPC algorithms with robustness properties. The first method is to minimize a nominal performance index while imposing tightened state and terminal constraints, provided that each admissible disturbance is fulfilled [Chisci *et al.* (2001); Marruedo *et al.* (2002)]. The second method is to solve a minimax optimization problem where the performance index under the worst-case effect of uncertainties is minimized [Bemporad *et al.* (2003); Kerrigan and Maciejowski (2004); Lee and Yu (1997); Magni *et al.* (2003); Raimondo *et al.* (2009); Scokaert and Mayne (1998)]. Specifically, in an open-loop formulation, a sequence of control actions is obtained. In a closed-loop formulation, a sequence of feedback control policies is computed. The third method is to calculate disturbance invariant sets to ensure that the evolution of closed-loop trajectory lie in an invariant tube [Langson *et al.* (2004); Mayne *et al.* (2005); Pin *et al.* (2009)].

In this chapter, MPC approaches to nonlinear systems with unmodeled dynamic are presented. By using Jacobian linearization, a nonlinear model can be decomposed to an affine model, an unknown high-order term, and umodeled dynamics. Consequently, the nonlinear MPC problem is formulated as a sequential quadratic programming with unknown parameters. When bounded additive uncertainties are explicitly taken into account, the robust MPC problem is formulated as a sequential minimax optimization problem with unknown parameters, which is further reformulated as a convex optimization problem with nonlinear inequality constraints. The unknown terms are then modeled by using a feedforward neural network called the Extreme Learning Machine (ELM) [Huang *et al.* (2005)] via supervised learning. To obtain optimal control signals, a simplified dual neural network is applied for solving the quadratic programming, and a two-layer recurrent neural network is applied for solving the nonlinear convex optimization problem. The contribution of this chapter is twofold. First, computational efficient algorithms are provided. Via online Jacobian linearization, nonlinear and robust MPC problems are formulated as convex optimization problems which can be solved by using neurodynamic optimization models in real time. Moreover, modeling errors resulting from linearization residual and unmodeled dynamics are numerically compensated by using an ELM. As a result, the computational burden is reduced, whereas the optimality is less compromised. Second, the presented approaches are viable for solving large-scale multi-variable control problems. In view of the inherent parallelism of neural networks, large scale optimization problems would not degrade the real time computational efficiency. In contrast, most existing MPC algorithms are incompetent for complex large-scale industrial processes.

The rest of this chapter is organized as follows. In Section 7.2, nonlinear and robust MPC problems are formulated. In Section 7.3, neural networks are applied for modeling the unknown parameters and solving the optimization problems. In Section 7.4, simulation results are provided. Finally, Section 7.5 concludes this chapter.

7.2 Problem Formulation

Consider a discrete-time nonlinear system with unmodeled dynamics

$$\begin{aligned} x(k+1) &= f\left(x(k), u(k)\right) + w(x(k), u(k)) \\ y(k) &= Cx(k) \end{aligned} \tag{7.1}$$

subject to input and output constraints

$$u_{\min} \leq u(k) \leq u_{\max}, \; y_{\min} \leq y(k) \leq y_{\max} \tag{7.2}$$

where $x(k) \in \Re^n$ is the state vector, $u(k) \in \Re^m$ is the control input vector, $y(k) \in \Re^p$ is the output vector, $f(\cdot)$ is a nonlinear function, and $w(x(k), u(k))$ represents the unmodeled dynamics of the system, which can be unindentified structures or unknown parameters. $u_{\min} \leq u_{\max}$ and $y_{\min} \leq y_{\max}$ are lower and upper bounds. The following assumptions will be used in this chapter.

Assumption 7.1. All state variables are available at each time k.

Assumption 7.2. $f(\cdot)$ is Lipschitz continuous and differentiable for all (x, u), with $f(0,0) = 0$.

Assumption 7.3. $w(x, u)$ is continuous at the origin with $w(0,0) = 0$.

MPC of (7.1) can be formulated as a constrained finite-horizon optimal control problem at each time instant k:

$$\begin{aligned}\min_{\Delta u(k)} J(\Delta u(k), x(k), y(k)) = & \sum_{j=1}^{N} \|y(k+j|k) - r(k+j)\|_{Q_j}^2 \\ & + \sum_{j=0}^{N_u-1} \|\Delta u(k+j|k)\|_{R_j}^2 + F(x(k+N|k)) \\ \text{subject to} \quad & \Delta u_{\min} \le \Delta u(k+j|k) \le \Delta u_{\max},\ j = 0, 1, \ldots, N_u - 1; \\ & u_{\min} \le u(k+j|k) \le u_{\max},\ j = 0, 1, \ldots, N_u - 1; \\ & y_{\min} \le y(k+j|k) \le y_{\max},\ j = 1, 2, \ldots, N; \end{aligned} \tag{7.3}$$

where N and N_u are respectively prediction horizon ($1 \le N$) and control horizon ($0 < N_u \le N$), $r(k+j)$ denotes the reference vector, $y(k+j|k)$ denotes the predicted output vector, $\Delta u(k+j|k)$ denotes the control input increment vector defined as $\Delta u(k+j|k) = u(k+j|k) - u(k-1+j|k)$, $x(k+N|k)$ denotes the predicted terminal state within the prediction horizon, and F is a terminal cost for stability purpose.

7.2.1 *Nominal stability*

Research on stability of MPC of constrained nonlinear systems has reached a relative mature stage. Essential ingredients that ensure nominal closed loop stability include the terminal cost $F(\cdot)$, a terminal constraint $x(k+N|k) \in \Omega$ where Ω is a region in the neighborhood of the origin, and a local stabilizing control law $\kappa(\cdot)$. $F(\cdot)$ and Ω can appear explicitly in MPC formulation with flexible choices, whereas $\kappa(\cdot)$ is merely implicit. Several approaches to design $F(\cdot)$ and Ω have been proposed [Limon *et al.* (2006)]. Based on the theorems in Lazar *et al.* (2008); Limon *et al.* (2006); Mayne *et al.* (2000), the nominal stability conditions can be stated as follows:

Suppose that for system (7.1), there exist a locally asymptotically stabilizing controller $u = \kappa(x)$ and a positive definite function $F(x)$ such that $F(\kappa(x)) - F(x) \le (x^\mathsf{T} Qx + u^\mathsf{T} Ru)$ for all $x \in \Omega$, and the constrained finite horizon optimal control problem (7.3) admits a feasible solution at time $k = 0$. Then the closed-loop system is nominally asymptotically stable. Ω is called the terminal region. Let $\Xi \subseteq \Omega$ be the set of all initial states for which the constrained optimal control problem has a feasible solution at time $k = 0$, Ξ defines a domain of attraction for the closed-loop system. Limón *et al.* (2006) shows that when the terminal constraint is removed, the stability can be preserved by weighting the terminal cost F. In other words, suitable choice of F implies the fulfillment of the terminal constraint $x(k+N|k) \in \Omega$. In this chapter, the terminal cost is designed as $F(x) = x^\mathsf{T} Px$ where P is a positive semi-definite weighting matrix.

7.2.2 *Quadratic programming formulation*

If the nonlinear model (7.1) is directly used for prediction, (7.3) will generally become a nonconvex optimization. By using Taylor series, the nonlinear function

f can be decomposed about an operating point $[x_o(k), u_o(k)]$ into an affine system plus an unknown term at each time instant k:

$$\begin{aligned} x(k+1) &= f(x_o(k), u_o(k)) + \frac{\partial f(x,u)}{\partial x}|_{x_o(k),u_o(k)}(x(k) - x_o(k)) \\ &\quad + \frac{\partial f(x,u)}{\partial u}|_{x_o(k),u_o(k)}(u(k) - u_o(k)) + \varepsilon_f(k) + w(x(k), u(k)) \end{aligned} \tag{7.4}$$

where $\partial f/\partial x$ and $\partial f/\partial u$ are Jacobians of f with respect to x and u, respectively; and $\varepsilon_f(k)$ is the higher-order residual corresponding to Taylor series. By denoting $A(k) = \partial f(x_o(k), u_o(k))/\partial x$, $B(k) = \partial f(x_o(k), u_o(k))/\partial u$, $\delta x(k) = x(k) - x_o(k)$, $\delta u(k) = u(k) - u_o(k)$, and $\delta y(k) = y(k) - Cx_o(k)$, system (7.1) can be reformulated as:

$$\begin{aligned} \delta x(k+1) &= A(k)\delta x(k) + B(k)\delta u(k) + \varepsilon(k) \\ \delta y(k) &= C\delta x(k) \end{aligned} \tag{7.5}$$

where $\varepsilon(k)$ denotes the combination of linearization residue and unmodeled dynamics. Define the following vectors:

$$\begin{aligned} \delta\bar{y}(k) &= [\delta y(k+1|k) \ldots \delta y(k+N|k)]^{\mathsf{T}} \in \Re^{Np} \\ \delta\bar{u}(k) &= [\delta u(k|k) \ldots \delta u(k+N_u-1|k)]^{\mathsf{T}} \in \Re^{N_u m} \\ \bar{\delta r}(k) &= [r(k+1) \ldots r(k+N)]^{\mathsf{T}} \in \Re^{Np} \\ \delta\bar{x}(k) &= [\delta x(k+1|k) \ldots \delta x(k+N|k)]^{\mathsf{T}} \in \Re^{Nn} \\ \Delta\bar{u}(k) &= [\Delta u(k|k) \ldots \Delta u(k+N_u-1|k)]^{\mathsf{T}} \in \Re^{N_u m} \\ \bar{\varepsilon}(k) &= [\varepsilon(k|k) \ldots \varepsilon(k+N-1|k)]^{\mathsf{T}} \in \Re^{Nn}. \end{aligned} \tag{7.6}$$

As the linearization is performed once for a given sampling time instant k, $A(k)$ and $B(k)$ are written as A and B for brevity. Using (7.5) as the prediction model, it follows

$$\begin{aligned} \delta y(k+1|k) &= CA\delta x(k|k) + CB\delta u(k-1) + CB\Delta u(k|k) + C\varepsilon(k), \\ \delta y(k+2|k) &= CA^2\delta x(k|k) + C(A+I)B\delta u(k-1) + C(A+I)B\Delta u(k|k) \\ &\quad + CB\Delta u(k+1|k) + CA\varepsilon(k) + C\varepsilon(k+1), \\ &\vdots \\ \delta y(k+N|k) &= CA^N\delta x(k|k) + C(A^{N-1} + A^{N-2} + \cdots + I)B\delta u(k-1) \\ &\quad + C(A^{N-1} + A^{N-2} + \cdots + I)B\Delta u(k|k) \\ &\quad + C(A^{N-2} + A^{N-3} + \cdots + I)B\Delta u(k+1|k) \\ &\quad + \cdots + C(A^{N-N_u} + A^{N-N_u-1} + \cdots + I) + B\Delta u(k+N_u-1|k) \\ &\quad + CA^{N-1}\varepsilon(k) + CA^{N-2}\varepsilon(k+1) + \cdots + C\varepsilon(k+N-1). \end{aligned}$$

The predicted output $\delta\bar{y}(k)$ can be written in a compact form as follows:

$$\delta\bar{y}(k) = S\delta x(k) + V\delta u(k-1) + M\Delta\bar{u}(k) + H\bar{\varepsilon}(k) \tag{7.7}$$

where

$$S = \begin{bmatrix} CA \\ CA^2 \\ \vdots \\ CA^N \end{bmatrix} \in \Re^{Nn\times n}, \quad V = \begin{bmatrix} CB \\ C(A+I)B \\ \vdots \\ C(A^{N-1}+\ldots+I)B \end{bmatrix} \in \Re^{Nn\times m},$$

$$M = \begin{bmatrix} CB & \cdots & 0 \\ C(A+I)B & \cdots & 0 \\ \vdots & \ddots & \vdots \\ C(A^{N-1}+\ldots+I)B & \cdots & C(A^{N-N_u}+\ldots+I)B \end{bmatrix} \in \Re^{Nn\times N_u m},$$

$$H = \begin{bmatrix} I & 0 & \cdots & 0 \\ CA & I & \cdots & 0 \\ \vdots & \vdots & \ddots & \vdots \\ CA^{N-1} & CA^{N-2} & \cdots & C \end{bmatrix} \in \Re^{Nn\times Nn}.$$

The predicted terminal state within the prediction horizon N can be expressed as

$$\delta x(k+N|k) = \tilde{S}\delta x(k) + \tilde{V}\delta u(k-1) + \tilde{M}\Delta\bar{u}(k) + \tilde{H}\bar{\varepsilon}(k) \tag{7.8}$$

where

$$\begin{aligned} &\tilde{S} = A^N \in \Re^{n\times n}, \quad \tilde{V} = (A^{N-1}+\cdots+I)B \in \Re^{n\times m} \\ &\tilde{M} = [(A^{N-1}+\cdots+I)B \cdots (A^{N-N_u}+\cdots+I)B] \in \Re^{n\times N_u m} \\ &\tilde{H} = [A^{N-1}\ A^{N-2}\cdots I] \in \Re^{n\times Nn}. \end{aligned}$$

Therefore, in light of (7.7) and (7.8), the optimization problem (7.3) becomes:

$$\begin{aligned} \min_{\Delta\bar{u}(k)} \quad & \|S\delta x(k) + V\delta u(k-1) + M\Delta\bar{u}(k) + H\bar{\varepsilon}(k)\|_Q^2 + \|\Delta\bar{u}(k)\|_R^2 \\ & + \|\tilde{S}\delta x(k) + \tilde{V}\delta u(k-1) + \tilde{M}\Delta\bar{u}(k) + \tilde{H}\bar{\varepsilon}(k)\|_P^2 \end{aligned} \tag{7.9}$$

$$\begin{aligned} \text{s.t.} \quad & \Delta\bar{u}_{\min} \le \Delta\bar{u}(k) \le \Delta\bar{u}_{\max}, \\ & \delta\bar{u}_{\min} \le \delta\bar{u}(k-1) + \tilde{I}\Delta\bar{u}(k) \le \delta\bar{u}_{\max}, \\ & \delta\bar{y}_{\min} \le S\delta x(k) + V\delta u(k-1) + M\Delta\bar{u}(k) + H\bar{\varepsilon}(k) \le \delta\bar{y}_{\max} \end{aligned} \tag{7.10}$$

where

$$\begin{aligned} &\delta u_{\min} = u_{\min} - u_o(k), \quad \delta u_{\max} = u_{\max} - u_o(k), \\ &\delta y_{\min} = y_{\min} - y_o(k), \quad \delta y_{\max} = y_{\max} - y_o(k), \end{aligned}$$

$$\tilde{I} = \begin{bmatrix} I & 0 & \dots & 0 \\ I & I & \dots & 0 \\ \vdots & \vdots & \ddots & \vdots \\ I & I & \dots & I \end{bmatrix} \in \Re^{N_u m \times N_u m}. \tag{7.11}$$

Denote $\Delta\bar{u} = \Delta\bar{u}(k)$ for simplicity, problem (7.10) can be rewritten as a quadratic programming (QP) as follows

$$\min \frac{1}{2} \Delta\bar{u}^{\mathsf{T}} W \Delta\bar{u} + c^{\mathsf{T}} \Delta\bar{u}, \ \text{ s.t. } l \le E\Delta\bar{u} \le h \tag{7.12}$$

where

$$\begin{aligned} W &= 2(M^{\mathsf{T}} Q M + \tilde{M}^{\mathsf{T}} P \tilde{M} + R) \in \Re^{N_u m \times N_u m}, \\ c &= 2M^{\mathsf{T}} Q(S\delta x(k) + V\delta u(k-1) + H\bar{\varepsilon}(k)) \\ &\quad + 2\tilde{M}^{\mathsf{T}} P(\tilde{S}\delta x(k) + \tilde{V}\delta u(k-1) + \tilde{H}\bar{\varepsilon}(k)) \in \Re^{N_u m}, \\ E &= \begin{bmatrix} -\tilde{I} & \tilde{I} & -M & M & I \end{bmatrix}^{\mathsf{T}} \in \Re^{(3N_u m + 2Np) \times N_u m}, \\ b &= \begin{bmatrix} -\delta\bar{u}_{\min} + \delta\bar{u}(k-1) \\ \delta\bar{u}_{\max} - \delta\bar{u}(k-1) \\ -\delta\bar{y}_{\min} + S\delta x(k) + V\delta u(k-1) + H\bar{\varepsilon}(k) \\ \delta\bar{y}_{\max} - S\delta x(k) - V\delta u(k-1) - H\bar{\varepsilon}(k) \end{bmatrix} \in \Re^{2N_u m + 2Np}, \\ l &= \begin{bmatrix} -\infty \\ \Delta\bar{u}_{\min} \end{bmatrix} \in \Re^{3N_u m + 2Np}, \\ h &= \begin{bmatrix} b \\ \Delta\bar{u}_{\max} \end{bmatrix} \in \Re^{3N_u m + 2Np}. \end{aligned}$$

Since W is positive definite, the objective function in (7.12) is strictly convex. As the feasible region defined by the constraints is a closed convex set, (7.12) is shown to be a strictly convex quadratic programming. The solution to the QP (7.12) provides the optimal control input.

7.2.3 *Minimax problem formulation*

Consider the nonlinear model (7.1) effected by additive uncertainties

$$\begin{aligned} x(k+1) &= f(x(k), u(k)) + w(x(k), u(k)) + v(k) \\ y(k) &= Cx(k) \end{aligned} \tag{7.13}$$

where $v \in \Re^n$ denotes a vector of additive bounded uncertainties. In this chapter, v satisfies the following assumption.

Assumption 7.4. The additive uncertainty v is limited in a compact set which contains the origin with known bound, i.e., $v \in \mathcal{V}$, $\|v\|_2 \le \zeta$.

When bounded uncertainties are explicitly considered, the MPC problem can be formulated as a minimax optimization problem:

$$\begin{aligned}
\min_{\Delta u(k)} \max_{v(k)} \; & J(\Delta u(k), v(k), x(k), y(k)) \\
&= \sum_{j=1}^{N} \|y(k+j|k) - r(k+j)|\,_{Q_j}^{2} \\
&+ \sum_{j=0}^{N_u-1} \|\Delta u(k+j|k)\|_{R_j}^2 + F(x(k+N|k))
\end{aligned}$$

$$\begin{aligned}
\text{subject to} \quad & \Delta u_{\min} \le \Delta u(k+j|k) \le \Delta u_{\max}, j = 0, 2, \ldots, N_u - 1; \\
& u_{\min} \le u(k+j|k) \le u_{\max}, j = 0, 1, \ldots, N_u - 1; \\
& v_{\min} \le v(k+j|k) \le v_{\max}, j = 1, 2, \ldots, N; \\
& y_{\min} \le y(k+j|k) \le y_{\max}, j = 1, 2, \ldots, N; \qquad (7.14)
\end{aligned}$$

where $v_{\min} \le v_{\max}$ are the lower and upper bounds of the additive uncertainty vector.

In the presence of v, Jacobian linearization results in the following input affine model:

$$\begin{aligned}
\delta x(k+1) &= A(k)\delta x(k) + B(k)\delta u(k) + \varepsilon(k) + v(k) \\
\delta y(k) &= C\delta x(k). \qquad (7.15)
\end{aligned}$$

Correspondingly, the future output vector within the prediction horizon can be predicted as

$$\delta\bar{y}(k) = S\delta x(k) + V\delta u(k-1) + M\Delta\bar{u}(k) + H\bar{\varepsilon}(k) + H\bar{v}(k) \qquad (7.16)$$

where $\bar{v}(k) = [v(k|k)\ldots v(k+N-1|k)]^{\mathsf{T}} \in \Re^{Nn}$. The predicted terminal state within the prediction horizon N can be expressed as

$$\delta x(k+N|k) = \tilde{S}\delta x(k) + \tilde{V}\delta u(k-1) + \tilde{M}\Delta\bar{u}(k) + \tilde{H}\bar{\varepsilon}(k) + \tilde{H}\bar{v}(k). \qquad (7.17)$$

In light of (7.16) and (7.17), the minimax optimization problem (7.14) can be transformed as

$$\begin{aligned}
\min_{\Delta\bar{u}(k)} \max_{\bar{v}(k)} \; & ||S\delta x(k) + V\delta u(k-1) + M\Delta\bar{u}(k) + H\bar{\varepsilon}(k) + H\bar{v}(k)) - \delta\bar{r}(k)||_Q^2 \\
& + \|\Delta\bar{u}(k)\|_R^2 + ||\tilde{S}\delta x(k) + \tilde{V}\delta u(k-1) + \tilde{M}\Delta\bar{u}(k) + \tilde{H}\bar{\varepsilon}(k) + \tilde{H}\bar{v}(k)||_P^2, \\
\text{s.t.} \quad & \Delta\bar{u}_{\min} \le \Delta\bar{u}(k) \le \Delta\bar{u}_{\max}, \\
& \bar{v}_{\min} \le \bar{v}(k) \le \bar{v}_{\max}, \\
& \delta\bar{u}_{\min} \le \delta\bar{u}(k-1) + \tilde{I}\Delta\bar{u}(k) \le \delta\bar{u}_{\max}, \\
& \delta\bar{y}_{\min} \le \delta\bar{y}(k) \le \delta\bar{y}_{\max}. \qquad (7.18)
\end{aligned}$$

Write $\Delta\bar{u}(k)$ and $\bar{v}(k)$ as $\Delta\bar{u}$ and $\bar{v}$ for brevity, the problem (7.18) can be rewritten as

$$\begin{aligned}\min_{\Delta\bar{u}}\max_{\bar{v}} \begin{bmatrix}\Delta\bar{u}\\ \bar{v}\end{bmatrix}^{\mathsf{T}} W_1 \begin{bmatrix}\Delta\bar{u}\\ \bar{v}\end{bmatrix} + \begin{bmatrix}c_1\\ c_2\end{bmatrix}^{\mathsf{T}} \begin{bmatrix}\Delta\bar{u}\\ \bar{v}\end{bmatrix}& \\ \text{s.t.}\quad \Delta\bar{u}_{\min} \le \Delta\bar{u} \le \Delta\bar{u}_{\max},& \\ \bar{v}_{\min} \le \bar{v} \le \bar{v}_{\max},& \\ E_1\Delta\bar{u} \le b_1 &\end{aligned} \tag{7.19}$$

where

$$W_1 = \begin{bmatrix}(\tilde{C}M)^{\mathsf{T}}Q\tilde{C}M + \tilde{M}^{\mathsf{T}}P\tilde{M} + R & (\tilde{C}M)^{\mathsf{T}}Q\tilde{C}H + \tilde{M}^{\mathsf{T}}P\tilde{H}\\ (\tilde{C}H)^{\mathsf{T}}Q\tilde{C}M + \tilde{H}^{\mathsf{T}}P\tilde{M} & (\tilde{C}H)^{\mathsf{T}}Q\tilde{C}H + \tilde{H}^{\mathsf{T}}P\tilde{H}\end{bmatrix} \in \Re^{(N_u m+Nn)\times(N_u m+Nn)},$$

$$\begin{aligned} c_1 &= 2(\tilde{C}M)^{\mathsf{T}}Q(\tilde{C}(S\delta x(k) + V\delta u(k-1) + H\bar{\varepsilon}(k)) - \delta\bar{(r)}(k)) \\ &\quad + 2\tilde{M}^{\mathsf{T}}P(\tilde{S}\delta x(k) + \tilde{V}\delta u(k-1) + \tilde{H}\bar{\varepsilon}(k)) \in \Re^{N_u m}, \\ c_2 &= 2(\tilde{C}H)^{\mathsf{T}}Q(\tilde{C}(S\delta x(k) + V\delta u(k-1) + H\bar{\varepsilon}(k)) - \delta\bar{(r)}(k)) \\ &\quad + 2\tilde{H}^{\mathsf{T}}P(\tilde{S}\delta x(k) + \tilde{V}\delta u(k-1) + \tilde{H}\bar{\varepsilon}(k)) \in \Re^{Nn}, \\ E_1 &= \begin{bmatrix}-\tilde{I} & \tilde{I} & -M & M\end{bmatrix}^{\mathsf{T}} \in \Re^{2(N_u m+Nn+Np)\times N_u m}, \\ b_1 &= \begin{bmatrix}-\delta\bar{u}_{\min} + \delta\bar{u}(k-1)\\ \delta\bar{u}_{\max} - \delta\bar{u}(k-1)\\ -\delta\bar{y}_{\min} + S\delta x(k) + V\delta u(k-1) + H\bar{\varepsilon}(k) + H\bar{v}_{\min}\\ \delta\bar{y}_{\max} - S\delta x(k) - V\delta u(k-1) - H\bar{\varepsilon}(k) - H\bar{v}_{\max}\end{bmatrix} \in \Re^{2N_u m+2Nn+2Np}. \end{aligned}$$

Let $\theta = \max_{\bar{v}} \mu^{\mathsf{T}}W_1\mu + c_l^{\mathsf{T}}\mu$, where $\mu = [\Delta\bar{u}^{\mathsf{T}}, \bar{v}^{\mathsf{T}}]^{\mathsf{T}}$ and $c_l = [c_1^{\mathsf{T}}, c_2^{\mathsf{T}}]^{\mathsf{T}}$. The optimization problem (7.19) is then equivalent to:

$$\begin{aligned}&\min\ \theta \\ &\text{s.t.}\ \begin{bmatrix}\mu\\ \theta\end{bmatrix}^{\mathsf{T}} \begin{bmatrix}W_1 & 0\\ 0 & 0\end{bmatrix} \begin{bmatrix}\mu\\ \theta\end{bmatrix} + \begin{bmatrix}c_l\\ -1\end{bmatrix}^{\mathsf{T}} \begin{bmatrix}\mu\\ \theta\end{bmatrix} \le 0, \\ &\quad \begin{bmatrix}E_2 & 0\end{bmatrix} \begin{bmatrix}\mu\\ \theta\end{bmatrix} \le b_2\end{aligned} \tag{7.20}$$

where

$$E_2 = \begin{bmatrix}E_1 & 0\\ I & 0\\ 0 & I\\ -I & 0\\ 0 & -I\end{bmatrix} \in \Re^{(4N_u m+4Nn+2Np)\times(N_u m+Nn)},$$

$$b_2 = \begin{bmatrix}b_1\\ \Delta\bar{u}_{\max}\\ \bar{v}_{\max}\\ -\Delta\bar{u}_{\min}\\ -\bar{v}_{\min}\end{bmatrix} \in \Re^{4N_u m+4Nn+2Np}.$$

Furthermore, the minimization problem (7.20) can be rewritten as the following linear minimization problem subject to quadratic and affine inequality constraints:

$$\begin{aligned} &\min \eta^{\mathsf{T}}\xi \\ &\text{s.t.} \quad \xi^{\mathsf{T}}\tilde{W}\xi + \tilde{c}^{\mathsf{T}}\xi \le 0, \;\; \tilde{E}\xi - \tilde{b} \le 0 \end{aligned} \tag{7.21}$$

where

$$\begin{aligned} \eta &= [0, 0, \ldots, 0, 1]^{\mathsf{T}} \in \Re^{N_u m + Nn + 1}, \\ \xi &= \begin{bmatrix} \mu \\ \theta \end{bmatrix} \in \Re^{N_u m + Nn + 1}, \\ \tilde{W} &= \begin{bmatrix} W_1 & 0 \\ 0 & 0 \end{bmatrix} \in \Re^{(N_u m + Nn + 1)\times(N_u m + Nn + 1)}, \\ \tilde{c} &= \begin{bmatrix} c_l \\ -1 \end{bmatrix} \in \Re^{N_u m + Nn + 1}, \\ \tilde{E} &= [E_2 \quad 0] \in \Re^{(4N_u m + 4Nn + 2Np)\times(N_u m - Nn + 1)}. \end{aligned} \tag{7.22}$$

As W_1 in (7.19) is symmetric and positive definite, $\tilde{W}$ in (7.21) are symmetric and positive semi-definite as well. Thus, the quadratic constraint in (7.21) is convex and twice differentiable. Problem (7.21) is shown to be a convex optimization problem. The solution to (7.21) gives the optimal control input increment vector $\Delta\bar{u}(k)$ that minimizes the worst-case performance index.

7.3 Neural Network Approaches

7.3.1 *Extreme learning machine*

ELM [Huang *et al.* (2005); Liang *et al.* (2006); Yang *et al.* (2012); Zhang *et al.* (2012)] is a single hidden-layer feedforward neural network. The connection weights from inputs to hidden neurons are randomly generated, whereas the connection weights from hidden neurons to outputs are computed analytically. The hidden layer of ELM does not need tuning. For $\mathcal{N}$ arbitrary distinct samples (x_i, d_i), where $\mathrm{x}_i = \left[\mathrm{x}_{i1}, \mathrm{x}_{i2}, \cdots, \mathrm{x}_{ip}\right]^{\mathsf{T}} \in \Re^{\mathrm{p}}$ and $d_i = [d_{i1}, d_{i2}, \cdots, d_{i\mathrm{m}}]^{\mathsf{T}} \in \Re^{\mathrm{m}}$, ELM with L hidden nodes and activation function $G(\mathrm{x})$ are mathematically modeled as:

$$\sum_{i=1}^{L} \beta_i G_i(\mathrm{x}_j) = \sum_{i=1}^{L} \beta_i G(a_i^{\mathsf{T}}\mathrm{x}_j + b_i) = o_j, \quad j = 1, \ldots, \mathcal{N} \tag{7.23}$$

where $a_i = \left[a_{i1}, a_{i2}, \cdots, a_{i\mathrm{p}}\right]^{\mathsf{T}}$ is the weight vector connecting the i-th hidden node and the input nodes, $\beta_i = [\beta_{i1}, \beta_{i2}, \cdots, \beta_{i\mathrm{m}}]^{\mathsf{T}}$ is the weight vector connecting the i-th hidden node and the output nodes, and b_i is the threshold of the i-th hidden node. $a_i^{\mathsf{T}}\mathrm{x}_j$ denotes the inner product of a_i and x_j. $G(\mathrm{x})$ can be any infinitely differential activation function.

Let $d = \left[d_1^{\mathsf{T}}, \cdots, d_N^{\mathsf{T}}\right]^{\mathsf{T}} \in \Re^{\mathcal{N} \times \mathtt{m}}$, $\beta = \left[\beta_1^{\mathsf{T}}, \cdots, \beta_L^{\mathsf{T}}\right] \in \Re^{L \times \mathtt{m}}$ and

$$\mathcal{H} = \begin{bmatrix} G(a_1 \cdot \mathtt{x}_1 + b_1) & \cdots & G(a_L \cdot \mathtt{x}_1 + b_L) \\ \vdots & \cdots & \vdots \\ G(a_1 \cdot \mathtt{x}_{\mathcal{N}} + b_1) & \cdots & G(a_L \cdot \mathtt{x}_{\mathcal{N}} + b_L) \end{bmatrix} \in \Re^{\mathcal{N} \times L}$$

where d is target vector, β is output weight matrix, and $\mathcal{H}$ is called hidden layer output matrix. The learning algorithm of ELM is to find a least-squares solution $\tilde{\beta}$ of the linear system $\mathcal{H}\beta = d$:

$$\left\|\mathcal{H}\tilde{\beta}\right\| = \min_{\beta} \|\mathcal{H}\beta - d\| . \tag{7.24}$$

The minimized norm least-squares solution of the above linear system is:

$$\tilde{\beta} = \mathcal{H}^{+} d \tag{7.25}$$

where $\mathcal{H}^{+}$ is the Moore-Penrose generalized inverse of matrix $\mathcal{H}$. Thus, ELM learning can be summarized as follows: Given a training set $\aleph = \{(\mathtt{x}_i, d_i) \mid \mathtt{x}_i \in \Re^{\mathtt{p}}, d_i \in \Re^{\mathtt{m}},\ i = 1, \ldots, \mathcal{N}\}$, hidden node activation function G and hidden node number L,

(1) Randomly generate hidden node weights a_i and thresholds b_i, $i = 1, \ldots, L$.
(2) Calculate the hidden layer output matrix $\mathcal{H}$.
(3) Calculate the output weight vector β: $\beta = \mathcal{H}^{+} d$.

7.3.2 *Simplified dual network*

Based on the Karush–Kuhn–Tucker conditions and duality theorems, a one-layer recurrent network called the simplified dual network was developed for solving quadratic programming problems by utilizing dual variables [Liu and Wang (2006)]. The simplified dual network is competent for solving (7.12) in real time, whose dynamical model is described as follows:

- State equation

$$\epsilon \frac{dz}{dt} = -E\Delta u + g(E\Delta u - z).$$

- Output equation

$$\Delta u = W^{-1}(E^{\mathsf{T}} z - c) \tag{7.26}$$

where ϵ is a positive constant, $z \in \Re^{3N_u m + 2Np}$ is the neural network state vector, Δu is the output vector, which corresponds to the decision variable in the quadratic programming (7.12), and $g(\cdot)$ is a piecewise linear activation function defined as:

$$g(\xi_i) = \begin{cases} l_i, & \xi_i < l_i; \\ \xi_i, & l_i \le \xi_i \le h_i; \\ h_i, & \xi_i > h_i. \end{cases}$$

The simplified dual network has a single-layer structure with totally $3N_u m+2Np$ neurons. According to the convergence analysis in Liu and Wang (2006), it is Lyapunov stable and globally convergent to the optimal solution of any strictly convex quadratic programming problem.

7.3.3 *A two-layer recurrent neural network*

Xia and Wang (2004b) presented a recurrent neural network for nonlinear convex optimization, which is capable of handling nonlinear inequality constraints. Write (7.21) in form of a general nonlinear convex optimization as follows:

$$\begin{aligned} &\text{minimize} \quad J(\xi) \\ &\text{subject to} \quad s(\xi) \le 0 \end{aligned} \tag{7.27}$$

where $J(\xi) = \eta^{\mathsf{T}}\xi$ is the objective function, $s(\xi) = [s_1(\xi), s_2(\xi)]^{\mathsf{T}}$ with $s_1(\xi) = \xi^{\mathsf{T}}\tilde{W}\xi + \tilde{c}^{\mathsf{T}}\xi$ and $s_2(\xi) = \tilde{E}\xi - \tilde{b}$. The functions J, s_1 and s_2 have been shown to be convex and twice differentiable. The Karush–Kuhn–Tucker conditions for (7.27) can be casted as

$$\begin{aligned} &\lambda \ge 0, \\ &s(\xi) \le 0, \\ &\nabla J(\xi) + \nabla s(\xi)\lambda \ge 0, \\ &\lambda^{\mathsf{T}} s(\xi) = 0 \end{aligned} \tag{7.28}$$

where λ is a vector of Lagrangian multipliers, $\nabla s(\xi) = [\nabla s_1(\xi), \nabla s_2(\xi)]$ is the gradient of $s(\xi)$, and $\nabla J(\xi)$ is the gradient of $J(\xi)$. Based on to the saddle point theorem [Bazaraa *et al.* (1993)] and the projection theorem [Kinderlehrer and Stampacchia (1980)], the dynamic equations of the two-layer recurrent neural network for solving (7.27) are described as follows [Xia and Wang (2004b)]

$$\epsilon \frac{d}{dt}\begin{pmatrix} \xi \\ \lambda \end{pmatrix} = -\begin{pmatrix} \nabla J(\xi) + \nabla s(\xi)\lambda \\ \lambda - (\lambda + s(\xi))^{+} \end{pmatrix} \tag{7.29}$$

where ϵ is a positive constant, $\lambda^{+} = [(\lambda_1)^{+}, (\lambda_2)^{+}]^{\mathsf{T}}$ and $[\lambda_i]^{+} = \max\{0, \lambda_i\}$. According to the theoretical analysis in Xia and Wang (2004b), the two-layer recurrent neural network (7.29) is Lyapunov stable and globally convergent to the unique optimal solution to the convex program (7.27).

7.3.4 *Overall MPC approach*

The presented neurodynamic optimization models are suitable for real time optimization of the quadratic programming (7.12) and the nonlinear convex optimization (7.21) for synthesizing nonlinear and robust MPC. However, it is worth noting that the term $\bar{\varepsilon}(k)$ is still unknown so far. Consequently, the optimization problems

(7.12) and (7.21) are with unknown parameters, and they can not be solved unless $\bar{\varepsilon}(k)$ can be estimated. Representing modeling errors resulting from the unmodeled dynamics and the linearization residue, $\bar{\varepsilon}(k)$ does not only depend on the current state and input, but also the future evolution of the system states and inputs. Computing the explicit form of this term is intractable. Thanks to the universal approximation and fast learning capability, ELM can be used in black-box identification of nonlinear functions from input and output data. From a pair of input-state sample data randomly generated within the operational domain, the state response $\hat{x}$ can be measured from the original plant. The errors resulting from linearization and unmodeled dynamics can be computed as:

$$\varepsilon(x, u, x_o, u_o) = \hat{x} - \frac{\partial f(x_o)}{\partial x}(x - x_o) - \frac{\partial f(u_o)}{\partial u}(u - u_o).$$

An ELM can be applied for mapping the relation between the two data sets by treating $[x\ u\ x_o\ u_o]^{\mathsf{T}}$ as the input vector and ε as the target vector. A well-trained ELM from sample data sets is capable of estimating the numerical value of $\varepsilon(k + j|k)$ $j = 1, \ldots, N - 1$ at each time instant k.

The nonlinear and robust MPC scheme for uncertain nonlinear systems with unmodeled dynamics based on supervised learning and neurodynamic optimization is summarized as follows:

1. Let $k = 1$. Set control time terminal T, prediction horizon N, control horizon N_u, sampling period τ, weight matrices Q and R. Compute the terminal cost matrix P.
2. Model ε via supervised learning by using an ELM.
3. Estimate $\bar{\varepsilon}(k)$ using the trained ELM in Step 2. Calculate the optimization problem parameters using the estimated value of $\bar{\varepsilon}(k)$.
4. Solve the quadratic programming (7.12) by applying the simplified dual network (7.26) or the nonlinear convex optimization problem (7.18) by applying the two-layer neural network (7.29) to obtain the optimal control increment vector $\Delta\bar{u}(k)$.
5. Compute the optimal control vector $\bar{u}(k)$ and implement $u(k|k)$.
6. If $k < T$, set $k = k + 1$, go to Step 3; otherwise end.

7.4 Simulation Results

In this section, simulation results on nonlinear control problems are provided to demonstrate the performance of the proposed MPC approaches.

7.4.1 *Example 1*

Consider a cement milling circuit presented in Piche *et al.* (2000) as modeled as follows:

$$
\begin{aligned}
0.3\dot{y}_f &= -y_f + (1 - \alpha(l, v_s, d)\hat{\varphi}(l, d)) + w_1, \\
\dot{l} &= -\hat{\varphi}(l, d) + u_f + y_r + w_2, \\
0.01\dot{y}_r &= -y_r + \alpha(l, v_s, d)\hat{\varphi}(l, d) + w_3
\end{aligned} \tag{7.30}
$$

where y_f is the product flow rate (tons/h), l is the load in the mill (tons), y_r is the tailing flow rate (tons/h), $\varphi(l, d)$ is the output flow rate of the mill (tons/h), u_f is the feed flow rate (tons/h), v_s is the classifier speed (rpm), d is the hardness of the material inside the mill. The nonlinear function φ is

$$\varphi(l, d) = \max\{0; -0.1116dl^2 + 16.5l\}.$$

But due to the model mismatch, only an estimated nonlinear function $\hat{\varphi}$ is available:

$$\hat{\varphi}(l, d) = \max\{0; -0.1dl^2 - 16l\}.$$

The nonlinear function α is

$$\alpha(l, v_s, d) = \frac{\hat{\varphi}^{0.8} v_s^4}{3.56 \times 10^{10} + \hat{\varphi}^{0.8} v_s^4}.$$

The unmodeled dynamics results from functional errors is represented as $w = [w_1\ w_2\ w_3]^{\mathsf{T}}$. Define $x = [y_f\ l\ y_r]^{\mathsf{T}}$, $u = [u_f\ v_s]^{\mathsf{T}}$, and $y = [y_f\ y_r]^{\mathsf{T}}$. The reference set point at $t < 3h$ is $r = [120\ 450]^{\mathsf{T}}$, and at $t \geq 3h$ is $r = [110\ 425]^{\mathsf{T}}$. The initial state is $x(0) = [115\ 60\ 455]^{\mathsf{T}}$, and the initial input is $u(0) = [130\ 170]^{\mathsf{T}}$. Let $N = 5$, $N_u = 3$, $Q = I$, $R = 0.1I$ and sampling period be 2 minutes. The constraints are: $80 \leq u_1 \leq 150$, $165 \leq u_2 \leq 180$. The terminal cost weighting parameter P is computed as

$$P = \begin{bmatrix} 0.33 & -0.09 & -0.03 \\ -0.09 & 1.31 & 0.17 \\ -0.03 & 0.17 & 0.81 \end{bmatrix}.$$

To apply the presented MPC schedule, we first model ε by using an ELM via supervised learning. The learning results based on 2000 training data and 2000 testing data are summarized in Table 7.1. To compare the performance, we also applied classical multilayer perceptron (MLP-BP) [Hornik *et al.* (1989)] and least-squares supporting vector machines (LS-SVM) [Suykens *et al.* (2002)] to model ε whose results are also shown in Table 7.1. As ELM randomly generates hidden

Table 7.1 Comparison of learning results in Example 1

	ELM	MLP-BP	LS-SVM
Hidden neurons/features	300	300	300
Training time	$0.4212s$	$12.38s$	$353.61s$
Training MSE	1.3×10^{-7}	3.9×10^{-7}	5.1×10^{-8}
Testing MSE	4.6×10^{-6}	7.3×10^{-7}	2.2×10^{-7}

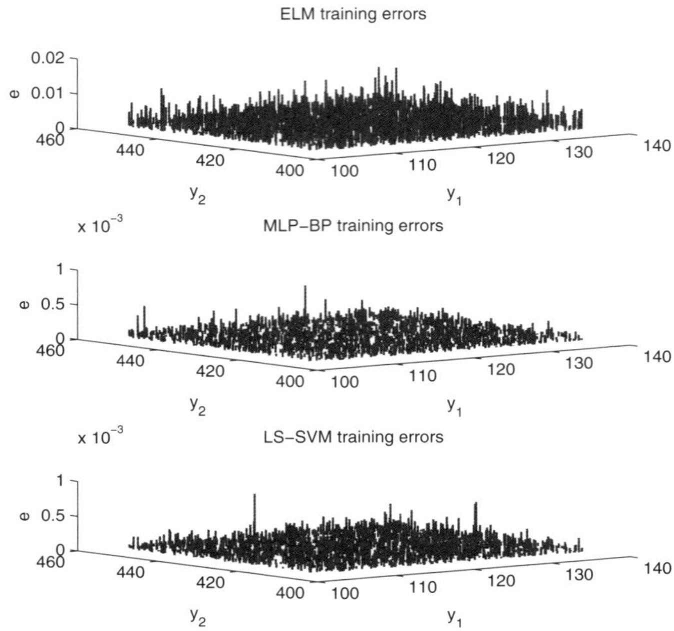

Fig. 7.1 L_2-norm of training errors in Example 1

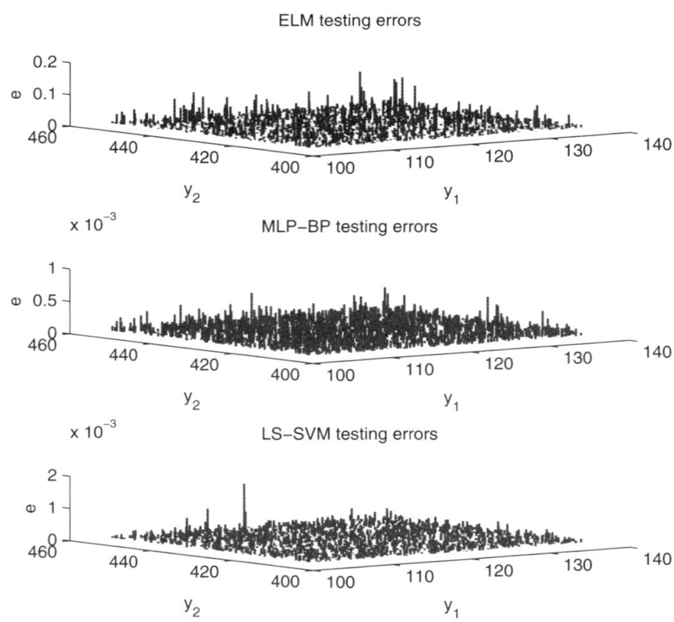

Fig. 7.2 L_2-norm of testing errors in Example 1

weights and MLP randomly generates initial weights, we run ten times for each fixed number of hidden nodes and take the mean square errors. The L_2-norm of training errors and testing errors are depicted in Figs. 7.1–7.2. From the learning results, we observe that ELM can learn thousands of times faster and provide comparable accuracy. It is reasonable to choose ELM as the modeling tool for MPC design.

MPC of (7.30) is then synthesized by solving the quadratic programming (7.12) using the simplified dual network. The control results are illustrated in Figs. 7.3-7.5. To show the performance, MPC on the exact model of the cement milling

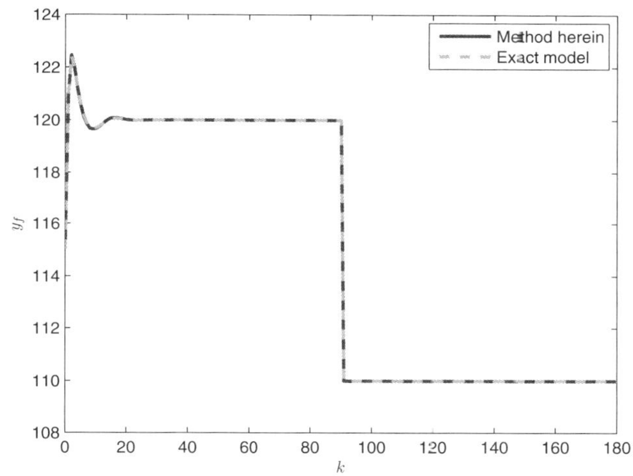

Fig. 7.3 Product flow rate in Example 1

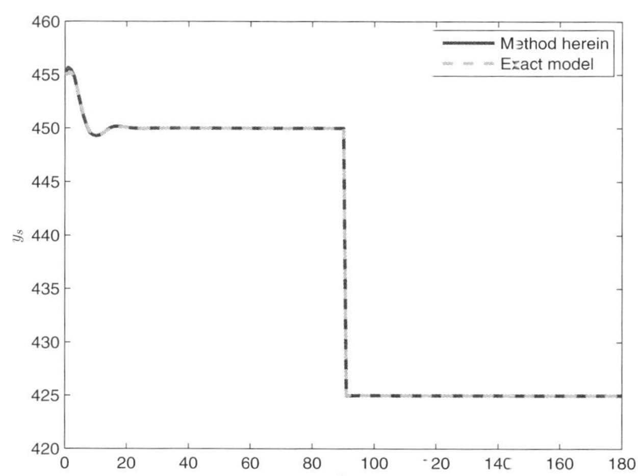

Fig. 7.4 Tailing flow rate in Example 1

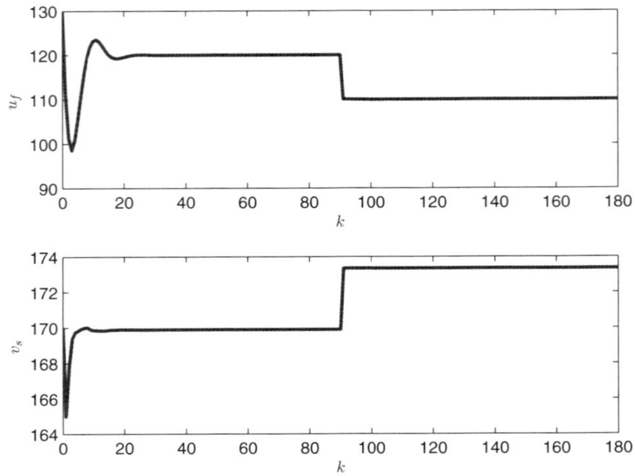

Fig. 7.5 Control inputs in Example 1

circuit is also applied. The presented nonlinear MPC results in superior set points tracking despite that the model is only partially known.

Now consider that the model (7.30) is affected by an bounded additive uncertainty $v = [v_1, v_2, v_3]^{\mathsf{T}}$, where v is assumed to satisfy $-2 \leq v \leq 2$. To obtain robust MPC against the uncertainty, the nonlinear optimization problems (7.21) is solved by using the two-layer neural network (7.29) in real time. The simulation results for this case are shown in Figs. 7.6–7.8. The system outputs reach the reference set points in the presence of bounded uncertainty.

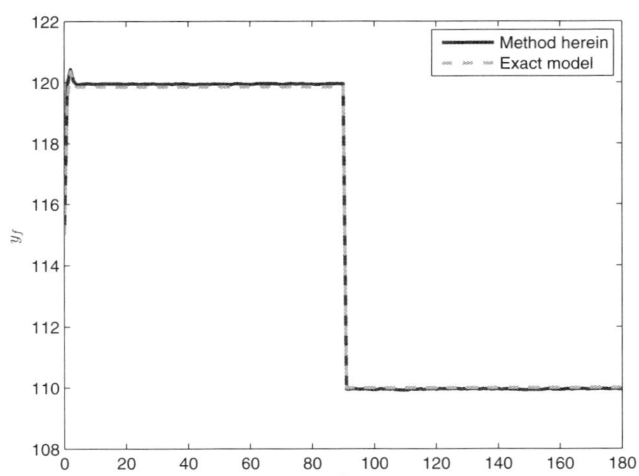

Fig. 7.6 Product flow rate in presence of uncertainty in Example 1

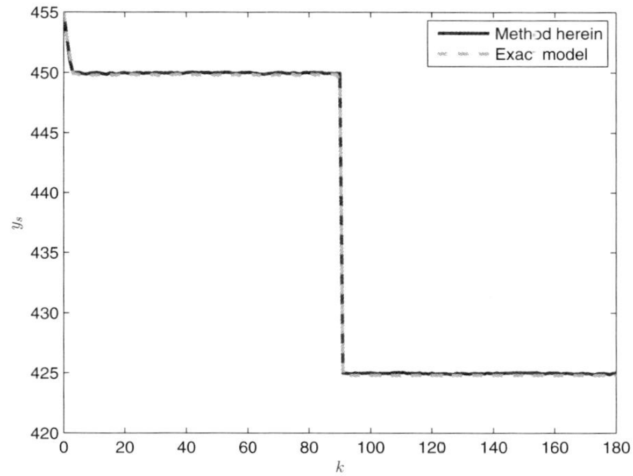

Fig. 7.7 Tailing flow rate in presence of uncertainty in Example 1

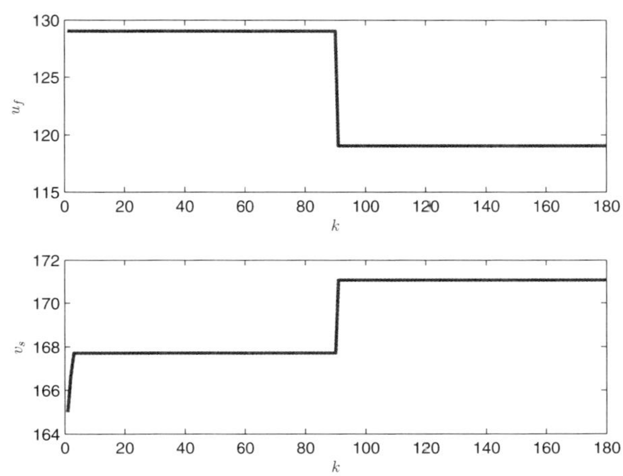

Fig. 7.8 Control inputs in presence of uncertainty in Example 1

7.4.2 *Example 2*

Consider a nonlinear plant model presented in Ge *et al.* (1999):

$$\begin{aligned}\dot{x}_1 &= x_2 + w_1,\\ \dot{x}_2 &= x_1^2 + 0.15u^3 + 0.1(1 + x_2^2)u + w_2,\\ y &= x_1.\end{aligned} \tag{7.31}$$

Assume $w = [w_1, w_2]^\mathsf{T} = [0, \sin(0.1u)]^\mathsf{T}$ is the unmodeled dynamics of the plant, The tracking objective is to force the output y to follow a reference signal $r =$

Table 7.2 Comparison of learning results in Example 2

	ELM	MLP-BP	LS-SVM
Hidden neurons/features	300	300	300
Training time	$0.5616s$	$47.65s$	$267.35s$
Training MSE	1.03×10^{-7}	1.98×10^{-7}	8.56×10^{-8}
Testing MSE	3.07×10^{-7}	4.80×10^{-7}	2.85×10^{-7}

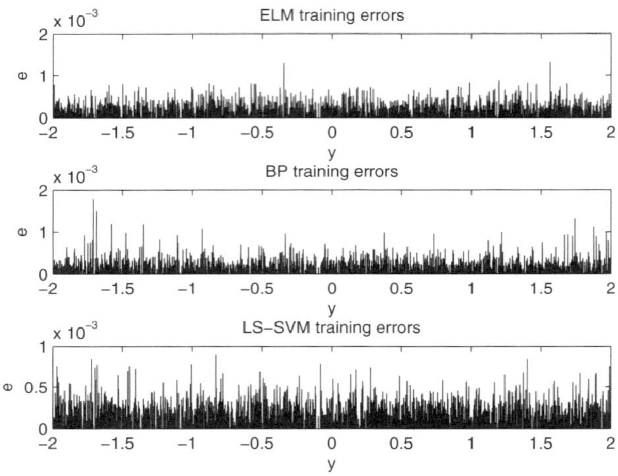

Fig. 7.9 L_2-norm of training errors in Example 2

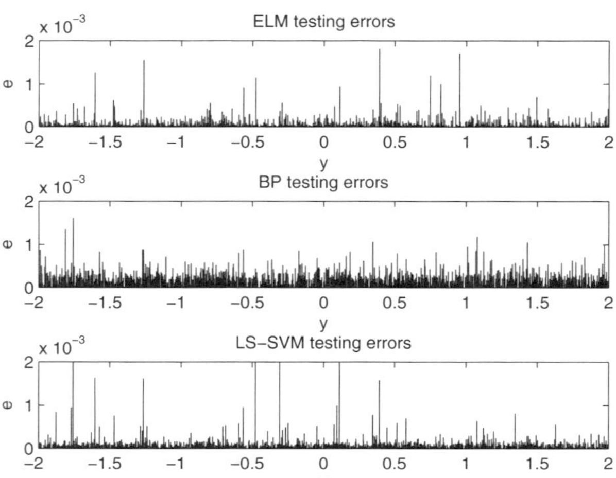

Fig. 7.10 L_2-norm of testing errors in Example 2

$\sin t + \cos(0.5t)$. The initial condition is $x(0) = [0.5,\ 0]^{\mathrm{T}}$. Set $N = 5$, $N_u = 2$, $Q = 50I$, $R = 0.1I$, and $\tau = 0.1s$. The input and output constraints are $-4 \leq u \leq 5$ and $-2 \leq y \leq 2$. The terminal weighting parameter is P=diag$(0.067, 0.148)$. The supervised learning results of ε based on 2000 sampling data by using ELM, MLP, and LS-SVM are summarized in Table 7.2. The training and testing errors are depicted in Figs. 7.9-7.10.

Using the trained ELM to estimate the unknown parameters in (7.12) enables the MPC of (7.31) in the absence of additive uncertainties to be solved by the neurodynamic optimization model (7.26) in real time. The controlled output is depicted in Fig. 7.11. The control input is shown in Fig 7.12. To illustrate the

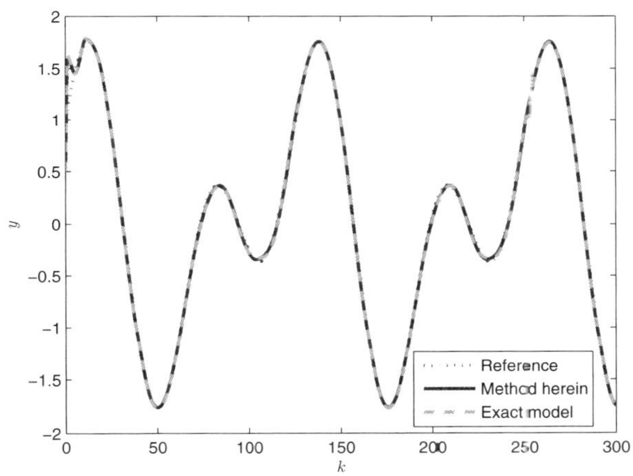

Fig. 7.11 Output in Example 2

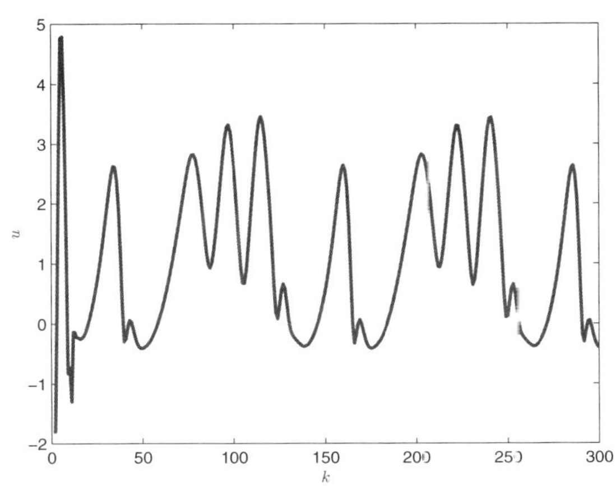

Fig. 7.12 Control input in Example 2

performance, we also applied MPC on the exact model of the plant. The results show that the presented nonlinear MPC approach results in superior tracking performance in the presence of unmodeled dynamics.

Now consider the case where the system model is affected by an additive bounded uncertainty $v = [v_1, v_2]^\mathsf{T}$ with $-0.1 \leq v \leq 0.1$. The reformulated optimization problem (7.21) is then repeatedly solved by using the recurrent neural network (7.29). The output tracking results is depicted in Fig. 7.13. The optimal control input is shown in Fig. 7.14. The tracking results shown that the presented robust MPC can sustain bounded uncertainties.

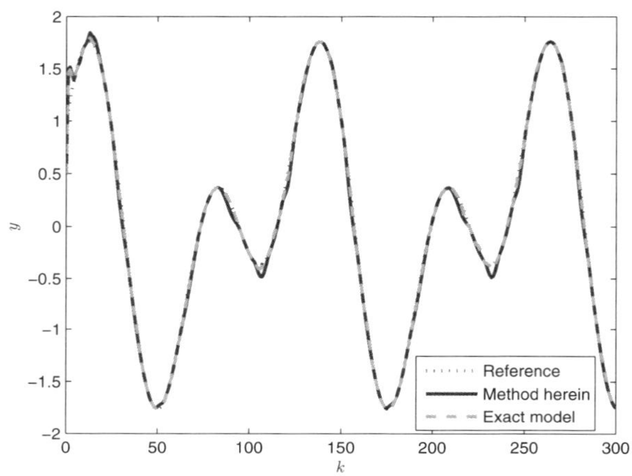

Fig. 7.13 Output in presence uncertainty in Example 2

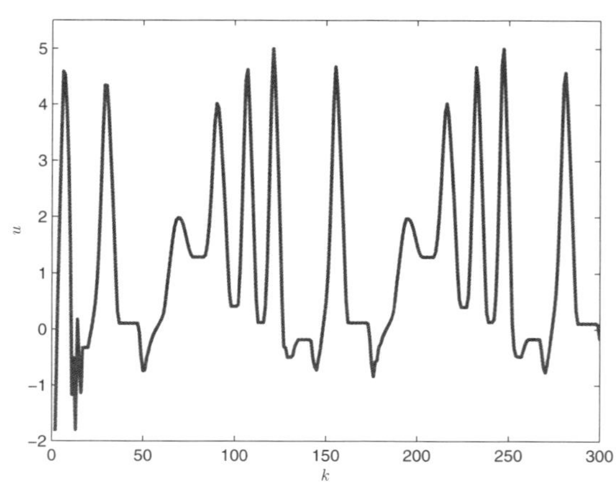

Fig. 7.14 Control input in presence uncertainty in Example 2

7.5 Conclusions

In this chapter, model predictive control approaches to nonlinear systems with unmodeled dynamics are presented based on supervised leanring and neurodynamic optimization. Via online Jacobian linearization, the nonlinear model predictive control problem is formulated as a quadratic programming in the absence of additive uncertainties. When additive bounded uncertainties are explicitly taken into account, the minimax method is applied for achieving robustness. The robust model predictive control problem is reformulated as a convex optimization problem subject to nonlinear constraints. Modeling errors resulting from linearization residues and unmodeled dynamics are compensated by using an extreme learning machine. Optimal control signals can be generated by using recurrent neural networks via real time optimization. Simulation results substantiated the effectiveness of the presented approaches.

Acknowledgement

This work was supported in part by the Research Grants Council of the Hong Kong Special Administrative Region under Grants CUHK416811E and Grant CUHK416812E, and in part by the National Natural Science Foundation of China under Grant 61273307.

References

Akesson, B. M. and Toivonen, H. T. (2006). A neural network model predictive controller, *Journal of Process Control* **16**, 9, pp. 937–946.

Bazaraa, M., Sherali, H. and Shetty, C. (1993). *Nonlinear Programming: Theory and Algorithms* (Wiley).

Bemporad, A., Borrelli, F. and Morari, M. (2003). Min-max control of constrained uncertain discrete-time linear systems, *IEEE Transactions on Automatic Control* **48**, 9, pp. 1600–1606.

Cavagnari, Magni, L. and Scattolini, R. (1999). Neural network implementation of nonlinear receding horizon control, *Neural Computing and Applications* **8**, 1, pp. 86–92.

Cheng, L., Hou, Z. G., Lin, Y., Tan, M., Zhang, W. C. and Wu, F. (2011). Recurrent neural network for nonsmooth convex optimization problems with applications to the identification of genetic regulatory networks, *IEEE Transactions on Neural Networks* **22**, 5, pp. 714–726.

Cheng, L., Hou, Z. G. and Tan, M. (2007). Constrained multi-variable generalized predictive control using a dual neural network, *Neural Computing and Applications* **16**, 6, pp. 505–512.

Chisci, L., Rossiter, J. A. and Zappa, G. (2001). Systems with persistent disturbances: Predictive control with restricted constraints, *Automatica* **37**, 7, pp. 1019–1028.

Colin, G., Chamaillard, Y., Bloch, G. and Corde, G. (2007). Neural control of fast nonlinear systems-application to a turbocharged SI engine with VCT, *IEEE Transactions on Neural Networks* **18**, 4, pp. 1101–1114.

Draeger, A., Engell, S. and Ranke, H. (1995). Model predictive control using neural networks, *IEEE Control Systems* **15**, 5, pp. 61–66.

Forti, M., Nistri, P. and Quincampoix, M. (2004). Generalized neural network for nonsmooth nonlinear programming problems, *IEEE Transactions on Circuits and Systems I: Regular Papers* **51**, 9, pp. 1741–1754.

Ge, S. S., Hang, C. C. and Zhang, T. (1999). Adaptive neural network control of nonlinear systems by state and output feedback, *IEEE Transactions on Systems, Man, and Cybernetics, Part B: Cybernetics* **29**, 6, pp. 818–828.

Guo, Z., Liu, Q. and Wang, J. (2011). A one-layer recurrent neural network for pseudoconvex optimization subject to linear equality constraints, *IEEE Transactions on Neural Networks* **22**, 12, pp. 1892–1900.

Hornik, K., Stinchcombe, M. and White, H. (1989). Multilayer feedforward networks are universal approximators, *Neural Networks* **2**, 5, pp. 359–366.

Hu, X. and Wang, J. (2007). A recurrent neural network for solving a class of general variational inequalities, *IEEE Transactions on Systems, Man, and Cybernetics, Part B: Cybernetics* **37**, 3, pp. 528–539.

Hu, X. and Wang, J. (2008). An improved dual neural network for solving a class of quadratic programming problems and its k-winners-take-all application, *IEEE Transactions on Neural Networks* **19**, 12, pp. 2022–2031.

Huang, G. B., Zhu, Q. Y. and Siew, C. K. (2005). Extreme learning machine: Theory and application, *Neurocomputing* **70**, 1, pp. 489–501.

Huang, J. and Lewis, F. (2003). Neural network predictive control for nonlinear dynamic systems with time-delay, *IEEE Transactions on Neural Networks* **14**, 2, pp. 377–389.

Kerrigan, E. C. and Maciejowski, J. M. (2004). Feedback min-max model predictive control using a single linear program: Robust stability and the explicit solution, *International Journal of Robust and Nonlinear Control* **14**, 4, pp. 395–413.

Kinderlehrer, D. and Stampacchia, G. (1980). *An Introduction to Variational Inequalities and Their Applications* (Academic Press, Inc.).

Langson, W., Chryssochoos, I., Rakovic, S. V. and Mayne, D. Q. (2004). Robust model predictive control using tubes, *Automatica* **40**, 1, pp. 125–133.

Lawrynczuk, M. (2010). Explicit neural network-based nonlinear predictive control with low computational complexity, in S. Tsumoto, R. Slowiński, J. Komorowski, and J. W. Grzymala-Busse (eds.), *Rough Sets and Current Trends in Computing* (Springer, Berlin), pp. 649–658.

Lawrynczuk, M. and Tatjewski, P. (2010). Nonlinear predictive control based on neural multi-models, *International Journal of Applied Mathematics and Computer Science* **20**, 1, pp. 7–21.

Lazar, M., Muñoz de la Peña, D., Heemels, W. and Alamo, T. (2008). On input-to-state stability of min-max nonlinear model predictive control, *Systems and Control Letters* **57**, 1, pp. 39–48.

Lee, J. H. and Yu, Z. (1997). Worst-case formulation of model predictive control for systems with bounded parameters, *Automatica* **33**, 5, pp. 763–781.

Liang, N. Y., Huang, G. B., Saratchandran, P. and Sundararajan, N. (2006). A fast and accurate online sequential learning algorithm for feedforward networks, *IEEE Transactions on Neural Networks* **17**, 6, pp. 1411–1423.

Limón, D., Alamo, T., Salas, F. and Camacho, E. F. (2006). On the stability of constrained MPC without terminal constraint, *IEEE Transactions on Automatic Control* **51**, 5, pp. 832–836.

Limón, D., Alamo, T., Salas, F. and Camacho, E. F. (2006). Input to state stability of min-max MPC controllers for nonlinear systems with bounded uncertainties, *Automatica* **42**, 5, pp. 797–803.

Liu, G. P. and Kadirkamanathan, V. (1998). Predictive control for non-linear systems using neural networks, *International Journal of Control* **71**, 6, pp. 1119–1132.

Liu, Q., Guo, Z. and Wang, J. (2012). A one-layer recurrent neural network for constrained pseudoconvex optimization and its application for dynamic portfolio optimization, *Neural Networks* **26**, pp. 99–109.

Liu, Q. and Wang, J. (2008). A one-layer recurrent neural network with a discontinuous hard-limiting activation function for quadratic programming, *IEEE Transactions on Neural Networks* **19**, 4, pp. 558–570.

Liu, Q. and Wang, J. (2011). A one-layer recurrent neural network for constrained nonsmooth optimization, *IEEE Transactions on Systems, Man, and Cybernetics, Part B: Cybernetics* **41**, 5, pp. 1323–1333.

Liu, Q. and Wang, J. (2013). A one-layer projection neural network for nonsmooth optimization subject to linear equalities and bound constraints, *IEEE Transactions on Neural Networks and Learning Systems* **24**, 5, pp. 812–824.

Liu, S. and Wang, J. (2006). A simplified dual neural network for quadratic programming with its KWTA application, *IEEE Transactions on Neural Networks* **17**, 6, pp. 1500–1510.

Magni, L., De Nicolao, G., Scattolini, R. and Allgöwer, F. (2003). Robust model predictive control of nonlinear discrete-time systems, *International Journal of Robust and Nonlinear Control* **13**, 3–4, pp. 229–246.

Marruedo, D. L., Alamo, T. and Camacho, E. F. (2002). Input-to-state stable MPC for constrained discrete-time nonlinear systems with bounded additive uncertainties, in *Proceedings of the 41st IEEE Conference on Decision and Control* (Atlantis, Bahamas), pp. 4619–4624.

Mayne, D. Q., Rawlings, J. B., Rao, C. V. and Scokaert, P. O. M. (2000). Constrained model predictive control: Stability and optimality, *Automatica* **36**, 6, pp. 789–814.

Mayne, D. Q., Seron, M. M. and Raković, S. V. (2005). Robust model predictive control of constrained linear systems with bounded disturbances, *Automatica* **41**, 2, pp. 219–224.

Mu, J., Rees, D. and Liu, G. P. (2005). Advanced controller design for aircraft gas turbine engines, *Control Engineering Practice* **13**, 8, pp. 1001–1015.

Pan, Y. and Wang, J. (2012). Model predictive control of unknown nonlinear dynamical systems based on recurrent neural networks, *IEEE Transactions on Industrial Electronics* **59**, 8, pp. 3089–3101.

Piche, S., Sayyar-Rodsari, B., Johnson, D. and Gerules, M. (2000). Nonlinear model predictive control using neural networks, *IEEE Control Systems* **20**, 3, pp. 53–62.

Pin, G., Raimondo, D. M., Magni, L. and Parisini, T. (2009). Robust model predictive control of nonlinear systems with bounded and state-dependent uncertainties, *IEEE Transactions on Automatic Control* **54**, 7, pp. 1681–1687.

Raimondo, D. M., Limon, D., Lazar, M., Magni, L. and Camacho, E. F. (2009). Min-max model predictive control of nonlinear systems: A unifying overview on stability, *European Journal of Control* **15**, 1, pp. 5–21.

Scokaert, P. O. M. and Mayne, D. Q. (1998). Min-max feedback model predictive control for constrained linear systems, *IEEE Transactions on Automatic Control* **43**, 8, pp. 1136–1142.

Song, Y., Chen, Z. and Yuan, Z. (2007). New chaotic PSO-base neural network predictive control for nonlinear process, *IEEE Transactions on Neural Networks* **18**, 2, pp. 595–601.

Suykens, J., Gestel, T., Brabanter, J., DeMoor, B. and Vandewalle, J. (2002). *Least Squares Support Vector Machines* (World Scientific).

Tatjewski, P. (2003). *Advanced Control of Industrial Processes, Structures and Algorithms* (Springer).

Wang, L. X. and Wan, F. (2001). Structured neural networks for constrained model predictive control, *Automatica* **37**, 8, pp. 1235–1243.

Werbos, P. J., McAvoy, T. and Su, T. (1992). Neural networks, system identification, and control in the chemical process industries, in White D. and Sofge D. (eds.), *Handbook of Intelligent Control: Neural, Fuzzy, and Adaptive Approaches* (Van Nostrand Reinhold), pp. 283–356.

Xia, Y. S., Feng, G. and Wang, J. (2005). A primal-dual neural network for online resolving constrained kinematic redundancy in robot motion control, *IEEE Transactions on Systems, Man, and Cybernetics, Part B: Cybernetics* **35**, 1, pp. 54–64.

Xia, Y., Feng, G. and Wang, J. (2008). A novel recurrent neural network for solving nonlinear optimization problems with inequality constraints, *IEEE Transactions on Neural Networks* **19**, 8, pp. 1340–1353.

Xia, Y. and Wang, J. (2004a). A general projection neural network for solving monotone variational inequalities and related optimization problems, *IEEE Transactions on Neural Networks* **15**, 2, pp. 318–328.

Xia, Y. and Wang, J. (2004b). A recurrent neural network for nonlinear convex optimization subject to nonlinear inequality constraints, *IEEE Transactions on Circuits and Systems I: Regular Papers* **51**, 7, pp. 1385–1394.

Yan, Z. and Wang, J. (2012a). Model predictive control for tracking of underactuated vessels based on recurrent neural networks, *IEEE Journal of Oceanic Engineering* **37**, 4, pp. 717–726.

Yan, Z. and Wang, J. (2012b). Model predictive control of nonlinear systems with unmodeled dynamics based on feedforward and recurrent neural networks, *IEEE Transactions on Industrial Informatics* **8**, 4, pp. 746–756.

Yang, Y., Wang, Y. and Yuan, X. (2012). Bidirectional extreme learning machine for regression problem and its learning effectiveness, *IEEE Transactions on Neural Networks and Learning Systems* **23**, 9, pp. 1498–1505.

Zhang, R., Lan, Y., Huang, G. B. and Xu, Z. (2012). Universal approximation of extreme learning machine with adaptive growth of hidden nodes, *IEEE Transactions on Neural Networks and Learning Systems* **23**, 2, pp. 365–371.

Chapter 8

Packet-Based Communication and Control Co-Design for Networked Control Systems

Yun-Bo Zhao and Guo-Ping Liu

Abstract A packet-based communication and control co-design framework is proposed for Networked Control Systems (NCSs). This framework takes advantage of the characteristic of the packet-based transmission in the networked control environment, which enables a sequence of control signals to be sent over the network simultaneously within one data packet. This consequently makes it possible to actively compensate for the communication constraints in NCSs with specially designed compensation mechanisms, which can not be achieved by conventional control approaches. These compensated communication constraints include all the major ones brought by the communication network to NCSs, i.e., the network-induced delay, data packet dropout and data packet disorder, thus making the packet-based co-design approach a unified framework for NCSs. Following the design of the packet-based framework, the resulting control system is mathematically formulated, its closed-loop stability is analyzed, and a receding horizon controller is designed to implement the scheme. Finally, the effectiveness of the co-design scheme is verified by numerical examples as well as an Internet-based test rig. It is believed that this packet-based communication and control co-design framework is an important step towards the convergence of control, communication and computation in the new era of the information technology.

8.1 Introduction

Networked control systems (NCSs) are control systems that are closed via communication networks, whose typical structure is depicted in Fig. 8.1. Distinct from conventional control systems where the data exchange between sensors, controllers, actuators, etc., is assumed to be costless, NCSs can contain a large number of control devices interconnected through some form of communication network over

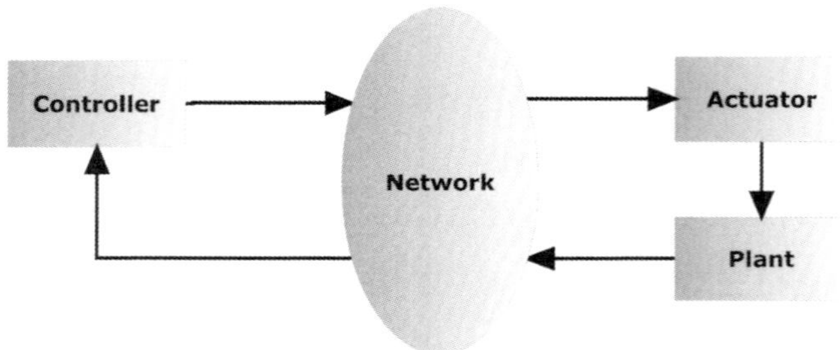

Fig. 8.1 The block diagram of general networked control systems

which data is exchanged. This system structure thus introduces the so-called communication constraints to the control systems, e.g., network-induced delay, data packet dropout, data packet disorder, data rate constraint, etc., making the design and analysis of NCSs a challenging task [Baillieul and Antsaklis (2007); Gupta and Chow (2010); Hespanha *et al.* (2007); Tipsuwan and Chow (2003); Zhang *et al.* (2013)].

As an emerging interdisciplinary research field in recent years, NCSs has attracted much attention from both the communication engineers and the control theory specialists. It has been so widely studied mainly due to its extensive practical applications, existing ones of which include smart home, remote surgery, smart transportation, just name a few, and a lot of potentials are expected in the near future [Hespanha *et al.* (2007)]. These promising applications, generally speaking, benefit from two essential advantages brought by NCSs, that is, the capability of both remote and distributed control. These capabilities are evidently due to the introduction of the communication network to NCSs, thus making the communication network to be essential in NCSs. An increasing trend is also noticed that more and more network-based applications are now configured over the Internet, mainly due to the low cost, easy maintenance, remote control capability, etc., brought by the Internet. However, unlike those conventional control networks such as ControlNet, DeviceNet, etc., that have been specially optimized for control applications [Lian *et al.* (2001)], the Internet is essentially a data network rather than a real-time network, meaning that the Internet is difficult to meet the critical real-time requirement of control systems. This fact implies that worse communication conditions in the Internet such as larger delay, more data packet dropout and disorder, etc., require even more careful treatment before the Internet can be eventually served as a reliable communication media for most control applications [Li *et al.* (2010); Tipsuwan and Chow (2004); Zeng and Chow (2012)].

To date many efforts have been made to deal with these so-called communication constraints in NCSs, ranging from the mathematical modeling and analysis from conventional control theory perspective [Donkers *et al.* (2012); Gao *et al.* (2008); Kruszewski *et al.* (2012); Postoyan and Nesic (2012); Vatanski *et al.* (2009)], to controller design and performance evaluation by taking advantage of the characteristics of the communication network in NCSs [Chen and Qiu (2013); Colandairaj *et al.* (2007); Greco *et al.* (2012); Guan *et al.* (2013); Moayedi *et al.* (2013); Walsh *et al.* (2002); Zhao *et al.* (2009a)], and further to control-oriented communication protocol design from the communication technology perspective [Cea and Goodwin (2013); Katewa and Gupta (2013); Lian *et al.* (2005); Longo *et al.* (2012); Tipsuwan and Chow (2004)], and so forth. Whatever specific methods are used in NCSs, a consensus is always held that the communication constraints are critical in NCSs and a promising approach ought to reduce the negative effects brought by the communication constraints as much as possible. Such approaches can be divided into the following two categories. One is to dynamically schedule the communication resources among different control tasks, in order to make full use of the communication resources efficiently, as done in Walsh and Ye (2001); Zhao *et al.* (2008b). The other way is to reduce the necessary data transmissions while maintaining the system performance at an acceptable level, that is, to find the trade-off between the use of the communication resource and the system performance [Otanez *et al.* (2002); Zhao *et al.* (2011b, 2010b)]. These two means are not fungible but rather, have different focuses. The former is applied to the scenario where the communication network is occupied by multiple NCSs so that the efficient allocation of the communication resources is vital while the latter more focuses on the reduction of the dependence on the communication network for a single NCS.

In this chapter, we exploit the fact that in most communication networks, data is transmitted in "packet" and within its effective load sending a single bit or several hundred bits consumes the same amount of network resources. This makes it possible in NCSs to actively compensate for the communication constraints by sending a sequence of control predictions in one data packet and then selecting the appropriate one corresponding to current network condition. This observation motivates the design of the so called "packet-based control" approach to NCSs in this chapter [Zhao *et al.* (2009a)]. Due to the active compensation process in the packet-based control approach, a better performance can be expected than those from conventional implementations where no characteristics of the network are specifically considered in the design. In a sense this communication and control co-design scheme falls into the first category of the approaches to the efficient usage of the communication resources, i.e., improving the system performance by taking full advantage of the communication networks. On the other hand, the packet-based control approach is flexible to be applied to various system settings, thus making it a unified framework for NCSs. In this chapter we present the approach for the basic system setting. For more applications of the approach, please refer to Zhao *et al.* (2008a,b,c,d,e, 2009b, 2010a, 2011a, 2012a,b,c).

The remainder of the chapter is organized as follows. The characteristic of the packet-based transmission of data in NCSs is first discussed in Section 8.2, to give a general context of the work to be done in this chapter. The packet-based communication and control co-design approach is then presented in detail in Section 8.3, which leads to a novel controller that can compensate for network-induced delay, data packet dropout and data packet disorder simultaneously. The stability criteria for the corresponding closed-loop system are investigated in Section 8.4, from different perspectives of switched system theory and delay-dependent analysis, respectively. The packet-based control framework generally allows any appropriate controller design method to be used, and a Generalized Predictive Control (GPC) based controller is designed in Section 8.5 for an example. Numerical and experimental examples to illustrate the effectiveness of the proposed approach are presented in Section 8.6 and Section 8.7 concludes the chapter.

8.2 Packet-Based Transmission in Networked Control Systems

In conventional control systems data is usually assumed to be transmitted costlessly, while in practical NCSs, the data exchange has to be completed by the communication network, in the form of "data packets". This packet-based transmission is one of the most important characteristics of NCSs that are distinct from conventional control systems [Antsaklis and Baillieul (2007); Baillieul and Antsaklis (2007)]. This characteristic can mean that the perfect data transmission as assumed in conventional control systems is absent in NCSs, thus deriving the most challenging issue in NCSs. The communication constraints caused by the packet-based transmission in NCSs include the network-induced delay, data packet dropout, data packet disorder, etc.. We discuss these in detail as follows.

A. Network-induced delay

With the network being inserted into the control loop in NCSs, network-induced delays are introduced in both the forward and backward channels, which are well known to significantly degrade the performance of the control systems.

Two types of network-induced delays are present according to where they occur.

- τ_{sc}: Network-induced delay from the sensor to the controller, i.e., backward channel delay;
- τ_{ca}: Network-induced delay from the controller to the actuator, i.e., forward channel delay.

The two types of network-induced delays may have different characteristics [Nilsson *et al.* (1998)]. In most cases, however, these delays are not treated separately and only the round trip delay is of interest [Fan *et al.* (2006); Hespanha *et al.* (2007)].

According to the types of the communication networks being used in NCSs, the characteristics of the network-induced delay vary as follows [Lian (2001); Tipsuwan and Chow (2003)].

- Cyclic service networks (e.g., Toking-Ring, Toking-Bus): Bounded delays and can be regarded as constant for certain occasions;
- Random access networks (e.g., Ethernet, CAN): Random and unbounded delays;
- Priority order networks(e.g., DeviceNet): Bounded delays for the data packets with higher priority and unbounded delays for those with lower priority.

Network-induced delay is one of the most important characteristics of NCSs which has been widely addressed in the literature to date, see, e.g., in He (2004); Liu and Shen (2006); Schenato *et al.* (2007); Yue and Han (2005).

B. Data packet dropout

It is well known that the transmission error in communication networks is inevitable, which in the case of NCSs then produces a situation called "data packet dropout". Data packet dropout can occur either in the backward or forward channel, and it makes either the sensing data or the control signals unavailable to NCSs, thus significantly degrading the performance of NCSs.

In communication networks, two different strategies are applied when a data packet is lost, that is, either to send the packet again or simply discard it. Using the terms from communication networks, these two strategies are called Transmission Control Protocol (TCP) and User Datagram Protocol (UDP) respectively [Stallings (2000)]. It is readily seen that with TCP, all the data packets will be received successfully, although it may take a considerably long time for some data packets; while with UDP, some data packets will be lost forever.

As far as NCSs is concerned, UDP is used in most applications due to the real-time requirement and the robustness of control systems. As a result, the effect of data packet dropout in NCSs has to be explicitly considered, as done in, e.g., Azimi-Sadjadi (2003); Imer *et al.* (2006); Xiong and Lam (2007) and the packet-based control approach in this chapter.

C. Data packet disorder

In most communication networks, different data packets suffer different delays, as mentioned above; it therefore produces a situation where a data packet sent earlier may arrive at the destination later or vice versa, that is, data packet disorder. This characteristic in NCSs can mean that a newly arrived control signal in NCSs may not be the latest, which never occurs in conventional control systems. Therefore, the effect of data packet disorder has to be specially dealt with. This, however, has rarely been touched to date. We will show later that this effect can be effectively overcome by using the comparison process in the packet-based control approach.

8.3 Packet-Based Control for Networked Control Systems

The NCS setup considered in this chapter is shown in Fig. 8.2, where $\tau_{sc,k}$ and $\tau_{ca,k}$ are the backward and forward channel delays, respectively. The plant is linear in discrete-time, represented by

$$\mathcal{S}_d\colon x(k+1) = Ax(k) + Bu(k)$$
$$y(k) = Cx(k) \tag{8.1}$$

with the system states $x(k) \in \mathbb{R}^n$, the control signals $u(k) \in \mathbb{R}^m$, and the system matrices $A \in \mathbb{R}^{n\times n}$, $B \in \mathbb{R}^{n\times m}$ and $C \in \mathbb{R}^{r\times n}$.

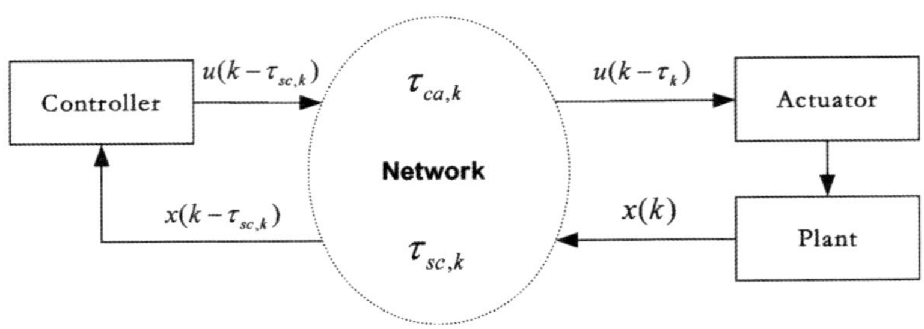

Fig. 8.2 The block diagram of networked control systems in discrete time

8.3.1 *Packet-based control for NCSs: A unified model*

For the system in (8.1), the conventional state feedback law is usually obtained as follows without considering the communication constraints in NCSs,

$$u(k) = Kx(k) \tag{8.2}$$

where the feedback gain K is time-invariant.

However, when the network-induced delay is considered, the state feedback law can not be simply defined as in (8.2) due to the unavailability of the current state information. The resulting control law using conventional approaches in time delay systems theory would have the following form

$$u(k) = Kx(k - \tau_k) \tag{8.3}$$

where the effect of the delay is not been specially treated in the design. Furthermore, when data packet dropout is also present, it can be seen from Fig. 8.2 that no matter where data packet dropout occurs, a certain control input will be unavailable to the actuator. In conventional time delay systems theory, there are mainly two ways to deal with this situation, either use the previous control input or adopt zero control [Richard (2003)]. For example, in Wu and Chen (2007), the last step of the control signal is used in the case of an unsuccessful transmission, as follows,

$$u(k) = \begin{cases} \bar{u}(k) & \text{if transmitted successfully,} \\ u(k-1) & \text{otherwise,} \end{cases} \tag{8.4}$$

where $\bar{u}(k)$ is the newly arrived control signal at time k.

Although the conventional control strategies in (8.3) and (8.4) are simple to implement, they are conservative in that they overlook the potential of providing an active prediction for the unavailable control input using available information of the system dynamics and previous system trajectory.

On the contrary, the control law based on the packet-based control approach is obtained as follows with explicit compensation for the communication constraint (see Algorithm 8.1 to be given later)

$$u(k) = K(\tau^*_{sc,k}, \tau^*_{ca,k})x(k - \tau^*_{sc,k} - \tau^*_{ca,k}), \tag{8.5}$$

or simply (see Algorithm 8.2 which will be given later)

$$u(k) = K(\tau^*_k)x(k - \tau^*_k) \tag{8.6}$$

where $\tau^*_{sc,k}$ and $\tau^*_{ca,k}$ are the network-induced delays of the control action that is actually applied to the plant at time k and $\tau^*_k = \tau^*_{sc,k} + \tau^*_{ca,k}$.

It is seen from the control laws in (8.5) and (8.6) that in the packet-based control approach, different feedback gains apply for different network conditions. This is why we call it a "Varying Feedback Gain" (VFG) scheme for NCSs. As will be presented later, these packet-based control laws can actively deal with the network-induced delay, data packet dropout and data packet disorder simultaneously, and therefore can be regarded as a unified model for NCSs. This control strategy can be compared with the conventional approach as in (8.3) and (8.4) where no active compensation is available.

Remark 8.1. In Zhang *et al.* (2005), the authors noticed the unavailability of the forward channel delay $\tau_{ca,k}$ and a controller was designed with the following form

$$u(k) = K(\tau_{sc,k}, \tau_{ca,k-1})x(k - \tau_{sc,k} - \tau_{ca,k}) \tag{8.7}$$

where the forward channel delay of the last step $\tau_{ca,k-1}$ was used instead. However, actually even $\tau_{ca,k-1}$ is generally unavailable for the controller in NCSs since in the case of a arbitrary forward channel delay, $\tau_{ca,k-1}$ can not be known to the controller until the controller receives information of $\tau_{ca,k-1}$ from the actuator. Therefore, it is seen that $\tau_{ca,k-1}$ can not be available for the controller earlier than time $k - 1 + \tau_{ca,k-1}$ even if an additional delay-free channel exists to send the information of $\tau_{ca,k-1}$ from the actuator to the controller. As a result, the above model in (8.7) is inappropriate in practice unless a special control structure is designed for the networked control environment as done in this chapter.

8.3.2 *Design of the packet-based control scheme*

For the design of the packet-based control scheme, the following assumptions are required.

Assumption 8.1. The control components in the considered NCS including the sensor, the controller and the actuator, are time-synchronized and the data packets sent from both the sensor and the controller are time-stamped.

Assumption 8.2. The sum of the maximum forward (backward) channel delay and the maximum number of continuous data packet dropout is upper bounded by $\bar{\tau}_{ca}$ ($\bar{\tau}_{sc}$ accordingly) and

$$\bar{\tau}_{ca} \leq \frac{B_p}{B_c} - 1 \tag{8.8}$$

where B_p is the size of the effective load of the data packet and B_c is the bits required to encode a single step control signal.

Remark 8.2. From Assumption 8.1, the network-induced delay that each data packet experiences is known by the controller and the actuator on its arrival.

Remark 8.3. Assumption 8.2 is required due to the need of packing the forward control signals and compensating for the network-induced delay in the packet-based control approach, which will be detailed later. The constraint in (8.8) is easy to be satisfied, e.g., B_p = 368bit for an Ethernet IEEE 802.3 frame which is often used [Stallings (2000)], while an 8-bit data (i.e., B_c = 8bit) can encode $2^8 = 256$ different control actions which is ample for most control implementations; In this case, 45 steps of forward channel delay is allowed by (8.8) which can actually meet the requirements of most practical control systems.

The block diagram of the packet-based control structure is illustrated in Fig. 8.3. It is distinct from a conventional control structure in two aspects: the specially designed packet-based controller and the corresponding Control Action Selector (CAS) at the actuator side.

In order to implement the control law in (8.5) and (8.6), we take advantage of the packet-based transmission of the network to design a packet-based controller

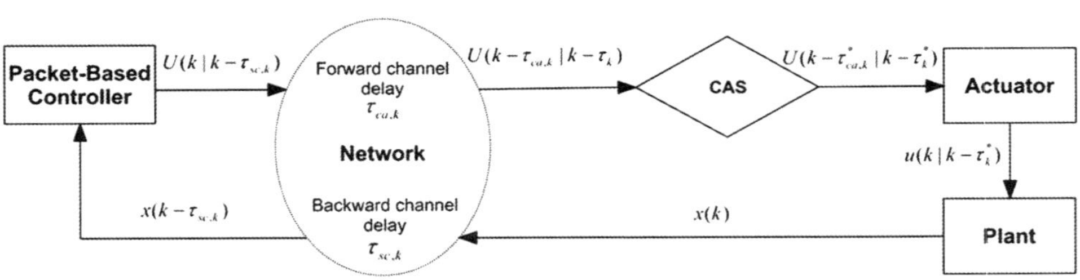

Fig. 8.3 Packet-based networked control systems in discrete time (with time synchronization)

instead of trying to obtain directly the current forward channel delay as this is actually impossible in practice. As for the control law in (8.5), the packet-based controller determines a sequence of forward control actions (called "Forward Control Sequence" (FCS)) as follows and sends them together in one data packet to the actuator,

$$U_1(k|k-\tau_{sc,k}) = [u(k|k-\tau_{sc,k}) \ \ldots \ u(k+\bar{\tau}_{cc}|k-\tau_{sc,k}]^{\mathrm{T}} \tag{8.9}$$

where $u(k+i|k-\tau_{sc,k}), i = 0, 1, \ldots, \tau_{ca,k}$ are the forward control action predictions based on information up to time $k - \tau_{sc,k}$.

When a data packet arrives at the actuator, the designed CAS compares its time stamp with the one already in CAS and only the one with the latest time stamp is saved. Denote the forward control sequence already in CAS and the one just arrived by $U_1(k_1 - \tau_{ca,k_1}|k_1 - \tau_{k_1})$ and $U_1(k_2 - \tau_{ca,k_2}|k_2 - \tau_{k_2})$ respectively, then the chosen sequence is determined by the following comparison rule

$$U_1(k-\tau^*_{ca,k}|k-\tau^*_k) = \begin{cases} U_1(k_2-\tau_{ca,k_2}|k_2-\tau_{k_2}), & \text{if } k_1-\tau_{k_1} < k_2-\tau_{k_2}; \\ U_1(k_1-\tau_{ca,k_1}|k_1-\tau_{k_1}), & \text{otherwise.} \end{cases} \tag{8.10}$$

The comparison process is introduced to deal with data packet disorder. After the comparison process, only the latest available information is used and the effect of data packet disorder is effectively overcome.

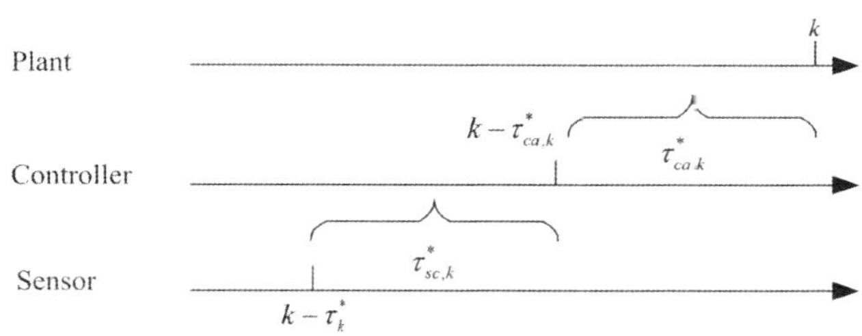

Fig. 8.4 Timeline in packet-based networked control systems

CAS also determines the appropriate control action from the FCS $U_1(k-\tau^*_{ca,k}|k-\tau^*_k)$ at each time instant as follows

$$u(k) = u(k|k-\tau^*_k). \tag{8.11}$$

The timeline of the packet-based control approach is illustrated in Fig. 8.4. It is necessary to point out that the appropriate control action determined by (8.11) is always available provided Assumption 8.2 holds.

The packet-based control algorithm under Assumptions 8.1 and 8.2 can now be summarized as follows.

In practice, it is often the case that we do not need to identify separately the forward and backward channel delays since it is normally the round trip delay that affects the system performance. In such a case, the simpler control law in (8.6)

Algorithm 8.1 Packet-based control with the control law in (8.5)

1: At time k, if the packet-based controller receives the delayed state data $x(k - \tau_{sc,k})$, it then:
(1a) Reads the current backward channel delay $\tau_{sc,k}$;
(1b) Calculates the FCS as in (8.9);
(1c) Packs $U_1(k|k - \tau_{sc,k})$ and sends it to the actuator in one data packet with time stamps k and $\tau_{sc,k}$.
If no data packet is received at time k, then let $k = k + 1$ and wait for the next time instant.
2: CAS updates its FCS by (8.10) once a data packet arrives;
3: The control action in (8.11) is picked out from CAS and applied to the plant.

instead of that in (8.5) is applied, for which the following assumption is required instead of Assumption 8.2.

Assumption 8.3. The sum of the maximum network-induced delay and the maximum number of continuous data packet dropout in round trip is upper bounded by $\bar{\tau}$ and

$$\bar{\tau} \le \frac{B_p}{B_c} - 1. \tag{8.12}$$

With the above assumption, the packet-based controller is modified as follows

$$U_2(k - \tau_{sc,k}|k - \tau_{sc,k}) = [u(k - \tau_{sc,k}|k - \tau_{sc,k}) \ \ldots \ u(k - \tau_{sc,k} + \bar{\tau}|k - \tau_{sc,k}]^{\mathsf{T}}. \tag{8.13}$$

It is noticed that in such a case the backward channel delay $\tau_{sc,k}$ is not required for the controller, since the controller simply produces $(\bar{\tau} + 1)$ step forward control actions whenever a data packet containing sensing data arrives. This relaxation implies that the time-synchronization between the controller and the actuator (plant) is not required any more and thus Assumption 8.1 can then be modified as follows.

Assumption 8.4. The data packets sent from the sensor are time-stamped.

The comparison rule in (8.10) and the determination of the actual control action in (8.11) remain unchanged since both of them are based on the round trip delay τ_k. The packet-based control algorithm with the control law in (8.6) can now be summarized as follows based on Assumptions 8.3 and 8.4.

The block diagram of the packet-based control approach in Algorithm 8.2 is illustrated in Fig. 8.5.

8.4 Stability of Packet-Based Networked Control Systems

In this section the stability criteria for the system in (8.1) using the aforementioned packet-based control approach with the control laws in (8.5) and (8.6) are investi-

Algorithm 8.2 Packet-based control with the control law in (8.6)

1: At time k, if the packet-based controller receives the delayed state data $x(k - \tau_{sc,k})$, then,
   (1a) Calculate the FCS as in (8.13);
   (1b) Pack $U_2(k - \tau_{sc,k}|k - \tau_{sc,k})$ and send it to the actuator in one data packet.
   If no data packet is received at time k, then let $k = k - 1$ and wait for the next time instant.
2: Steps 2 and 3 remain the same as in Algorithm 8.1.

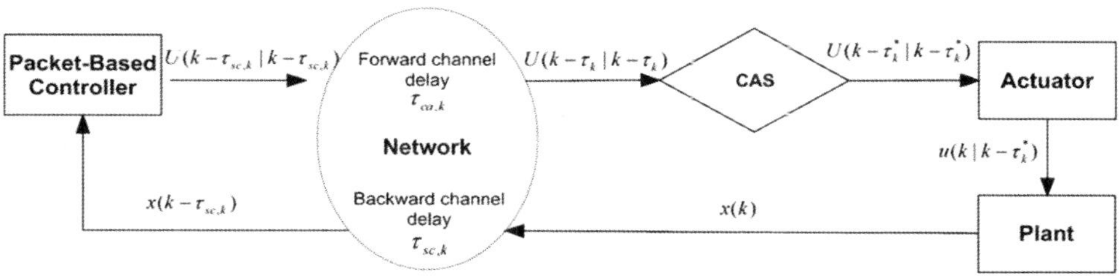

Fig. 8.5 Packet-based networked control systems in discrete time (without time synchronization)

gated. Two stability analysis approaches, i.e., results from switched system theory and delay-dependent analysis, are applied, by modeling the closed-loop system into different forms. Unless otherwise specified, all the stability related notions in this chapter are under the Lyapunov stability framework.

8.4.1 *A switched system theory approach*

An intuitive observation on the packet-based control approach is that, at every execution time, a specific control action is determined by the CAS according to the current network condition. Thus, regarding this selection process as "switches" among different subsystems, then yields the following analysis from the viewpoint of switched system theory.

Let $X(k) = [x(k) \; x(k-1) \; \cdots \; x(k-\bar{\tau})]$. The closed-loop formula for the system in (8.1) using the packet-based controllers in (8.5) and (8.6) can then be represented in augmented forms as

$$X(k+1) = \Xi_{\tau^*_{sc,k}, \tau^*_{ca,k}} X(k) \tag{8.14}$$

and

$$X(k+1) = \Xi_{\tau^*_k} X(k) \tag{8.15}$$

respectively, where

$$\Xi_{\tau^*_{sc,k},\tau^*_{ca,k}} = \begin{pmatrix} A & \cdots & BK_{\tau^*_{sc,k},\tau^*_{ca,k}} & \cdots & \cdots \\ I_n & & & & 0 \\ & I_n & & & 0 \\ & & \ddots & & \vdots \\ & & & I_n & 0 \end{pmatrix}$$

$$\Xi_{\tau^*_k} = \begin{pmatrix} A & \cdots & BK_{\tau^*_k} & \cdots & \cdots \\ I_n & & & & 0 \\ & I_n & & & 0 \\ & & \ddots & & \vdots \\ & & & I_n & 0 \end{pmatrix}$$

and I_n is the identity matrix with rank n.

With the closed-loop system model in (8.14), we then obtain the following stability criterion.

Theorem 8.1. *The closed-loop system in (8.14) is stable if there exists a positive definite solution $P = P^{\mathsf{T}} > 0$ for the following $(\bar{\tau}_{sc} + 1) \times (\bar{\tau}_{ca} + 1)$ LMIs*

$$\Xi^{\mathsf{T}}_{\tau^*_{sc,k},\tau^*_{ca,k}} P \Xi_{\tau^*_{sc,k},\tau^*_{ca,k}} - P < 0.$$

Proof. Let $V(k) = X^{\mathsf{T}}(k)PX(k)$ be a Lyapunov candidate, and then its increment along the system in (8.14) can be obtained as

$$\begin{aligned} \Delta V(k) &= V(k+1) - V(k) \\ &= X^{\mathsf{T}}(k)(\Xi^{\mathsf{T}}_{\tau^*_{sc,k},\tau^*_{ca,k}} P \Xi_{\tau^*_{sc,k},\tau^*_{ca,k}} - P)X(k) \\ &< 0 \end{aligned}$$

which completes the proof. □

The following stability criterion for the closed-loop system model in (8.15) readily follows from Theorem 8.1.

Proposition 8.1. *The closed-loop system in (8.15) is stable if there exists a positive definite solution $P = P^{\mathsf{T}} > 0$ for the following $(\bar{\tau} + 1)$ LMIs*

$$\Xi^{\mathsf{T}}_{\tau^*_k} P \Xi_{\tau^*_k} - P < 0.$$

8.4.2 *A delay dependent analysis approach*

In this subsection, the closed-loop stability is investigated using a delay dependent analysis approach as in He (2004).

Without augmenting the system states as done in the last subsection, the closed-loop formula for the system in (8.1) using the packet-based controller in (8.5) can be obtained as

$$x(k+1) = Ax(k) + BK(\tau^*_{ca,k}, \tau^*_{sc,k})x(k - \tau^*_k). \tag{8.16}$$

It is noticed that in practice there is at least a single step delay in both the forward and backward channels, and therefore we have $\tau^*_k = \tau^*_{sc,k} + \tau^*_{ca,k} \geq 2$. Let $\bar{K} = \max_{2 \leq \tau^*_k \leq \bar{\tau}_{ca} + \bar{\tau}_{sc}} ||K(\tau^*_{ca,k}, \tau^*_{sc,k})||$ where $|| \cdot ||$ denotes the Euclidean norm. Then $BK(\tau^*_{ca,k}, \tau^*_{sc,k})$ can be represented by

$$BK(\tau^*_{ca,k}, \tau^*_{sc,k}) = B_m \cdot K'(\tau^*_{ca,k}, \tau^*_{sc.k}) \tag{8.17}$$

where $B_m = \bar{K}B$ is a constant matrix and $K'(\tau^*_{ca,k}, \tau^*_{sc,k}) = \frac{K(\tau^*_{ca,k}, \tau^*_{sc,k})}{\bar{K}}$. It is readily to conclude that $||K'(\tau^*_{ca,k}, \tau^*_{sc,k})|| \leq 1, \ \forall\ 1 \leq \tau^*_{ca,k} \leq \bar{\tau}_{ca}, 1 \leq \tau^*_{sc,k} \leq \bar{\tau}_{sc}$.

Theorem 8.2. *If there exists* $P_i = P_i^{\mathsf{T}} > 0,\ i = 1, 2, 3,\ X = \begin{pmatrix} X_{11} & X_{12} \\ * & X_{22} \end{pmatrix} \geq 0$, $N_i, i = 1, 2$ *with appropriate dimensions and* $\gamma > 0$ *satisfying the following two LMIs,*

$$\begin{pmatrix} X_{11} & X_{12} & N_1 \\ * & X_{22} & N_2 \\ * & * & P_3 \end{pmatrix} \geq 0 \tag{8.18}$$

$$\begin{pmatrix} \Phi_{11} & \Phi_{12} & (A-I)^{\mathsf{T}}H & P_1 B_m \\ * & \Phi_{22} + \gamma I & 0 & 0 \\ * & * & -H & HB_m \\ * & * & * & -\gamma I \end{pmatrix} < 0 \tag{8.19}$$

where

$$\begin{aligned} \Phi_{11} &= (\bar{\tau} - 1)P_2 + P_1(A - I) + (A - I)^{\mathsf{T}} P_1 \\ &\quad + N_1 + N_1^{\mathsf{T}} + \bar{\tau} X_{11}, \\ \Phi_{12} &= N_2^{\mathsf{T}} - N_1 + \bar{\tau} X_{12}, \\ \Phi_{22} &= -P_2 - N_2 - N_2^{\mathsf{T}} + \bar{\tau} X_{22}, \\ H &= P_1 + \bar{\tau} P_3, \end{aligned}$$

then the closed-loop system in (8.16) is stable.

Proof. Let $d_1 = 2$, $d_2 = \bar{\tau}$, and $\Delta A_d(k) = BK(\tau^*_{ca,k}, \tau^*_{sc,k})$ in Theorem 7.3 in He (2004), then the above theorem can be obtained using the same techniques as in He (2004). □

Remark 8.4. Following the same procedure, the stability criterion for the system in (8.1) using the packet-based controller in (8.6) can be obtained analogously.

Remark 8.5. It is seen that the aforementioned stability criteria are simple propositions of existing results from switched system theory and delay-dependent analysis respectively. The former emphasizes on the "switch" property of packet-based control while the latter on the "time delay" property of the system. However, none of them is perfect: the switched system theory approach does not consider explicitly the time delay by augmenting the system states, whilst the delay-dependent analysis approach neglects the "switch" property which therefore leads to a stability criterion that is valid for any type of delay changes within an allowed upper bound. Hence, better stability analysis is still needed for the proposed packet-based control approach.

Up to now we have provided the packet-based control structure for NCSs whilst the controller design remains open. Indeed, under the packet-based control framework, any controller design approach can be used to obtain the VFGs as in (8.5) and (8.6) provided it can result in a satisfactory system performance. In the following section, a GPC-based controller is designed as an example.

8.5 Packet-Based Controller Design: A GPC-Based Approach

In GPC, an optimization process is repeated at every control instant to determine a sequence of forward control signals that optimize future open-loop plant behavior based on current system information. Different from conventional GPC implementations where only the first control prediction is actually applied to the plant, in this chapter the first $\bar{\tau}_{ca}+1$ (or $\bar{\tau}+1$ for the control law in (8.6)) forward control predictions are all used to implement the packt-based control approach proposed in the previous section.

Taking account of the communication constraints in NCSs which delay the sensing data, the objective function for open-loop optimization in GPC is therefore defined as follows,

$$J_{k,\tau_{sc,k}} = X^{\mathsf{T}}(k|k-\tau_{sc,k})QX(k|k-\tau_{sc,k}) + U'^{\mathsf{T}}(k|k-\tau_{sc,k})RU'(k|k-\tau_{sc,k}) \quad (8.20)$$

where $J_{k,\tau_{sc,k}}$ is the objective function at time k, $U'(k|k-\tau_{sc,k}) = [u(k-\tau_{sc,k}|k-\tau_{sc,k})\cdots u(k+N_u-1|k-\tau_{sc,k})]^{\mathsf{T}}$ is the FCS, $X(k|k-\tau_{sc,k}) = [x(k+1|k-\tau_{sc,k})\cdots x(k+N_p|k-\tau_{sc,k})]^{\mathsf{T}}$ is the predictive state trajectory, Q and R are constant weighting matrixes and N_p and N_u are the prediction horizon and the control horizon respectively.

The predictive states at time k based on the state at time $k-\tau_{sc,k}$ and the control sequences from $k-\tau_{sc,k}$ can be obtained by iteration as

$$\begin{aligned} x(k+j|k-\tau_{sc,k}) &= A^{j+\tau_{sc,k}}x(k-\tau_{sc,k}) \\ &\quad + \sum_{l=-\tau_{sc,k}}^{j-1} A^{j-l-1}Bu(k+l|k-\tau_{sc,k}), j=1,2,\cdots,N_p. \end{aligned}$$

Thus, we obtain

$$X(k|k-\tau_{sc,k}) = E_{\tau_{sc,k}}x(k-\tau_{sc,k}) + F_{\tau_{sc,k}}U'(k|k-\tau_{sc,k})$$

where $E_{\tau_{sc,k}} = [(A^{\tau_{sc,k}+1})^{\mathsf{T}} \cdots (A^{\tau_{sc,k}+N_p})^{\mathsf{T}}]^{\mathsf{T}}$ and $F_{\tau_{sc,k}}$ is a block lower triangular matrix with its non-null blocks defined by $(F_{\tau_{sc,k}})_{ij} = A^{\tau_{sc,k}+i-j}B$, $j - i \leq \tau_{sc,k}$.

The optimal control inputs can then be calculated by substituting the above equation to (8.20) and optimizing $J_{k,\tau_{sc,k}}$, which turns out to be state feedback control,

$$u(k+j|k-\tau_{sc,k}) = K_{\tau_{sc,k},j}x(k-\tau_{sc,k}), \quad j = 0, 1, 2, \ldots, \bar{\tau}_{ca}$$

where $K_{\tau_{sc,k}} = [K_{\tau_{sc,k},0}^{\mathsf{T}} \cdots K_{\tau_{sc,k},\bar{\tau}_{ca}}^{\mathsf{T}}]^{\mathsf{T}}$, can be calculated by

$$K_{\tau_{sc,k}} = -M_{\tau_{sc,k}}(F_{\tau_{sc,k}}^{\mathsf{T}} Q F_{\tau_{sc,k}} + R)^{-1} F_{\tau_{sc,k}}^{\mathsf{T}} Q E_{\tau_{sc,k}}$$

and $M_{\tau_{sc,k}} = [0_{m(\bar{\tau}_{ca}+1)\times m\tau_{sc,k}} \; I_{m(\bar{\tau}_{ca}+1)\times m(\bar{\tau}_{ca}+1)} \; 0_{m(\bar{\tau}_{ca}-1)\times m(N_u-\bar{\tau}_{ca})}]$. The FCS in (8.9) for Algorithm 8.1 can then be constructed by

$$U(k|k-\tau_{sc,k}) = K_{\tau_{sc,k}}x(k-\tau_{sc,k}). \tag{8.21}$$

The FCS in (8.13) for Algorithm 8.2 can also be constructed analogously.

Remark 8.6 (State observer). *If the state vector x is not available, an observer must be included*

$$\hat{x}(k+1|k) = A\hat{x}(k|k-1) + Bu(k) + L(y_m(k) - C\hat{x}(k|k-1)) \tag{8.22}$$

where $\hat{x}(k)$ is the observed state at time k, and $y_m(k)$ is the measured output. If the plant is subject to white noise disturbances affecting the process and the output with known covariance matrices, the observer becomes a Kalman filter and the gain L is calculated solving a Riccati equation.

Remark 8.7. In Liu *et al.* (2007), state feedback $u_k = K\hat{x}_{k|k-\tau_{sc,k}}$ was also used, where K was artificially chosen without consideration of the communication constraints and $\hat{x}_{k|k-\tau_{sc,k}}$ depends on "the state estimation $\hat{x}_{k-\tau_{sc,k}|k-\tau_{sc,k}-1}$, the past control input up to u_{k-1}, and the past output up to $y_{k-\tau_{sc,k}}$ of the system". However, under the compensation scheme in the forward channel in Liu *et al.* (2007), the whole sequence of the optimal forward control signals $U(k|k-\tau_{sc,k}^*)$ is sent to the actuator and only one of them is chosen to be applied to the plant. Thus, unless information from the actuator is received we have no idea which control prediction was really used if the data packets in the forward channel were arbitrarily delayed. Hence, the use of the previous control inputs implies an additional communication channel which can send the applied control inputs to the controller efficiently. Without such a channel, the approaches proposed in these publications are only applicable to such a situation where there is no delay or data packet dropout in the forward channel. This requirement is relaxed in this section by redesigning the controller where the objective function includes as part of it the previous control increment sequence from $k - \tau_{sc,k}$ to $k - 1$. As a result, the FCS at time k are only based on data up to time $k - \tau_{sc,k}$, which is always available in practice.

8.6 Numerical and Experimental Examples

In this section, numerical and experimental examples are presented to illustrate the effectiveness of the proposed packet-based control approach to NCSs.

8.6.1 *Numerical examples*

Example 8.1 (Validating the GPC-based controller). A second order system in (8.1) is adopted, which is open-loop unstable with the following system matrices,

$$A = \begin{pmatrix} 0.98 & 0.1 \\ 0 & 1 \end{pmatrix}, \quad B = \begin{pmatrix} 0.04 \\ 0.1 \end{pmatrix}, \quad C = \begin{pmatrix} 1 & 0 \\ 0 & 1 \end{pmatrix}.$$

In order to illustrate the effectiveness of the proposed packet-based control approach compared with conventional design approach, the Linear Quadratic Optimal (LQR) control method is used to design a state feedback law for this system without consideration of the communication constraints, which yields the time-invariant feedback gain $K_{LQR} = [0.7044\ 1.3611]$. In the simulation, the initial state $x_0 = [-1\ -1]^{\mathsf{T}}$, the upper bounds of the delays and continuous dropout (disorder) are $\bar{\tau} = 3, \bar{\tau}_{ca} = 2, \bar{\tau}_{sc} = 1$, and the control and prediction horizon in the GPC-based controller proposed in Section 8.5 are set as $N_u = 8, N_p = 10$ respectively. The delays in both channels are set to vary arbitrarily within their upper bounds.

The simulation results show that it is unstable using this LQR controller (Fig. 8.6) while it is stable using the packet-based control approach (Fig. 8.7) in the presence of communication constraints.

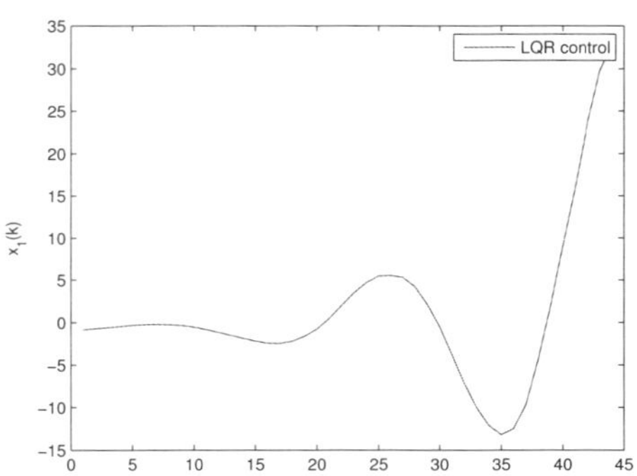

Fig. 8.6 System is unstable using LQR controller in Example 8.1

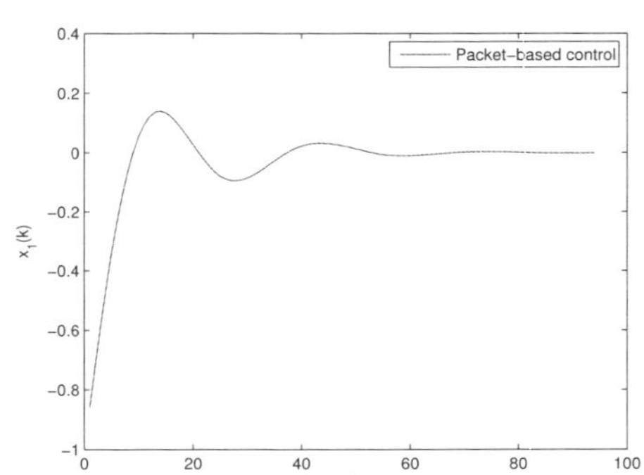

Fig. 8.7 System is stable using packet-based controller in Example 8.1

Example 8.2 (Example 1 in Liu *et al.* (2007)). The system matrices for the system in (8.1) are as follows,

$$A = \begin{pmatrix} 1.0100 & 0.2710 & -0.4880 \\ 0.4820 & 0.1000 & 0.2400 \\ 0.0020 & 0.3681 & 0.7070 \end{pmatrix}, \quad B = \begin{pmatrix} 5 & 5 \\ 3 & -2 \\ 5 & 4 \end{pmatrix}, \quad C = \begin{pmatrix} 1 & 2 & 3 \\ 4 & 3 & 1 \end{pmatrix}.$$

In Liu *et al.* (2007), the above system is illustrated to be stable with the observer in (8.22) where $\bar{\tau}_{sc} = 2$, $\bar{\tau}_{ca} = 1$, and

$$L = \begin{pmatrix} -0.3614 & 0.3326 \\ 0.0332 & 0.0576 \\ 0.2481 & -0.0750 \end{pmatrix}, \quad K = \begin{pmatrix} 0.5858 & -0.1347 & -0.4543 \\ -0.5550 & 0.0461 & 0.4721 \end{pmatrix}.$$

However, using the packet-based control approach, this system is unstable with the same $\bar{\tau}_{sc}$, $\bar{\tau}_{ca}$ and L (see Fig. 8.8. Other parameters: $N_u = 8$, $N_p = 10$.). This fact seems to mean the approach in Liu *et al.* (2007) is better than the approach in this chapter, but we need to remember that the approach in Liu *et al.* (2007) takes advantage of more information to design the predictive controller and some of the information used is not easy to obtain in practice (Remark 8.7). On the other hand, the simulation results do illustrate that the VFG scheme in this chapter is superior to the previous Fixed Feedback Gain (FFG) scheme in Liu *et al.* (2007), where the same system is stable using the approach in this chapter when $\bar{\tau}_{sc} = \bar{\tau}_{sc} = 1$ (Fig. 8.9) and yet is unstable using the same state feedback in (8.16) with the fixed K above (Fig. 8.10).

Example 8.3 (Validating Theorem 8.2). The system matrices for the system in (8.1) are set as

$$A = \begin{pmatrix} 0.7 & 0.2 \\ 0.3 & 0.5 \end{pmatrix}, \quad B = \begin{pmatrix} 0.05 \\ 0.2 \end{pmatrix}, \quad C = (1 \;\; 0).$$

This system can be shown using Theorem 8.2 to be stable under $\bar{\tau}_{sc} = 3$, $\bar{\tau}_{ca} = 2$, $N_u = 8$, $N_p = 10$. The simulation result is illustrated in Fig. 8.11.

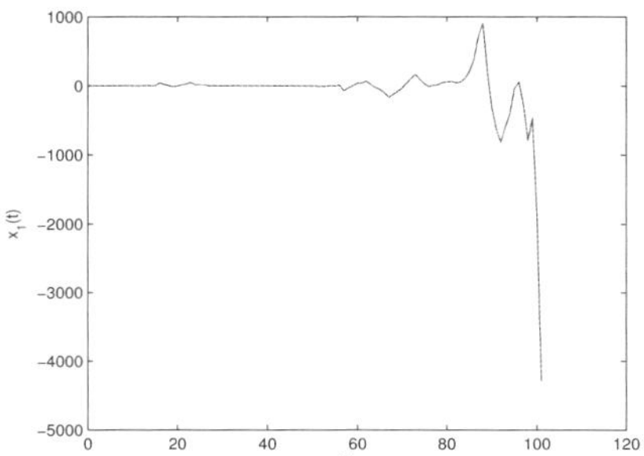

Fig. 8.8 Example 8.2. Packet-based control, unstable, $\bar{\tau}_{sc} = 2$, $\bar{\tau}_{ca} = 1$

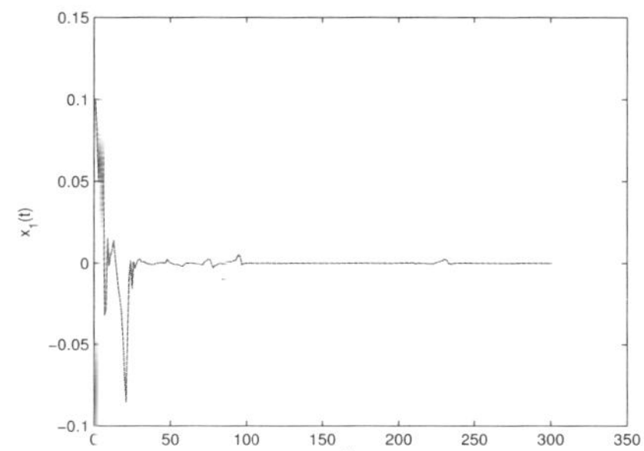

Fig. 8.9 Example 8.2. VFG scheme, stable, $\bar{\tau}_{sc} = 1$, $\bar{\tau}_{ca} = 1$

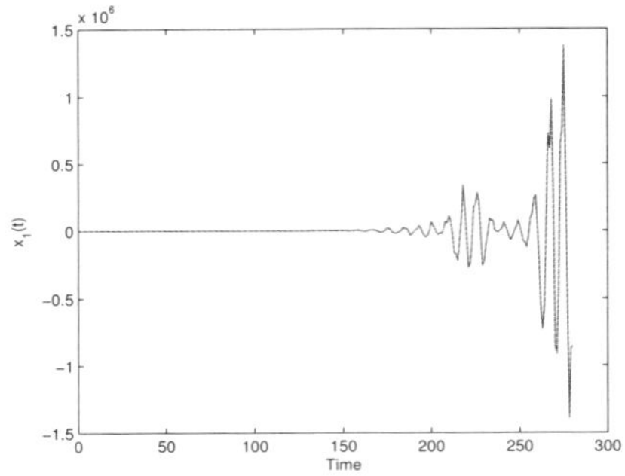

Fig. 8.10 Example 8.2. FFG scheme, unstable, $\bar{\tau}_{sc} = 1$, $\bar{\tau}_{ca} = 1$

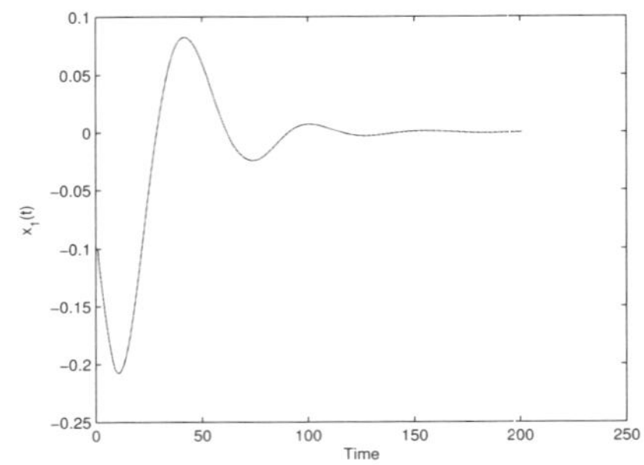

Fig. 8.11 Example 8.3. Packet-based control, stable, $\bar{\tau}_{sc} = 3$, $\bar{\tau}_{ca} = 2$

8.6.2 *Experimental example*

Example 8.4 (Experimental example). In this example, an Internet-based test rig is used to verify the effectiveness of the packet-based control approach. This test rig consists of a plant (DC servo system, see Fig. 8.12) which is located in the University of Glamorgan, Pontypridd, UK, and a remote controller which is located in the Institute of Automation, Chinese Academy of Sciences, Beijing, China (see Fig. 8.13). The plant and the controller are connected via the Internet. A web-based laboratory is also available at *www.ncslab.net* to implement experiments online. For further information of this test rig, the reader is referred to Hu (2008); Hu *et al.* (2007).

Fig. 8.12 The DC servo plant at the University of Glamorgan

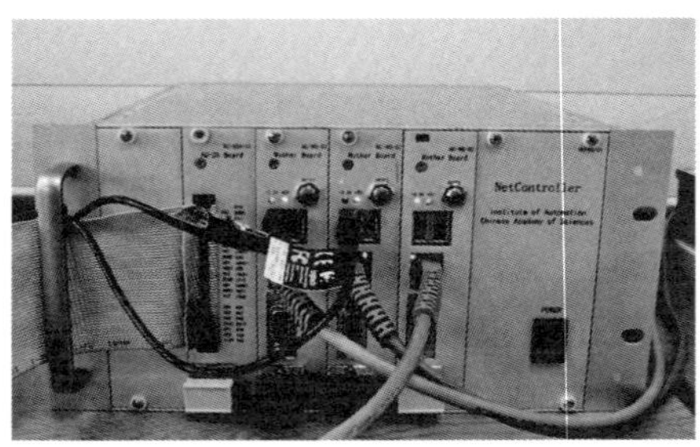

Fig. 8.13 The network controller at the Chinese Academy of Sciences

The DC servo system is identified by Hu (2008) to be a third-order system and in state-space description has the following system matrices:

$$A = \begin{pmatrix} 1.12 & 0.213 & -0.333 \\ 1 & 0 & 0 \\ 0 & 1 & 0 \end{pmatrix}, \quad B = \begin{pmatrix} 1 \\ 0 \\ 0 \end{pmatrix}, \quad C = \begin{pmatrix} 0.0541 & 0.0050 & 0.0001 \end{pmatrix}.$$

To enable the use of state feedback in the packet-based control approach, a state observer as in Remark 8.6 is designed with $L = [6\ 6\ 6]^{\mathsf{T}}$. The packet-based controller is calculated by using the GPC-based controller design approach in Section 8.5. To this end, the upper bounds of the network-induced delays (data packet dropout as well) in both forward and backward channels are assumed to be 4 steps of the sampling period (The sampling period is set as 0.04s and thus the delay bounds are 0.16s for both backward and forward channel delays.), since typically the round trip delay in the experiment is not larger than 0.32s. The packet-based controller can then be obtained as:

$$K = [K_0^{\mathsf{T}}\ \ K_1^{\mathsf{T}}\ \ K_2^{\mathsf{T}}\ \ K_3^{\mathsf{T}}\ \ K_4^{\mathsf{T}}]^{\mathsf{T}},\quad K_0 = \begin{pmatrix} -1.3217 & 0.1276 & 0.4296 \\ -0.1356 & 0.0306 & 0.0445 \\ 0.2688 & -0.0220 & -0.0816 \\ 0.1255 & -0.0096 & -0.0396 \\ 0.0610 & -0.0061 & -0.0190 \end{pmatrix},$$

$$K_1 = \begin{pmatrix} -0.2193 & 0.0219 & 0.0844 \\ 0.2177 & -0.0032 & -0.0662 \\ 0.1298 & -0.0087 & -0.0381 \\ 0.0621 & -0.0035 & -0.0198 \\ 0.0114 & -0.0014 & -0.0035 \end{pmatrix},\quad K_2 = \begin{pmatrix} 0.1120 & 0.0005 & -0.0201 \\ 0.1183 & 0.0032 & -0.0348 \\ 0.0726 & -0.0050 & -0.0201 \\ 0.0192 & -0.0007 & -0.0062 \\ 0.0035 & -0.0009 & -0.0010 \end{pmatrix},$$

$$K_3 = \begin{pmatrix} 0.0894 & 0.0021 & -0.0130 \\ 0.0832 & 0.0056 & -0.0239 \\ 0.0398 & -0.0028 & -0.0099 \\ 0.0106 & -0.0001 & -0.0035 \\ 0.0007 & -0.0007 & -0.0002 \end{pmatrix},\quad K_4 = \begin{pmatrix} 0.0721 & 0.0030 & -0.0076 \\ 0.0515 & 0.0073 & -0.0140 \\ 0.0267 & -0.0021 & -0.0058 \\ 0.0059 & 0.0001 & -0.0021 \\ 0.0005 & -0.0007 & -0.0001 \end{pmatrix},$$

where the subscripts of K_0, K_1, K_2, K_3 and K_4 are with respect to different backward channel delays.

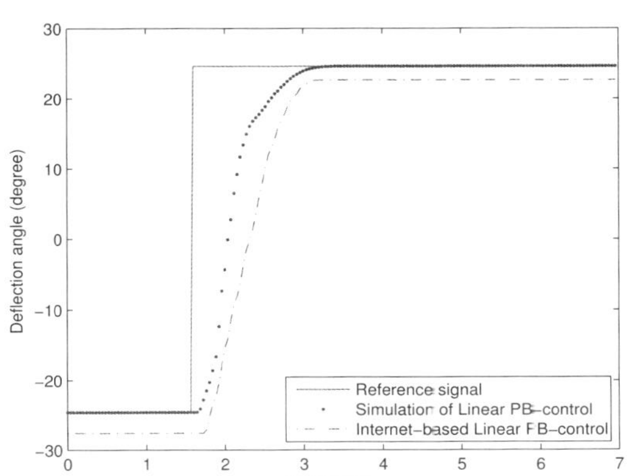

Fig. 8.14 Comparison between simulation and experimental results of linear packet-based control system in Example 8.4

The comparison between the simulation and experimental results is illustrated in Fig. 8.14, which shows that the packet-based control approach is effective in practice.

It is seen however that there is some difference between simulation and experimental results. Several possible reasons may contribute to this difference: 1) the identified model for the DC servo system may not be accurate enough; 2) the dead zone of the DC servo plant has not been considered; and 3) the measurement of the network-induced delays is not accurate in practice.

8.7 Conclusions

Networked control systems are the integration of conventional control systems and the communication networks. Therefore, a natural way to deal with the communication constraints in NCSs is to put the problem under the co-design framework–design with the integration of control theory and communication technology. Motivated by this co-design angle and benefited from the observation of the packet-based transmission in the networked control environment, we are able to propose the so-called packet-based communication and control co-design framework for networked control systems. By showing its effectiveness in dealing with the communication constraints including the network-induced delay, data packet dropout and data packet disorder simultaneously, theoretically, numerically, and experimentally, we are on our way towards a unified approach to networked control systems, and consequentely the convegence of control, communication and computation in such an exciting era featured by ubiquitous networks, ubiquitous control.

Acknowledgement

This work was supported in part by the National Natural Science Foundation of China under grant numbers 61304048, 61273104 and 61333003.

References

Antsaklis, P. and Baillieul, J. (2007). Special issue on technology of networked control systems, *Proceedings of the IEEE* **95**, 1, pp. 5–8.

Azimi-Sadjadi, B. (2003). Stability of networked control systems in the presence of packet, in *Proceedings of the 42nd IEEE Conference on Decision and Control* (Hawaii, USA), pp. 676–681.

Baillieul, J. and Antsaklis, P. J. (2007). Control and communcation challenges in networked real-time systems, *Proceedings of the IEEE* **95**, 1, pp. 9–28.

Cea, M. and Goodwin, G. C. (2013). Stabilization of systems over bit-rate-constrained networked control architectures, *IEEE Transactions on Industrial Informatics* **9**, 1, pp. 357–364.

Chen, W. and Qiu, L. (2013). Stabilization of networked control systems with multirate sampling, *Automatica* **49**, 6, pp. 1528–1537.

Colandairaj, J , Irwin, G. and Scanlon, W. (2007). Wireless networked control systems with QoS-based sampling, *IET Control Theory and Applications* **1**, 1, pp. 430–438.

Donkers, M., Heemels, W., Bernardini, D., Bemporad, A. and Shneer, V. (2012). Stability analysis of stochastic networked control systems, *Automatica* **48**, 5, pp. 917–925.

Fan, Y., Jiang, Z. P. and Zhang, H. (2006). Network flow control under capacity constraints: A case study, *Systems and Control Letters* **55**, 8, pp. 681–688.

Gao, H., Chen, T. and Lam, J. (2008). A new delay system approach to network-based control, *Automatica* **44**, 1, pp. 39–52.

Greco, L., Chaillet, A. and Bicchi, A. (2012). Exploiting packet size in uncertain nonlinear networked control systems, *Automatica* **48**, 11, pp. 2801–2811.

Guan, Z. H., Chen, C. Y., Feng, G. and Li, T. (2013). Optimal tracking performance limitation of networked control systems with limited bandwidth and additive colored white Gaussian noise, *IEEE Transactions on Circuits and Systems I: Regular Papers* **60**, 1, pp. 189–198.

Gupta, R. and Chow, M. Y. (2010). Networked control system: Overview and research trends, *IEEE Transactions on Industrial Electronics* **57**, 7, pp. 2527–2535.

He, Y. (2004). *Delay-Dependent Robust Stability and Stabilization Based on Free-Weighting Matrices (in Chinese)*, Ph.D. Dissertation, Institute of Information Science and Engineering, Central South University, Changsha, China.

Hespanha, J. P., Naghshtabrizi, P. and Xu, Y. (2007). A survey of recent results in networked control systems, *Proceedings of the IEEE* **95**, 1, pp. 138–162.

Hu, W. (2008). *Design of Networked Control Systems and Global Web-Based Control Laboratory*, Ph.D. Dissertation, Faculty of Advanced Technology, University of Glamorgan, Pontypridd, UK.

Hu, W., Liu, G. and Rees, D. (2007). Event-driven networked predictive control, *IEEE Transactions on Industrial Electronics* **54**, 3, pp 1603–1613.

Imer, O. C., Yuksel, S. and Basar, T. (2006). Optimal control of LTI systems over unreliable communication links, *Automatica* **42**, 9, pp. 1429–1439.

Katewa, V. and Gupta, V. (2013). On a rate control protocol for networked estimation, *Automatica* **49**, 5, pp. 1310–1317.

Kruszewski, A., Jiang, W., Fridman, E., Richard, J. and Toguyeni, A. (2012). A switched system approach to exponential stabilization through communication network, *IEEE Transactions on Control Systems Technology* **20**, 4, pp. 887–900.

Li, Y., Yang, S., Zhang, Z. and Wang, Q. G. (2010). Network load minimisation design for dual-rate internet-based control systems, *IET Control Theory and Applications* **4**, 2, pp. 197–205.

Lian, F. L. (2001). *Analysis, Design, Modeling, and Control of Networked Control Systems*, Ph.D. Dissertation, University of Michigan, USA.

Lian, F. L., Moyne, J. R. and Tilbury, D. M. (2001). Performance evaluation of control networks: Ethernet, ControlNet, and DeviceNet, *IEEE Control Systems* **21**, 1, pp. 66–83.

Lian, F. L., Moyne, J. R. and Tilbury, D. M. (2005). Network protocols for networked control systems, in D. Hristu-Varsakelis and W. S. Levine (eds.), *Handbook of Networked and Embedded Control Systems* (Birkhäuser, Boston), pp. 651–675.

Liu, G. P., Xia, Y., Chen, J., Rees, D. and Hu, W. (2007). Networked predictive control of systems with random network delays in both forward and feedback channels, *IEEE Transactions on Industrial Electronics* **54**, 3, pp. 1282–1297.

Liu, X. and Shen, J. (2006). Stability theory of hybrid dynamical systems with time delay, *IEEE Transactions on Automatic Control* **51**, 4, pp. 620–625.

Longo, S., Herrmann, G. and Barber, P. (2012). Robust scheduling of sampled-data networked control systems, *IEEE Transactions on Control Systems Technology* **20**, 6, pp. 1613–1621.

Moayedi, M., Foo, Y. and Soh, Y. (2013). Networked LQG control over unreliable channels, *International Journal of Robust and Nonlinear Control* **23**, 2, pp. 167–189.

Nilsson, J., Bernhardsson, B. and Wittenmark, B. (1998). Stochastic analysis and control of real-time systems with random time delays, *Automatica* **34**, 1, pp. 57–64.

Otanez, P., Moyne, J. and Tilbury, D. (2002). Using deadbands to reduce communication in networked control systems, in *Proceedings of the American Control Conference* (Anchorag, Alaska), pp. 3015–3020.

Postoyan, R. and Nesic, D. (2012). A framework for the observer design for networked control systems, *IEEE Transactions on Automatic Control* **57**, 5, pp. 1309–1314.

Richard, J. P. (2003). Time-delay systems: An overview of some recent advances and open problems, *Automatica* **39**, 10, pp. 1667–1694.

Schenato, L., Sinopoli, B., Franceschetti, M., Poolla, K. and Sastry, S. S. (2007). Foundations of control and estimation over lossy networks, *Proceedings of the IEEE* **95**, 1, pp. 163–187.

Stallings, W. (2000). *Data and Computer Communications*, 6th edn. (Prentice Hall).

Tipsuwan, Y. and Chow, M. Y. (2003). Control methodologies in networked control systems, *Control Engineering Practice* **11**, 10, pp. 1099–1111.

Tipsuwan, Y. and Chow, M. Y. (2004). Gain scheduler middleware: A methodology to enable existing controllers for networked control and teleoperation -part I: Networked control, *IEEE Transactions on Industrial Electronics* **51**, 6, pp. 1218–1226.

Vatanski, N., Georges, J. P., Aubrun, C., Rondeau, E. and Jämsä-Jounel, S. L. (2009). Networked control with delay measurement and estimation, *Control Engineering Practice* **17**, 2, pp. 231–244.

Walsh, G. C., Beldiman, O. and Bushnell, L. G. (2002). Error encoding algorithms for networked control systems, *Automatica* **38**, 2, pp. 261–267.

Walsh, G. C. and Ye, H. (2001). Scheduling of networked control systems, *IEEE Control Systems* **21**, 1, pp. 57–65.

Wu, J. and Chen, T. (2007). Design of networked control systems with packet dropouts, *IEEE Transactions on Automatic Control* **52**, 7, pp. 1314–1319.

Xiong, J. and Lam, J. (2007). Stabilization of linear systems over networks with bounded packet loss, *Automatica* **43**, 1, pp. 80–87.

Yue, D. and Han, Q. L. (2005). Delayed feedback control of uncertain systems with time-varying input delay, *Automatica* **41**, 2, pp. 233–240.

Zeng, W. and Chow, M. Y. (2012). Optimal tradeoff between performance and security in networked control systems based on coevolutionary algorithms, *IEEE Transactions on Industrial Electronics* **59**, 7, pp. 3016–3025.

Zhang, L., Gao, H. and Kaynak, O. (2013). Network-induced constraints in networked control systems: A survey, *IEEE Transactions on Industrial Informatics* **9**, 1, pp. 403–416.

Zhang, L., Shi, Y., Chen, T. and Huang, B. (2005). A new method for stabilization of networked control systems with random delays, *IEEE Transactions on Automatic Control* **50**, 8, pp. 1177–1181.

Zhao, Y. B., Kang, Y., Liu, G. P. and Rees, D. (2011a). Stochastic stabilization of packet-based networked control systems, *International Journal of Innovative Computing Information and Control* **7**, 5, pp. 2441–2455.

Zhao, Y. B., Kim, J. and Liu, G. P. (2011b). Error bounded sensing for packet-based networked control systems, *IEEE Transactions on Industrial Electronics* **58**, 5, pp. 1980–1989.

Zhao, Y. B., Kim, J. and Liu, G. P. (2012a). Offline model predictive control based gain scheduling for networked control systems, *IET Control Theory and Applications* **6**, 16, pp. 2585–2591.

Zhao, Y. B., Kim, J., Liu, G. P. and Rees, D. (2012b). Compensation and stochastic modeling of discrete-time networked control systems with data packet disorder, *International Journal of Control, Automation and Systems* **10**, 5, pp. 1055–1063.

Zhao, Y. B., Kim, J., Sun, X. M. and Liu, G. P. (2012c). On the delay effects of different channels in Internet-based networked control systems, *International Journal of Systems Science* **44**, 11, pp. 2119–2129.

Zhao, Y. B., Liu, G. P. and Rees, D. (2008a). Improved predictive control approach to networked control systems, *IET Control Theory and Applications* **2**, 8, pp. 675–681.

Zhao, Y. B., Liu, G. P. and Rees, D. (2008b). Integrated predictive control and scheduling co-design for networked control systems, *IET Control Theory and Applications* **2**, 1, pp. 7–15.

Zhao, Y. B., Liu, G. P. and Rees, D. (2008c). Networked predictive control systems based on Hammerstein model, *IEEE Transactions on Circuits and Systems II: Express Briefs* **55**, 5, pp. 469–473.

Zhao, Y. B., Liu, G. P. and Rees, D. (2008d). A predictive control based approach to networked Hammerstein systems: Design and stability analysis, *IEEE Transactions on Systems, Man, and Cybernetics, Part B: Cybernetics* **38**, 3, pp. 700–708.

Zhao, Y. B., Liu, G. P. and Rees, D. (2008e). A predictive control based approach to networked Wiener systems, *International Journal of Innovative Computing Information and Control* **4**, 11, pp. 2793–2802.

Zhao, Y. B., Liu, G. P. and Rees, D. (2009a). Design of a packet-based control framework for networked control systems, *IEEE Transactions on Control Systems Technology* **17**, 4, pp. 859–865.

Zhao, Y. B., Liu, G. P. and Rees, D. (2009b). Modeling and stabilization of continuous-time packet-based networked control systems, *IEEE Transactions on Systems, Man, and Cybernetics, Part B: Cybernetics* **39**, 6, pp. 1646–1652.

Zhao, Y. B., Liu, G. P. and Rees, D. (2010a). Actively compensating for data packet disorder in networked control systems, *IEEE Transactions on Circuits and Systems-II: Express Briefs* **57**, 11, pp. 913–917.

Zhao, Y. B., Liu, G. P. and Rees, D. (2010b). Packet-based deadband control for Internet-based networked control systems, *IEEE Transactions on Control Systems Technology* **18**, 5, pp. 1057–1067.

Chapter 9

Review of Some Approximate Privacy Measures of Multi-Agent Communication Protocols

Bhaskar DasGupta and Venkatakumar Srinivasan

Abstract Privacy preserving computational models have become an important research area due to the increasingly widespread usage of sensitive data in networked environments, as evidenced by distributed computing applications and game-theoretic settings (e.g., auctions). Over the years computer scientists have explored many quantifications of privacy in computation. Much of this research focused on designing perfectly privacy-preserving protocols, i.e., protocols whose execution reveals no information about the parties' private inputs beyond that implied by the outcome of the computation. Unfortunately, perfect privacy is often either impossible, or infeasibly costly to achieve (e.g., requiring impractically extensive communication steps). To overcome this, researchers have also investigated various notions of approximate privacy. In this chapter, we review a few such notions and known results for them.

9.1 Introduction

Consider the following standard two-agent communication model as originally proposed by Yao (1979). We have two agents, say Alice and Bob, interacting via a public channel as depicted in Fig. 9.1. Each of Alice and Bob holds a private input, $x_1 \in \mathbb{X}_1$ and $x_2 \in \mathbb{X}_2$ respectively, that is known to her and him, respectively, and they would like to compute a function $f\colon \mathbb{X}_1 \times \mathbb{X}_2 \mapsto \mathbb{R}$ of their two private inputs. Alice and Bob alternately execute "rounds" of a "communication protocol", where in each round they make available a small amount of information about their private inputs, such as an answer to a range query on their private inputs or a few bits of their private inputs, until each of them has seen enough information to be able to compute the value of $f(x_1, x_2)$. This setting can be generalized in an obvious manner to $d > 2$ agents computing a d-ary function

$f\colon \mathbb{X}_1 \times \mathbb{X}_2 \times \cdots \times \mathbb{X}_d \mapsto \mathbb{R}$ by allowing each agent to broadcast information about its private input via a public communication channel in a round-robin order; see Fig. 9.2 for an illustration. Without loss of generality and to simplify exposition, we may assume that $\mathbb{X}_1 = \mathbb{X}_2 = \cdots = \mathbb{X}_d = \{0, 1, \ldots, n\}$ for some positive integer n that is a power of two. With this assumption, the function f can also be visualized as $n \times n \times \cdots \times n$ d-dimensional real matrix $A_{f,d}$ in which the ith dimension represents the possible inputs of the ith agent, and each entry contains the value of f associated with a particular set of inputs from the d agents (i.e., $A_{f,d}[x_1, x_2, \ldots, x_d] = f(x_1, x_2, \ldots, x_d)$). We will denote $A_{f,2}$ simply as A_f.

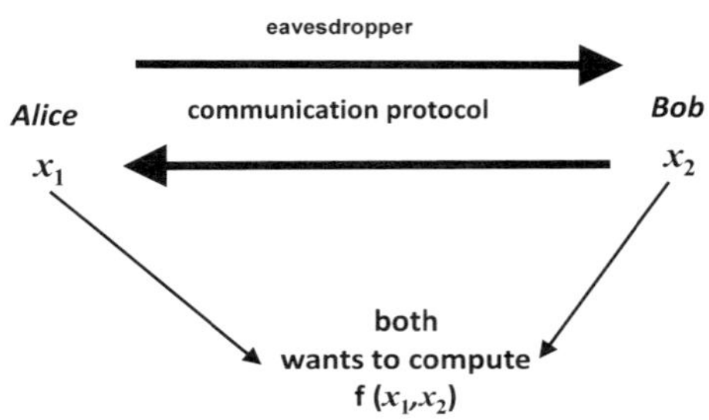

Fig. 9.1 The standard two-agent communication model [Yao (1979)]

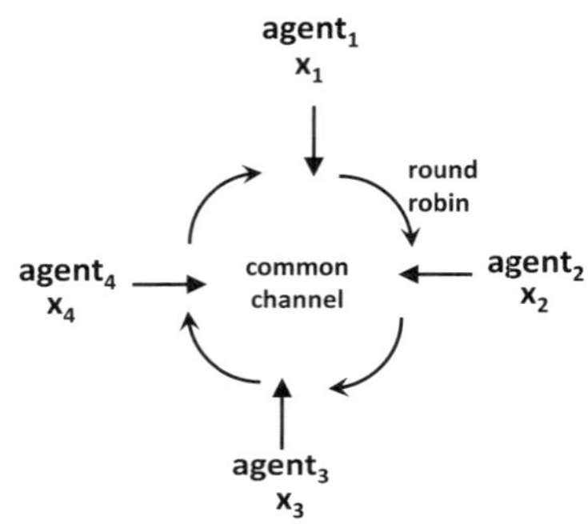

Fig. 9.2 The d-agent communication model for $d = 4$ computing $f(x_1, x_2, x_3, x_4)$

A typical line of research in the above two and multi-agent model of communication, starting with the seminal work of Yao (1979), lies in investigating the *communication complexity* issues, e.g., how many rounds of communications are necessary or sufficient to compute a given family of functions. Such investigations have resulted in a well-developed research area with many interesting results; the reader is referred to a textbook such as Kushilevitz and Nisan (1997) for an overview and basic results in this field, such as the number of bits that need to be exchanged in a two-agent communication protocol by Alice and Bob to compute a given function f is at least $\log_2(\text{rank}_f)$ where rank_f is the linear rank of the matrix A_f over the reals [Kushilevitz and Nisan (1997)]. Applications of these results and techniques have led to the famous AT^2 bound for VLSI networks and analysis of dynamic data structures, to mention a few.

The central question that is the topic of the chapter is however motivated differently from the previous line of research by recent increasingly widespread usage of sensitive data in networked environments, as evidenced by distributed computing applications, game-theoretic settings (e.g., auctions) and more. For example, one motivation, as explained in details in Feigenbaum *et al.* (2010), comes from privacy concerns in auction theory in Economics. An offline or online auction can obviously be viewed as an exchange of information between the bidders and the auctioneer where the goal is to the compute the function that determines the winner of the auction (see Fig. 9.3). Traditionally desired goals of designing auction mechanisms include maximizing revenues and ensuring that the designed mechanisms are indeed truthful, i.e., bidders fare best by letting their truthful bids known [Nisan *et al.* (2007)]. However, more recently, another complementary goal that has gained significant attention, specially in the context of online auctions such as administered by google and other similar companies, is to be able to preserve privacy of the bidders, i.e., bidders must not reveal more information that what is necessary to the auctioneer for optimal outcomes [Comi *et al.* (2011); Feigenbaum *et al.* (2010)]. Thus, for these types of multi-agent communication protocols, the alternate goal is to preserve the privacy of the agents as opposed to optimizing the communication complexity. Informally, the privacy question that is the focus of this chapter is the following: *given a communication protocol to compute a function via multi-agent communication, how can we quantify the amount of extra information about the agents private inputs, beyond what is necessary to compute the function value, that is revealed by the execution of the protocol*? Note that there are two conflicting constraints: the agents do need to communicate sufficient information for computing the function value, but would prefer not to communicate too much information about their private inputs.

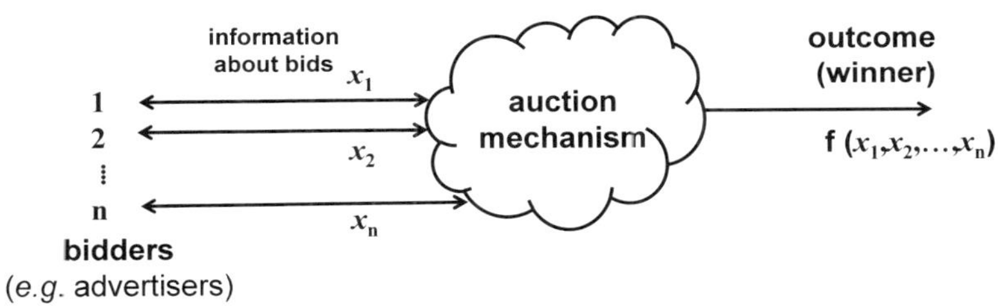

Fig. 9.3 An online auction mechanism viewed as a multi-agent communication problem

To give a concrete example, consider a sequential second-price auction[1] of an item via a protocol in which the price of the item is incrementally increased and publicly announced until the winner is determined. However, such a protocol publicly reveals more information about the bidders than what is absolutely necessary to determine the winner which could be detrimental for the bidders. For example, the protocol reveals the information about the identity of the winner (with revealing his/her bid) together with the bid of the second-highest bidder, and revealing such additional information could put the winner at a disadvantage in the bidding process of a similar item in the future since the auctioneer could set a lower reserve price. In this chapter, we will review a generalized geometric privacy framework that

[1] In such an auction, the winner is the bidder with the highest bid and the price paid by the winner is that of the second-highest bid.

captures applications of the above type as well as other applications in multi-agent computation.

9.1.1 *Perfect vs. approximately perfect privacy*

Unfortunately, even though perfect privacy is the most desirable goal, it is often either impossible, or infeasibly costly to achieve (e.g., requiring impractically extensive communication steps). For example, using the combinatorial characterization of privately computable functions put forth by Chor and Kushilevitz (1991) and Kushelvitz (1992), it is possible to show that the millionaires problem (defined in Section 9.3) is not perfectly privately computable [Chor and Kushilevitz (1991)] and the two-bidder second-price Vickrey auction (also defined in Section 9.3) is perfectly privately computable but only at the cost of and exponential amount of communication by the bidders [Brandt and Sandholm (2008)]. Thus, much of the current research works focus on quantifying and analyzing approximate notions of privacy [Ada *et al.* (2012); Comi *et al.* (2011); Dwork (2006); Feigenbaum *et al.* (2010)].

9.1.2 *Privacy analysis in other environments*

Besides the distributed function computation environment, privacy preserving methods have also been studied in other environments. For example, in the context of mining statistical databases a privacy preserving protocol is expected to provide reliable information about a record being queried while revealing as little information as possible about other records in the database. One model to address the privacy issue in such a data mining environment is the so-called *Differential Privacy Model* originally introduced by Dwork (2006). This model was introduced to investigate the issue of preserving privacy in statistical databases. For example, if the statistical database is a representative sample of some population, differential privacy model allows a user to learn the properties of the population while preserving the privacy of individuals in the population. More formally, a randomized query returning algorithm $\mathcal{A}$ provides ε-differential privacy if, for all pairs of data sets $\mathcal{D}_1$ and $\mathcal{D}_2$ differing on at most one element and for all subset $\mathcal{S}$ of answers provided by $\mathcal{A}$, we have[2]

$$\begin{aligned}&\Pr\left[\text{querying on } D_1 \text{ returns a member in } \mathcal{S}\right]\\&\quad\le \mathrm{e}^{\varepsilon}\Pr\left[\text{querying on } D_2 \text{ returns a member in } \mathcal{S}\right]\end{aligned}$$

where the probabilities are taken is over the coin tosses of the randomized algorithm $\mathcal{A}$ (i.e., informally, removing one record from the database does not make a query output too much more or less likely). Differential privacy is usually achieved by computing the correct answer to a query and adding a noise drawn from the so-

[2] e denotes the base of natural logarithm.

called $\mathcal{L}aplace(f(\varepsilon))$ distribution for some appropriate function f. This approach is sufficient to handle individual queries. In Dwork (2006), the author also provides a mechanism for ensuring differential privacy in case of adaptive queries.

Alternative frameworks, such as the k-anonymization approach [Sweeney (2002)], has also been used for this application.

9.2 Various Frameworks to Quantify Privacy of Protocols

In this section, we review several well-known frameworks to quantify exact or approximate privacy of communication protocols for distributed function computation, one of which is the main topic of this chapter.

9.2.1 *Communication complexity based approaches*

The origin of these frameworks can be traced back to the early works of Chor and Kushilevitz (1991) on characterizations of privately computable functions that can be computed in a perfect private manner, and that of Kushelvitz (1992) on communication complexity issues of privately computable functions. Based on these results, the following two privacy frameworks were independently developed by researchers (an exact characterization of the relationship between these two frameworks is still an open research question):

- Bar-Yehuda *et al.* (1993) provided a combinatorial framework to quantify the amount of privacy that can be maintained in the computation of a function, and the communication cost of achieving this amount of privacy.
- Geometric frameworks to quantify exact and approximate privacy in computing a function in two- and multi-agent communication settings were first formulated by Feigenbaum *et al.* (2010) and subsequently further analyzed by Comi *et al.* (2011). This geometric framework is the main framework that is reviewed in this chapter.

Notable among other results on this approach is the work of Brandt and Sandholm (2008) that, using the framework of Kushelvitz (1992), provided an exponential lower bound on the communication complexity of privacy-preserving second-price Vickrey auctions.

9.2.2 *Information-theoretic approaches*

The study of information-theoretically private protocols can be traced back to the works in Ben-Or *et al.* (1988); Chaum *et al.* (1988). An underlying assumption in these works was that a constant fraction of the agents are honest, i.e., these agents follow the protocol even if deviating from the protocol may benefit them.

9.2.3 *Cryptographic approaches*

Another approach to securing privacy in any multi-agent communication is to use cryptographic tools that rely on various (mostly unproven but always widely believed to be true) complexity-theoretic assumptions. The origin of this line of research can be traced back to the earlier works of Yao [Yao (1982, 1986)]. Usually these types of protocols are very communication intensive, though communication efficient cryptographic protocols have indeed been obtained in some recent papers in mechanism design problems [Dodis *et al.* (2000); Naor *et al.* (1999)].

9.2.4 *Two-agent differential privacy framework*

This framework, introduced by McGregor *et al.* (2010), attempts to extend the differential privacy model mentioned in Section 9.1.2 in the context of distributed function computation in a two-agent communication setting. In this setting the two agents, say agents A and B, want to find out the hamming distance between the n bit inputs that they hold. This setting is defined in the following manner:

- A *mechanism* M (on Σ^n) is a family of probability distributions $\{\mu_x : x \text{ is an input value}\}$ on $\mathbb{R}$. Such a mechanism M is ε-differentially private if and only if the following condition holds:
  - $\forall\, x, x' \in \Sigma^n : \left|x - x'\right|_H = 1$, and
  - for all measurable subsets S of $\mathbb{R}$, $\mu_x(S) \leq e^{\varepsilon} \mu_x(S)$.

  where $\left|x - x'\right|_H$ denotes the Hamming distance between x and x'.
- $\mathsf{VIEW}^A_{\mathcal{P}}(x, y)$ is the joint probability distribution over inputs x, y, the transcript of a protocol $\mathcal{P}$ and the private randomness of agent A (the probability space is private randomness for both agents). $\mathsf{VIEW}^B_{\mathcal{P}}(x, y)$ is defined in a similar manner with respect to agent B.

Then, a communication protocol $\mathcal{P}$ has ε-differential privacy if and only if both of the following conditions hold:

(a) For all input x, $\mathsf{VIEW}^A_{\mathcal{P}}(x, y)$ is ε-differential private.
(b) For all input y, $\mathsf{VIEW}^B_{\mathcal{P}}(x, y)$ is ε-differential private.

A major contrition of McGregor *et al.* (2010) is a *lower bound* on the *least additive error* of any differentially private protocol that is used to compute the hamming distance.

Theorem 9.1. [McGregor *et al.* (2010)] *Let $\mathcal{P}(x, y)$ be a randomized protocol with ε-differential privacy for inputs $x, y \in \{0, 1\}^n$, and let $\delta > 0$. Then, with probability at least $1 - \delta$ over $x, y \in \{0, 1\}^n$ and the coin tosses of $\mathcal{P}$, output of agent B differs from $\langle x, y\rangle$ by at least $\Omega\left(\frac{\sqrt{n}}{\log n} \times \frac{\delta}{e^{\varepsilon}}\right)$.*

An obvious research question for investigation is to see if the above lower bound can be improved or if an actual protocol with a matching upper bound can be found.

9.3 Benchmark Problems and Functions

Often the usefulness of a privacy definition in distributed function computation is checked by demonstrating its value for a class of interesting functions ("benchmark" functions). We mention a few such functions here.

Set-covering function $f_{\text{set-cover}}$: Suppose that the universe $\mathcal{U}$ consists of k elements $u_1, u_2, \ldots, u_k$, and the vectors $\vec{x} = (x_1, x_2, \ldots, x_k) \in \{0,1\}^k$ and $\vec{y} = (y_1, y_2, \ldots, y_k) \in \{0,1\}^k$ encode membership of the elements in two sets $S_{\vec{x}}$ and $S_{\vec{y}}$, i.e., x_i (respectively, y_i) is 1 if and only if $u_i \in S_{\vec{x}}$ (respectively, $u_i \in S_{\vec{y}}$). Then,

$$f_{\text{set-cover}}(\vec{x}, \vec{y}) \stackrel{\text{def}}{=} \bigwedge_{i=1}^{k} (x_i \vee y_i) = \begin{cases} 1, & \text{if } S_{\vec{x}} \cup S_{\vec{y}} = \mathcal{U} \\ 0, & \text{otherwise.} \end{cases}$$

Set-covering type of functions are useful for studying the differences between deterministic and non-deterministic communication complexities [Kushilevitz and Nisan (1997)].

Equality function $f_=$: For two boolean vectors $\vec{x} = (x_1, x_2, \ldots, x_k) \in \{0,1\}^k$ and $\vec{y} = (y_1, y_2, \ldots, y_k) \in \{0,1\}^k$:

$$f_=(\vec{x}, \vec{y}) \stackrel{\text{def}}{=} \begin{cases} 1, & \text{if } \forall i\colon x_i = y_i \\ 0, & \text{otherwise.} \end{cases}$$

The equality function provides a useful testbed for evaluating privacy preserving protocols [Bar-Yehuda *et al.* (1993); Kushilevitz and Nisan (1997)]

Set-disjointness function f_{disjoint}: We have two boolean vectors $\vec{x} = (x_1, x_2, \ldots, x_k) \in \{0,1\}^k$ and $\vec{y} = (y_1, y_2, \ldots, y_k) \in \{0,1\}^k$ (encoding set memberships of elements from an universe) and we wish to decide if they disagree on every coordinate or not, i.e.,

$$f_{\text{disjoint}}(\vec{x}, \vec{y}) \stackrel{\text{def}}{=} \begin{cases} 1, & \text{if } \forall i\colon x_i \neq y_i \\ 0, & \text{otherwise.} \end{cases}$$

The set-disjointness function plays an important role in the theory and application of communication complexity.

Set-intersection function $f_{\text{intersect}}$: We have two boolean vectors $\vec{x} = (x_1, x_2, \ldots, x_k) \in \{0,1\}^k$ and $\vec{y} = (y_1, y_2, \ldots, y_k) \in \{0,1\}^k$ (encoding set memberships of elements from an universe) and we wish to determine the coordinates in which both of them have a 1, i.e.,

$$f_{\text{intersect}}(\vec{x}, \vec{y}) \stackrel{\text{def}}{=} (z_1, z_2, \ldots, z_n) \text{ where, for each } 1 \leq j \leq k,\ z_j = x_j \wedge y_j$$

The set-intersection function has motivated the study of privacy-preserving computation for many years. A typical application of the set-intersection function is when two organizations wish to compute the set of common members without disclosing the members of only one of the organizations [Feigenbaum *et al.* (2010)].

Millionaires problem $f_{\text{millionaire}}$: In this problem, the two agents are two millionaires, each knowing his/her own wealth as own private information, and the goal is to discover the identity of the richer millionaire while preserving the privacy of both agents. Formally, for two boolean vectors $\vec{x} = (x_0, x_1, \ldots, x_{k-1}) \in \{0,1\}^k$ and $\vec{y} = (y_0, y_1, \ldots, y_{k-1}) \in \{0,1\}^k$:

$$f_{\text{millionaire}}(\vec{x}, \vec{y}) \stackrel{\text{def}}{=} \begin{cases} 0, & \text{if } \sum_{j=0}^{k-1} 2^j x_j \geq \sum_{j=0}^{k-1} 2^j y_j \\ 1, & \text{otherwise.} \end{cases}$$

Privacy-preserving protocols for the millionaires problem was investigated in papers such as Chor and Kushilevitz (1991); Feigenbaum *et al.* (2010); Yao (1979).

Second-price Vickrey auction: In a 2nd-price Vickrey auction [Vickrey (1961)] involving one item and two bidders, each having a private value of the item, the goal is to declare the bidder with the higher value as the winner (breaking ties arbitrarily) and reveal the identity of the winner as well as the value of the losing bidder. Formally, for two boolean vectors $\vec{x} = (x_0, x_1, \ldots, x_{k-1}) \in \{0,1\}^k$ and $\vec{y} = (y_0, y_1, \ldots, y_{k-1}) \in \{0,1\}^k$:

$$f_{\text{Second-priceVickrey}}(\vec{x}, \vec{y}) \stackrel{\text{def}}{=} \begin{cases} (0, y_0, y_1, \ldots, y_{k-1}), & \text{if } \sum_{j=0}^{k-1} 2^j x_j \geq \sum_{j=0}^{k-1} 2^j y_j \\ (1, x_0, x_1, \ldots, x_{k-1}), & \text{otherwise.} \end{cases}$$

Second-price Vickrey auction is a fundamental technique in mechanism design for inducing truthful behavior in one-item auctions [Nisan *et al.* (2007)].

9.4 Examples of Standard Communication Protocols

A protocol $\mathcal{P}$ will refer to (a priori) fixed set of rules for communication, and the transcript of $\mathcal{P}$ is the total information (e.g., bits) exchanged during an execution of $\mathcal{P}$. By a "run" of the protocol, one refers to the entire execution of the protocol for a specific set of private inputs of the two agents. For simplicity, we illustrate these protocols for two agents only, but they are easily generalized for the case of $d > 2$ agents in an obvious manner. Typically, one assumes that in each communication round of a protocol $\mathcal{P}$, except the very last one, one of the agents alternately sends out a bit that is computed as a function of that agents' input and communication history. The last message sent in $\mathcal{P}$ is assumed to contain the actual value of the function and therefore may require a larger number of bits. The final outcome of

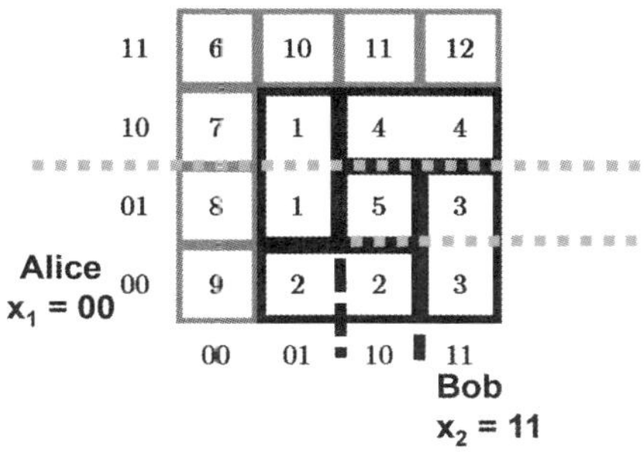

Fig. 9.4 Illustration of one run of the bisection protocol when Alice and Bob has private inputs 00 and 11, respectively. The recursive partitioning induced by the execution of the protocol is shown by thick dashed lines.

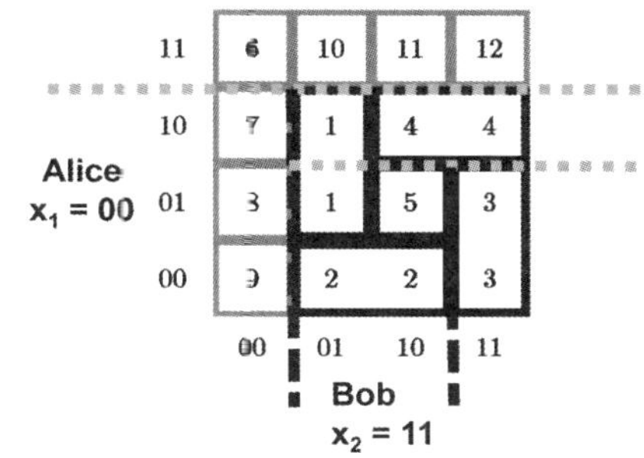

Fig. 9.5 Illustration of one run of the dissection protocol when Alice and Bob has private inputs 00 and 11, respectively. The recursive partitioning induced by the execution of the protocol is shown by thick dashed lines.

the protocol $\mathcal{P}$ is denoted by the function $t_{\mathcal{P}}$. Viewed in this manner, each run of a protocol recursively induces a partition of the associated matrix A_f of a function f. Three such well-known communication protocols studied in the literature are as follows:

α-bisection protocol: For a constant $\alpha \in \left[\frac{1}{2}, 1\right)$, a protocol is a α-bisection protocol provided the following two conditions hold:

- at each communication step, the maintained subset of inputs of each agent is a set of contiguous integers.
- at each communication step, the communicating agent partitions its input space of size z into two halves of size $\alpha\, z$ and $(1-\alpha)\, z$.

Bisection protocol: A $\frac{1}{2}$-bisection protocol is simply called a bisection protocol (see Fig. 9.4 for an illustration).

Bounded-bisection protocol: For an integer valued function $g(k)$ such that $0 \leq g(k) \leq k$, bounded-bisection$_{g(k)}$ is the protocol that runs a bisection protocol with $g(k)$ bisection operations followed by a protocol (if necessary) in which each agent repeatedly partitions its input space into two halves one of which is of size exactly one.

Sealed-bid auction protocol: This straightforward protocol is applicable for functions that represent the auction of an item. Here the auctioneer receives sealed bids from all bidders and computes the outcome based on this information.

Ascending-price English auction protocol: This straightforward protocol is applicable for functions that represent the auction of an item. Suppose that we have two bidders only. We start with a price of zero for the item and, in each discrete time step increment, we increase the price by one until one of the two bidders indicates that his/her value for the item is less than the current price, say a. Then, we allocate the item to the other bidder for a price of $a - 1$.

Comi et al. [Comi *et al.* (2011)] considered a more general version of the α-bisection protocol in the following manner. When designing protocols for environments in which the input of each agent has a natural ordering (e.g., the set of input of an agent from $\{0,1\}^k$ can represent the numbers $0, 1, 2, \ldots, 2^k - 1$, as is in the case when computing the maximum/minimum of two inputs, in the millionaires problem, in second-price auctions, and more), a natural restriction is to allow protocols such that each agent asks questions of the form "*Is your input between a and b (in this natural order over possible inputs)?*", where $a, b \in \{0,1\}^k$. Notice that such a protocol divides the input space into two (not necessarily equal) halves (see Fig. 9.5). Such protocols were termed as the dissection protocol in Comi *et al.* (2011) and were useful in analyzing average loss of privacy.

9.5 A Geometric Approach to Quantify Privacy

In this section, we review a recent interesting geometric approach to privacy is based on communication complexity that was initiated by Feigenbaum et al. [Feigenbaum *et al.* (2010)] and subsequently followed up in Ada *et al.* (2012); Comi *et al.* (2011). Although originally motivated by agents' privacy in mechanism design, the definitions and tools can be easily applied to distributed function computation in general. This framework allows one to quantify approximate privacy as well as study the trade-off between privacy preservation and communication complexity. For simplicity of exposition, we discuss the framework first for two agents and later comment on how to generalize it when $d > 2$ agents communicate. As mentioned in the introduction, we have two agents and a function $f : \mathbb{X} \times \mathbb{X} \mapsto \mathbb{R}$ of two arguments to compute, where $\mathbb{X} = \{0, 1, \ldots, n\}$ with $n = 2^k$ for some positive integer k, and such a function f can be visualized via the associated two-dimensional matrix A_f. For convenience, we will view the elements in $\mathbb{X}$ in binary as a k-bit number whenever required.

Intuitively, a quantification of (exact or approximate) privacy should satisfy the objective that any observer of the protocol $\mathcal{P}$ should not be able to distinguish the private inputs of the two communicating agents from as large a set as possible of other possible private inputs. To capture this intuition, Feigenbaum et al. [Feigenbaum *et al.* (2010)] makes use of the machinery of communication-complexity theory to introduce the so-called *Privacy Approximation Ratio* (PAR) via a geometric and combinatorial interpretation of protocols. To define PAR, we first need to state some basic communication complexity definitions for a two-agent communication model [Kushilevitz and Nisan (1997)].

Definition 9.1 (see Fig. 9.6 for an illustration).

(I) A region R of A_f is any subset of entries in A_f. R is monochromatic if all entries in R are of the same value. A monochromatic region R of A_f is max-

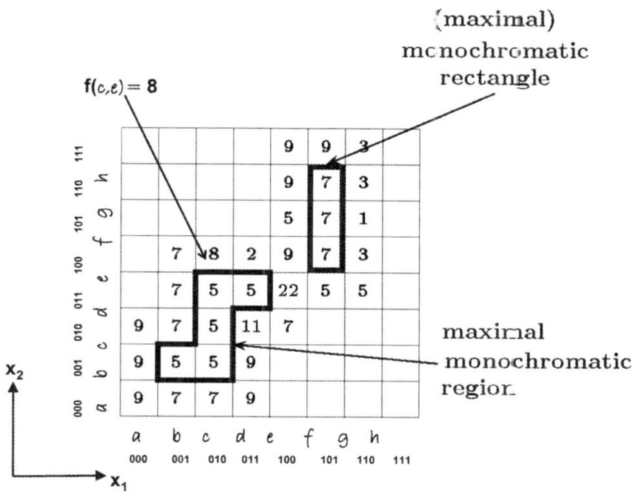

Fig. 9.6 Illustration of concepts in Definition 9.1

imal if no other monochromatic region in A_f properly contains it. The ideal monochromatic partition of A_f is made up of the maximal monochromatic regions.

(II) A partition of A_f is a collection of disjoint regions in A_f whose union equals to A_f. A monochromatic partition of A_f is a partition all of whose regions are monochromatic.

(III) A rectangle in A_f is a sub-matrix of A_f. A tiling of A_f is a partition of A_f into rectangles. A tiling T_1 of A_f is said to be a refinement of another tiling T_2 of A_f if every rectangle in T_1 is contained in some rectangle in T_2.

(IV) A protocol $\mathcal{P}$ achieves perfect privacy if, for every pair of inputs (x_1, x_2) and (x_1', x_2') such that $f(x_1, x_2) = f(x_1', x_2')$, it holds that $t_{\mathcal{P}}(x_1,\ x_2) = t_{\mathcal{P}}(x'_1,\ x'_2)$.

(V) A communication protocol $\mathcal{P}$ for f is perfectly privacy-preserving if the monochromatic tiling induced by P is the ideal monochromatic partition of A_f.

(VI) Let $R^P(x_1, x_2)$ be the monochromatic rectangle induced by protocol P for $(x_1, x_2) \in \{0,1\}^k \times \{0,1\}^k$ and $R^I(x_1, x_2)$ be the monochromatic region containing $A_f[x_1, y_1]$ in the ideal monochromatic partition of A_f. Then P has a worst-case privacy-approximation-ratio (PAR) of Δ_{worst} if

$$\Delta_{\text{worst}} = \max_{(x_1,x_2)} \left[\frac{|\, R^I(x_1, x_2) \,|}{|\, R^P(x_1, x_2) |} \right].$$

See Fig. 9.7 for an illustration.

(VII) Let $\mathcal{D}$ be a probability distribution over the space of inputs. The average case privacy-approximation-ratio (PAR) of a communication protocol P under distribution $\mathcal{D}$ for function f is

$$\Delta_{\mathcal{D}} = \mathbb{E}_{\mathcal{D}}\left[\frac{|R^I(x_1, x_2)|}{|R^P(x_1, x_2)|}\right]$$

where $\mathbb{E}_{\mathcal{D}}$ denotes the expectation with respect to the distribution $\mathcal{D}$.

(VIII) The worst case PAR for a function f is the minimum, over all protocols P for f, of the worst case PAR of P.

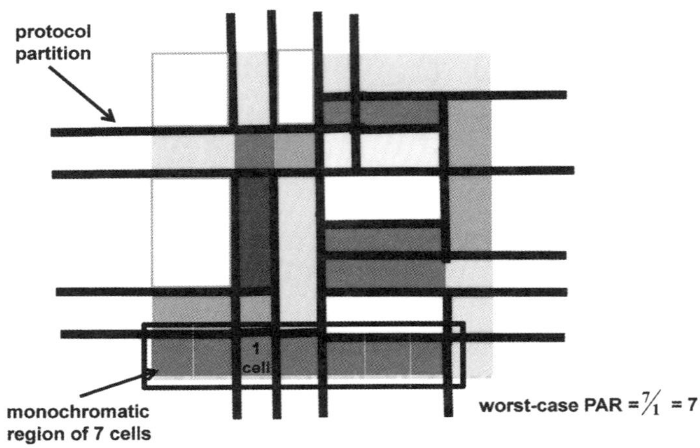

Fig. 9.7 Illustration of the calculation of the worst-case PAR Δ_{worst}

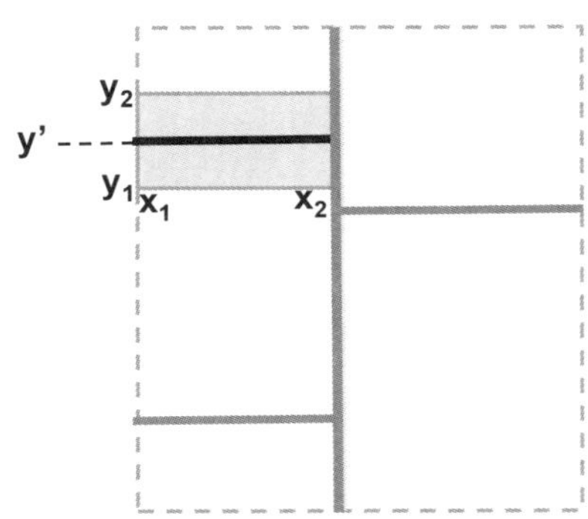

Fig. 9.8 Partitioning an ideal monochromatic rectangle leads to loss of privacy

In Definition 9.1 (VI)–(VIII), the underlying assumption is that partitioning an ideal monochromatic rectangle results in loss of privacy. The intuition behind this is as follows. Consider the situation depicted in Fig. 9.8 where the shaded ideal monochromatic rectangle is partitioned into two rectangles by a protocol. Note that the value of $f(x, y)$ is the same for all $x_1 \leq x \leq x_2$ and

Table 9.1 List of bounds on average and worst-case PAR for a few functions as derived in Feigenbaum *et al.* (2010). $\mathcal{D}$ is assumed to be the uniform distribution.

Protocol	Computed function		
	$f_{\text{millionaire}}$	$f_{\text{Second-priceVickrey}}$	
	$\Delta_{\mathcal{D}}$	Δ_{worst}	$\Delta_{\mathcal{D}}$
arbitrary	$\geq 2^k - \frac{1}{2} - \frac{1}{2^{k+1}}$		$\leq \frac{2}{3}2^k + \frac{1}{3\,2^k}$
bisection	$= \frac{3}{2}2^k - \frac{1}{2}$	$= 2^{\frac{k}{2}}$	$= \frac{k}{2} + 1$
α-bisection		$= 2^{\frac{k}{2}}$ $\left(\text{assuming } \alpha > \frac{1}{2^k}\right)$	
bounded-bisection$_{g(k)}$			$= \frac{g(k)+3}{2} - \frac{2^{g(k)}}{2^{k+1}} + \frac{1}{2^{k+1}} - \frac{1}{2^{g(k)+1}}$
sealed-bid auction			$= \frac{2^{k+1}}{3} + \frac{1}{3\,2^k}$
Ascending-price English auction			$= 1$

$y_1 \leq y \leq y_2$ since the shaded rectangle is monochromatic. But, observing the protocol allows one to distinguish between subsets of these inputs, namely inputs in the subset $\{(x,y) \mid x_1 \leq x \leq x_2, y_1 \leq y < y'\}$ from inputs in the subset $\{(x,y) \mid x_1 \leq x \leq x_2, y' \leq y < y_2\}$, thereby revealing extra information.

Using the above framework and definitions, Feigenbaum et al. [Feigenbaum *et al.* (2010)] provided calculations of worst-case and average PAR values for a number of functions as summarized in Table 9.1.

9.5.1 *Tiling functions and dissection protocols*

Comi et al. [Comi *et al.* (2011)] further investigated this geometric approach by defining a special class of functions called the "tiling" functions, and analyzing the power of the more general dissection protocol in computing these tiling functions. The dissection protocol was defined in Section 9.4. To illustrate the main ideas more clearly, here we consider a slightly simplified version of the definition of a tiling function in which we assume that the index of a row (respectively, a column) of A_f is the same as the private value held by the first (respectively, second) agent. Then, a function f is a tiling function [Comi *et al.* (2011)] if the monochromatic regions in A_f form a tiling; the number of monochromatic regions in this tiling is denote by ∇_f. See Fig. 9.9 for illustrations. Comi et al. [Comi *et al.* (2011)] proved the following results.

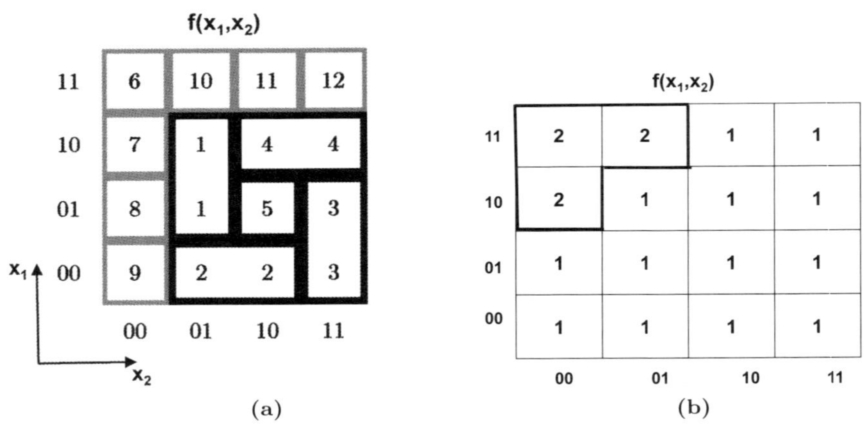

Fig. 9.9 Example of (a) tiling and (b) non-tiling functions

Theorem 9.2.

(*a*) *Every boolean tiling function can be computed in a perfectly privacy-preserving manner.*

(*b*) *There exists a tiling function* $f : \{0,1\}^k \times \{0,1\}^k \mapsto \{0,1\}^3$ *such that every dissection protocol* $\mathcal{P}$ *for* f *has* $\Delta_{\text{worst}} = \Omega\left(2^{\frac{k}{2}}\right)$.

(*c*) *Let $\mathcal{D}$ be the uniform distribution. Then, the following results hold.*

(*i*) *For any tiling function f, there is a dissection protocol $\mathcal{P}$ using at most $O(\nabla_f)$ communication rounds such that*

- $\Delta_{\mathcal{D}} \leq 4$.
- *$\mathcal{P}$ can be computed in $O\left(k4^k\right)$ time.*

(*ii*) *There exists a tiling function f such that for every dissection protocol we have $\Delta_{\mathcal{D}} \geq \frac{11}{9}$.*

A proof of Theorem 9.2$(c)(i)$ was obtained in Comi *et al.* (2011) via a connection between protocols and the binary space partitions (BSP). BSPs present a way to implement a geometric divide-and-conquer strategy and is an extremely popular approach in numerous applications such as hidden surface removal, visibility problems, and motion planning [Tóth (2005)]. A BSP for a collection of disjoint rectangles in the two-dimensional plane can be defined in the following manner. The plane is divided into two parts by cutting rectangles with a horizontal or vertical line if necessary. The two resulting parts of the plane are divided recursively in a similar manner and the process continues until at most one fragment of the original rectangles remains in any part of the plane. This division process can be naturally represented as a binary tree (BSP-tree) where a node represents a part of the plane and stores the cut that splits the plane into two parts that its two children represent. Each leaf of the BSP-tree then represents the final partitioning of the plane by storing at most one fragment of an input rectangle; see Fig. 9.10 for an illustration. The following result on BSP was shown in d'Amore and Franciosa (1992):

> [d'Amore and Franciosa (1992)] For any set of disjoint axis-parallel rectangles in the plane, there is a BSP such that every rectangle is partitioned into at most 4 rectangles due to BSP.

The above result can be used to prove Theorem 9.2$(c)(i)$ by identifying BSPs with dissection protocols.

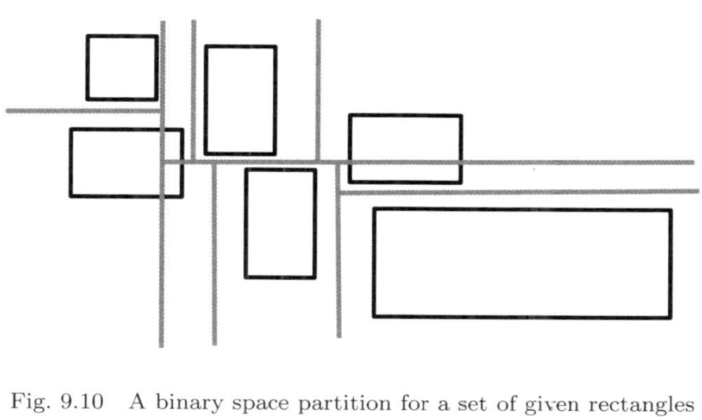

Fig. 9.10 A binary space partition for a set of given rectangles

9.5.2 *Generalization for d > 2 agents*

Comi et al. [Comi *et al.* (2011)] showed that the average PAR is very high for dissection protocols even for 3 agents and uniform distribution, thereby suggesting that this quantification of privacy may not provide good bounds for three or more agents. More precisely, they proved the following result.

Theorem 9.3. *Let $\mathcal{D}$ denote the uniform distribution. Then, there exists a tiling function $f\colon \{0,1\}^k \times \{0,1\}^k \times \{0,1\}^k \mapsto \{0,1\}^{3k}$ such that every dissection protocol must have $\Delta_{\mathcal{D}} = \Omega\left(2^k\right)$.*

9.6 Conclusions

In this chapter, we have provided an overview to privacy preserving computing in a distributed function computation setup that includes game-theoretic settings. In particular, we have reviewed in greater details a recently developed geometric approach to quantifying loss of privacy. Future research questions of interest include identifying other non-tiling classes of functions for which good approximate-privacy preserving protocols are possible and relating the geometric privacy model to other privacy models.

References

Ada, A., Chattopadhyay, A., Cook, S., Fontes, L., Koucky, M. and Pitassi, T. (2012). The hardness of being private, in *IEEE 27th Annual Conference on Computational Complexity* (Porto, Portugal), pp. 192–202.

Bar-Yehuda, R., Chor, B., Kushilevitz, E. and Orlitsky, A. (1993). Privacy, additional information, and communication, *IEEE Transactions on Information Theory* **39**, 6, pp. 1930–1943.

Ben-Or, M., Goldwasser, S. and Wigderson, A. (1988). Completeness theorems for non-cryptographic, fault-tolerant computation, in *Proceedings of 20th ACM Symposium on Theory of Computing* (Chicago, USA), pp. 1–10.

Brandt, F. and Sandholm, T. (2008). On the existence of unconditionally privacy preserving auction protocols, *ACM Transactions on Information and Systems Security* **11**, 2, pp. 1–21.

Chaum, D., Crépeau, C. and Damgard, I. (1988). Multiparty, unconditionally secure protocols, in *Proceedings of 20th ACM Symposium on Theory of Computing* (Chicago, USA), pp. 11–19.

Chor, B. and Kushilevitz, E. (1991). A zero-one law for boolean privacy, *SIAM Journal on Discrete Mathematics* **4**, pp. 36–47.

Comi, M., DasGupta, B., Schapira, M. and Srinivasan, V. (2011). On communication protocols that compute almost privately, *Algorithmic Game Theory* **6982**, pp. 44–56.

d'Amore, F. and Franciosa, P. G. (1992). On the optimal binary plane partition for sets of isothetic rectangles, *Information Processing Letters* **44**, 5, pp. 255–259.

Dodis, Y., Halevi, S. and Rabin, T. (2000). A cryptographic solution to a game theoretic problem, in M. Bellare (ed.), *Advances in Cryptology–CRYPTO 2000* (Springer, Berlin), pp. 112–130.

Dwork, C. (2006). Differential privacy, in *Proceedings of 33rd International Colloquium on Automata, Languages and Programming* (Venice, Italy), pp. 1–12.

Feigenbaum, J., Jaggard, A. and Schapira, M. (2010). Approximate privacy: Foundations and quantification, in *Proceedings of 11th ACM Conference on Electronic Commerce* (Cambridge, USA), pp. 167–178.

Kushelvitz, E. (1992). Privacy and communication complexity, *SIAM Journal on Discrete Mathematics* **5**, 2, pp. 273–284.

Kushilevitz, E. and Nisan, N. (1997). *Communication Complexity* (Cambridge University Press).

McGregor, A., Mironov, I., Pitassi, T., Reingold, O., Talwar, K. and Vadhan, S. (2010). The limits of two-party differential privacy, in *IEEE 51st Annual Symposium on Foundations of Computer Science* (Las Vegas, USA), pp 81–90.

Naor, M., Pinkas, B. and Sumner, R. (1999). Privacy preserving auctions and mechanism design, in *Proceedings of 1st ACM Conference on Electronic Commerce* (Denver, USA), pp. 129–139.

Nisan, N., Roughgarden, T., Tardos, E. and Vazirani, V. (2007). *Algorithmic Game Theory* (Cambridge University Press).

Sweeney, L. (2002). k-anonymity: A model for protecting privacy, *International Journal on Uncertainty, Fuzziness and Knowledge-Based Systems* **10**, 5, pp. 557–570.

Tóth, C. D. (2005). Binary space partitions: Recent developments, in J. E. Goodman, J. Pach and E. Welzl (eds.), *Combinatorial and Computational Geometry* (Cambridge University Press), pp. 529–556.

Vickrey, W. (1961). Counterspeculation, auctions and competitive sealed tenders, *The Journal of Finance* **16**, 1, pp. 8–37.

Yao, A. C. (1979). Some complexity questions related to distributive computing, in *Proceedings of 11th Annual ACM Symposium on Theory of Computing* (Atlanta, USA), pp. 209–213.

Yao, A. C. (1982). Protocols for secure computation, in *Proceedings of 23rd IEEE Symposium on Foundations of Computer Science* (Chicago, USA), pp. 160–164.

Yao, A. C. (1986). How to generate and exchange secrets, in *IEEE 27th Annual Symposium on Foundations of Computer Science* (Toronto, Canada), pp. 162–167.

Chapter 10

Encoding-Decoding Machines for Online Concept-Drift Detection on Datastreams

Cesare Alippi, Giacomo Boracchi, Li Bu, and Dongbin Zhao

Abstract The ability to detect online the occurrence of concept drift in a datastream is a hot research topic with striking implications. In addition to applications such as network traffic monitoring and credit-card fraud detection, the issue is relevant in several monitoring-based and big data applications. This chapter describes a change-detection approach based on encoding-decoding machines, which are computationally simple, yet effective mechanisms for detecting concept drift in a datastream. As an illustrative example we describe a linear machine based on SVD and contrast the change-detection capabilities with a linear predictive modeling-based inspection. The proposed solution can be easily embedded in intelligent sensors and adopted in sensor networks where the change-detection outcomes of several units can be aggregated in a cognitive level.

10.1 Introduction

The push towards virtual sensors on the one hand and system-on-chip solutions on the other has made it possible the design of intelligent sensors characterized by not trivial capabilities. While the discussion is more obvious for virtual sensors where computation availability is generally not an issue, it is less immediate for physical sensors. Recent advances in electronics led to a new generation of sensors mounting, on the same chip, a thermistor and a programmable embedded finite-state machine in addition to the accelerometer sensor [1]. The thermistor acts as a secondary auxiliary sensor providing the temperature value for thermal compensation purposes. Sensor programmability by using a custom defined instruction set grants a new

[1]See, for instance, the `LIS3DSH` digital MEMS sensor released by ST `http://www.st.com/web/en/resource/technical/document/datasheet/DM00040962.pdf`

level of processing ability for the incoming datastream as well as the possibility to integrate in the future more sophisticated forms of intelligence close to the probe.

The nature of datastream acquired from intelligent sensors (virtual and real) depends on the specific application. For instance, datastreams for virtual sensors can be found in network traffic monitoring, credit-card fraud detection, web click streams [Chawla *et al.* (2004)] while a plethora of applications are associated with raw datastreams coming from analog or digital sensors.

Borrowing the terminology typically used in the classification scenario, we denote by concept a specific stationary state of the data-generating process. Then, concept drift indicates a change that affects the data-generating process, which is not anymore stationary and moves, abruptly or gradually, into a different state. Concept drift can be associated with a change in the monitored environment (non-stationarity/time variance), a model bias at the detecting agent, faults and aging effects affecting physical sensors. Therefore, in all the above applications, the ability to deal with concept drift is of paramount importance and it would be beneficial for intelligent sensors to autonomously detect and undertake suitable countermeasures to deal with concept-drift.

The literature concerning concept drift is mostly focused on classification applications, with few works facing different applications e.g., see Ilin *et al.* (2004). Depending on the learning mechanisms adopted [Elwell and Polikar (2011)], we distinguish existing approaches into passive or active. In passive approaches the application undergoes a continuous training and adaptation, without explicitly determining whether (and when) concept drift has occurred. Differently, active approaches exploit some sort of triggering mechanism, namely, a Change-Detection Test (CDT) [Basseville and Nikiforov (1993)], and the application evolves and adapts to new operating conditions solely once the change has been detected.

Passive methods are very popular in the classification scenario, and the best known solution to adapt to concept drift consists in an ensembles of classifiers. Ensembles of classifiers allow the incremental learning of concept drift: the algorithm in Elwell and Polikar (2011), named Learn++.NSE, learns the current concept from consecutive batches of data without making assumptions about the nature of concept drift. A different approach has been suggested in Grinblat *et al.* (2011) where an ensemble of reference classifiers based on SVMs has been considered. An adaptive Bayesian-inspired algorithm has been considered in De Oliveira and Caticha (2010), which incorporated changes in the data distribution to solve the classification problem. The relevant problem of classifying skewed datastreams has been addressed in Gao *et al.* (2008), while Minku and Yao (2012) showed that one should consider different levels of ensemble diversity before and after the drift to get a good generalization on new concepts. An online datastream-oriented classification algorithm has been suggested in Abdulsalam *et al.* (2011). The algorithm, which can handle multiple classes, adjusts its parameters to respond to changing class boundaries in the datastream. He and Chen (2008) proposes an incremental multiple-object learning method, which can automatically and adaptively learn from continuous video streams; its results are independent from the specific classifier employed.

The peculiarity of active approaches is the use of some form of CDT to detect concept drift. In classification applications CDTs are introduced to either monitor the classification error (measured on supervised samples) or the distribution of incoming observations and/or features derived from them. The inspection of the classification error, e.g., see Gama *et al.* (2004); Nishida and Yamauchi (2009); Widmer and Kubat (1996), is meant to detect changes affecting the classes distribution and implicitly assumes that the error would increase in case of concept drift. For this reason, some authors propose the use of a fixed threshold on the classification accuracy as in Widmer and Kubat (1996), while others weakened such assumption and introduced an adaptive thresholding mechanism [Gama *et al.* (2004)]. The solution presented in Nishida and Yamauchi (2009) relies on an exponentially weighted moving average chart applied to the classification error. In Alippi *et al.* (2013a) the average classification error computed over subsequence of data is treated as a binomial distribution to detect concept drift over time. A second use of a CDT sees the triggering mechanism applied to inspect raw data or features extracted from them, and this strongly relates to this work. The Mann-Kendall [Kendall (1975)], the CUmulative SUMs CUSUM test [Manly and MacKenzie (2000)] and its extension, the CI-CUSUM [Alippi and Roveri (2008)], went in this direction and proposed automatic sequential methods for change detection. Another approach, which has shown to be particularly effective for scalar datastreams, is the ICI-based CDT suggested in Alippi *et al.* (2011a), and its very performing hierarchical extension [Alippi *et al.* (2011b)], which introduces a validation phase that is activated after each detection. An advantage of active approaches is that, once the change has been detected, the classifier can react, discarding its obsolete knowledge and undergoing a reconfiguration phase taking advantage only of recent supervised samples, which represent the current state of the process [Alippi and Roveri (2008)]. Of course, after each detection, it is convenient to reconfigure also the CDT, to detect a further drift which might stem from the new state.

That said, it is obvious that the use of a CDT is application independent, and can be used for any application beside adaptive classifiers. However, most of the online CDTs operating on the observations (or on features extracted from them), typically require the strong and hardly satisfied assumption that observations (or features) are independent and identically distributed (i.i.d.). Examples of features are the distribution moments [Alippi and Roveri (2008); Alippi *et al.* (2011a)] or the parameters of a suitable dynamic model designed to track the evolution of the incoming signal [Basseville and Nikiforov (1993)].

When observations are not i.i.d., e.g., show a clear temporal dynamic, most of CDTs can not be directly applied. In these cases, the most straightforward solution to detect concept drift consists in computing the residuals with respect to a suitable approximating model $\mathcal{M}_0$, and then perform a CDT phase there [Alippi *et al.* (2013b); Gustafsson (2000)]. The residuals, i.e., the discrepancy between the observations and their forecast/reconstruction provided by $\mathcal{M}_0$, can be often treated as i.i.d. realizations of a random variable, hence enabling the use of the above presented CDTs at the residuals level. The machine $\mathcal{M}_0$ is learnt from an initial training phase, which can be regarded as a characterization of the stationary conditions of the data-generating process. The CDT is then configured in cascade, analyzing the residuals computed by $\mathcal{M}_0$ over additional observations generated in stationary conditions. Computing an effective approximation model $\mathcal{M}_0$ is not a trivial task since, often, a system model for the data-generating process is not provided. In these situations, the model complexity must be suitably identified for the problem and $\mathcal{M}_0$ then configured [Alippi *et al.* (2012)].

Here, we propose a $\mathcal{M}_0$ based on an auto-associative [Kramer (1992)] machine that approximates the incoming observations by acting on an encoder-decoder philosophy. We constrain the machine to be a linear function of the last observation and of a state vector, which represents the machine memory to keep low the computational complexity. Reconstruction residuals provided by such machine constitute the input of a CDT. In the following, the encoding-decoding CDT denotes the combination of the encoding-decoding machine with a CDT. Thanks to their flexibility and reduced computational burden, the encoding-decoding CDT can very appealing solution for intelligent sensors.

The structure of the chapter is organized as follows. Section 10.2 introduces the general formulation of the considered encoding-decoding machine and their use for change-detection purposes, while Section 10.3 and Section 10.4 introduce a specific machine based on the Singular Value Decomposition (SVD). Experiments are provided in Section 10.5, while Section 10.6 presents concluding remarks.

10.2 A CDT Based on the Encoding-Decoding Machine

In what follows we provide a formal and general description of the class of machines we consider, and show how to combine them with a CDT.

10.2.1 *The observation model*

We consider a very general observation model, where observations $x_t \in \mathbb{R}$ are scalar

$$x_t = \chi_t + \eta, \quad \forall t. \tag{10.1}$$

χ_t is the unknown noise-free observation at time t and η is the noise, which is modeled as a random variable following an unknown probability density function (pdf) of zero expectation (i.e., $E[\eta] = 0$). The noise-free observations χ_t are assumed to be generated from a stationary data-generating process $\mathcal{P}_0$, which is also assumed to be unknown.

Concept drift may affect both $\mathcal{P}_0$, which may shift to a different process $\mathcal{P}_1$, as well as the noise η, which may change its distribution. Of course, the detection of both situations is of primary importance in several application scenarios, as these might indicate abnormal states for the data-generating process, possibly induced by faults.

10.2.2 *The recurrent encoding-decoding machine*

In the sequel, we consider $\mathcal{M}_0$ to be the linear, recurrent machine as in Voegtlin (2005) and Sperduti (2006)

$$y_t = W_x x_t + W_y y_{t-1}, \quad \forall t \tag{10.2}$$

where $y_{t-1} \in \mathbb{R}^s$ is a column vector describing the state of the machine at time $t-1$, while the column vector $W_x \in \mathbb{R}^s$ and the matrix $W_y \in \mathbb{R}^{s \times s}$ weight the contributions of the current input and previous state vector, respectively. (10.2) describes how the current observation x_t is encoded in the memory of the machine, i.e., in the state vector y_t.

The estimate of the current noise-free observation χ_t is extracted from the state vector y_t by means of a linear projection

$$\widehat{x}_t = E^{\mathsf{T}} y_t \tag{10.3}$$

being E^{T} a row vector, which is also part of the machine $\mathcal{M}_0$. (10.3) describes the decoding phase of the machine, where an estimate of the current observation is extracted from the state vector.

The residual sequence associated with machine $\mathcal{M}_0$ is

$$e_t = x_t - \widehat{x}_t, \quad \forall t \tag{10.4}$$

and represents the discrepancy between the observations and the result of encoding/decoding operation carried out by of $\mathcal{M}_0$.

Both matrices W_x and W_y, as well as the projection vector E have to be leant from the training sequence $\text{TS} = \{x_t, t < L\}$, which contains L observations generated in stationary conditions (i.e., belonging to $\mathcal{P}_0$). Clearly, the combination of (10.2) and (10.3) should not provide the identity otherwise the machine degenerates in a trivial configuration. We recall that residuals inspection with a CDT allows us for detecting potential concept drift in the data-generating process, and this will be detailed in what follows.

10.2.3 *The encoding-decoding CDT*

In the ideal case, where the machine $\mathcal{M}_0$ is always able to correctly estimate χ_t, the residuals e_t in (10.4) would correspond to i.i.d. realizations of a random variable (η), hence enabling the use of any CDT [Basseville and Nikiforov (1993)]. In this case, a CDT configured on the residuals computed in the training sequence TS would be able, in principle, to perceive any concept drift.

Denote by CDT(t) the outcome of any CDT at time t, which has been executed on the residuals computed by means of $\mathcal{M}_0$, i.e.,

$$\mathrm{CDT}(t) = \begin{cases} 0, & \text{if } \{e_i, i < t\} \text{ is stationary} \\ 1, & \text{otherwise.} \end{cases} \tag{10.5}$$

We refer to encoding-decoding CDT as the combination of $\mathcal{M}_0$ followed by any CDT (which is executed on the approximation residuals from $\mathcal{M}_0$). Given the nonlinear structure of a generic datastream and the linear structure of the machine $\mathcal{M}_0$ we should expect e_t to be therefore affected by model bias. However, if such model bias is not too relevant, or it is constant, or can be modeled as a stochastic component, the described approach remains viable.

We point out that it is not advisable to both learn $\mathcal{M}_0$ and configure the CDT from the same data in TS. In fact, the residuals computed from data used to learn $\mathcal{M}_0$ may not be representative of the actual distribution of the residuals in the operational life, and therefore a cross-validation procedure is recommended. A portion of TS should be used to learn $\mathcal{M}_0$ (which in turn may require a further, internal partitioning in training, test and validation sets), and the leaned machine is then used to compute the residuals on the remaining part of TS: only these latter residuals should be considered to configure the CDT.

10.3 An Encoding-Decoding Machine Based on SVD

In the following, we consider the linear machine suggested in Sperduti (2006), there designed to embed discrete structures (graphs) into vectorial spaces. In the following, we review the basics behind the machine and we detail its practical use.

10.3.1 *The SVD-based encoding-decoding machine*

In Sperduti (2006), a linear machine of the form (10.2) is presented, where the coefficient matrices are defined as

$$W_y = \begin{bmatrix} \mathbf{0}_{1,s-1} & 0 \\ \mathbf{1}_{s-1,s-1} & \mathbf{0}_{s-1,1} \end{bmatrix} \quad \text{and} \quad W_x = \begin{bmatrix} 1 \\ \mathbf{0}_{s-1,1} \end{bmatrix} \tag{10.6}$$

where $\mathbf{0}_{r,c}$ is a rectangular zero matrix of r rows and c columns and $\mathbf{1}_{r,r}$ is the identity matrix of order r. The parameter $s \in \mathbb{N}$ determines the dimension of the state vector. We comment that, when these matrices are considered in (10.2), W_y operates as a one-step right shift, whereas W_x builds a vector of $\mathbb{R}^s$ having the current observation x_t in the first component and zeros elsewhere.

In practice, the machine defined by (10.2) and (10.6) corresponds to the matrix formulation of the customary operation of selecting a sliding window over the most recent s observations. It follows that the state vector $y(t) \in \mathbb{R}^s$ becomes:

$$y_t = \begin{cases} [x_t, x_{t-1}, \ldots, x_{t-s}], & \text{if } t \geq s \\ [x_t, x_{t-1}, \ldots, x_1, \mathbf{0}_{1,s-t}], & \text{if } t < s \end{cases} \tag{10.7}$$

where zero padding is meant to be employed at the beginning of the dataset, i.e., when $t < s$. In this specific situation, a straightforward solution to define the projection vector $E^\top$ consists in $E^\top = W_x^\top$, such that

$$\widehat{x}_t = W_x^\top y_t \quad \forall t, \tag{10.8}$$

which, of course, induces a lossless reconstruction of the input x_t, i.e., $\widehat{x}_t = x_t \ \forall t$. Therefore, (10.8) reconstructs also the noise and the consequent residual $e_t = 0 \, \forall t$, would not highlight any concept drift. However, this machine provides an algebraic framework to select a sliding window, and it is always possible to modify the projection step (10.8), to prevent perfect reconstruction and learn $\mathcal{M}_0$ such that it can approximate $\mathcal{P}_0$, for instance, by means of a SVD decomposition.

Consider the SVD of the matrix associated with the state vectors y_t as follows. Let $\{x_0, \ldots, x_{\tau+s}\} \subset \text{TS}$ be a subsequence of the training sequence TS (i.e., $\tau + s < L$), generated in stationary conditions. Compute all the state vectors from this subsequence and build the matrix $Y_{\tau,s}$ whose columns are vectors y_t, $t \leq \tau$. Therefore, the SVD of $Y_{\tau,s}$ yields

$$Y_{\tau,s} = U_{\tau,\tau} \Sigma_{\tau,s} V_{s,s}^\top, \tag{10.9}$$

and build the projection in the subspace of the first p singular values of $\Sigma_{\tau,s}$. Basic algebra shows that the projection matrix is $\tilde{V}\tilde{V}^\top \in \mathbb{R}^{s \times s}$, where $\tilde{V} = V_{s,p}$, is a matrix whose columns contain the eigenvectors of $V_{s,s}^\top$ associated with the p largest eigenvalues. The approximating machine $\mathcal{M}_0$ is obtained by projecting state y_t into the space of the first p eigenvectors by means of $\tilde{V}\tilde{V}^\top \in \mathbb{R}^{s \times s}$. We thus define the estimate of the observation at time t as

$$\widehat{x}_t = W_x^\top (\tilde{V}\tilde{V}^\top) y_t. \tag{10.10}$$

When $p < s$, the machine given by (10.2), (10.6) and (10.10) does not yield perfect reconstruction since it retains only the components of the state vector belonging to the subspace of the first p eigenvectors. In practice, such a machine implements a standard preprocessing step based on projection over the first p SVD components. Of course, the projection of the state in the subspace of the first p eigenvectors can be included in the encoding phase, replacing in (10.2) the matrices W_x and W_y with $\tilde{W}_x = (\tilde{V}\tilde{V}^\top) W_x$ and $\tilde{W}_y = (\tilde{V}\tilde{V}^\top) W_y$, while the encoding remains that of (10.8).

10.3.2 *The residuals*

The quality of the reconstruction provided by the machine defined from (10.6) and (10.10) can be evaluated by inspecting the reconstruction residuals which, given the particular structure of the machine, can be rewritten as

$$e_t = (x_t - \widehat{x}_t) = W_x^{\mathsf{T}} \left(\mathbf{1}_{s\times s} - \tilde{V}\tilde{V}^{\mathsf{T}} \right) y(t). \tag{10.11}$$

The error can be then expressed as a linear combination of the state vector, which corresponds to the linear filter over a sliding window

$$e_t = \sum_{i=1}^{s} [a]_i [y_t]_i = a^{\mathsf{T}} y_t \tag{10.12}$$

where $[a]_i$ stands for the i-th entry of the column vector a, and similarly $[y_t]_i$. These coefficients depend on the entries of the matrix $\left(\mathbf{1}_{s\times s} - \tilde{V}\tilde{V}^{\mathsf{T}}\right)$ and after basic calculus

$$\alpha_1 = 1 - \sum_{j>p}^{s} [V]_{1,j} \left[V^{\mathsf{T}}\right]_{j,1} = 1 - \sum_{j>p}^{s} [V]_{1,j}^2$$

while for $i \geq 2$

$$\alpha_i = -\sum_{j>p}^{s} [V]_{1,j} [V]_{i,j}$$

where $[V]_{i,j}$ denotes the element at the i-th row and j-th column. In what follows we provide a practical criteria to select p from a training sequence.

10.3.3 *Learning $\mathcal{M}_0$*

Determining the most appropriate machine $\mathcal{M}_0$ for a given problem requires definition of s and parameter estimation of p. In particular, p controls the level of reconstruction (from 0 to 100%) of the observations and can be selected through cross validation procedure as described in the sequel. Differently, s characterizes the reconstruction ability of the machine.

When no a priori information is available about the nature of the noise, we recommend to select p as the parameter maximizing the whiteness of the residuals in (10.4) over a validation set. More specifically, if we denote by $e_{t,p}$ the residuals at time t, defined as in (10.4), when the matrix $\tilde{V}$ in (10.10) is built from the first p eigenvectors, we select p minimizing the autocorrelation of $e_{t,p}$ at $t_{\max}$ lags, over a suitable validation interval $[L_1, L_2] \subset [0, L]$, i.e.,

$$p = \underset{p=1,\dots,s}{\operatorname{argmin}} \sum_{w=1}^{t_{\max}} \left| \sum_{t=L_1}^{L_2} e_{t,p} e_{t-w,p} \right|. \tag{10.13}$$

This specific criterion encourages choices of p that make the residuals sequence less correlated, thus easing the CDT task in detecting concept drift.

10.3.4 *Extensions*

Possible extensions of the machine $\mathcal{M}_0$ consist in augmenting the state vector y_t to include also nonlinear features extracted from the most recent observations. In this case, the basic functioning of the machine as a sliding window (10.2) can still be preserved, though the machine has to be applied in the space of feature vectors rather than on the observations themselves. Such a feature extraction phase allows the machine $\mathcal{M}_0$ to learn high-order relations of the observations, if needed.

For instance, following the approach used to introduce nonlinear decision boundaries to linear classifiers [Hastie *et al.* (2009)], one could define the operator $\mathcal{F}\,:\,\mathbb{R}^s \rightarrow \mathbb{R}^{ks}$ mapping each state vector y_t into a feature vector $\mathcal{F}(y_t)$ that includes nonlinear transformation of each element of y_t: as an illustrative example, we could define $\mathcal{F}(y_t) = \left[y_t,\ y_t^2,\ y_t^3, y_{t-1}, \ldots, y_{t-s}, y_{t-s}^2, y_{t-s}^3\right]^{\mathsf{T}}$, yielding $k = 3$. When such augmented state vectors are used, W_x and W_y in (10.6) become

$$W_y = \begin{bmatrix} \mathbf{0}_{k,k(s-1)} & \mathbf{0}_{k,k} \\ \mathbf{1}_{k(s-1),k(s-1)} & \mathbf{0}_{k(s-1),k} \end{bmatrix} \quad \text{and} \quad W_x = \begin{bmatrix} \mathbf{1}_{k,1} \\ \mathbf{0}_{k(s-1),1} \end{bmatrix} \tag{10.14}$$

and therefore the SVD in (10.9) has to be computed from the matrix $X_{\tau,ks}$, obtained by stacking $\mathcal{F}(y_t)\,,\ t < \tau$. Such a machine $\mathcal{M}_0$ could exploit linear relations among powers of the observations.

10.4 An ICI-Based Encoding-Decoding CDT

We show here how the encoding-decoding machine described in Section 10.3 can be used to build a sophisticated CDT operating at the residual level.

10.4.1 *Designing the encoding-decoding CDT*

The encoding-decoding CDT is composed of the recurrent SVD-based machine presented in Section 10.3, which provides the residual datastream to a CDT that performs the concept-drift detection. In particular, we consider the ICI-based hierarchical CDT suggested in Alippi *et al.* (2011b) for its reduced computational complexity and effectiveness.

The ICI-based hierarchical CDT operates at two functional levels:

- At the first level, the ICI-based CDT [Alippi *et al.* (2011a)] monitors features extracted out of the input data: features have been designed in such a way to approach a Gaussian distribution. The core of the CDT is the Intersection of Confidence Intervals (ICI) rule [Goldenshluger and Nemirovski (1997)], an adaptive filtering technique to regularize data through polynomial regression, which is used to detect changes in the expected value of features. Unfortunately, this CDT suffers from a structural generation of false positives over time.

- The second level has been designed to mitigate the false positives introduced by the first level CDT, and is activated every time the first level detects a change. The goal of the second level is to validate/reject the detections by means of an hypothesis test. This is carried out by considering the Hotelling T^2 statistic [Burr (1976)] to assess whether there is a change in the mean of the extracted features before and after the perspective change or not (the null hypothesis of the test being "the difference between the mean of feature vectors is zero"). The first level detection is confirmed when the test statistic provides enough statistical evidence for rejecting the null hypothesis, and the parameters of the first level CDT are reconfigured.

The ICI-based CDT is very effective and has interesting properties, which makes it the adequate tool to be coupled with our encoding-decoding machine. First, any persistent concept drift that affects the residuals of $\mathcal{M}_0$ can be detected, eventually at the expenses of an increased detection latency. Second, it is characterized by a low computational burden, thus it can be effectively combined with $\mathcal{M}_0$ maintaining a reduced computational complexity of the overall method. The ICI-based CDT also provides a refined estimate of the change time-instant (once the change has been detected); such information allows to partition the features in two sets referring to the *no concept drift* and *potential concept drift* states. This information is essential for setting up the hypothesis test at the second level, which validates/rejects the detections of the first level.

Here, the residuals in (10.4) are fed as input of the ICI-based hierarchical CDT, which computes and inspects two features: the sample mean and the power-law transform [Mudholkar and Trivedi (1981)] of the sample variance. More specifically, the features are extracted from non-overlapping windows of ν residuals. The sample mean $M(j)$ and the transformed variance $S(j)$ are computed on the j-th window as

$$M(j) = \frac{1}{\nu} \sum_{t=(j-1)\nu+1}^{\nu j} e_t \tag{10.15}$$

and

$$S(j) = \left[\frac{1}{\nu - 1} \sum_{t=(j-1)\nu+1}^{\nu j} (e_t - M(j))^2 \right]^{h_0} \tag{10.16}$$

where h_0 is estimated on the training sequence as suggested in Mudholkar and Trivedi (1981) and yields values that can be considered approximately Gaussian distributed. Note that the windows are not overlapping, thus the estimates can be treated as i.i.d. Gaussian realizations.

These two features reveal relevant classes of perturbations: in particular, $S(j)$ is needed to detect concept drift caused by precision degradation (i.e., an increase in the variance of η) or a stack-at fault (the observations becomes frozen to a constant value), while $M(j)$ addresses other relevant situation such as the abrupt perturbation and the gradual drift cases. Both features are needed to detect more sophisticated concept drift.

10.4.2 *Adaptation*

The CDT combining the encoding-decoding machine and the ICI-based hierarchical CDT can be integrated within an adaptive mechanism to allow the application/intelligent sensor to track the data-generating process in a nonstationary environment. We recall that changes can be associated with different situations:

- Time variance, i.e., the data-generating process is not stationary (concept drift affects the process).
- Model bias, i.e., the machine $\mathcal{M}_0$ is not able to globally explain the incoming observations.
- Soft and hard faults affecting the sensor (concept drift affecting the sensors).

Without any extra or a priori information, it is not possible to distinguish which of the above situations gave rise to the concept drift. However, when we may exclude the fault at sensors, any detection indicates either that the environment has changed or that the machine was no longer adequate to track the process, thus the machine $\mathcal{M}_0$ needs to be retrained, the hierarchical ICI-based CDT reconfigured, and an alarm raised so that the application/intelligent sensor can react accordingly (e.g., in classification applications, the knowledge base of the classifier has to be refreshed).

Summarizing, the encoding-decoding CDT operates as illustrated in Fig. 10.1 and discussed in what follows. Once suitably trained from TS, the encoding-decoding machine $\mathcal{M}_0$ is ready to operate and provides, at each time instant, a residual e_t (10.12) to the feature extraction module of the ICI-based CDT. The j-th window of residuals is then populated, and features $M(j)$ and $S(j)$ are computed as soon as ν residuals are gathered together. These features are then analyzed by the ICI-rule, thus implementing the ICI-based CDT test at the lower level. As soon as a change is detected (at time $\widehat{T}$), the refinement procedure is activated yielding $\widehat{T}_{\text{ref}}$, a refined estimate of the change-time instant, which is then provided to the upper level of the hierarchical CDT for change validation. An hypothesis test on the Hotelling T^2 statistic is then used to assess whether the mean of the features over TS and within the interval $[\widehat{T}_{\text{ref}}, \widehat{T}]$ are statistically different at a given confidence α or not (the null hypothesis of the test is "the difference between the feature mean is null"). When the detection is validated, concept drift is confirmed, and the encoding-decoding CDT must be retrained by using data in interval $[\widehat{T}_{\text{ref}}, \widehat{T}]$. In particular, such a re-training phase involves both the machine $\mathcal{M}_0$ and the ICI-based hierarchical CDT. Conversely, when the detection is not validated, it is considered to be a false positive, and only the ICI-based CDT at the lower level needs to be reconfigured.

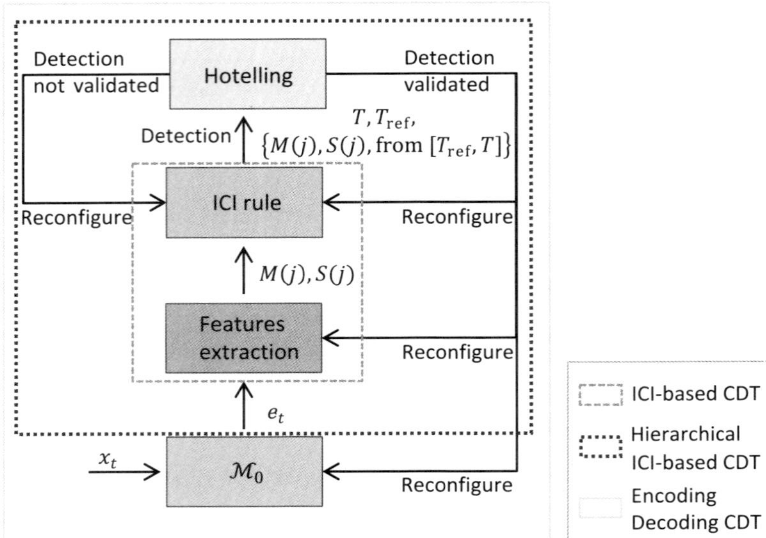

Fig. 10.1 An illustrative scheme of the proposed encoding-decoding CDT. The scheme also details the components of the encoding-decoding CDT and the way how these are integrated

10.4.3 *Single vs. multiple sensors*

We saw that it is not possible to distinguish between a sensor fault and a change in the environment when analyzing only data coming from a single datastream in a context free of a-priori information. However, when a set of sensors is deployed in the same environment (as it typically happens in sensor-network scenario), the situation becomes different. In fact, it is possible to design a cognitive Fault-Diagnosis System (FDS), which takes advantage of spatial and temporal relationships among sensors. In this direction, Alippi *et al.* (2013d) proposed a FDS that, by relying on a dependency graph that models the functional dependencies (in terms of transfer functions) among data acquired by network units, is able to discriminate between faults, changes in the environment and false positives induced by model bias. The proposed encoding-decoding CDT could be naturally integrated within such a framework for fault detection and isolation.

10.5 Experiments

Experiments show the performance of the encoding-decoding CDT on both synthetically generated datastreams and measurements coming from real world applications. The encoding-decoding CDT (combining $\mathcal{M}_0$ with the hierarchical ICI-based CDT) is compared against an ideal model-based counterpart, where the hierarchical ICI-based CDT is executed on residuals provided by a suitbale predictive model.

$\mathcal{M}_0$ was always configured setting $s = 5$, providing $L = 500$ samples as training sequence and using $\tau = 200$ of them for computing the SVD. We learned the p parameter following the procedure described in Section 10.3.3, analyzing 100 training samples (i.e., $L_1 = \tau$, $L_2 = 300$, $t_{\max} = 20$). To configure the model-based counterpart, we follow a classical system identification procedure [Ljung (1999)], fitting several autoregressive models to the initial 300 samples, and selecting the best one according to the Akaike information criteria [Akaike (1974)]. The remaining 200 samples in TS are used to compute the residuals (both using the encoding-decoding machine and the fitted autoregressive model), to configure the hierarchical ICI-based CDT. The hierarchical ICI-based CDT operates on subsequences of $\nu = 20$ residuals, and has been configured using $\Gamma = 2$ and $\alpha = 0.05$; further details can be found in Alippi *et al.* (2011b).

The concept-drift detection capabilities are assessed by means of the following figures of merit.

- Detection Latency (DL): the average detection latency over N runs, which is an estimate of the expected delay $E[\widehat{T} - T^*]$, conditioned to the fact that $\widehat{T} \geq T^*$. Since the distribution of the detections is particularly skewed (detections are always nonnegative), we also report the $25\%, 50\%$ and 75% quantiles of the empirical distribution. These values illustrate, better than the mean and the standard deviation, the actual distribution of these values.
- False positives rate (FPR): the percentage of detections (over N runs) that do not correspond to a concept drift.
- False negative rate (FNR): measured as the percentage of times (over N runs) where the CDT is does not detect concept drift.
- The Signal-To-Noise ratio of the change: this corresponds to the ratio between the energy of the residuals before and after the change:

$$\mathrm{SNR}_{\mathrm{CHANGE}} = \left[\frac{\sum_{t=T^*+1}^{N} e_t^2}{\sum_{t=L+1}^{T^*} e_t^2} \frac{T^* - L}{N - (T^* - L)} \right]. \tag{10.17}$$

  This quantity is not representative of the actual performance of the CDT, since it does not even take into account $\widehat{T}$. On the contrary, it indicates how perceivable the concept drift is from residuals $\{e_t, L < t \leq N\}$. The larger the value, the easier is to detect concept drift. Since the distribution of the $\mathrm{SNR}_{\mathrm{CHANGE}}$ is particularly skewed ($\mathrm{SNR}_{\mathrm{CHANGE}}$ is nonnegative) and suffers from very large outliers, beside the average over the dataset, we report the $25\%, 50\%$ and 75% quantiles. These values illustrate better than the mean and the standard deviation the actual distribution of the values of $\mathrm{SNR}_{\mathrm{CHANGE}}$.

It is important to mention some limitations of the comparison between the encoding-decoding CDT and the ICI-based hierarchical CDT applied to the residuals from a suitable predictive model. On the one hand, the encoding-decoding CDT does not enforce any analytical model of the observations, while in contrast the family of predictive models considered in the fitting often includes the process that generated the data. In this sense, the residuals obtained when fitting predictive models have to be considered sort of ideal ones. On the other hand, predictive models does not consider the current observation x_t to compute $\widehat{x}_t$, while the encoding-decoding machine does consider x_t: the encoding-decoding machine performs regularization rather than prediction on the datastream.

10.5.1 *The "ARMA" dataset*

We consider $\mathcal{P}_0$ as the autoregressive moving-average (ARMA) process

$$x_t = \sum_{i=1}^{l} \phi_i x_{t-i} + \sum_{i=1}^{m} \psi_i \eta_{t-i} + \eta_t \tag{10.18}$$

where $\theta_0 = [\phi_1, \dots, \phi_l, \psi_1, \dots, \psi_m]$ represents the parameter vector, $\eta_t \sim \mathcal{N}(0, \sigma^2)$ denotes the innovation at time t, l and m corresponds to the orders of the autoregressive and moving-average terms, respectively. Each data sequence is composed of 6000 samples and contains an abrupt and permanent shift in the parameter vector θ_0 at time $T^* = 4000$: the observations before T^* are generated from an ARMA process parameterized by θ_0, while after T^* the data-generating process becomes an ARMA parameterized by $\theta_1 \neq \theta_0$. Such a change mimics a concept drift affecting the process.

A dataset containing 10000 sequences was prepared by randomly generating the parameter vectors θ_0 and θ_1, including only those yielding stable systems. In each sequence, the orders (l, m) are also randomly selected within their range $l \in [1, 4]$ and $m \in [0, 2]$. The standard deviation of η is set to $\sigma = 0.1$. To choose the best fitting ARMA model to the observations we tested all the possible pairs (l, m) in the above described range. Results are reported in Table 10.1. The figures of merit reported in Table 10.1 indicate that the encoding-decoding CDT is, as expected, less effective than its ideal model-based counterpart. The statistics of detection latency and of $\text{SNR}_{\text{CHANGE}}$ indicate that the residuals computed using the machine reveal

Table 10.1 The performance of the two CDTs on the ARMA dataset

	Figures of Merit			
CDT Considered	**FNR**	**FPR**	**DL**	$\textbf{SNR}_{\text{CHANGE}}$
Encoding Decoding CDT	19.18%	2.63%	376.61 (100 220 480)	8×10^4 (0.09 1.17 12.60)
ARMA-based CDT	10.37%	4.81%	276.81 (60 130 340)	1.6×10^6 (0.40 3.81 39.67)

randomly generated. The numbers within brackets in the DL and $\text{SNR}_{\text{CHANGE}}$ columns indicates the 25%, 50% and 75% quantiles of the corresponding empirical distribution. These values are far more meaningful than the mean and standard deviations since the distributions are very skewed (DL and $\text{SNR}_{\text{CHANGE}}$ are not negative) and may contain outliers (in particular $\text{SNR}_{\text{CHANGE}}$).

the concept drift less effectively than the residuals from a suitable predictive model. However, the encoding-decoding CDT is able to properly operate on these data, without either enforcing any parametric model of the observations or requesting model completeness.

10.5.2 *The "hairdryer" application*

The application refers to a single input/single output (SISO) system described in Ljung (1999), modeling the relationship between the power of a heating device (the input) and the air temperature (the output). As described in Ljung (1999), this system can be successfully modeled by means of an autoregressive moving average with exogenous input (ARMAX) having 9 time lags on the regressive parts, 6 on the input and 2 on the innovators. Both input and output signals contain 1000 data samples; the sampling time is 0.08s. The dataset is taken from file *dryer2.mat* belonging to the *System Identification* toolbox in Matlab. As pointed out in Ljung (1999), the process has simple dynamics with small disturbances.

The dataset and the results are reported in Fig. 10.2 and Fig. 10.3: from $T^* = 800$ onwards we artificially corrupted the observations introducing an additive noise term having standard deviation $\sigma = 0.1$. Such a concept drift mimics the precision degradation due, for instance, to a faulty sensor. The performance of the encoding-decoding CDT on this real world data are in line with the results and the comments in Section 10.5.1. Numerical results are reported in the captions of Fig. 10.2 and Fig. 10.3.

10.5.3 *The mountain temperature application*

This datastream is acquired from an hybrid wired-wireless monitoring system deployed on the Alps to monitor a potential rock collapse [Alippi *et al.* (2013c)]. Temperature is measured every 5 minutes and data are sent to a remote control station for subsequent processing (the deployment also considers tiltmeter, accelerometer and strain gauge sensors). Here, we analyze a short sequence (1000 measurements), and run the two CDTs configured over the initial 500 measurements. There are no studies concerning which model describes these data at best. Therefore, as in Section 10.5.1, we exhaustively tested all the configuration of ARMA models having AR order $l \in [1, 4]$ and MA order $m \in [0, 4]$, selecting the best one using the Akaike information criteria. In this case, as in Section 10.5.2, we introduce a precision degradation at $T^* = 800$, where the additive Gaussian noise has standard deviation $\sigma = 1$.

Figs. 10.4 and 10.5 illustrate the performance of the two CDTs: the bumps in the residuals show that neither the machine $\mathcal{M}_0$ nor an ARMA model are adequate to approximate these data. In this case, the detection of the encoding-decoding CDT is prompter, though the detectability in terms of $\text{SNR}_{\text{CHANGE}}$ indicates that the residuals from ARMA highlights better the change.

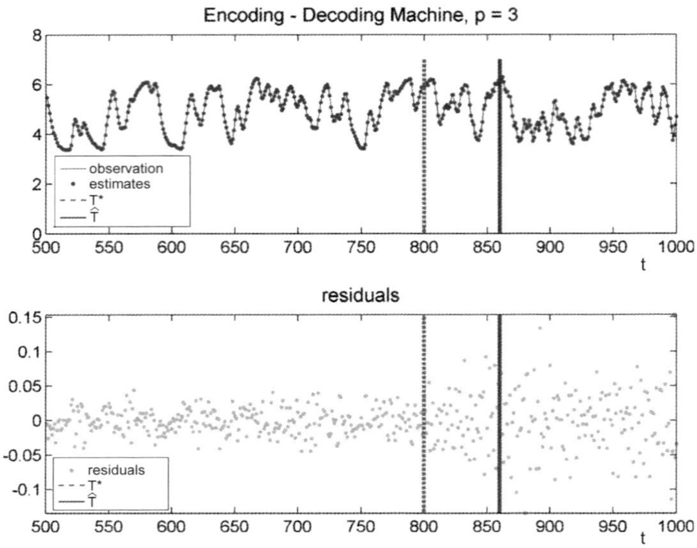

Fig. 10.2 Hairdryer Dataset corrupted by a subtle precision degradation ($\sigma = 0.1$), occurring at time $T^* = 800$ (dashed line). The Encoding-decoding CDT ($p = 3$) detects the change at $\widehat{T} = 840$ (solid line). In this case, $\text{SNR}_{\text{CHANGE}} = 14.6$. The initial training sequence ($L = 500$) is not reported.

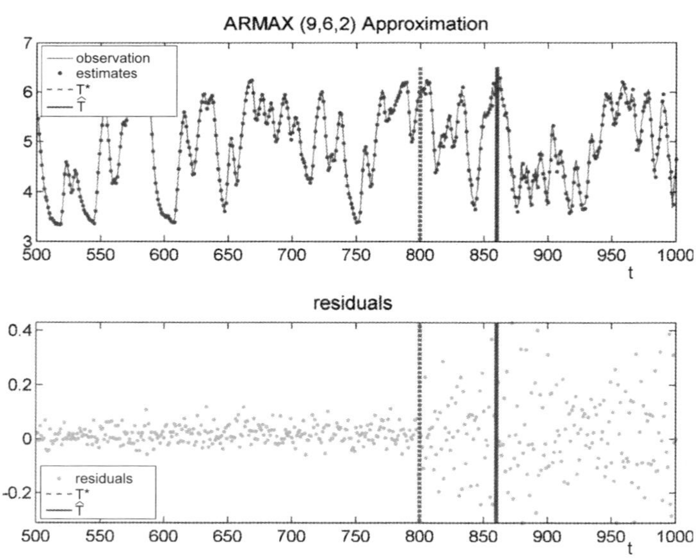

Fig. 10.3 Hairdryer Dataset, a subtle precision degradation ($\sigma = 0.1$), introduced at time $T^* = 800$ (dashed line). The CDT executed on the residuals from an ARMAX (9,6,2) detects the change at $\widehat{T} = 840$ (solid line). In this case, $\text{SNR}_{\text{CHANGE}} = 75.7$: though the detection coincides with that of the encoding-decoding CDT, the change is easier to perceive here than in Fig. 10.2. The initial training sequence ($L = 500$) is not reported.

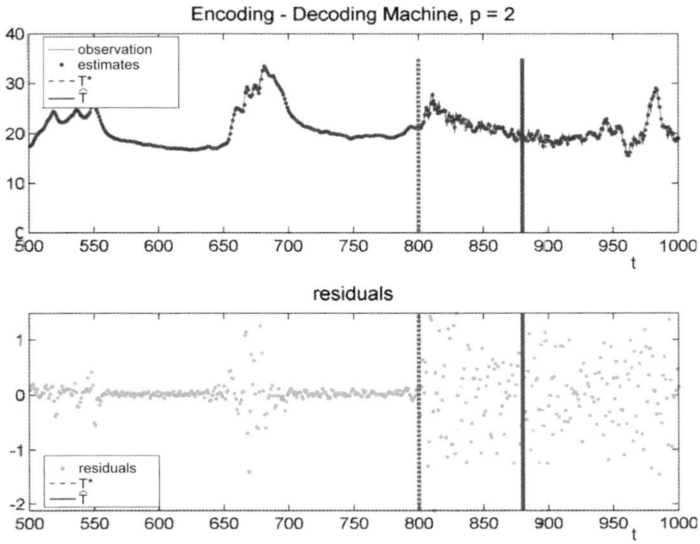

Fig. 10.4 Mountain Dataset, a precision degradation (σ = 1), introduced at time T^* = 800 (dashed line). The Encoding-decoding CDT (p = 2) detects the change at $\widehat{T}$ = 880 (solid line), while $\mathrm{SNR}_{\mathrm{CHANGE}}$ = 4.82. The initial training sequence (L = 500) is not reported.

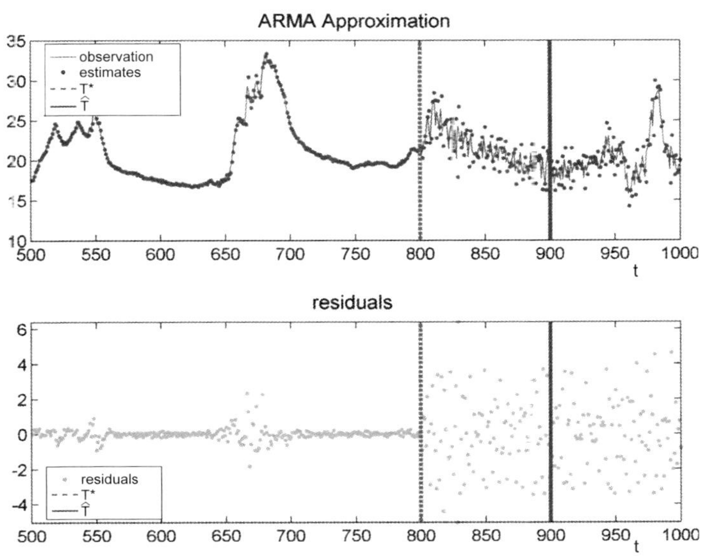

Fig. 10.5 Mountain Dataset, a precision degradation occurring at time T^*= 800 (dashed line). The hierarchical ICI-based CDT executed on the residuals from the fitting ARMA model detects the concept drift at time T^* = 900, in this case $\mathrm{SNR}_{\mathrm{CHANGE}}$ = 46.5. The initial training sequence (L= 500) is not reported.

10.6 Conclusions

We introduced a novel approach to detect concept drift in processes characterized by a temporal or spatial dynamic. Peculiarity of the proposed solution is the use of an encoding-decoding machine, learned to characterize the data-generating process within an initial training sequence. The learned machine is then used to process online incoming observations, computing residuals between the actual data the and the machine output. Such a residuals can be monitored by any CDT, to detect concept drift affecting the process itself, changes in the noise distribution (due for instance to faults in the sensing apparatus) as well as the obsolescence of the machine to express the current state of the process (model bias). These machines could be then employed in situations where no valid parametric model to approximate the observation is known.

The family of the proposed machines is general, and includes also recurrent ones that deserves further investigation. All these machines exploit simple linear operations, thus are particularly suited for online processing datastreams. As a meaningful example we combined a very simple machine based on SVD of the observations with the hierarchical ICI-based CDT: the resulting encoding-decoding CDT was able to properly operate on both synthetically generated data and measurements from real sensors.

The suggested method can be easily integrated within a recurrent concept framework or a fault-diagnosis scheme. In fact, when each sensor in a network mounts its own encoding-decoding CDT, the concept-drift detections can be analyzed at an abstract, cognitive, level where functional spatial relationships between sensors are exploited. Within such diagnostic framework we can differentiate among changes in the environment, model bias and sensor faults provided that enough information is given.

Acknowledgement

This work was supported by the FP7 EU project i-Sense, Making Sense of Nonsense, Contract No: INSFO-ICT-270428, the National Natural Science Foundation of China (Nos. 61273136, 61273140, 60921061, 61034002), and the visiting professorship of Chinese Academy of Sciences.

References

Abdulsalam, H., Skillicorn, D. B. and Martin, P. (2011). Classification using streaming random forests, *IEEE Transactions on Knowledge and Data Engineering* **23**, 1, pp. 22–36.

Akaike, H. (1974). A new look at the statistical model identification, *IEEE Transactions on Automatic Control* **19**, 6, pp. 716–723.

Alippi, C., Boracchi, G. and Roveri, M. (2011a). A just-in-time adaptive classification system based on the intersection of confidence intervals rule, *Neural Networks* **24**, 8, pp. 791–800.

Alippi, C., Boracchi, G. and Roveri, M. (2012). On-line reconstruction of missing data in sensor/actuator networks by exploiting temporal and spatial redundancy, in *Proceedings of International Joint Conference on Neural Networks* (Birsbane, Australia), pp. 1–8.

Alippi, C., Boracchi, G. and Roveri, M. (2013a). Just-in-time classifiers for recurrent concepts, *IEEE Transactions on Neural Networks and Learning Systems* **24**, 4, pp. 620–634.

Alippi, C., Boracchi, G. and Roveri, M. (2013b). Ensembles of change-point methods to estimate the change point in residual sequences, *Soft Computing* **17**, 11, pp. 1971–1981.

Alippi, C., Boracchi, G. and Roveri, M. (2011b). A hierarchical, nonparametric sequential change-detection test, in *Proceedings of the International Joint Conference on Neural Networks* (California, USA), pp. 2889–2896.

Alippi, C., Camplani, R., Galperti, C., Marullo, A. and Roveri, M. (2013c). A high frequency sampling monitoring system for environmental and structural applications. Part A: The hardware architecture, *ACM Transactions on Sensor Networks* **9**, 4, pp. 1–32.

Alippi, C., Ntalampiras, S. and Roveri, M. (2013d). A cognitive fault diagnosis system for distributed sensor networks, *IEEE Transactions on Neural Networks and Learning Systems* **24**, 8, pp. 1213–1226.

Alippi, C. and Roveri, M. (2008). Just-in-time adaptive classifiers–part I: Detecting nonstationary changes, *IEEE Transactions on Neural Networks* **19**, 7, pp. 1145–1153.

Alippi, C. and Roveri, M. (2008). Just-in-time adaptive classifiers–part II: Designing the classifier, *IEEE Transactions on Neural Networks* **19**, 12, pp. 2053–2064.

Basseville, M. and Nikiforov, I. V. (1993). *Detection of Abrupt Changes: Theory and Application* (Prentice-Hall).

Burr, I. W. (1976). *Statistical Quality Control Methods* (CRC Press).

Chawla, N. V., Japkowicz, N. and Kotcz, A. (2004). Editorial: Special issue on learning from imbalanced data sets, *SIGKDD Explorations* **6**, 1, pp. 1–6.

De Oliveira, E. A. and Caticha, N. (2010). Inference from aging information, *IEEE Transactions on Neural Networks* **21**, 6, pp. 1015–1020.

Elwell, R. and Polikar, R. (2011). Incremental learning of concept drift in nonstationary environments, *IEEE Transactions on Neural Networks* **22**, 10, pp. 1517–1531.

Gama, J., Medas, P., Castillo, G. and Rodrigues, P. (2004). Learning with drift detection, in J. G. Carbonell and J. Siekmann (eds.), *Advances in Artificial Intelligence SBIA 2004* (Springer, Berlin), pp. 66–112.

Gao, J., Ding, B., Fan, W., Han, J. and Yu, P. S. (2008). Classifying data streams with skewed class distributions and concept drifts, *IEEE Internet Computing* **12**, 6, pp. 37–49.

Goldenshluger, A. and Nemirovski, A. (1997). On spatial adaptive estimation of nonparametric regression, *Mathematical Methods of Statistics* **6**, 2, pp. 135–170.

Grinblat, G. L., Uzal, L. C., Ceccatto, H. A. and Granitto, P. M. (2011). Solving nonstationary classification problems with coupled support vector machines, *IEEE Transactions on Neural Networks* **22**, 1, pp. 37–51.

Gustafsson, F. (2000). *Adaptive Filtering and Change Detection* (Wiley).

Hastie, T., Tibshirani, R. and Friedman, J. (2009). *The Elements of Statistical Learning: Data Mining, Inference and Prediction*, 2nd edn. (Springer).

He, H. and Chen, S. (2008). IMORL: Incremental multiple-object recognition and localization, *IEEE Transactions on Neural Networks* **19**, 10, pp. 1727–1738.

Ilin, A., Valpola, H. and Oja, E. (2004). Nonlinear dynamical factor analysis for state change detection, *IEEE Transaction on Neural Networks* **15**, 3, pp. 559–575.

Kendall, M. (1975). *Rank Correlation Methods* (Griffin).

Kramer, M. A. (1992). Autoassociative neural networks, *Computers and Chemical Engineering* **16**, 4, pp. 313–328.

Ljung, L. (1999). *System Identification: Theory for the User*, 2nd edn. (Prentice Hall).

Manly, B. F. J. and MacKenzie, D. I. (2000). A cumulative sum type of method for environmental monitoring, *Environmetrics* **11**, 2, pp. 151–166.

Minku, L. and Yao, X. (2012). DDD: A new ensemble approach for dealing with concept drift, *IEEE Transactions on Knowledge and Data Engineering* **24**, 4, pp. 619–633.

Mudholkar, G. S. and Trivedi, M. C. (1981). A Gaussian approximation to the distribution of the sample variance for nonnormal populations, *Journal of the American Statistical Association* **76**, 374, pp. 479–485.

Nishida, K. and Yamauchi, K. (2009). Learning, detecting, understanding, and predicting concept changes, in *Proceedings of International Joint Conference on Neural Networks* (Atlanta, USA), pp. 2280–2287.

Sperduti, A. (2006). Exact solutions for recursive principal components analysis of sequences and trees, in *Proceedings of International Conference on Artificial Neural Networks* (Athens, Greece), pp. 349–356.

Voegtlin, T. (2005). Recursive principal components analysis, *Neural Networks* **18**, 8, pp. 1051–1063.

Widmer, G. and Kubat, M. (1996). Learning in the presence of concept drift and hidden contexts, *Machine Learning* **23**, 1, pp. 69–101.

Chapter 11

Recognizing sEMG Patterns for Interacting with Prosthetic Manipulation

Zhaojie Ju, Gaoxiang Ouyang, Marzena Wilamowska-Korsak, and Honghai Liu

Abstract It is a challenge to achieve a satisfactory rate for the sEMG pattern recognition, which is becoming the main focus of the on-going research in rehabilitation and prosthetics. This chapter introduces nonlinear feature extraction and nonlinear classification approaches to efficiently identify different human hand manipulations based on surface electromyography (sEMG) signals. The recurrence plot is employed to represent dynamical characteristics of sEMG during hand movements as nonlinear features. Fuzzy Gaussian Mixture Models (FGMMs) are proposed and employed as a nonlinear classifier to recognise these hand grasps and in-hand manipulations captured from different subjects. Results from a variety of experiments comparing 14 individual features, 19 multi-features and 4 classifiers demonstrate the proposed nonlinear measures provide essential supplemental information to the good performance in multi-features. It also proves that FGMMs with capacity of modelling nonlinear datasets outperform commonly used approaches including Linear Discriminant Analysis (LDA), Gaussian Mixture Models (GMMs) and Support Vector Machine (SVM). Specially, the best performance with the recognition rate of 96.7% is achieved by FGMMs with the multi-feature combining Willison Amplitude (WAMP) and Determinism (DET).

11.1 Introduction

The electromyogram (EMG) signal is a measure of the summed activity of a number of motor unit action potentials (MUAP) lying in the vicinity of the recording electrode [Farina *et al.* (2004); Karlsson *et al.* (2009)], which contains rich information about users motion intention and thus can be utilised in the control of rehabilitation devices and assistive robots, such as the prosthetic hand or exoskeleton [Artemiadis and Kyriakopoulos (2010); Yin *et al.* (2012)], as shown in Fig. 11.1. Thus, the analysis of the EMG signals is particularly attractive in that it can provide the useful information to recognise EMG patterns for different hand gestures [Al-Timemy *et al.* (2013); Cipriani *et al.* (2009)]. However, there are inherent difficulties in deriving a general model on the relationship between the recorded sEMG and hand gestures when humans perform contractions [Liu (2011)]. In order to control a prosthetic device, the fundamental challenge is to efficiently handle sEMG signals and identify the intention of the users [Kwon and Kim (2011)]. Recent advances in technologies of signal processing and pattern recognition have made it possible to develop advanced EMG detection and analysis techniques [Farfan *et al.* (2010)]. The prosthetic control scheme based on pattern recognition usually includes two major parts: feature extraction and classification [Chen *et al.* (2010)].

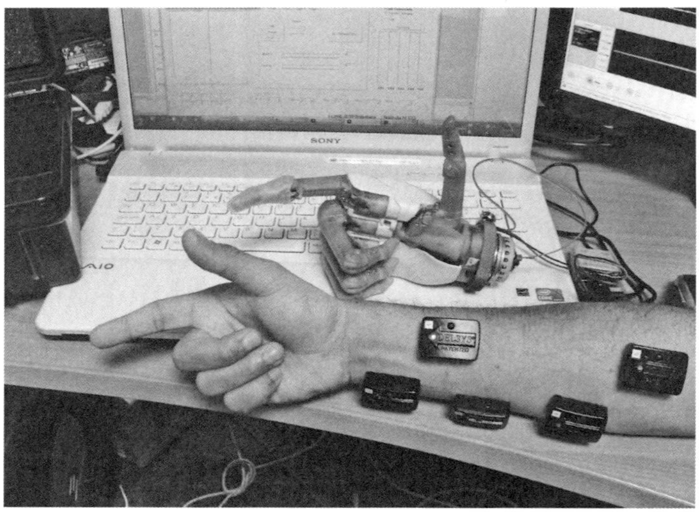

Fig. 11.1 Controlling prosthetic hand, i-limb, from Touch Bionics, by sEMG signals

Various methods have been widely used for feature extraction ranging from traditional linear methods such as time and frequency analysis to nonlinear methods. In the time domain, traditional methods of EMG amplitude analysis mainly consist of the mean absolute value, number of zero crossings, waveform length, number of slope sign changes and autoregressive model [Hudgins *et al.* (1993); Li *et al.* (2010)]. Graupe *et al.* (1978) used an autoregressive (AR) model to represent EMG signals, and motions were determined based on the parameters of the AR model. Oskoei and Hu (2008) evaluated six preselected single features and four multi-features and compared them in terms of the recognition rate and the multi-feature achieved the best performance. In the frequency domain, the methods include Fourier transform coefficients [Peleg *et al.* (2002)]. More recently, instead of analysing the sEMG signal independently in the time or frequency domain, time-frequency methods, such as short time Fourier transform, wavelet, and wavelet package, have been introduced for feature extraction [Englehart *et al.* (1999)]. Chu *et al.* (2006) used a wavelet packet transform to extract a feature vector form EMG signals for recognising the hand motions. To some extent, these traditional methods of analysing EMG have the capability of tracking muscular changes [Li *et al.* (2010)]. However, traditional methods of sEMG amplitude and spectral analysis are not effective in analysing complex transient signals [Duchene and Goubel (1993)]. In particular, the spectral analysis is based on an assumption that the observed variations of electrical field of muscle activity are time-invariant (stationary) processes [Xie *et al.* (2010)]. In practice, the sEMG signal is a non-stationary signal and stems from a highly non-linear system, which is a complex signal embedded in noise [Akay (2000)]. It is thus promising to develop new methods to characterize sEMG changes in different neural activation of muscles based on nonlinear methods. Filligoi and Felici (1999) employed a recurrence quantification analysis (RQA) to analyse the non-linear dynamical characteristics of sEMG data. Ju *et al.* (2013b) investigated and evaluated nonlinear approaches to extract EMG signal features and to identify different types of manipulation motions including different hand grasps and in-hand manipulations. The experimental results demonstrated the proposed nonlinear measures provide important supplemental information and the combination of linear and nonlinear measures of EMG signals could achieve the highest recognition rate.

In terms of classifiers, lots of methodologies have been proposed and applied to process and discriminate sEMG signals, including Neural Networks [Chen *et al.* (2010)], such as Multilayer Perceptron Network [Tenore *et al.* (2009)], statistic approaches such as Hidden Markov Models (HMMs) [Chan and Englehart (2005)] and Gaussian Mixture Models (GMMs) [Huang *et al.* (2005)], and fuzzy methods [Khezri and Jahed (2007)]. For example, Tenore *et al.* (2009) proposed a system using traditional time-domain features of EMG signals and a multilayer perceptron as a classifier, which can identify ten finger movements with the recognition greater than 90%. In addition, a few studies compared several methods [Castellini and van der Smagt (2009); Ju *et al.* (2011b)], e.g., Castellini and van der Smagt (2009) reported that SVM achieved a higher recognition rate, about 90% than Neural Networks (NN) and Locally Weighted Projection Regression on five grasp motions, while Liu *et al.* (2007) proposed Cascaded Kernel Learning Machine (CKLM) compared to other classifiers such as k-nearest neighbours, multi-layer NN and SVM. Khezri and Jahed (2007) employed an adaptive neuro-fuzzy inference system to identify hand motion commands by a hybrid back propagation and a subtractive-clustering algorithm. Subasi (2012) compared multi-layer perceptron neural networks (MLPNN), dynamic fuzzy neural network (DFNN) and adaptive neuro-fuzzy inference system (ANFIS) based classifiers in relation to their accuracies in the classification of EMG signals. Tang *et al.* (2012) proposed a cascaded-classifier that divides the classification procedure into several levels and different sEMG features were employed individually in the different levels.

Most of above recognition methods more concern about individual muscle contractions, and the natural movements with combined multiple muscle contractions have not yet been fully tested and addressed. In addition, the uncertainties embedded in the sEMG signals, arising from the non-stationary nature, different subjects, muscle fatigue, etc., have not been considered. This chapter investigates nonlinear approaches for sEMG feature extraction and classification using recurrence plot and Fuzzy Gaussian Mixture Models (FGMMs). FGMMs are used to identify sEMG based natural hand motions in order to control prosthetic hands, with a novel recognition algorithm to improve the recognition efficiency. Different combinations of the linear and nonlinear features are also investigated. Natural hand motions from different subjects, which consist of combinations of finger flexions, extensions, adductions and abductions, are addressed to test various linear/nonlinear features and different classifiers.

11.2 Recurrence Plot and Quantification Analysis

In the nonlinear dynamics theory [Packard *et al.* (1980)], phase space of the signal needs to be reconstructed, during which time delay methods are usually used to embed a scalar time series into a m-dimensional space:

$$e_k = (u_k, u_{k+\tau}, \ldots, u_{k+(m-1)\tau}) \tag{11.1}$$

where $k = 1, 2, \ldots, L-(m-1)\tau$, τ is the time delay and m is the embedding dimension, $m \geq 2$. Taken's theorem states that if the data is infinite and noise-free, the time delay τ can be chosen almost arbitrarily. However, real EMG signals are finite and associated with noises, and the time delay needs to be carefully considered: if τ is too small, the reconstructed vector is too close to serve as independent coordinates; if it is too large, the vector becomes independent and loses the connection with each other [Kim *et al.* (1999)]. The most common method to choose a proper time delay is based on detection of the first local minimum of the mutual information (MI) function, since the first minimum portrays the time delay where the signal $u_{t+\tau}$ adds maximal information to the knowledge obtained from u_t [Fraser and Swinney (1986)]. The MI method produces non-linear characteristics of time series, so it is better to estimate time delay than linear autocorrelation functions. For choosing the parameter m, there are a number of different criteria. Cao proposed a method that determines the minimum embedding dimension; it can overcome some shortcomings of false nearest neighbours [Cao (1997)]. In this study, the MI method [Fraser and Swinney (1986)] and Cao's method [Cao (1997)] are employed to estimate the time delay and the embedding dimension respectively.

Recurrence plots (RP), proposed by Eckmann *et al.* (1987), describes the recurrence property of a deterministic dynamical system, i.e., visualizing the time dependent behaviour of orbits in a phase space. The key step of RP is to calculate the following matrix:

$$R_{i,j}(\varepsilon) = \Theta(\varepsilon - \|e_i - e_j\|), \quad i,j = 1,\ldots,N \tag{11.2}$$

where $N = L - (m-1)\tau$, ε is a predefined cutoff distance, $\|\cdot\|$ is the norm (e.g., the Euclidean norm) and $\Theta(x)$ is a Heaviside function. The phase space vector e_i can be reconstructed using Takens time delay method, $e_i = (u_i, u_{i+\tau}, \ldots, u_{i+(m-1)\tau})$ [Takens (1981)], based on the observations. The cutoff distance ε defines a sphere centred at e_j: if e_i falls within this sphere, i.e., the state is close to e_j, then $R_{i,j} = 1$; otherwise $R_{i,j} = 0$. The binary values of $R_{i,j}$ can be simply visualized with the colours black for 1 and white for 0. Thereby, RP can be considered as a visual inspection of a high dimensional phase space trajectory: in other words, RP indicates the time evolution of a trajectory. In short, RP can describe the characteristics of large-scale and small-scale patterns of a dynamical system [Marwan *et al.* (2002)]. One example of the RP for one sEMG signal can been seen in Fig. 11.2.

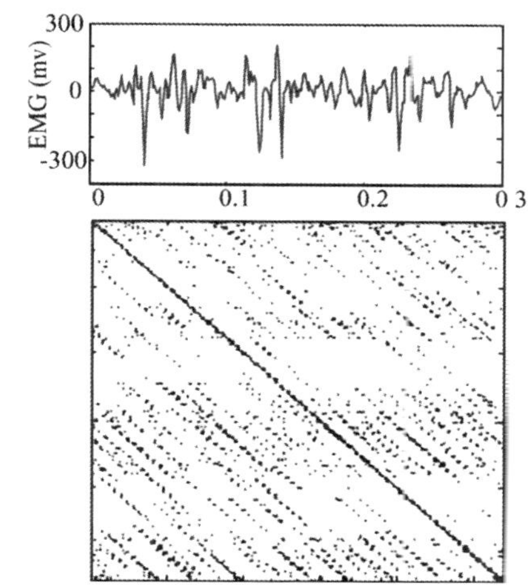

Fig. 11.2 An example of the RP for one sEMG signal

In order to further investigate properties of RP, several measures of complexity that quantify the small-scale structures in RP called recurrence quantification analysis (RQA), have been proposed [Marwan *et al.* (2002); Trulla *et al.* (1996)]. Here, we introduced two measure variables that are determinism (DET) and entropy (ENTR). They are based on the recurrence point density and diagonal line structures of the RP. For more details please refer to Marwan *et al.* (2007).

The classical measure of RQA is the recurrence rate (RR)

$$\mathrm{RR}(\varepsilon) = \frac{1}{N^2}\sum_{i,j=1}^{N} R_{i,j}(\varepsilon), \tag{11.3}$$

RR is a measure of the density of recurrence points and counts the black dots in the RP.

The frequency distribution of the lengths l of diagonal structures in the RP is $P^{\varepsilon}(l) = \{l_i;\ i = 1, 2, \ldots, N\}$. Processes with stochastic behaviour cause none or very short diagonals, whereas the deterministic processes cause longer diagonals and less single, isolated recurrence points. Therefore, the ratio of recurrence points on the diagonal structures (of at least length $l_{\min}$) to all recurrence points is called the DET, and is introduced as a determinism (or predictability) measure of the system. It is given as:

$$\mathrm{DET} = \frac{\sum_{l=l_{\min}}^{N} lP^{\varepsilon}(l)}{\sum_{i,j}^{N} R_{i,j}(\varepsilon)} = \frac{\sum_{l=l_{\min}}^{N} lP^{\varepsilon}(l)}{\sum_{l=1}^{N} lP^{\varepsilon}(l)} \tag{11.4}$$

where $l_{\min}$ is the threshold, which excludes the diagonal lines formed by the tangential motion of a phase space trajectory, and in this study we fixed at $l_{\min} = 2$.

ENTR is the Shannon entropy of the frequency distribution of the diagonal line lengths,

$$\mathrm{ENTR} = -\sum_{l=l_{\min}}^{N} p(l) \ln p(l) \tag{11.5}$$

where $p(l) = P(l)/\sum_{l=l_{\min}}^{N} P(l)$. ENTR measures the complexity of the deterministic structure in a dynamical system, and ENTR becomes larger if the structure is more complex.

A parameter specific to the RP is the cutoff distance ε. If it is too large, almost every point is a neighbour of each other point, which produces a saturation of the RP including irrelevant points; on the contrary, if it is too small, there may be almost no recurrence points, which loses information of the underlying system [Marwan *et al.* (2007)]. Several criteria for the choice of the cutoff distance have been advocated in the literature [Matassini *et al.* (2002); Thiel *et al.* (2002)]. One approach uses a fixed number of neighbours, N_n, for every point of the trajectory, called fixed amount of nearest neighbours (FAN) [Trulla *et al.* (1996)]. In this approach, the cutoff distance ε_i changes for each state e_i to ensure all columns of the RP have the same recurrence density. Using this neighbourhood criterion, ε_i can be adjusted in such a way that the recurrence rate has a fixed predetermined value (i.e., $RR = N_n/N$) [Marwan *et al.* (2007)]; and the RP is invariant under enlarging or reducing of the amplitude of time series.

11.3 Nonlinear Recognition Method

11.3.1 *Gaussian mixture models*

As one of the most statistically mature methods in pattern recognition and machine learning [Ba and Odobez (2009); Ju *et al.* (2009)], Gaussian Mixture Models (GMMs) have been successfully implemented to identify high frequency signals such as speech and EMG signals with good performance [Huang *et al.* (2005); Lu *et al.* (2011)].

The probability density function for a Gaussian distribution is given by

$$p(x|\theta) = \frac{1}{(2\pi)^{\frac{d}{2}}\sqrt{|\Sigma|}} \exp\left(-\frac{(x-\mu)^{\mathsf{T}}\Sigma^{-1}(x-\mu)}{2}\right) \tag{11.6}$$

where the set of parameters has $\theta = (\mu, \Sigma)$, μ is the mean, Σ is the covariance matrix of the Gaussian, d is the dimension of vector x, and exp denotes the exponential function.

Let $\mathcal{X} = \{x_1, \ldots, x_n\}$ be a d-dimensional observed dataset of n vectors. If the distribution of $\mathcal{X}$ can be modelled by a mixture of k Gaussians, the density of each vector is

$$p(x_t|\Theta) = \sum_{i=1}^{k} \alpha_i p_i(x_t|\theta_i) \tag{11.7}$$

where the parameters are $\Theta = (\alpha_1, \ldots, \alpha_k, \theta_1, \ldots, \theta_k)$ and $(\alpha_1, \ldots, \alpha_k)$ are the k mixing coefficients of the k mixed components such that $\sum_{i=1}^{k} \alpha_i = 1$; each p_i is a density function parameterized by θ_i. The resulting density for the samples is

$$p(\mathcal{X}|\Theta) = \prod_{t=1}^{n} p(x_t|\Theta) = \mathcal{L}(\Theta|\mathcal{X}). \tag{11.8}$$

The function $\mathcal{L}(\Theta|\mathcal{X})$ is called the likelihood of the parameters given the data, or the likelihood function. The likelihood is considered as a function of the parameters Θ where the data $\mathcal{X}$ is fixed. In the maximum likelihood problem, the objective is to estimate the parameters set Θ that maximizes $\mathcal{L}$. That is to find Θ^* where

$$\Theta^* = \arg\max_{\Theta} \mathcal{L}(\Theta|\mathcal{X}). \tag{11.9}$$

Usually, the $\log(\mathcal{L}(\Theta|\mathcal{X}))$ is maximized instead because it is analytically easier. The log-likelihood expression is given by

$$\begin{aligned} \log(\mathcal{L}(\Theta|\mathcal{X})) &= \log\left(\prod_{t=1}^{n} p(x_t|\Theta)\right) \\ &= \sum_{t=1}^{n} \log\left(\sum_{i=1}^{k} \alpha_i p_i(x_t|\theta_i)\right). \end{aligned} \tag{11.10}$$

Directly maximizing the log-likelihood is difficult, hence an auxiliary objective function Q is taken into account

$$Q = \sum_{t=1}^{n} \sum_{i=1}^{k} w_{it} \log\left[\alpha_i p_i(x_t|\theta_i)\right] \tag{11.11}$$

where w_{it} is a posteriori probability for individual class i, $i = 1, \ldots, k$, and it satisfies

$$w_{it} = \frac{\alpha_i p_i(x_t|\theta_i)}{\sum_{s=1}^{k} \alpha_s p_s(x_t|\theta_s)} \tag{11.12}$$

and

$$\sum_{i=1}^{k} w_{it} = 1. \tag{11.13}$$

Maximizing (11.11) guarantees that $p(\mathcal{X}|\Theta)$ is maximized if it is performed by an EM algorithm (e.g., [Bilmes (1998); Huang *et al.* (1990)]). The iteration of an EM algorithm estimating the new parameters in terms of the old parameters is given as follows:

- E-step: compute "expected" classes of all data points for each class using (11.12).
- M-step: compute maximum likelihood given the data's class membership distributions according to (11.14)–(11.16).

$$\alpha_i^{new} = \frac{1}{n}\sum_{t=1}^{n} w_{it} \tag{11.14}$$

$$\mu_i^{new} = \frac{\sum_{t=1}^{n} w_{it}x_t}{\sum_{t=1}^{n} w_{it}} \tag{11.15}$$

$$\Sigma_i^{new} = \frac{\sum_{t=1}^{n} w_{it}(x_t - \mu_i^{new})(x_t - \mu_i^{new})^{\mathsf{T}}}{\sum_{t=1}^{n} w_{it}}. \tag{11.16}$$

When training GMMs, k-means is employed for initialization before EM starts. The iteration of EM algorithm stops when the change value of log-likelihood is below a preset threshold. An example of the GMMs trained result with six components is shown in Fig. 11.3.

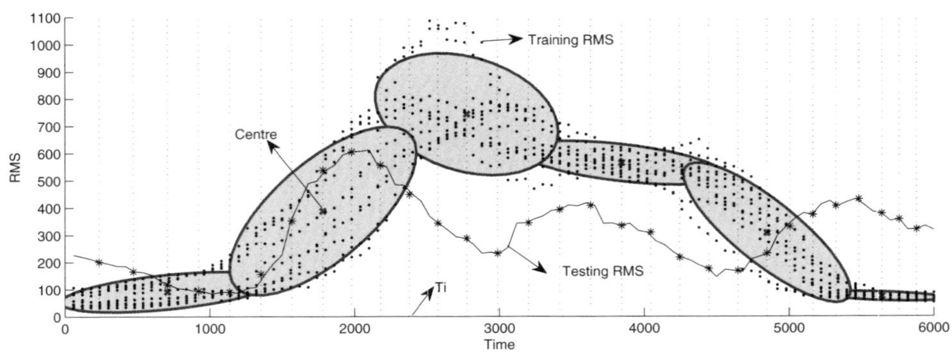

Fig. 11.3 An example of the GMMs trained result with six components (circle) on the root mean square (RMS) feature (dots). The black line is the RMS from the testing motion; dots on the black line are the re-sampled points for the testing RMS at T_i time instance.

11.3.2 *Fuzzy Gaussian mixture models*

In order to equip GMMs with nonlinear fitting capabilities, fuzzy Gaussian mixture models (FGMMs) were proposed in our previous work [Ju and Liu (2012, 2011a)], which were proved to have better fitting performance and a faster convergence speed than conventional GMMs. In this chapter, the nonlinear method, FGMMs, will be employed to recognize hand motions including different hand grasps and in-hand manipulations via sEMG signals. In this section, we will firstly revisit the expectation-maximization (EM) algorithm for the FGMMs and then propose its recognition method. In Ju and Liu (2012), two types of FGMMs were proposed, including the distance based FGMMs and the probability based FGMMs. In this chapter, the distance based FGMMs are chosen and referred as FGMMs, since the performance of the former is better than the latter. The processing of training with FGMMs is summarized as follows.

Let $\mathcal{X} = \{x_1, x_2, \ldots, x_n\}$ be the d dimensional observed dataset with n samples; $k \geq 2$ be the number of the components; n be the number of the sampling points; $m > 1$ be the degree of fuzziness; $\varepsilon > 0$ be a small preset real positive number. The initialisation of FGMMs is achieved by Fuzzy C-Means (FCM). The iteration of the EM algorithm for the FGMMs is

- E-step: Compute 'expected' classes of all data points for each class.

$$u_{it} = \left[\sum_{j=1}^{k} \left(\frac{d_{it}}{d_{jt}} \right)^{\frac{2}{m-1}} \right]^{-1} \tag{11.17}$$

where u_{it} is the degree of membership of x_t in the i-th cluster; d_{jt} is the dissimilarity between point x_t and j-th cluster; d_{it} is the dissimilarity between point x_t and i-th cluster, which can be archived by

$$\mathrm{d}_{\mathrm{it}}^2 = \begin{cases} \dfrac{\exp\left(\frac{(x_t-\mu_i)^\mathsf{T}\Sigma_i^{-1}(x_t-\mu_i)}{2}\right)}{\left(\alpha_{\mathrm{i}}(2\pi)^{-\frac{d}{2}}|\Sigma_i|^{-\frac{1}{2}}\right)^{\frac{m-1}{m}}} & (|a_i| < \varepsilon) \\[2em] \dfrac{1}{\alpha_i^{\frac{m-1}{m}} p_i(x_t|\theta_i)} & (|a_i| \geq \varepsilon) \end{cases} \tag{11.18}$$

where μ_i is the mean and Σ_i is the covariance matrix of the ith Gaussian component; α_i is the weight of ith component; a is first parameter of the standard $y = ax^2 + b$ which is used to shape the principle component axis; $p_i(x_t|\theta_i)$ is the probability density function of point x_t to the ith component and it has

$$p(x_t|\theta) = \left(\sum_{j=1}^{J_t} \prod_{s=1}^{2} \frac{\exp\left(\frac{-l_j^2(v_{st})m}{2\Sigma_s(m-1)}\right)}{\sqrt{2\pi|\Sigma_s|}} \prod_{s=3}^{d} \frac{\exp\left(\frac{-v_{st}^2 m}{2\Sigma_s(m-1)}\right)}{\sqrt{2\pi|\Sigma_s|}} \right)^{\frac{m-1}{m}} \tag{11.19}$$

where $l_j(v_{1t})$ is the arc length of the j-th projected coordinate $z = [z_{1j}, z_{2j}]$, which is transferred from point x_t, on the standard curve principle axis; $l_j(v_{2t})$ is the distance between the transferred point $[v_{1t}, v_{2t}]$ and its projected point z. More details about how to get these projected points or transferred points can be found in Ju and Liu (2012).

- M-step: Compute maximum likelihood given the data's class membership distributions.

  If $|a_i| < \varepsilon$

$$\mu_i^{new} = \frac{\sum\limits_{t=1}^{n} u_{it}^m x_t}{\sum\limits_{t=1}^{n} u_{it}^m} \tag{11.20}$$

$$\Sigma_i^{new} = \frac{\sum\limits_{t=1}^{n} u_{it}^m (x_t - \bar{\mu}_i)(x_t - \bar{\mu}_i)^{\mathsf{T}}}{\sum\limits_{t=1}^{n} u_{it}^m}. \tag{11.21}$$

  If $|a| \geq \varepsilon$

$$(C_i^{new}, T_i^{new}, Q_i^{new}) = \mathrm{LSFM}(\mathrm{PCA}(\mathcal{X} \cdot U_i)) \tag{11.22}$$

  $U_i = [u_{i1}, \ldots, u_{in}]$; PCA$(\cdot)$ is the principal component analysis function for estimating the translation matrix T_i^{new} and rotation matrix U_i^{new}. LSFM$(\cdot)$ is least-squares fitting method for estimating control parameters $C_i^{new} = (a, c)$ which shapes the curve axis with standard curve $y = ax^2 + b$; and the new estimated mean and covariance are

$$\mu_i^{new} = \frac{\sum\limits_{t=1}^{n} u_{it}^m x_t}{\sum\limits_{t=1}^{n} u_{it}^m} + (Q_i^{new})^{-1} \underbrace{[0, b, 0, \ldots, 0]^{\mathsf{T}}}_{d} + T_i^{new} \tag{11.23}$$

$$\Sigma_{ie}^{new} = \frac{\sum\limits_{t=1}^{n} u_{it}^m \bar{L}_{te}^{(i)}}{\sum\limits_{t=1}^{n} u_{it}^m} \quad (e = 1, 2) \tag{11.24}$$

$$\Sigma_{i(3-d)}^{new} = \frac{\sum\limits_{t=1}^{n} u_{it}^m (x_t - \mu_i^{new})_{(3-d)} (x_t - \mu_i^{new})_{(3-d)}^{\mathsf{T}}}{\sum\limits_{t=1}^{n} u_{it}^m}. \tag{11.25}$$

The inputs of the FGMMs are the EMG features, number of components k, degree of fuzziness m and threshold ε. Then EM algorithm for FGMMs has been utilised to find optimised centres of the components μ, their covariance Σ and the

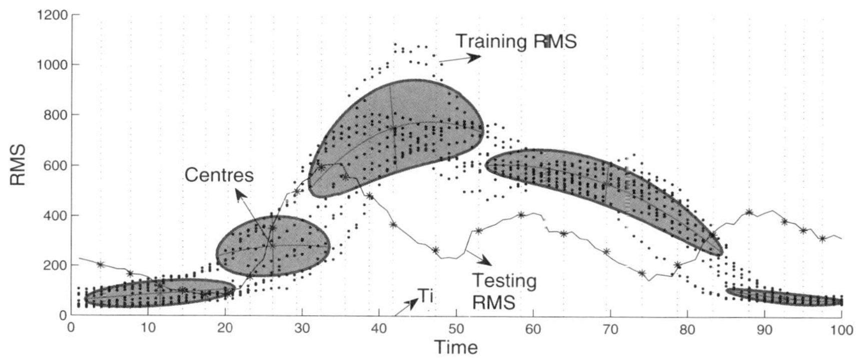

Fig. 11.4 An example of the FGMMs trained result with six components (circle) on the root mean square (RMS) feature (dots). The black line is the RMS from the testing motion; dots on the black line are the re-sampled points for the testing RMS at T_i time instance.

control parameters C, T, Q, which are the outputs of the FGMMs and will be used in the recognition process. Details to implement the EM algorithm of FGMMs can be found in the appendix in Ju and Liu (2012). Fig. 11.4 gives an example of the trained model via FGMMs, where there are six components to fit the sEMG RMS feature.

11.3.3 *Recognizing sEMG signals using FGMMs*

Consider that the FGMMs have k components/patterns. To recognise the testing motion, a similarity function is proposed in (11.26). The similarity of the testing motion and the trained model of FGMMs is defined by the normalised log-likelihood between the re-sampled testing points and the FGMMs components as

$$Si = \frac{1}{5k} \sum_{j=1}^{5k} \log \left(\sum_{i=1}^{k} \alpha_i p_i(x_{T_j} | \theta_i) \right) \tag{11.26}$$

where α_i is the mixing coefficient of the i-th component, if the component's curvature parameter $a_i < \varepsilon$, the $p(x|\theta)$ will be calculated by

$$p(x|\theta) = \frac{1}{(2\pi)^{\frac{d}{2}} \sqrt{|\Sigma|}} \exp \left(-\frac{(x-\mu)^{\mathsf{T}} \Sigma^{-1} (x-\mu)}{2} \right) \tag{11.27}$$

if its curvature parameter $a_i \geq \varepsilon$, $p(x|\theta)$ is achieved by (11.19). x_{T_j} is the selected points from the testing data $\mathbf{x}$ at the time instance of T_j, which can be achieved by (11.28).

$$T_j = \mu_f - \frac{\eta_j(\mu_f - \mu_{f-1})}{3} + \frac{\gamma_j(\mu_{f+1} - \mu_f)}{3} \tag{11.28}$$

where T_j is the time sampling points for the testing data; $j \in (1, \ldots, 5 \cdot k)$; μ_f is the time label of the fth component centre; $f = \lfloor (j-1)/5 \rfloor$; the parameters, η and γ, are achieved by (11.29) and (11.30):

$$\eta_j = \begin{cases} 2 - [(j-1) \mod 5] & \text{if } [(j-1) \mod 5] < 2, \\ 0 & \text{else.} \end{cases} \tag{11.29}$$

$$\gamma_j = \begin{cases} [(j-1) \mod 5] - 2 & \text{if } [(j-1) \mod 5] > 2, \\ 0 & \text{else.} \end{cases} \tag{11.30}$$

where mod is the module operation to find the remainder of division of one number by another.

The resampling process, shown in Fig. 11.4, is utilized to reduce the computation cost. The re-sampled points will normally be much fewer than the original data points (reducing data points from n to $5k$, given $n >> k$), and thus calculation of the similarity between the testing motion and the trained model will be more efficient. On the other hand, the re-sampling process locates five re-sampling points around each component, which means the re-sampling points will cover the major distribution of the components even if some of them are relatively small.

11.4 Experimental Results

11.4.1 *Data collection and description*

The sEMG of 5 forearm muscles shown in Fig. 11.5, i.e., flexor carpi radialis, flexor carpi ulnaris, flexor pollicis longus, flexor digitorum profundus and extensor digitorum, were measured. To obtain clearer signals, subjects were scrubbed with alcohol and shaved if necessary and then electrodes were applied over the body using the die cut medical grade double-sided adhesive tape. Five electrodes locations were selected according to the musculoskeletal systems related these five muscles and confirmed by muscle specific contractions, which include manually resisted finger flexion, extension and abduction. The EMG capture system using Trigno Wireless Sensors was employed, the resolution is 16 bits and the sampling rate is 4000 Hz.

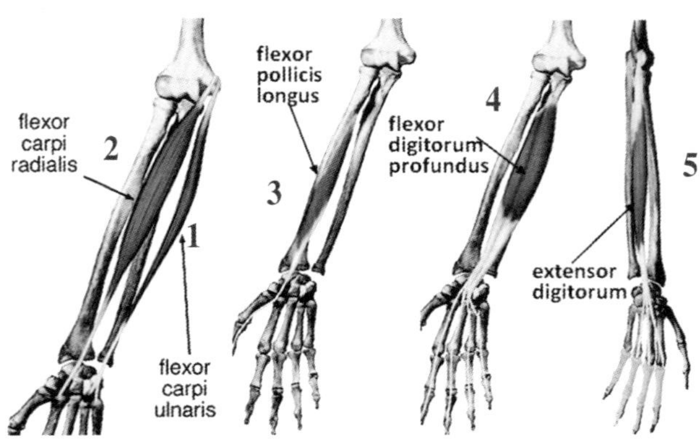

Fig. 11.5 Muscle positions for the electrode locations

Its size is 37mm × 26mm × 15mm and the range of its guaranteed performance is 40 meters. The real time sEMG signals were visualized on a computer screen giving participants feedback to choose electrode locations with stronger sEMG signals.

Eight healthy right-handed subjects including 2 females and 6 males volunteered for the study. Their ages varies from 23 to 40 and the average is 32.5 years old; body height average is 175.5cm; body mass average is 70kg. All participants gave informed consent prior to the experiments and the ethical approval for the study was obtained from University of Portsmouth CCI Faculty Ethics Committee. All subjects were trained to manipulate different objects. Participants had to perform ten grasps or in-hand manipulations which are shown in Fig. 11.6 and the motions are listed as follows

(1) Grasp and lift a book using five fingers with the thumb abduction.
(2) Grasp and lift a can full of rice using thumb, index finger and middle finger only.
(3) Grasp and lift a can full of rice using five fingers with the thumb abduction.
(4) Grasp and lift a big ball using five fingers.
(5) Grasp and lift a disc container using thumb and index finger only.
(6) Uncap and cap a marker pen using thumb, index finger and middle finger.
(7) Open and close a pen box using five fingers.
(8) Pick up a pencil using five fingers, flip it and place it on the table.
(9) Hold and lift a dumbbell.
(10) Grasp and lift a cup using thumb, index finger and middle finger.

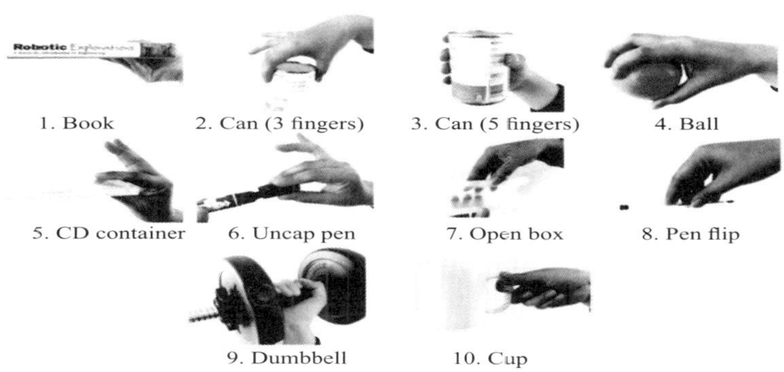

Fig. 11.6 Ten different hand motions including grasps and in-hand manipulations

The way to grasp or manipulate objects had been shown to the participants in the demonstration before they performed and every motion lasted about 2 to 4 seconds. Each motion was repeated 10 times. Between every two repetitions, participants had to relax the hands for 2 seconds in the intermediate state that is opening hand naturally without any muscle contraction. These intermediate states were used to segment the motions. Once one motion with ten repetitions was finished, participants had to relax the hand for 2 minutes before the next motion started. This was designed to overcome the effects of muscle fatigue.

11.4.2 *Feature extraction and parameter setting*

Balancing the algorithm's effectiveness and efficiency is critical in realtime system. Data segmentation is important for both the validity of sEMG features and time delay of the system. It has been shown that the time delay between the onset of muscle contraction made by a subject and the corresponding motion in a device should be less than 300ms for the real time operation; and the minimum time interval between two distinct contractions is about 200ms to contain enough information to estimate a motion state of the hand [Oskoei and Hu (2008); Stark (1968)]. In our experiments, we choose the segment length of 300ms and the increment of 50ms.

In this chapter, various EMG features have been used and compared with the nonlinear features: DET and ENTR, including the time domain features: integrated EMG (IEMG), mean absolute value (MAV), root mean square (RMS), waveform length (WL), zero crossing (ZC), slope sign change (SSC), willison amplitude (WAMP) and frequency domain features: autoregressive coefficients (AR), power spectral density (PSD).

In the reconstruction of the m-dimensional phase space, the time delay τ and the embedding dimension m of the sEMG epochs are determined by using the MI method and the Cao's method, and one example could be seen in Fig. 11.7.

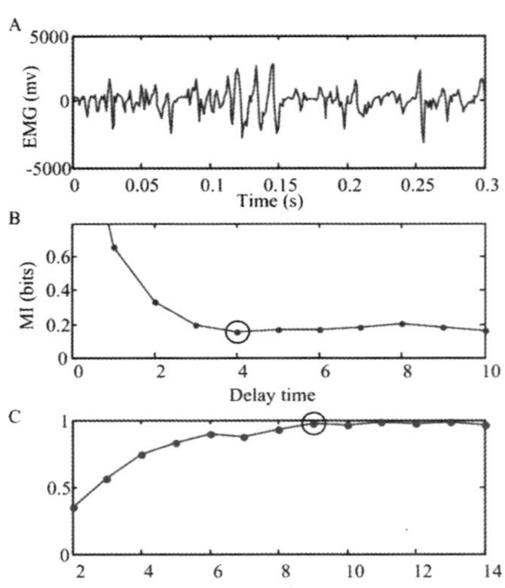

Fig. 11.7 An example of the phase construction for one sEMG epoch. (A) An example of sEMG signal; (B) mutual information (MI) versus delay time, the circle indicates the vicinity of τ, where the first local minimum of MI is; (C) embedding dimension, the circle shows E1 stop significant changing around $m = 9$.

From the results of MI and Cao methods applied to all EMG epochs, the optimum time delay values of τ, based on detection of the first local minimum of the MI function, ranged from 2 to 8 samples (mean and standard deviation is 3.68 ± 1.41) for different EMG epochs. So $\tau = 4$ is selected for the phase space reconstruction. The optimum embedding dimension m ranged from 5 to 11 (mean and standard deviation are 8.35 ± 1.31) for different EMG segments. Therefore, $m = 9$ is suitable for the topologically proper reconstruction of the EMG data. The fixed number of neighbours N_n is 15 for the every 300ms EMG segment, and the recurrence rate RR is about 0.05. The performance of every recognition algorithm in the following chapter is evaluated by leave-one-subject-out cross-validation.

11.4.3 *Recognition with one single feature*

In this section, the performance of different single features is compared using a standard recognition algorithm i.e., GMMs. The recognition results are shown in Fig. 11.8 including both the average and the standard deviation. The WAMP has the highest accuracy while the HIST has the lowest in the time domain features. The MAV, IEMG, WL and RMS have similar performance around 87% and ZC has similar performance with SSC around 77.5%. In the frequency features, the AR2 (the second order of the regression model) has the best accuracy and the AR1 (the first order of the regression model) has the lowest. Compared with features in the time domain and the frequency domain, the nonlinear features, DET and ENTR, have lower accuracies, which are around 72%. The AR2 has the highest recognition rate, 91.4%, among all features. The standard deviation results show that AR2

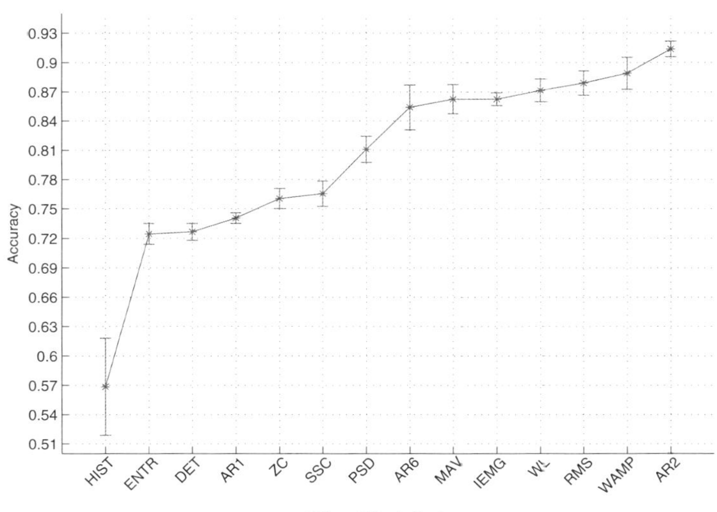

Fig. 11.8 Comparative results in terms of recognition rates with different single features using GMMs

with the best performance is very stable with small deviation while HIST has the largest deviation.

Fig. 11.9 presents the confusion matrix for the ten different motions using AR2 feature and GMMs. Motion 1, which is grasping and lifting a book using five fingers, has the lowest rate of only 71%. Motion 5 and motion 8, grasping a CD container and flipping a pen, have full accuracies, and motion 9, lifting a dumbbell has also achieved a high rate of 99%. Among all the motions, motions 6, 7 and 8 can be regarded as in-hand manipulations and others are grasps motions. The in-hand manipulations have relatively high rates than grasps.

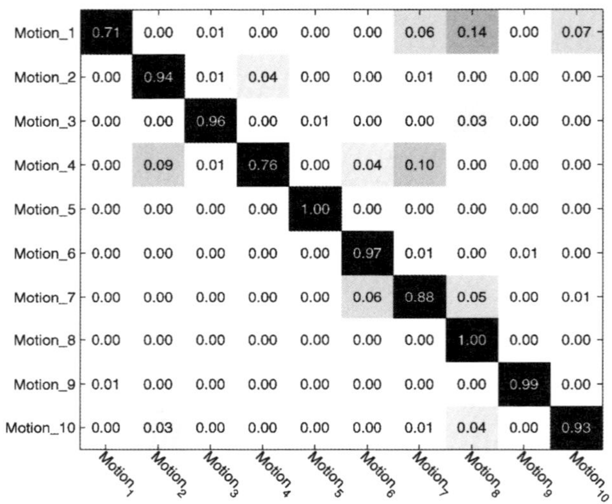

Fig. 11.9 Confusion matrix for the ten hand motions with the single feature AR2 using GMMs

11.4.4 *Recognition with multiple features*

The literature indicates the multi-features combining different single features may have better performance than the single features due to the rich information contained[Huang *et al.* (2005); Oskoei and Hu (2008)]. According to the work in by Chen *et al.* (2010); Huang *et al.* (2005); Oskoei and Hu (2008), some popular multi-features are evaluated in this section. Due to the large amount of possible combination of different single features, only the selections with relatively good performance are discussed here. Fig. 11.10 shows the comparison result of 19 different multi-features using GMMs including the average rates and their standard deviations. Since ENTR has the similar performance with DET, only results of DET have been discussed here. The results show that multi-features achieved overall better and more stable performance compared with the single features, and many of them are above 92%, such as MAV+WL, MAV+DET, and WAMP+DET. These features are better to work together than individually. However, it is not always true for

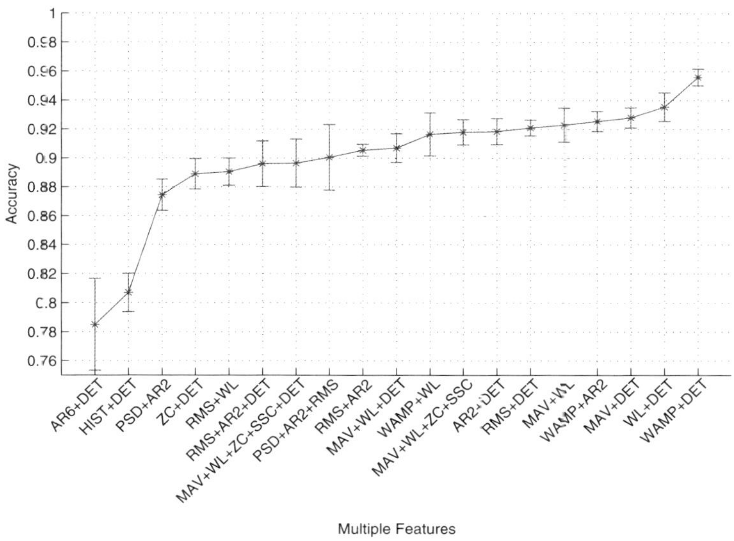

Fig. 11.10 Comparative results in terms of recognition rates with different multi-features using GMMs

every combination. For example, AR2 has a recognition rate of 91.4% when used alone, as shown in Fig. 11.8, but the multi-features, RMS+AR2+DET, have a lower accuracy, which is 89.6%, as displayed in Fig. 11.10. This may happen due to the redundancy in the multi-features. Among these multi-features, AR6+DET has the lowest recognition rate, and WAMP+DET has the highest rate of 95.7%. When using WAMP alone, the recognition rate is 88.5%, while WAMP+DET achieves 95.75% rate, which is much higher than using WAMP alone. The accuracy of WAMP+DET has improved the best performance from 91.4% achieved by the single features by more than 4 percent. When combined WAMP with other features in the time domain or frequency domain, such as AR2, WL, AR6, MAV, etc., the best rates achieved are 92.6% for WAMP+AR2 and 91.7% for WAMP+WL, which are both lower than combined with DET feature. The results of the standard deviations show that the best multi-feature still has stable performance with small standard deviation. Features of AR6+DET and PDS+AR2+RMS have large standard deviations. Generally, the standard deviations of the above multi-features are larger than the single features.

The confusion matrix for the different motions by the WAMP+DET feature and GMMs classifier is shown in Fig. 11.11. Most of the motions have above 90% recognition rates, except the motion 4 has the lowest of 84% recognition rate. This may be due to the similarity between motion 4 and motion 2.The results have demonstrated that the combined features with nonlinear feature can have much better performance than single features. In addition, RR, as a simplest measure in RQA, has also been used as a nonlinear feature in some work such as Chandrasekaran

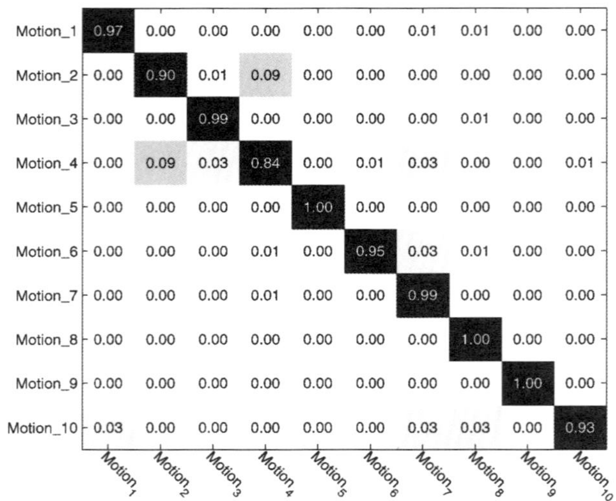

Fig. 11.11 Confusion matrix for the ten hand motions with the multi-feature WAMP+DET using GMMs

(2007); Rissanen *et al.* (2008). Here, it has also been investigated as supplemental information and compared with DET and ENTR. Its highest recognition rate is around 93%, which is achieved by the combination of WAMP+RR. It has weaker performance than both DET and ENTR.

11.4.5 *Recognition with different classifiers*

This section is to compare the effectiveness and efficiency of different classifiers, including Linear Discriminant Analysis (LDA), GMMs, Support Vector Machine (SVM) and FGMMs, with the identified best multi-features. LDA is a linear classifier seeking directions in space that has maximum discriminability and has also been employed to analyse biosignals [Ouyang *et al.* (2009)]. To have a fare comparison, the recognition process for GMMs is the same as FGMMs which has been proposed in Section 11.3.3. The parameter for GMMs and FGMMs is the number of the components ranging from 2 to 20 with increments of one. It is chosen with the best performance for each trained motion model. In this chapter, radial basis function for the SVM classifier has been employed, which has been demonstrated to have satisfactory performance in pattern recognition tasks [Chen *et al.* (2010); Steinwart and Christmann (2008)]. The parameters for SVM are the kernel parameter ranging from 1 to 10 with increments of one and penalty cost whose range is from 1 to 501 with increments of 50 achieved by using LIBSVM [Chang and Lin (2012)] package. These parameters are selected with their best performance.

All the features including the single features and multiple features have been evaluated using above different classifiers, and it was found that the combined

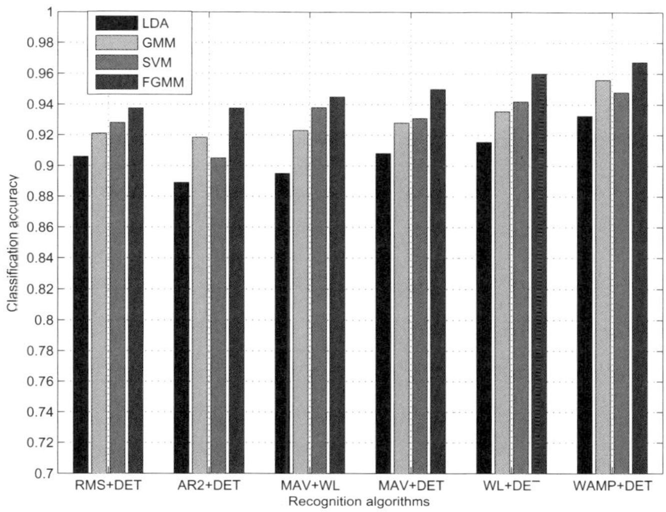

Fig. 11.12 Recognition results with various multi-features using different classifiers

feature of WAMP+DET has the best performance for all four classifiers. Since we are more interested in features with better performance, only the multi-features with better performance, such as RMS+DET, AR2+DET, MAV+WL, MAV+DET, WL+DET and WAMP+DET, are reported here. The recognition results by different classifiers are shown in Fig. 11.12. Among all the features and classifiers, FGMMs together with the feature WAMP+DET have the highest recognition rate, which is 96.7%.

To show their efficiency, Fig. 11.13 presents the box plot of different classifiers for all the selected multi-features and all different subjects. LDA has the lowest accuracies for all the multi-features and the average is 90.77%. GMMs and SVM generally have similar recognition performance and their averages are 93.16% and 93.24%, respectively. FGMMs have the highest accuracies, whose average reaches as high as 95.09%.

As we are more interested in the best performance of the best classifier and the best feature, the confusion matrix of the highest accuracy with WAMP+DET and FGMMs is shown in Fig. 11.14. The recognition rate of the motion 4 has been improved to 88% though there are still 5% of them which have been mistaken as motion 2. All the other motions have satisfactory results with an average of 97.67%, where the lowest is 93% for motion 2 and motion 6. The experiment results have demonstrated that the nonlinear classifier, FGMM, has better performance than others including GMMs, SVM and LDA, especially with the multi-feature AR2+DET.

The time costs by the single feature extraction and the recognition of different classifiers have also been investigated, as shown in Fig. 11.15, as this must be finished in the time interval of 300ms. All of these time costs are estimated in MATLAB running in a 2.3GHz intel Core i7 computer. The single features are from the multiple features in Fig. 11.12, where the combined features have relatively good performance. From the results, we can see the DET feature takes the most time cost, which is 11.72ms. AR2 takes around 2ms and the others, including RMS, MAV, WL and WAMP, all take less than 1ms. Compared with the time costs by different classifiers, FGMM takes 2.42ms to recognise a single motion, which is slightly more than others. According to the above results, the classifier, FGMMs, and the feature, DET+AR2, will take the most computational time to recognise

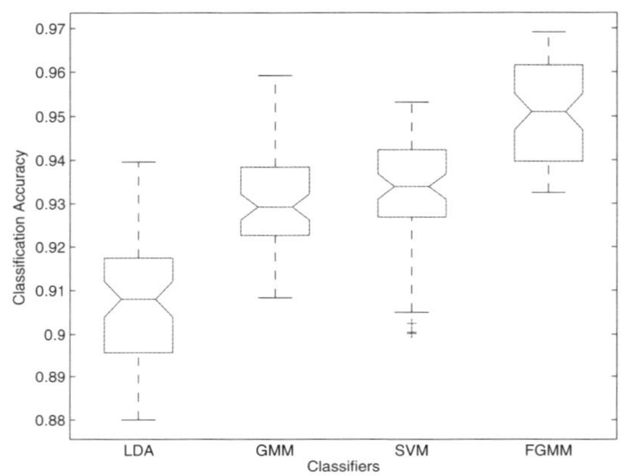

Fig. 11.13 Box plot results for the different classifiers in terms of different features and subjects

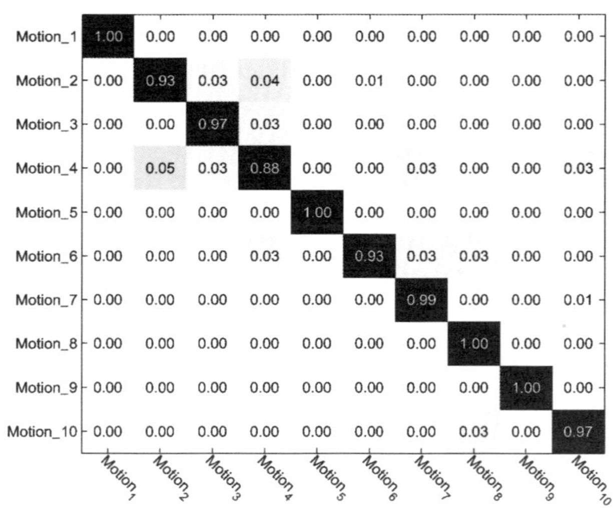

Fig. 11.14 Confusion matrix for the ten hand motions with the multi-feature WAMP+DET using FGMMs

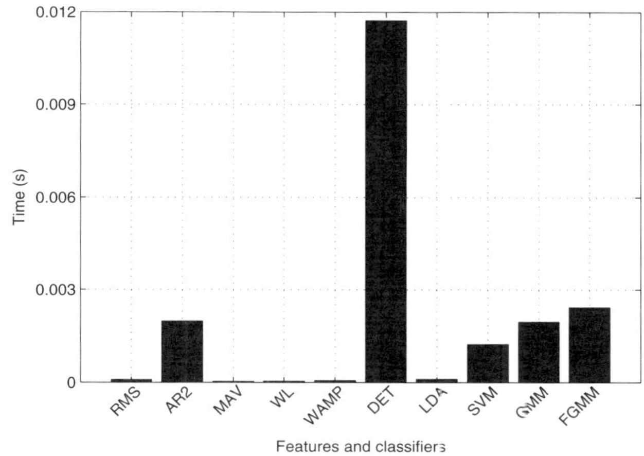

Fig. 11.15 The time cost by the single feature extraction and the recognition of different classifiers

a single motion, and the time cost by the feature extraction and recognition will be around 16.14ms (11.72+2+2.42). If a window length of 200ms is considered, FGMM with DET+AR2 feature will take around 220ms, which will satisfy the time requirement (the time delay is better to be less than 300ms) of the real prosthetic hand system.

11.5 Conclusions

In this chapter, nonlinear approaches to extract sEMG signal features and to identify different types of manipulation motions including different hand grasps and in-hand manipulations have been investigated and evaluated. Various examples have been conducted to demonstrate their performance.

Nonlinear measures, DET and ENTR, were proposed and extracted by the recurrence plot and quantification analysis to represent dynamical characteristics of sEMG during movements. Their performance was compared with a variety of time and frequency features, which demonstrated that the multi-features, WAMP+DET, outperformed all the 14 single features and 18 other multi-features, though the nonlinear features achieved low recognition rates when used individually. The multi-features combining both the nonlinear feature and other features usually have improved performance, since the nonlinear features provide important supplemental information, which can effectively represent the dynamical characteristics of sEMG for hand movements.

In addition, FGMMs, which are capable of modelling nonlinear datasets, were used as a nonlinear classifier. The proposed recognition algorithm also improved the efficiency of computational cost. Three other commonly used classifiers: LDA, GMMs and SVM have also been employed and compared with FGMMs, and the results showed that FGMMs outperform others in terms of the recognition rate. Especially, the highest recognition rate, 96.7%, was attained by using the multi-feature WAMP+DET and the nonlinear classifier FGMMs. It is evident that the proposed nonlinear features and classifier are essential to improve the recognition rate. The computational cost of the features and classifiers show that the proposed nonlinear methods potentially satisfy the time requirement of the real-time system.

In the future, more complex daily hand motions will be tested and automatic detection of different hand states in such complex motions may also be of great help to improve the system performance. Moreover, it will also be interesting to test the proposed algorithms to real-time control the prosthetic robotic hands worn by the amputees [Ju and Liu (2013a); Liu (2011)].

Acknowledgement

The authors would like to acknowledge financial support from UK National Engineering and Physical Scientific Research Council under Grant No. EP/G041377/1, from the Royal Society Grant under grant No. IJP08/R2, from the Leverhulme Trust under grant No. 13754 and from British Council for an Overseas Research Scholarship Award.

References

Akay, M. (2000). *Nonlinear Biomedical Signal Processing: Dynamic Analysis and Modeling* (Wiley-IEEE Press).

Al-Timemy, A., Bugmann, G., Escudero, J. and Outram, N. (2013). Classification of finger movements for the dexterous hand prosthesis control with surface electromyography, *IEEE Journal of Biomedical and Health Informatics* **17**, 3, pp. 608–618.

Artemiadis, P. K. and Kyriakopoulos, K. J. (2010). An EMG-based robot control scheme robust to time-varying EMG signal features, *IEEE Transactions on Information Technology in Biomedicine* **14**, 3, pp. 582–588.

Ba, S. and Odobez, J. (2009). Recognizing visual focus of attention from head pose in natural meetings, *IEEE Transactions on Systems, Man, and Cybernetics, Part B: Cybernetics* **39**, 1, pp. 16–33.

Bilmes, J. A. (1998). A gentle tutorial on the EM algorithm and its application to parameter estimation for gaussian mixture and hidden markov models, *International Computer Science Institute* ICSI-TR-97-0, pp. 1–13.

Cao, L. (1997). Practical method for determining the minimum embedding dimension of a scalar time series, *Physica D: Nonlinear Phenomena* **110**, 1, pp. 43–50.

Castellini, C. and van der Smagt, P. (2009). Surface EMG in advanced hand prosthetics, *Biological Cybernetics* **100**, 1, pp. 35–47.

Chan, A. D. and Englehart, K. B. (2005). Continuous myoelectric control for powered prostheses using hidden markov models, *IEEE Transactions on Biomedical Engineering* **52**, 1, pp. 121–124.

Chandrasekaran, S. (2007). A nonlinear dynamic modelling for speech recognition using recurrence plot-a dynamic bayesian approach, in *Proceedings of IEEE International Conference on Signal Processing and Communications* (Dubai, United Arab Emirates), pp. 516–519.

Chang, C. and Lin, C. (2012). LIBSVM: A library for support vector machines, *ACM Transactions on Intelligent Systems and Technology* **2**, 3, pp. 1–39.

Chen, X., Zhu, X. and Zhang, D. (2010). A discriminant bispectrum feature for surface electromyogram signal classification, *Medical Engineering and Physics* **32**, 2, pp. 126–135.

Chu, J. U., Moon, I. and Mun, M. S. (2006). A real-time EMG pattern recognition system based on linear-nonlinear feature projection for a multifunction myoelectric hand, *IEEE Transactions on Biomedical Engineering* **53**, 11, pp. 2232–2239.

Cipriani, C., Antfolk, C., Balkenius, C., Rosén, B., Lundborg, G., Carrozza, M. C. and Sebelius, F. (2009). A novel concept for a prosthetic hand with a bidirectional interface: A feasibility study, *IEEE Transactions on Biomedical Engineering* **56**, 11, pp. 2739–2743.

Duchene, J. and Goubel, F. (1993). Surface electromyogram during voluntary contraction: Processing tools and relation to physiological events, *Critical Reviews in Biomedical Engineering* **21**, 4, pp. 313–397.

Eckmann, J., Kamphorst, S. and Ruelle, D. (1987). Recurrence plots of dynamical systems, *Europhysics Letters* **4**, pp. 973–977.

Englehart, K., Hudgins, B., Parker, P. A. and Stevenson, M. (1999). Classification of the myoelectric signal using time-frequency based representations, *Medical Engineering and Physics* **21**, 6–7, pp. 431–438.

Farfan, F., Politti, J. and Felice, C. (2010). Evaluation of EMG processing techniques using information theory, *Biomedical Engineering Online* **9**, pp. 1–18.

Farina, D., Merletti, R. and Enoka, R. M. (2004). The extraction of neural strategies from the surface EMG, *Journal of Applied Physiology* **96**, 4, pp. 1486–1495.

Filligoi, G. and Felici, F. (1999). Detection of hidden rhythms in surface EMG signals with a nonlinear time-series tool, *Medical Engineering and Physics* **21**, 6, pp. 439–448.

Fraser, A. and Swinney, H. (1986). Independent coordinates for strange attractors from mutual information, *Physical Review A* **33**, 2, pp. 1134–1140.

Graupe, D., Magnussen, J. and Beex, A. (1978). A microprocessor system for multifunctional control of upper-limb prostheses via myoelectric signal identification, *IEEE Transactions on Automatic Control* **23**, 4, pp. 538–544

Huang, X., Ariki, Y. and Jack, M. (1990). *Hidden Markov Models for Speech Recognition* (Edinburgh University Press).

Huang, Y., Englehart, K. B., Hudgins, B. and Chan, A. D. (2005). A Gaussian mixture model based classification scheme for myoelectric control of powered upper limb prostheses, *IEEE Transactions on Biomedical Engineering* **52**, 11, pp. 1801–1811.

Hudgins, B., Parker, P. and Scott, R. N. (1993). A new strategy for multifunction myoelectric control, *IEEE Transactions on Biomedical Engineering* **40**, 1, pp. 82–94.

Ju, Z. and Liu, H. (2011a). A unified fuzzy framework for human hand motion recognition, *IEEE Transactions on Fuzzy Systems* **19**, 5, pp. 901–913.

Ju, Z. and Liu, H. (2012). Fuzzy gaussian mixture models, *Pattern Recognition* **45**, 3, pp. 1146–1158.

Ju, Z. and Liu, H. (2013a). Human hand motion analysis with multisensory information, *IEEE/ASME Transactions on Mechatronics* **19**, 2, pp. 456–466.

Ju, Z., Liu, H., Zhu, X. and Xiong, Y. (2009). Dynamic grasp recognition using time clustering, gaussian mixture models and hidden markov models, *Journal of Advanced Robotics* **23**, 10, pp. 1359–1371.

Ju, Z., Ouyang, G., Wilamowska-Korsak, M. and Liu, H. (2013b). Surface EMG based hand manipulation identification via nonlinear feature extraction and classification, *IEEE Sensor Journal* **13**, 9, pp. 3302–3311.

Ju, Z., Zhu, X. and Liu, H. (2011b). Empirical copula-based templates to recognize surface EMG signals of hand motions, *International Journal of Humanoid Robotics* **8**, 4, pp. 725–741.

Karlsson, J. S., Roeleveld, K., Grönlund, C., Holtermann, A. and Östlund, N. (2009). Signal processing of the surface electromyogram to gain insight into neuromuscular physiology, *Philosophical Transactions of the Royal Society A: Mathematical, Physical and Engineering Sciences* **367**, 1887, pp. 337–356.

Khezri, M. and Jahed, M. (2007). Real-time intelligent pattern recognition algorithm for surface EMG signals, *Biomedical Engineering Online* **6**, pp. 1–12.

Kim, H., Eykholt, R. and Salas, J. (1999). Nonlinear dynamics, delay times, and embedding windows, *Physica D: Nonlinear Phenomena* **127**, 1–2, pp. 48–60.

Kwon, S. and Kim, J. (2011). Real-time upper limb motion estimation from surface electromyography and joint angular velocities using an artificial neural network for human-machine cooperation, *IEEE Transactions on Information Technology in Biomedicine* **15**, 4, pp. 522–530.

Li, G., Schultz, A. E. and Kuiken, T. A. (2010). Quantifying pattern recognitionbased myoelectric control of multifunctional transradial prostheses, *IEEE Transactions on Neural Systems and Rehabilitation Engineering* **18**, 2, pp. 185–192.

Liu, H. (2011). Exploring human hand capabilities into embedded multifingered object manipulation, *IEEE Transactions on Industrial Informatics* **7**, 3, pp. 389–398.

Liu, Y., Huang, H. and Weng, C. (2007). Recognition of electromyographic signals using cascaded kernel learning machine, *IEEE/ASME Transactions on Mechatronics* **12**, 3, pp. 253–264.

Lu, L., Ghoshal, A. and Renals, S. (2011). Regularized subspace gaussian mixture models for speech recognition, *IEEE Signal Processing Letters* **18**, 7, pp. 419–422.

Marwan, N., Carmen Romano, M., Thiel, M. and Kurths, J. (2007). Recurrence plots for the analysis of complex systems, *Physics Reports* **438**, 5, pp. 237–329.

Marwan, N., Wessel, N., Meyerfeldt, U., Schirdewan, A. and Kurths, J. (2002). Recurrence-plot-based measures of complexity and their application to heart-rate-variability data, *Physical Review E* **66**, 2, p. 026702.

Matassini, L., Kantz, H., Holyst, J. and Hegger, R. (2002). Optimizing of recurrence plots for noise reduction, *Physical Review E* **65**, 2, pp. 021102.1–21102.6.

Oskoei, M. and Hu, H. (2008). Support vector machine-based classification scheme for myoelectric control applied to upper limb, *IEEE Transactions on Biomedical Engineering* **55**, 8, pp. 1956–1965.

Ouyang, G., Dang, C. and Li, X. (2009). Multiscale entropy analysis of EEG recordings in Epileptic rats, *Biomedical Engineering: Applications, Basis and Communications* **21**, 3, pp. 169–176.

Packard, N., Crutchfield, J., Farmer, J. and Shaw, R. (1980). Geometry from a time series, *Physical Review Letters* **45**, 9, pp. 712–716.

Peleg, D., Braiman, E., Yom-Tov, E. and Inbar, G. (2002). Classification of finger activation for use in a robotic prosthesis arm, *IEEE Transactions on Neural Systems and Rehabilitation Engineering* **10**, 4, pp. 290–293.

Rissanen, S., Kankaanpaa, M., Meigal, A., Tarvainen, M., Nuutinen, J., Tarkka, I., Airaksinen, O. and Karjalainen, P. (2008). Surface EMG and acceleration signals in Parkinsons disease: Feature extraction and cluster analysis, *Medical and Biological Engineering and Computing* **46**, 9, pp. 849–858.

Stark, L. (1968). *Neurological Control Systems: Studies in Bioengineering* (Springer).

Steinwart, I. and Christmann, A. (2008). *Support Vector Machines* (Springer).

Subasi, A. (2012). Classification of EMG signals using combined features and soft computing techniques, *Applied Soft Computing* **12**, 8, pp. 2188–2198.

Takens, F. (1981). Detecting strange attractors in turbulence, *Dynamical Systems and Turbulence, Warwick 1980* pp. 366–381.

Tang, X., Liu, Y., Lv, C. and Sun, D. (2012). Hand motion classification using a multichannel surface electromyography sensor, *Sensors* **12**, 2, pp. 1130–1147.

Tenore, F. V., Ramos, A., Fahmy, A., Acharya, S., Etienne-Cummings, R. and Thakor N. V. (2009). Decoding of individuated finger movements using surface electromyography, *IEEE Transactions on Biomedical Engineering* **56**, 5, pp. 1427–1434.

Thiel, M., Romano, M., Kurths, J., Meucci, R., Allaria, E. and Arecchi, F. (2002). Influence of observational noise on the recurrence quantification analysis, *Physica D: Nonlinear Phenomena* **171**, 3, pp. 138–152.

Trulla, L., Giuliani, A., Zbilut, J. and Webber, C. (1996). Recurrence quantification analysis of the logistic equation with transients, *Physics Letters A* **223**, 4, pp. 255–260.

Xie, H. B., Guo, J. Y. and Zheng, Y. P. (2010). Fuzzy approximate entropy analysis of chaotic and natural complex systems: Detecting muscle fatigue using electromyography signals, *Annals of Biomedical Engineering* **38**, 4, pp. 1483–1496.

Yin, Y. H., Fan, Y. J. and Xu, L. D. (2012). EMG and EPP-integrated human-machine interface between the paralyzed and rehabilitation exoskeleton, *IEEE Transactions on Information Technology in Biomedicine* **16**, 4, pp. 542–549.

Chapter 12

Energy Demand Management Through Uncertain Data Forecasting: A Hybrid Approach

Marco Severini, Stefano Squartini and Francesco Piazza

Abstract Although Smart Grids may represent the solution to the limits of nowadays Power Grid, the turnover may not occur in the next future yet due the complex nature of energy distribution. Thus, as a more short term effort, to improve the responsiveness of the energy demand to the power grid load, more and more energy providers apply dynamic pricing schemes for grid users. Believing that dynamic pricing policies may be an effective asset even at a micro-grid level, an hybrid energy management scheme is proposed in this contribution. While the nonlinear nature of a micro grid, involving the task allocation and the thermal constraint satisfaction, can be modeled as a mixed integer nonlinear programming problem, neural-network forecasting abilities can provide a sustainable support under realistic operating conditions. Based on the forecast of solar energy production and grid energy prices and outdoor temperature, the optimization of tasks allocation is aimed to lower both the user costs and the grid burden while accounting the thermal comfort of the user. Through computer simulations, whose degree of realism is enhanced by the adoption of forecast data, the shift of the grid load towards low energy price hours is confirmed.

12.1 Introduction

Smart Grids are reckoned to be the next generation power grid technology, intended to overcome many limits of nowadays infrastructures [Cecati *et al.* (2010); Ipakchi and Albuyeh (2009)]. In fact, with the assistance of information technology and automatic routing of power, they provide the means to improve energy distribution. While this technology may provide a major step ahead in power distribution, production and consumption may also require attention. More precisely, the actual energy storage technology is not able to support the power distribution in a cost-wise fashion. Thus, to maintain the service parameters within standards, energy production must always match the energy consumption. This requirement, in turn, affects the energy market, which therefore requires a specially strict regulation.

To overcome this limit several solutions are nowadays object of research. The most active seem to focus on distributed storage through grid connected electrical vehicles [Liu *et al.* (2013); Saber and Venayagamoorthy (2012); Venayagamoorthy (2011)] and demand response [Chakrabarti *et al.* (2012); Chapman *et al.* (2013); Wang *et al.* (2013)].

The former, while appealing and promising, still does not account two major aspect of the energy storage. The first one is the efficiency of the storage system. In fact, even though in an highly efficient system conversion losses may be lower than 5%, on a large scale system they can easily amount to the order of mega watt hours per day. The second aspect is the life expectation of the energy storage itself. Since charge and discharge shorten a battery life, grid connected electrical vehicles may most likely require early battery replacement. Therefore, a grid connected vehicle solution may be unappealing to the owners of the vehicle, and may prevent them to include their vehicle in the storage pool.

The latter approach, also quite interesting, promotes the regulation of the electricity demand through prices. It relies however on the user, it also requires consumer profiling [Ghosh *et al.* (2012)] and data disaggregation [Makonin *et al.* (2013)]. While it may still too early to draw conclusion, the reliance on the end user may not lead to the expected performances.

In the authors' opinion, a more promising solution is represented by Micro Grids [Kirthiga *et al.* (2013)]. While micro grids represent the next evolutionary step of smart grids, they may also lead the demand response approach to the next level. By accounting production and consumption within the power grid, the entire energy flow can be managed as a whole, improving efficiency, reliance, and reducing the need of energy storage systems.

In this perspective, domestic tasks and energy management [Squartini *et al.* (2013b)] may have a major role, since it can respond automatically to a dynamic pricing scheme. While it can also account local energy storage and production, its strong point is the ability to manage the energy consumption in a transparent way with respect to the user.

On a large scale therefore, such approach may provide the means to regulate the energy demand over the grid. As such, not only the energy balance across the grid can be maintained more efficiently, but also a less strictly regulated energy market may be possible. Also, although electrical vehicles may still be an integral part of the domestic energy management, the storage included in domestic solar power plants may be used instead, with less costs, since it is already part of the plant, and more efficiency, since it is used as extrema ratio.

Obviously enough, the automation of domestic tasks, being also a process transparent to the user, is an highly constrained decisional problem. Therefore, among the available solving techniques [De Angelis *et al.* (2013a); Squartini *et al.* (2013a)], many, such as Particle Swarm Optimization (PSO) [Soares *et al.* (2013)], Fuzzy Logic [Cui and Dai (2011)], Adaptive Dynamic Programming [Fuselli *et al.* (2013, 2012); Boaro *et al.* (2013)] but also Linear Programming [Tham and Luo (2013)], have been deemed not particularly suited to deal with the task scheduling, whereas Mixed Integer Programming ones [Bozchalui and Sharma (2012); De Angelis *et al.* (2013b); Nagasaka *et al.* (2012)] have shown to be quite promising.

As in our previous work [Severini *et al.* (2013)], extending the concept of task automation, although not strictly a task, the thermal management of the domestic environment is accounted. On purpose, the thermal behaviour of the domestic environment is modelled by means of the European standards [ISO (2003, 2007a,b)]. Due to the highly nonlinear nature of the problem, a mixed computational scheme, combining a genetic algorithm (GA) and a deterministic minimization approach, is used to apply the thermal requirement of the user and compute the energy demand of the environment.

The resulting report is integrated within the energy balance and a MILP problem, modelling the energy management, is implemented to select the optimal task schedule. An hybrid computational approach, mixing deterministic and heuristic computing technique, is therefore employed to minimize the energy bill through optimal scheduling.

In the present work data uncertainty is also included [Ciabattoni *et al.* (2013a,b)]. In order to select the optimal execution time table for the assigned tasks, the knowledge of the future values of the scheduling parameters is a key factor. However, data uncertainty may lead to sub-optimal results. To evaluate this aspect a few predictors, based on Artificial Neural Networks [Hernandez *et al.* (2013)], are implemented to provide a day ahead forecast of solar irradiation, outdoor temperature and electricity price.

The model at the core of the task scheduling and energy cost minimization is presented in Section 12.2, whereas in Section 12.3, the thermal model at the base of the heating/cooling system is proposed. The forecasters are described in Section 12.4. In Section 12.5, a brief overview on the solving algorithm part of the framework is reported. In Section 12.6, the simulated scenario is presented and the simulation results are discussed. Section 12.7 draws the work conclusions.

12.2 Home Energy Management Problem: The Model

The purpose of automated task scheduling is the selection of an optimal time table, so that the tasks are executed within the time frame provided by the user, while lowering the energy bill to its minimum. Since the energy price rise when the energy demand increase, the optimal time table will most likely contain the energy purchase during peak hours, thus shaping the energy demand of the building. In this perspective, a dynamic pricing scheme may provide an almost real-time feedback to the scheduler. On a large scale, not only the energy cost is contained, but even the grid burden can be equalized over the daily time frame.

Clearly, if the thermal regulation is accounted, since the thermal comfort is not supposed to depend on the energy price, the energy demand of the heat pump cannot be regulated on a price basis. However, if local energy storage and production are available, some sort of demand shaping may still be possible. In conclusion, to adjust the energy demand of the building, the scheduling problem can be modelled by evaluating the monetary balance of energy consumption and production. In the following:

$$Q = \sum_{t=1}^{slots} \left\{ \left[\sum_{j=1}^{houses} \sum_{i=1}^{tasks} \left(E_{j,i} tb_{j,i,t}\right) + \left(E_t^c - E_t^d\right) - E_t^{re} + E_t^{so} + E_t^{he} \right] C_t + -E_t^{so} S_t^{price} \right\} \tag{12.1}$$

$tb_{j,i,t}$ is the task binary variable that defines the activity state (ON/OFF) of the i-th task of j-th building during the t-th time slot. The variable $E_{j,i}$ represents the energy demand of the i-th task of j-th building in each time slot. Therefore the sum over i and j indices returns the total energy demand of the assigned tasks at the t-th time slot. The amount $\left(E_t^c - E_t^d\right)$ represents the energy transferred to or from the storage at the t-th time slot, while E_t^{re} is the renewable energy production, E_t^{so} accounts the sold energy amount and E_t^{he} describes the heater energy demand. Therefore, the quantity within brackets accounts the net energy demand at each time slot. On the other hand, since C_t represents the energy purchase price at the t-th time slot, the amount within braces describes the total energy cost minus the total energy income at each time slot, that is the energy balance at the t-th time slot.

Said monetary balance shall be complemented by the constraints binding together the unknowns of the equation, that have been already presented in our previous work [Severini *et al.* (2013)].

12.3 Modelling the Thermal Optimization Sub-Problem

Since the thermal regulation is not affected by the energy price, it may be more convenient to address this aspect as a separate problem. The core of the thermal regulation system is an heat pump. Since this matter has been already discussed in a previous work [Severini *et al.* (2013)], it will only be presented briefly.

12.3.1 *Notations*

Indices:

t time slot index

j	house index
k	room index

Parameters

$surf$	number of surfaces of the house thermal model
E_t^{he}	electrical energy demand at the t-th time slot
$P_{j,k,t}^{th}$	thermal power at the t-th time slot to the k-th room of the j-th building
P_t^{th}	thermal power at the t-th time slot (single room case)
Δ	time slot duration (in hours)
θ_{hp}	heat pump output temperature (°C)
θ_i	indoor temperature (°C)
θ_o	outdoor temperature (°C)
$\theta_t(t)$	target temperature at the t-th time slot (°C)
ε	temperature tolerance (0.5) (°C)
c_p	air heat capacity at NTP
M_{air}	air mass in the house (Kg)
k_l	house heat loss factor (W/°C)

12.3.2 *Heat pump constraints*

The electrical energy, required at each time slot, is obtained as the thermal energy, required by each room in the same time frame, on the Coefficient of Performance (COP) of the heat pump when heating, whereas the Energy Efficiency Ratio (EER) is used if cooling:

$$E_t^{he} = \frac{1}{COP} \sum_{j=1}^{houses} \sum_{k=1}^{rooms} (\Delta P_{j,k,t}^{th}) \quad \forall t : t = 1, \ldots, slots. \tag{12.2}$$

In turn, the thermal power $P_{j,k,t}^{th}$ of the k-th room of the j-th building at the t-th time slot is the amount that satisfies the temperature constraints. These constraints are defined with the heat balance of the buildings at its base [Fux *et al.* (2014); Kazanavičius *et al.* (2006); Qela and Mouftah (2010)]. If the overdot notation is used to represent the time derivative, the heat fluxes can be defined as

$$\dot{Q}_{hp} = P^{th} c_p (\theta_{hp} - \theta_i) \tag{12.3}$$

$$\dot{Q}_{loss} = k_l (\theta_i - \theta_o) \tag{12.4}$$

where (12.3) expresses the thermal energy provided by the heat pump, while (12.4) represents the thermal energy escaping through the walls. On the other hand, the indoor temperature variation, over time, can be calculated as the net heat flux on the thermal capacity of the air mass inside the room, which leads to the following:

$$\dot{\theta}_i = \frac{1}{M_{air}c_p}(\dot{Q}_{hp} - \dot{Q}_{loss}). \tag{12.5}$$

To account the discrete time domain required by the scheduler, (12.5) can be rewritten as a difference equation:

$$\theta_i(t+1) = \theta_i(t) + \frac{\Delta}{M_{air}c_p}\left[P_t^{th}c_p(\theta_{hp} - \theta_i(t)) - k_l(\theta_i(t) - \theta_o(t))\right] \tag{12.6}$$

being t the discrete time variable. By algebraic manipulation the temperature is then be expressed as

$$\theta_i(t+1) = \theta_i(t)\left[1 - \frac{P_t^{th}\Delta}{M_{air}} - \frac{k_l\Delta}{M_{air}c_p}\right] + \frac{P_t^{th}\theta_{hp}}{M_{air}} + \frac{k_l\theta_o(t)}{M_{air}c_p} \tag{12.7}$$

leading to the representation of the indoor temperature as a function of thermal fluxes, indoor temperature, and outdoor temperature in the previous time slot. As a result, the constraints set is obtained by requiring that

$$\begin{gathered}\forall t,:\ t = 1, \ldots, slots \\ \theta_t(t) - \varepsilon \leq \theta_i(t) \leq \theta_t(t) + \varepsilon.\end{gathered} \tag{12.8}$$

The coefficients M_{air} and k_l, that appear in the equations, represent, respectively, the air mass within each room and the heat loss factor of each room. Their value is obtained by means of a simple building geometry accounting several rooms. Regarding k_l, it is worth to mention that its value is estimated in accordance to the EN 12831, EN ISO 13370, EN ISO 13789 standards. This matter will be discussed in Section 12.6.

12.4 Neural Networks for Uncertain Data Forecasting

In order to compute the energy balance, the information concerning energy price, solar production and outdoor temperature is required. Therefore, to select the optimal time table to be applied in the future, information on the future values of energy price, solar production and outdoor temperature shall be used.

Although forecast data can be retrieved from many providers, such as weather forecast services and independent system operators, the prediction error and its effect on the scheduling process are to be taken into account. On purpose, to better assess the prediction error, for each of the required information set a forecaster has been developed. Since absolute prediction accuracy is now paramount in the current work, the implementation has been designed with forecaster simplicity in mind.

In particular, by means of the Neural Network toolbox, provided by the MathWorks MatLab[1] 2012a framework, three multi-layered perceptrons (MLP) have been used as forecaster. The neural networks training is carried out over a data set covering two years, namely, 2009 and 2010. The National Climatic Data Center[2] (NCDC) has been selected for the meteorological records, whereas the ISO of New England[3] (ISO-NE) provided information about the energy price.

The three forecaster share the same MLP structure: 49 neurons are used in the input layer, 48 in the hidden layer and 24 in the output layer. For every neuron the hyperbolic tangent is used as activation function. A day ahead prediction approach is used, thus 24 samples, one per hour, are generated at a time from the samples of the previous day. The data normalization is left to the toolbox and it maps the input to the interval $[-1, 1]$. The performance function of choice is the Mean Absolute Error (MAE). To reduce the training time, given the number of neurons, the gradient descent with momentum and adaptive learning rate is used. To avoid a premature completion of the training process, the number of validation check is fixed equal to 500, and the number of epoch has been increased accordingly.

Concerning temperature and irradiation forecasting, for each predicted set, the input data is represented by the 24 temperature hourly samples, the 24 solar irradiation hourly samples, and the day index within the year, which is coded through the minus cosine function. Regarding the price forecast, on the other hand, the energy price hourly samples are used as input in place of the solar irradiation hourly samples. The temperature samples are still used as input in this case. Clearly, since the output data is composed by 24 samples representing the hourly data of the next day, recurring prediction and error propagation have been avoided.

Given the simple structure of the network and its input set, the prediction accuracy is usually high if little to no changes are recorded from a day to the next. On the other hand, when highly variable conditions are encountered, the prediction accuracy drops. In this regard, even though a deep analysis of the results has not been carried out, a simple evaluation reveals that a proper data preprocessing would greatly improve the performance of the forecaster. Additionally, the nature of the forecast error appears to be systemic, and probably due to the lack of input data. As a result, the performance is not on par with the performance of the state of the art. Nonetheless the obtained forecast provides an ample set of cases, that are more than adequate to test the error propagation through the scheduler. In a real life scenario, an accurate predictor would be mandatory. As such, the improvement of the forecaster issues will be addressed in future works.

---

[1] http://www.mathworks.com

[2] http://www1.ncdc.noaa.gov
http://www1.ncdc.noaa.gov/pub/data/nsrdb-solar/station-data-2010/

[3] http://www.iso-ne.com/aboutiso/index.html
http://www.iso-ne.com/markets/hstdata/hourly/index.html

12.5 Optimization Algorithms

By means of the suggested objective function and constraints, the scenario of interest is modeled. Based on historical data, the forecast of outdoor temperature, solar irradiation and energy prices are generated by means of three ANN forecaster.

Whether forecast or historical data are used, energy prices are directly assigned to the model. Solar irradiation values are used to compute the energy production of the solar power plant, whereas the outdoor temperature is used to compute the energy demand of the heat pump through the thermal model.

For instance, the thermal load model is instantiated and, by means of the outdoor temperature, the energy demand of the heat pump is computed. This information is then integrated with the solar energy production and the energy rates to provide the input to the energy balance model. Then the main sub-problem, implementing the energy balance, is instantiated.

After the model instantiation and set up, the MILP problem is generated and the MILP solver is invoked. At the end of the solving routine the task time table and the expected energy monetary balance are obtained. The resulting algorithmic framework is presented in Fig. 12.1.

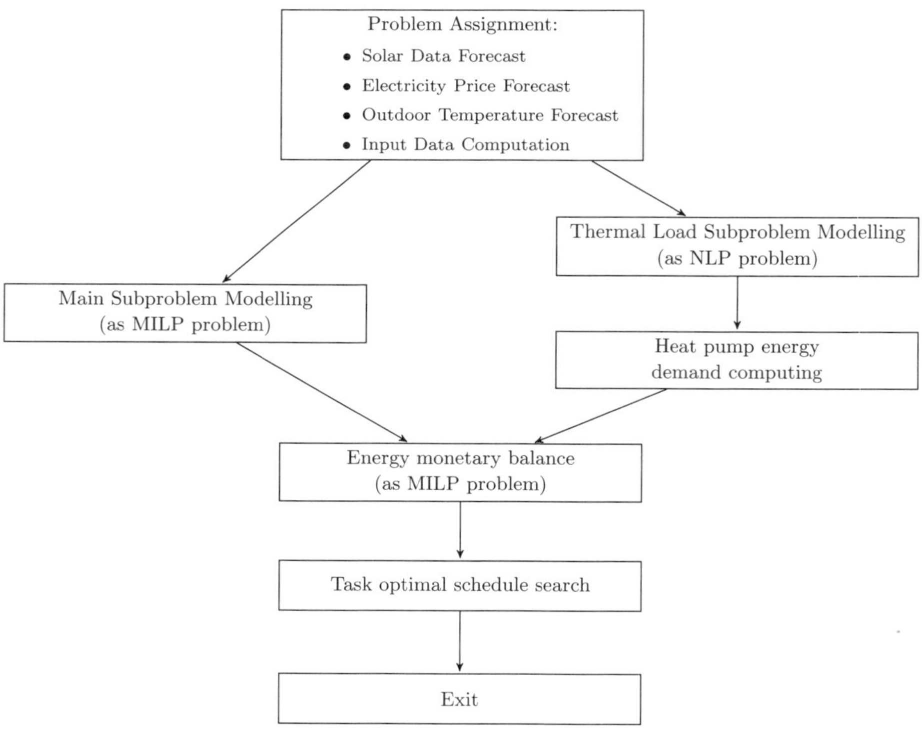

Fig. 12.1 Algorithmic framework flow chart

The first step is represented by the data input assignment, in which the scenario configuration is set up. Then, the sub-problem focused on the heat pump energy demand is formulated.

The next step requires to integrate the main scheduling sub-problem, involving the task scheduling and the energy management, with the heat pump energy demand, thus a MILP problem, modelling the energy monetary balance, is obtained. The main programming problem is then solved to obtain the optimal task and the energy resource schedule.

After the time table is retrieved, if said time table is based on forecast data, the same framework is used as a simulator to compute the actual energy balance resulting from the task execution and energy management. In contrast, in the search phase, the returned energy balance is the expected, or predicted one. In other words, the two balances coincide if the prediction error is zero, but they differ when a prediction error occurs.

To run the framework as a simulator, the forecast data is replaced by the corresponding historical data, and the time table obtained in the previous phase is assigned as an input. The framework, running as a simulator, computes the activity of the energy storage, the energy production and consumption, and therefore computes the electricity balance based on the historical data and the provided task schedule.

12.6 Smart Home Energy Management: Case Studies and Simulation Results

The scheduling process is carried out by means of the MatLab environment. The MILP solver, namely, GLPK, is invoked through the Opti Toolbox[4] 1.81 interface for MatLab. The Hybrid Genetic Algorithm and the Artificial Neural Networks, on the other hand, are part of their respective toolboxes provided by Mathworks. The framework is hosted by a Laptop PC based on the Intel Core i7 CPU series, with 8GB of ram, running on the Microsoft Windows 8 64-bit OS.

12.6.1 *Thermal model characterization*

To properly evaluate the performance of the thermal regulation, a simple building structure is employed as a target. While the structure is simplified enough to reduce the complexity of the model, the characterization is based on the European Standards EN 12831:2003, EN 13370:2007 and EN 13789:2007, so that all the most important thermal loss contributions are included.

A schematic representation of the building is reported in Fig. 12.2.

The building parameters are shown in Table 12.1.

[4]http://www.i2c2.aut.ac.nz/Wiki/OPTI/index.php/Main/HomePage

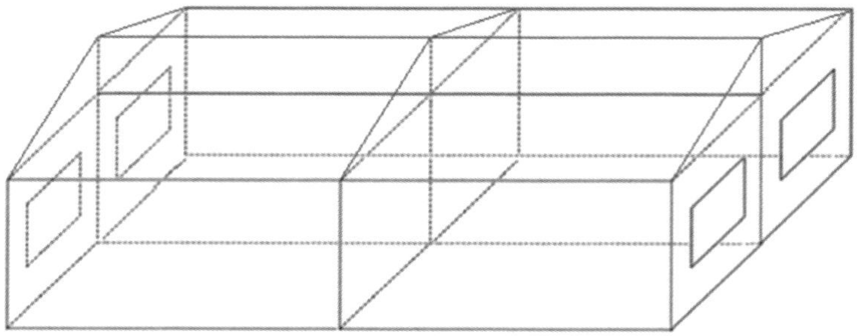

Fig. 12.2 Building model schematic representation

Table 12.1 Building geometry and parameters

Parameters	Value	Unit
Building's length	25	m
Building's width	10	m
Rooms' length	12.2	m
Rooms' width	4.7	m
Building's height	4	m
Roof's pitch	35	°
Windows' count	4	-
Windows' height	1	m
Windows' width	1	m
Windows' thickness	0.005	m
Walls' thickness	0.2	m
Slab on grade thickness	0.3	m
Windows' thermal conductivity	0.78	$W/(m^2 \cdot °C)$
Walls' thermal conductivity	0.30	$W/(m^2 \cdot °C)$
Slab on grade thermal conductivity	0.82	$W/(m^2 \cdot °C)$

This set of parameters and the air density at NTP (1.204 kg/m^3) are used to compute the air mass within each room, referred as M_{air} in the model description in Section 12.3. The parameter referred as k_l that is, the heat loss factor, is computed accordingly to the procedure stated in the aforementioned European Standards.

Concerning the heat pump, a 6 kW device is used. The COP factor is equal to 3.4 whereas the EER factor is chosen equal to 3. In the current scenario the outdoor temperature spans from about 15°C to about 35°C, thus no additional adjustment to either the COP or the EER is deemed necessary. The output temperature of the heat pump is assumed equal to 50°C when heating, and equal to 10°C when cooling. The heat pump is assumed able to switch from heating to cooling and vice versa depending on the outdoor temperature.

In order to compute the energy demand of the thermal regulation process, the MatLab GA solver setup assumes a population size limited to 20 candidates, a generation stall limited to 10, whereas the tolerance is set to 1E−10, the mutation function is set to *@mutationdadaptfeasible* and the hybrid function is set to

@*fmincon*. The genetic algorithm is therefore used to locate the area around the global minimum, whereas the *fmincon* operator locates the actual minimum within that area, improving the convergence time of the algorithm.

12.6.2 *Task scheduling and energy cost accounting*

The scheduling interval spans over a time frame of 24 hours, and requires the management of a set of appliances composed by a washing machine, a drying machine, a oven and a dish washer. The TV is also considered in the scenario. The tasks are described in Table 12.2.

Table 12.2 Assigned tasks and their respective timings

	Allowed time window		Forbidden time window		Tasks Parts	Energy Demand	ID	Master ID	Duration (Hrs)
	Begin	End	Begin	End					
Washing machine	8:00	22:00	14:00	16:00	3	800	task 1	-	1
						1000	task 2	task 1	1
						900	task 3	task 2	1
Drying Machine	8:00	22:00	14:00	16:00	1	2500	task 4	task 3	2
Oven	10:00	13:00			1	2000	task 5	-	2
Dishwasher	14:00	22:00			3	800	task 6	-	1
						1000	task 7	task 6	1
						500	task 8	task 7	1
TV	8:00	18:00			1	100	task 9	-	10

Regarding the storage system, a single set of batteries is considered. The parameters are reported in Table 12.3.

Table 12.3 Storage system parameters

η_c	η_d	Chl^{MIN}	Chl^{MAX}	P^c_{MAX}	P^d_{MAX}
		kWh		kW	
0.85	0.85	1	5	2	2

Concerning the energy rates, a dynamic pricing scheme is emulated by mean of either historical or forecast market prices. On this regards, since the values refers to the wholesale locational marginal price, whereas the retail price includes taxes and ancillary costs, to obtain a realistic retail price, a multiplicative factor of ten is also accounted. As a result, in the current scenario, the retail energy price is assumed to be ten times the wholesale price. In the same scenario, the energy selling price has been assumed equal to 1 cent per kWh.

12.6.3 *Solar production*

The current simulation scenario assumes a solar power plant as part of the domestic environment, its panel area being 20 squared meters, and its efficiency index being equal to 20%. As previously stated, the energy production is computed using the hourly solar irradiance as an input.

12.6.4 *Data forecasting*

To evaluate the scheduling framework performance, a forecast data set is used as an input to the scheduler. Although the forecast time frame spans over almost two years, which allowed several tests to be carried out, only the forecasts corresponding to a few days are presented in the current manuscript as a sample.

Also, to evaluate the forecast accuracy in a normalized fashion, for each day, the MAE of the forecast, against the corresponding historical data, is divided by the absolute mean value of the historical data. This approach has been devised to avoid the issues, due to negative and null values, that usually affect the Mean Absolute Percentage Error (MAPE).

This characterization was used to select three forecast set of samples. The set are referred to as "day 1" (02/05/2010), "day 2" (10/08/2010) and "day 3" (24/09/2010). The normalized MAE for these forecasts lies, respectively in the intervals [0, 0.1], (0.1, 0.2] and (0.2, 0.3].

In Figs. 12.3, 12.4 and 12.5, the input data profiles corresponding, respectively, to solar energy production, outdoor temperature and energy prices, are presented comparing the forecast data against the historical counterpart for each of the selected days.

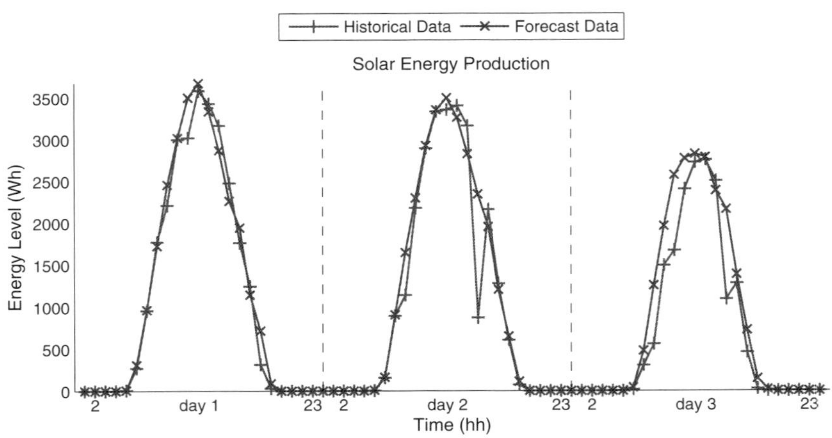

Fig. 12.3 Solar Energy Production: Forecast against historical data

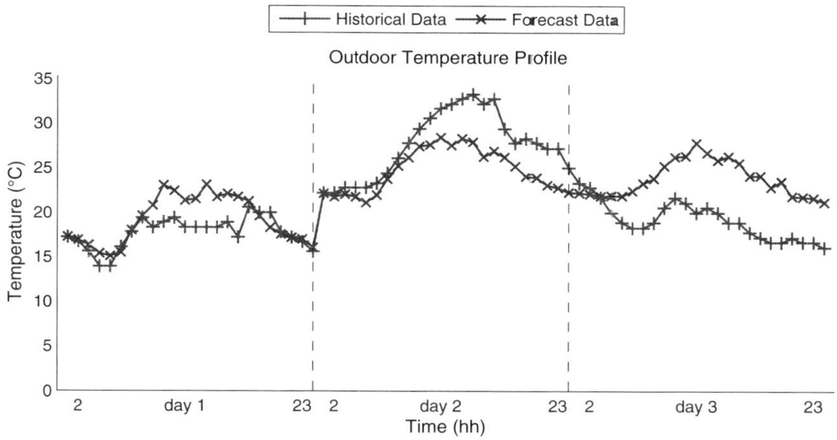

Fig. 12.4 Temperature Profile: Forecast against historical data

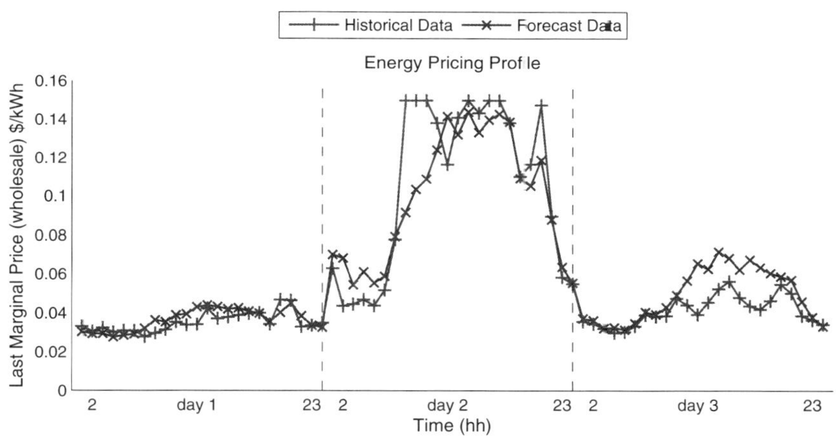

Fig. 12.5 Energy Pricing Profile: Forecast against historical data

12.6.5 *Energy management performance*

The suggested set-up is used to evaluate the framework performance. By means of the data forecast, a task time table is computed. Under these circumstances, the expected energy demand of the heat pump is the amount computed using the outdoor temperature forecast, whereas the expected energy production is computed with the solar irradiance forecast as a basis. After the time table is retrieved, the framework is used as a simulator. The returned energy monetary balance, which results from the execution of the task, accounts the actual energy demand of the heat pump, the actual energy production and the actual prices, thus it provides the actual energy cost, rather than the expected one. Since the forecast always include a prediction error, the result is almost certainly sub optimal, thus said result will be addressed as "Sub Optimal" in the reported plots and figures.

The performance reference is obtained from an ideal case that assumes an exact prediction as possible. Whenever the prediction error can be considered equal to zero, the data forecast will match the historical counterpart. In this scenario, then, it is possible to search for the task time table using the historical data rather than the forecast counterpart. As a result, the actual energy monetary balance obtained through simulation will match the expected cost, computed by the framework during the search of the task time table. As such, since the algorithm will provide the time table that minimize the energy cost, the optimal result is obtained. For this reason, the obtained result will be referred to as "Optimal" in the reported plot and figures.

The comparison of optimal and sub optimal results will provide the means to evaluate the relationship, between prediction error and scheduling performance, in term of energy net cost.

An additional reference case, addressed as "baseline scenario", consists of a domestic environment, without any energy management feature, nor energy storage facility and without task scheduling abilities. In this environment, the energy produced by the solar power plant is sold directly to the main grid, whereas the tasks are executed in the first available time slot. A time slot is assumed to be available if it falls within the time frame given to the task and if the task can be executed without exceeding the maximum power allowed by the building wirings. For this scenario no plot will be presented.

Concerning the realism of the scenario two aspects shall be accounted. The first one is related to the assumption that, within the environment, the thermal regulation depends on a thermostat. The thermal model, provided by the framework, is only meant to compute the energy demand. If historical data is used an input, the actual energy demand will be obtained. If forecast data is used instead, the expected energy demand is obtained.

The second aspect pertains the realism of the thermal regulation. In order to evaluate the framework performance, a strict thermal regulation is used to enforce the heat pump activity. In real life however, thermal regulation is usually not used all year around, thus a certain degree of realism is discarded in the current work.

The first test to be carried out is based on the data set referred to as "day 1". The error prediction of the data forecast, within this set, is less then 10%. As said, first the energy demand of the heat pump is computed. This routine depends on the outdoor temperature, the thermal constraints and the building thermal behaviour. For the optimal case, the temperature profile is reported in Fig. 12.6, whereas for the sub-optimal case the resulting indoor profile is reported in Fig. 12.7. By comparing the two, the effect of the prediction error is clear. The difference becomes even more apparent if the energy demand, reported in Fig. 12.8, is evaluated.

The comparison of the profiles highlights that the effects of the prediction error highly depend on the temperature constraints. Since no thermal regulation is required without thermal constraints, the prediction error is unable to propagate during the time slots where temperature requirements are not assigned. Also, the indoor temperature is bound to follow the outdoor temperature, and the thermal regulation is only required to maintain the indoor temperature within the given range. Therefore, if both the outdoor temperature and the corresponding forecast profile fall into the assigned range, no regulation will be necessary, thus the prediction error will not propagate. In the current scenario, a 1°C temperature range was chosen, thus the error propagation occurs, if a wider range where to be chosen, however, a different situation may present itself.

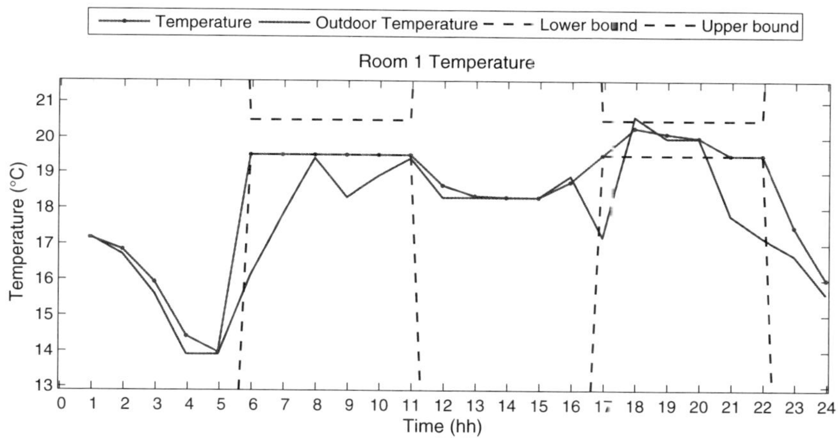

Fig. 12.6 Data set "day 1": Room temperature profile against outdoor temperature and constraints

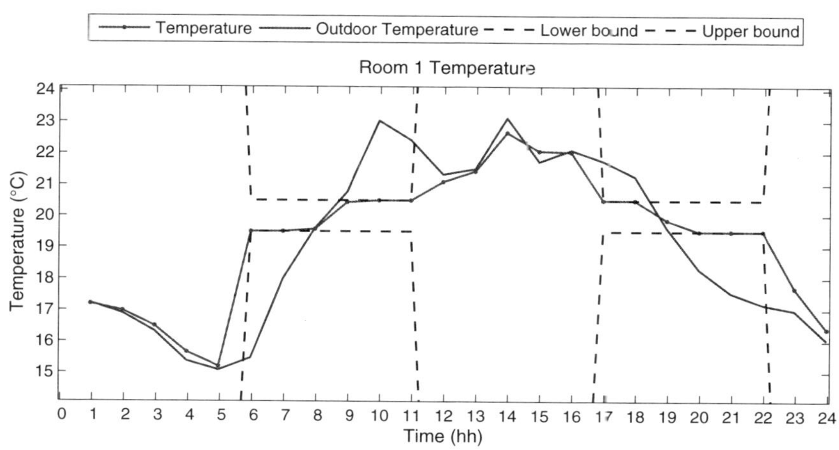

Fig. 12.7 Data set "day 1": Room temperature profile against forecast outdoor temperature and constraints

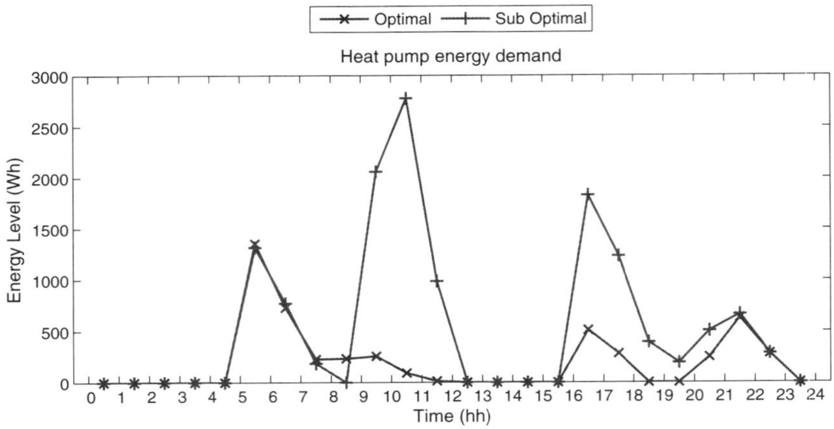

Fig. 12.8 Data set "day 1": Hourly energy amount consumed by the heat pump

Recalling that the task scheduling process accounts the heat pump activity, the energy price and local production, it is also possible to conclude that the prediction error, originating from the outdoor temperature forecast, affects the time table indirectly, whereas the prediction errors originating from both the solar irradiation and electricity price forecasts operate directly. On this subject, while a few educated guesses can be made in very specific cases, a general model of the interaction among the prediction errors does not exists. Due to their stochastic nature, in fact, it is not possible to assess beforehand the error amount, moreover in each time slot. In turn, then, the interaction among error, that is if the errors sum up or cancel each other out, is not known.

Although, from a general point of view, the way the task time table is distorted by the prediction errors is not predictable, it should also be noted that the time slot allocation is also subject to the system constraints. The maximum power allowed by the house wirings shall be accounted, and thus the fact that the number of tasks that can be executed at once is limited. The sequential nature of the task is also to be considered, and thus the fact that the tasks cannot be executed in any order. In addition, it shall be noted that each task has a given time frame. In other words, the robustness of the scheduling process, with respect to prediction errors, may increase the more tasks are to be scheduled, since the degrees of freedom of the allocation process are reduced. Nonetheless, it shall also be observed that, when energy demanding tasks are involved, even slightly distorted time schedules may lead to significant performance drop.

For instance, if the optimal time schedule, reported in Fig. 12.9, is to be compared to the sub-optimal time schedules, reported in Fig. 12.10, it is possible to observe that only the entries marked as "Task 3" and "Task 8" are actually affected by the forecast error.

By simulating the task execution using the computed task schedule, the actual

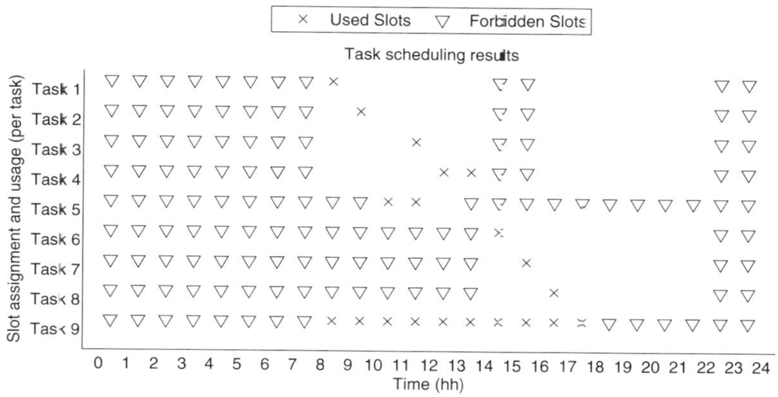

Fig. 12.9 Data set "day 1": Optimal task scheduling activity

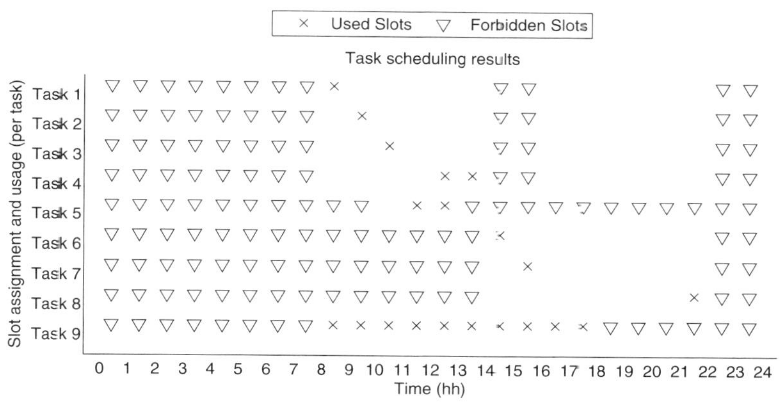

Fig. 12.10 Data set "day 1": Sub optimal task scheduling activity

energy management profiles can be evaluated. Concerning the reference case, being optimal, the energy allocation produce the minimum energy cost. When the sub optimal case is considered, obviously, a less efficient result is achieved. It may also worth to mention, in this case, that since the optimal case, being optimal, guarantees the best achievable performance, the sub optimal setup can, at most, achieve the same result. That is, under no circumstances better performances can be achieved.

Concerning the energy costs, reported in Fig. 12.11, the "Optimal" entry produces a total energy cost of \$0.21, whereas the "Sub Optimal" counterpart leads to a total energy cost of \$0.29. Concerning the energy income, reported in Fig. 12.12, the total income of \$0.1 is obtained in both cases. The energy minimum net cost (reference case) is thus \$0.106, the expected net cost for the forecast case is \$0.69, whereas the actual net cost for the forecast case is \$0.19.

Also the stored energy level can be evaluated (Fig. 12.13), and a better insight of the scheduling process can be gained. In particular, after 20.00 the battery is fully charged, meaning that the different slope in the energy cost plots, after 20.00, only depends on the task allocation. Also, at 8.00 the optimal time schedule allows the battery to recharge to full level, whereas the sub optimal time table prevent the recharging phase till 10.00 and requires an additional discharge at 13.00, thus increasing the battery stress.

As a further mean of comparison, for a baseline scenario, the energy cost will amount to \$5.31, whereas the energy income would amount to \$0.34. The net energy cost will amount to \$4.97, meaning that the optimal case net cost will amount to

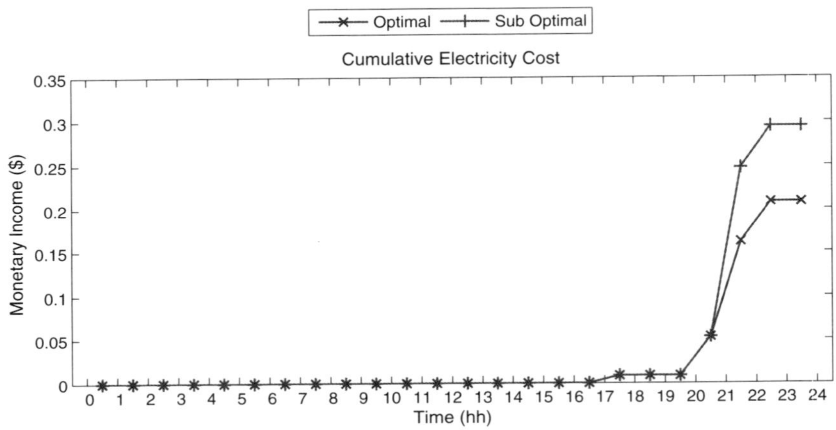

Fig. 12.11 Data set "day 1": Energy cost over time

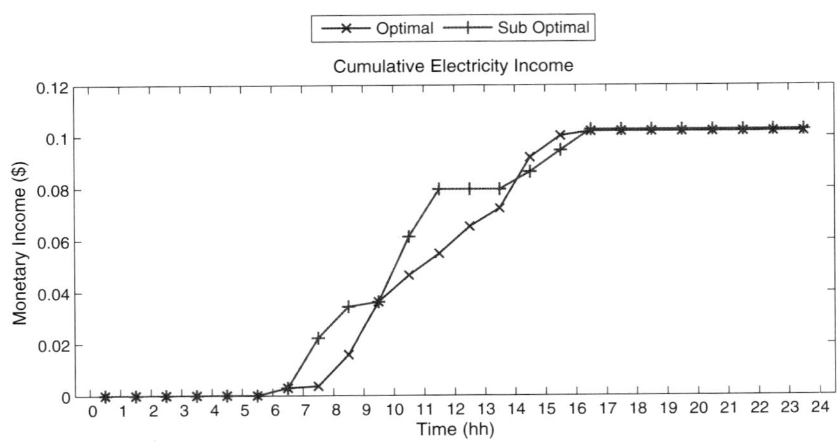

Fig. 12.12 Data set "day 1": Energy production income over time

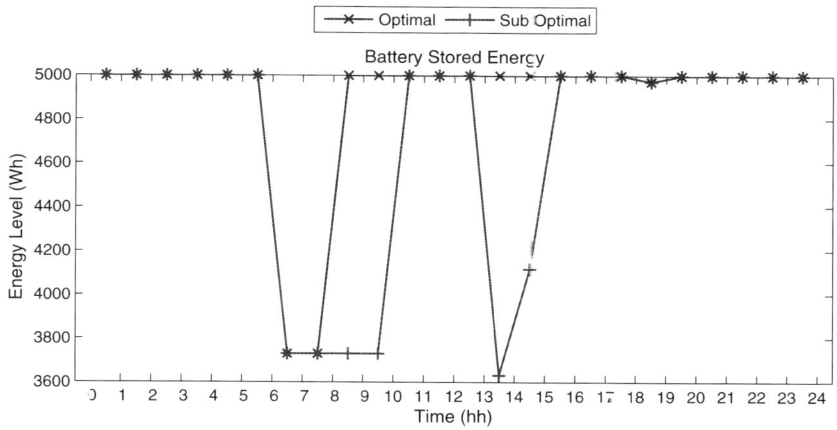

Fig. 12.13 Data set "day 1": Battery energy level over time

2.1% of the baseline net cost, whereas the sub-optimal net energy cost will amount to 3.8% of the baseline net cost.

By evaluating the scheduling process by means of the "day 3" set-up, a prediction error higher than 20% and lower than 30% will be accounted. Similarly to the previous case, concerning the heat pump expected energy demand, the forecast case will differ from the reference case (Fig. 12.14). The difference will be much more remarkable, since the prediction error is greater.

Pertaining the computed task schedules (Figs. 12.15 and 12.16), since the prediction errors originating from both the irradiance and price forecast are also greater than the previous case, the difference between the optimal case and the sub-optimal

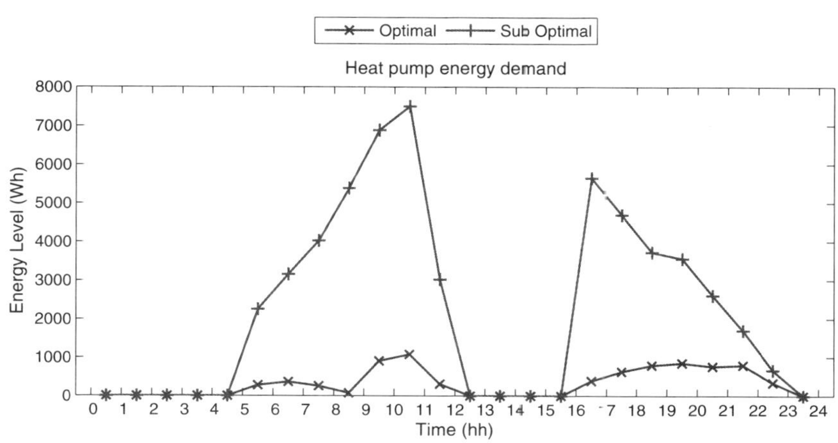

Fig. 12.14 Data set "day 3": Hourly energy amount consumed by the heat pump

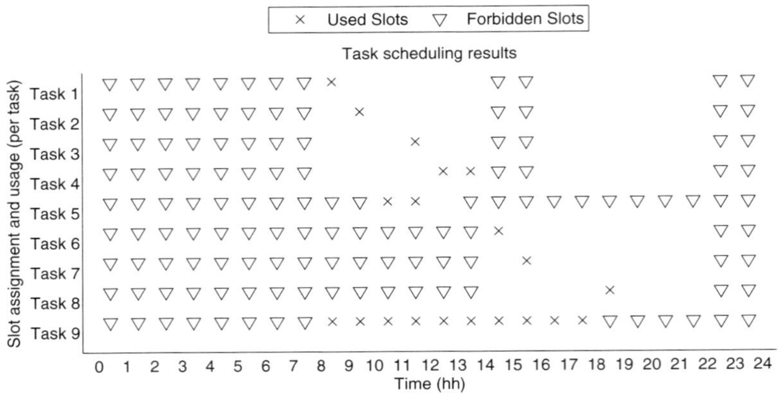

Fig. 12.15 Data set "day 3": Optimal task scheduling activity

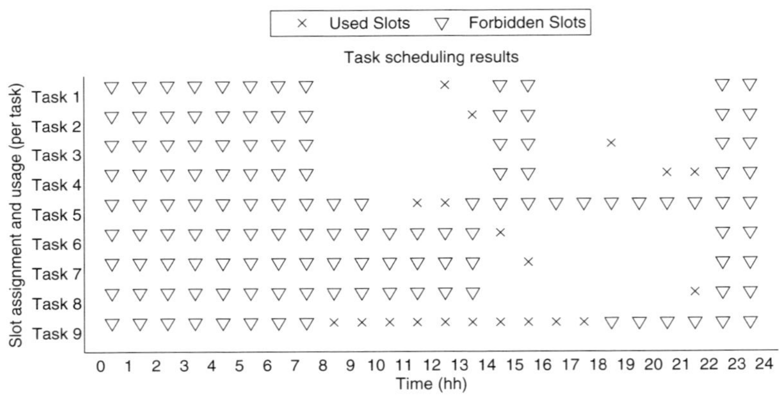

Fig. 12.16 Data set "day 3": Sub optimal task scheduling activity

counter part are much more evident. This result, also, suggests that the maximum and minimum errors, thus, accounted with their sign, may be a set of indices much more meaningful, towards the distortion of the task schedule, with respect to the MAE.

By simulating the task execution, based on each of the task schedule, the cumulative energy costs, depicted in Fig. 12.17 are obtained. The energy income amounts, on the other hand, are reported in Fig. 12.18, whereas the energy storage level over time is presented in Fig. 12.19.

Although the involved values may differ from the previous case, the same conclusion still holds. For instance, the net energy cost of the optimal case amounts to \$1.07, the expected net cost of the sub optimal case amounts, on the other hand, to \$11.72, while the actual net cost of the sub optimal scheduling amounts to \$1.73.

If the baseline scenario is taken into account, it is possible to observe that the total energy cost amounts to \$8.11 and the total energy income amounts to \$0.21. Since the net energy cost of the baseline scenario amounts to \$7.9, the net cost of the optimal case amounts to 13.5% of the baseline net energy cost. The net energy cost of the sub optimal case, on the other hand, equals the 21% of the baseline net cost.

Based on the proposed evaluations, it appears that, while the prediction errors may affect the scheduling process, thus impairing the allocation of the tasks and the energy management, the performance loss is not directly related to the amount of said errors nor their interaction. Clearly, the forecast values used by the scheduler may promote or prevent the allocation of each time slot to the given tasks.

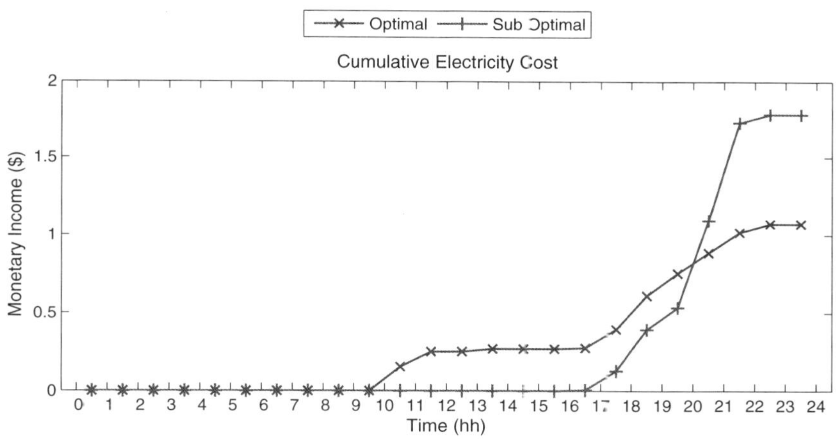

Fig. 12.17 Data set "day 3": Energy cost over time

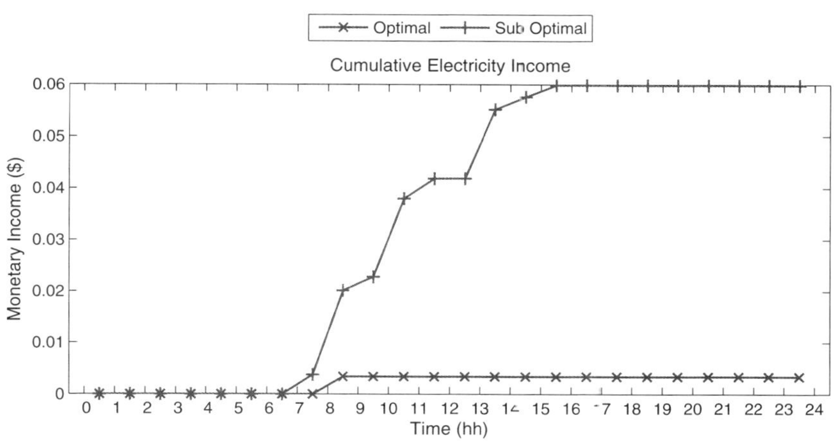

Fig. 12.18 Data set "day 3": Energy production income over time

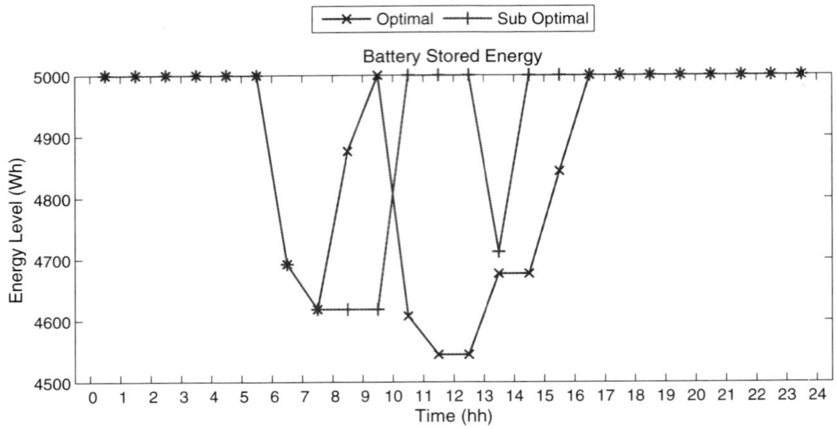

Fig. 12.19 Data set "day 3": Battery energy level over time

In this perspective, of course, the prediction errors may alter the time slots allocation in any way. However, if the optimal time slot for a given task can be addressed as the i-th time slot, and if a sub-optimal time slot assigned to the said task can be addressed as j-th time slot, it can be concluded that the performance loss actually depends on the difference between the i-th price and j-th price, that is the electricity price in the i-th time slot and the j-th time slot respectively.

Also, since the sub-optimal task allocation depends on both constraints and forecasts, due to the lack of correlation among the involved entities, it may be safe to assume that the sub-optimal allocation distortion is actually a random process, at least if a prediction error exists. On this subject, the "day 2" case, in which the prediction error is higher than 10% and lower than 20%, appears to be particularly meaningful.

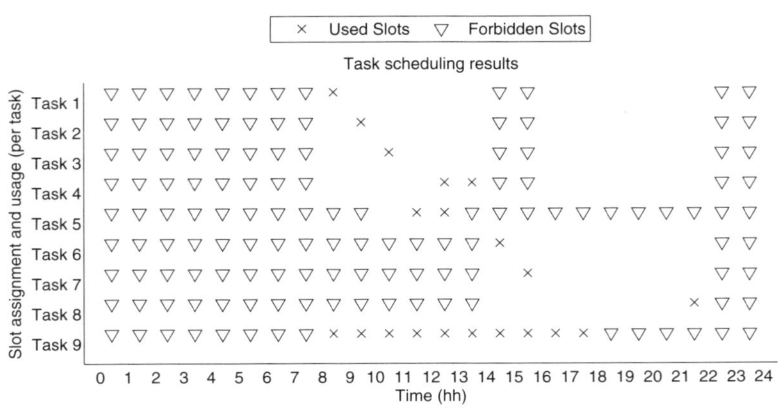

Fig. 12.20 Data set "day 2": Optimal task scheduling activity

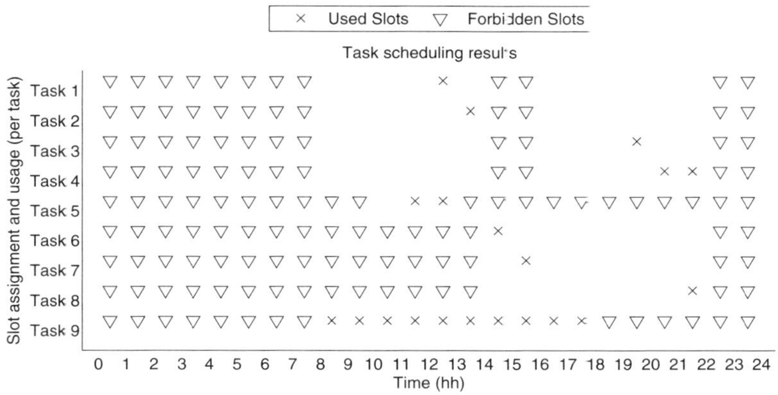

Fig. 12.21 Data set "day 2": Sub optimal task scheduling activity

In fact, if the optimal task schedule (Fig. 12.20) is compared against the sub optimal task schedule (Fig. 12.21), it is possible to notice that the "day 2" case is quite similar to the "day 3" one. However, in the "day 2" case, the net energy cost amounts to \$32.218 for the optimal and \$32.235 for the sub-optimal task schedule, thus there no meaningful difference in terms of energy cost between the optimal and the sub-optimal case.

This aspect may also be seen as the reason why the scheduling process seems to be fairly robust against prediction error, in the sense that the entity of the performance loss does not seem directly bounded to the entity of the prediction error. Moreover, even if the performance loss is to be accounted, the improvement over a baseline scheme still remains remarkable.

12.7 Conclusions

The potential shown by task scheduling and energy management solutions within smart home environments, and the theoretical benefits for both users and providers, pointed out that in order to take advantage of such technology, a time ahead scheduling approach is mandatory. In order to investigate this hypothesis, a performance evaluation of a scheduling process, based on a day ahead data forecast, is carried out.

The evaluation required, therefore, a few MLP based forecasters, that have been developed, implemented and trained. Then a forecast data set is generated. While the forecast data is used to select the time table under "realistic conditions", the historical data counterpart is used to simulate the task execution thus computing the resulting energy bill.

Also, the historical data are used to compute the ideal task schedule, under the assumption that the prediction error is zero. Under these circumstances the optimal energy bill is computed and the comparison between the sub optimal and the optimal task scheduling processes take place.

By reiterating the same process over several data set, the relationship between prediction error and scheduling performance is analyzed and evaluated, and then the conclusion are drawn. The investigation revealed that, while the issue at hand remains rather complex, a few key points can be highlighted.

Due to the interaction among prediction errors and constraints, the effects of prediction errors on the scheduling performance can hardly be esteemed. Even though indirectly, the historical data can be assumed to be responsible of the performance drop, whereas the performance drop does not appear to be bound to the entity of the prediction error.

While these results appear to be within reason, it also seems clear that, if more accurate forecasts are used, sub optimal task allocations are less likely to occur. Therefore, future works are aimed towards the improvement of data forecasts, to contain the prediction error, by providing additional information sources as the input of the ANN. Pertaining the irradiance and temperature forecasts, additional information on meteorological conditions can be used. Pertaining the price forecast also, grid load, fuel prices, and renewable source availability can be accounted.

In addition, although in the present work the point of view of the end user is taken into account, it shall be noted that the end user perspective is not the only one available. For instance, to investigate the power grid and the energy market behaviour in response to a large scale adoption of task schedulers may provide an interesting insight over power distribution.

Acknowledgement

This research was supported by the funding obtained within SMILING, an Italian Project for Technology Transfer (MISE - RIDITT Call 2009).

References

Boaro, M., Fuselli, D., Angelis, F., Liu, D., Wei, Q. and Piazza, F. (2013). Adaptive dynamic programming algorithm for renewable energy scheduling and battery management, *Cognitive Computation* **5**, 2, pp. 264–277.

Bozchalui, M. and Sharma, R. (2012). Analysis of electric vehicles as mobile energy storage in commercial buildings: Economic and environmental impacts, in *IEEE Power and Energy Society General Meeting* (San Diego, USA), pp. 1–8.

Cecati, C., Mokryani, G., Piccolo, A. and Siano, P. (2010). An overview on the smart grid concept, in *IECON 2010-36th Annual Conference on IEEE Industrial Electronics Society* (Glendale, USA), pp. 3322–3327.

Chakrabarti, B., Bullen, D., Edwards, C. and Callaghan, C. (2012). Demand response in the new zealand electricity market, in *IEEE 2012 PES Transmission and Distribution Conference and Exposition* (Orlando, USA), pp. 1–7.

Chapman, A. C., Verbic, G. and Hill, D. J. (2013). A healthy dose of reality for game-theoretic approaches to residential demand response, in *IEEE 2013 IREP Symposium on Bulk Power System Dynamics and Control-IX Optimization, Security and Control of the Emerging Power Grid* (Rethymno, Greece), pp. 1–13.

Ciabattoni, L., Cimini, G., Grisostomi, M., Ippoliti, G., Longhi, S. and Mainardi, E. (2013a). Supervisory control of PV-battery systems by online tuned neural networks, in *IEEE International Conference on Mechatronics* (Vicenza, Italy), pp. 99–104.

Ciabattoni, L., Ippoliti, G., Longhi, S., Pirro, M. and Cavalletti, M. (2013b). Solar irradiation forecasting for PV systems by fully tuned minimal RBF neural networks, in B. Apolloni, S. Bassis, A. Esposito, and F. C. Morabito (eds.), *Neural Nets and Surroundings* (Springer, Berlin), pp. 289–300.

Cui, H. and Dai, W. (2011). Multi-objective optimal allocation of distributed generation in smart grid, in *IEEE International Conference on Electrical and Control Engineering* (Yichang, China), pp. 713–717.

De Angelis, F., Boaro, M., Fuselli, D., Squartini, S. and Piazza, F. (2013a). A comparison between different optimization techniques for energy scheduling in smart home environment, in B. Apolloni, S. Bassis, A. Esposito, and F. C. Morabito (eds.), *Neural Nets and Surroundings* (Springer, Berlin), pp. 311–320.

De Angelis, F., Boaro, M., Fuselli, D., Squartini, S., Piazza, F. and Wei, Q. (2013b). Optimal home energy management under dynamic electrical and thermal constraints, *IEEE Transactions on Industrial Informatics* **9**, 3, pp. 1518–1527.

Fuselli, D., De Angelis, F., Boaro, M., Liu, D., Wei, Q., Squartini, S. and Piazza, F. (2012). Optimal battery management with ADHDP in smart home environments, in J. Wang, G. G. Yen, and M. M. Polycarpou (eds.), *Advances in Neural Networks–ISNN 2012* (Springer, Berlin), pp. 355–364.

Fuselli, D., De Angelis, F., Boaro, M., Squartini, S., Wei, Q., Liu, D. and Piazza, F. (2013). Action dependent heuristic dynamic programming for home energy resource scheduling, *International Journal of Electrical Power and Energy Systems* **48**, pp. 148–160.

Fux, S. F., Ashouri, A., Benz, M. J. and Guzzella, L. (2014). EKF based self-adaptive thermal model for a passive house, *Energy and Buildings* **68**, pp. 811–817.

Ghosh, S., Sun, X. and Zhang, X. (2012). Consumer profiling for demand response programs in smart grids, in *IEEE Innovative Smart Grid Technologies–Asia* (Tianjin, China), pp. 1–6.

Hernandez, L., Baladron, C., Aguiar, J., Carro, B., Sanchez-Esguevillas, A., Lloret, J., Chinarro, D., Gomez-Sanz, J. and Cook, D. (2013). A multi-agent system architecture for smart grid management and forecasting of energy demand in virtual power plants, *IEEE Communications Magazine* **51**, 1, pp. 106–113.

Ipakchi, A. and Albuyeh, F. (2009). Grid of the future, *IEEE Power and Energy Magazine* **7**, 2, pp. 52–62.

ISO (2003). EN 12831:2003 Heating Systems in Buildings–Method for Calculation of the Design Heat Load.

ISO (2007a). EN ISO 13370:2007 Thermal Performance of Buildings-Heat Transfer via the Ground–Calculation Methods.

ISO (2007b). EN ISO 13789:2007 Thermal Performance of Buildings-Transmission and Ventilation Heat Transfer Coefficients-Calculation Method.

Kazanavičius, E., Mikuckas, A., Mikuckienė, I. and Čeponis, J. (2006). The heat balance model of residential house, *Information Technology and Control* **35**, 4, pp. 391–396.

Kirthiga, M., Daniel, S. and Gurunathan, S. (2013). A methodology for transforming an existing distribution network into a sustainable autonomous micro-grid, *IEEE Transactions on Sustainable Energy* **4**, 1, pp. 31–41.

Liu, C., Chau, K., Wu, D. and Gao, S. (2013). Opportunities and challenges of vehicle-to-home, vehicle-to-vehicle, and vehicle-to-grid technologies, *Proceedings of the IEEE* **101**, 11, pp. 2409–2427.

Makonin, S., Popowich, F. and Gill, B. (2013). The cognitive power meter: Looking beyond the smart meter, in *26th Annual IEEE Canadian Conference on Electrical and Computer Engineering* (Regina, Canada), pp. 1–5.

Nagasaka, K., Ando, K., Xu, Y., Takamori, H., Wang, J., Mitsuta, A., Saito, O. and Go, E. (2012). A research on operation planning of multi smart micro grid, in *IEEE 2012 International Conference on Advanced Mechatronic Systems* (Tokyo, Japan), pp. 351–356.

Qela, B. and Mouftah, H. (2010). Simulation of a house heating system using C#–an energy conservation perspective, in *IEEE 23rd Canadian Conference on Electrical and Computer Engineering* (Calgary, Canada), pp. 1–5.

Saber, A. and Venayagamoorthy, G. (2012). Resource scheduling under uncertainty in a smart grid with renewables and plug-in vehicles, *IEEE Systems Journal* **6**, 1, pp. 103–109.

Severini, M., Squartini, S. and Piazza, F. (2013). Hybrid soft computing algorithmic framework for smart home energy management, *Soft Computing* **17**, 11, pp. 1983–2005.

Soares, J., Morais, H., Sousa, T., Vale, Z. and Faria, P. (2013). Day-ahead resource scheduling including demand response for electric vehicles, *IEEE Transactions on Smart Grid* **4**, 1, pp. 596–605.

Squartini, S., Boaro, M., De Angelis, F., Fuselli, D. and Piazza, F. (2013a). Optimization algorithm for home energy resource scheduling in presence of data uncertainty, in *4th International Conference on Intelligent Control and Information Processing* (Beijing, China), pp. 323–328.

Squartini, S., Fuselli, D., Boaro, M., De Angelis, F. and Piazza, F. (2013b). Home energy resource scheduling algorithms and their dependency on the battery model, in *IEEE Symposium on Computational Intelligence Applications in Smart Grid* (Singapore), pp. 122–129.

Tham, C. K. and Luo, T. (2013). Sensing-driven energy purchasing in smart grid cyber-physical system, *IEEE Transactions on Systems, Man, and Cybernetics: Systems* **43**, 4, pp. 773–784.

Venayagamoorthy, G. (2011). Dynamic, stochastic, computational, and scalable technologies for smart grids, *IEEE Computational Intelligence Magazine* **6**, 3, pp. 22–35.

Wang, Z., Paranjape, R., Sadanand, A. and Chen, Z. (2013). Residential demand response: An overview of recent simulation and modeling applications, in *26th Annual IEEE Canadian Conference on Electrical and Computer Engineering* (Regina, Canada), pp. 1–6.

Chapter 13

Many-Objective Evolutionary Algorithms and Hybrid Performance Metrics

Zhenan He and Gary G. Yen

Abstract In recent years, the design and evaluation of MOEAs, specifically for many-objective optimization problems, have been gaining an increasing attention because a large number of real-world problems involves simultaneously optimizing multiple conflicting objectives (in most cases, more than five) which are commonly referred to as many-objective optimization problems (MaOPs). Evolutionary algorithms have been successfully exploited to solve multiobjective optimization problems. In literature, a heuristic approach is often taken. For a chosen benchmark problem with specific problem characteristics, the performance of multiobjective evolutionary algorithms (MOEAs) is evaluated via some heuristic chosen performance metrics. The conclusion is then drawn based on statistical findings given the preferable choices of performance metrics. The conclusion, if any, is often indecisive and reveals no insight pertaining to specific problem characteristics that the underlying MOEA could perform the best. In this chapter, we introduce an ensemble method to compare six state-of-the-art MOEAs designed specifically for many-objective optimization problems under a number of carefully crafted benchmark problems, by combining a number of performance metrics using double elimination tournament selection. The double elimination design allows characteristically poor performance of a quality algorithm to still be able to win it all. Experimental results show that the proposed metrics ensemble can offer a more comprehensive comparison among various MOEAs than what could be obtained from single performance metric alone. The end result is a ranking order among all chosen MOEAs, but not quantifiable measures pertaining to the underlying MOEAs.

13.1 Introduction

Many real-world problems involve simultaneously optimizing multiple conflicting objectives (in most cases, more than five) which are commonly referred to as many-objective optimization problems (MaOPs). The subject has been gaining an increasing attention in recent years. Evolutionary algorithms (EAs) have established themselves as the approaches for exploring the Pareto-optimal fronts in multiobjective optimization problems (MOPs), which contain two or three objectives in general. Multiobjective Evolutionary Algorithms (MOEAs) do not guarantee to identify optimal tradeoffs, but attempt to find a good approximation. Although MOEAs have been effectively used to explore the Pareto-optimal fronts in MOPs, the performance of most MOEAs deteriorates severely in face of MaOPs [Ishibuchi *et al.* (2008)]. The main reason that MOEAs lose exploring capability in solving MaOPs is largely due to ineffective definition of Pareto optimality.

Compared with low-dimensional MOPs, the increasing number of objectives in MaOPs causes the Pareto optimality to lose the selection pressure during the evolution process. Over past few years, a number of state-of-the-art MOEAs based on Pareto optimality, e.g., NSGA-II [Deb *et al.* (2002a)] and SPEA2 [Zitzler *et al.* (2001)], perform very well to solve low-dimensional MOPs because Pareto optimality is effective to facilitate the convergence of the population in low-dimensional space. However, in high-dimensional space, the proportion of non-dominated individuals rises quickly with the number of objectives. When the number of objectives exceeds five, the proportion of non-dominated individuals in a randomly generated initial population is often higher than 90% [He and Yen (2013)]. This leads to significantly diminishing selection pressure during the evolutionary process no matter how the MOEA is designed, if the Pareto dominance relation is used.

In addition to the deficiency in the definition of Pareto dominance stated above, other complications such as visualization of high-dimensional objective spaces [Ishibuchi *et al.* (2008)] and a very high computational cost (due to a large number of individuals needed to obtain a good representation of the Pareto front) [Singh *et al.* (2011)] have contributed to the challenges of solving MaOPs.

From the above discussions, difficulties caused by a large number of objectives have rendered the existing MOEAs ineffective to solve MaOPs. The efforts in addressing this issue have led to the developments of new algorithms, often called Many Objective Evolutionary Algorithms. In literature, there are mainly four types of MOEAs that have been proposed to solve MaOPs.

First, algorithms that modify the Pareto dominance concept to adapt it to higher dimensional space include Pareto α-dominance [Batista *et al.* (2010)], Pareto ε-dominance [Batista *et al.* (2010)], and Pareto cone ε-dominance [Batista *et al.* (2010)]. Heuristically chosen parameters are incorporated into all of these methods. Each modified Pareto dominance design is a relaxing form of the Pareto dominance in that it makes one individual dominates others easier in high-dimensional space. Based on a similar idea, ε-domination based multi-objective evolutionary algorithm (ε-MOEA) [Deb *et al.* (2005)] is proposed and has been found to have a good performance for MaOPs [Hadka and Reed (2012)]. On the other hand, there are some studies incorporating fuzzy concept into Pareto dominance for new fitness evaluation mechanism to continuously differentiate individuals into different degrees of optimality beyond the classification of the original Pareto dominance. Based on this notion, a fuzzy Pareto dominance (FD) relation is defined and incorporated into the designs of NSGA-II, so called FD-NSGA-II [He and Yen (2013)]. It applies to search for a Pareto optimal set for many-objective optimization problems by maintaining the selection pressure towards the Pareto front throughout the entire evolutionary process.

The second class is based on the idea of performance metrics. For example, volume dominance (VD) [Hughes (2008)] is based on the volume of dominated objective space by the individual. Hypervolume estimation algorithm for multiobjective optimization (HypE) [Bader and Zitzler (2011)], is probably the most successful implementation of this class in that it has been shown to be more effective than other MOEAs for MaOPs. Also, there are some other designs in a similar spirit such as indicator-based evolutionary algorithm (IBEA) [Zitzler and Simon (2004)] and SMS-EMOA [Beume *et al.* (2007)].

The third class is decomposition based designs, such as multiobjective evolutionary algorithm based on decomposition (MOEA/D) [Zhang and Li (2007)] and reference-point based many-objective NSGA-II (MO-NSGA-II) [Deb and Jain (2012c)]. This type of methods decomposes a multiobjective optimization problem into a number of scalar optimization subproblems and optimizes them simultaneously. In the evolution process, two aggregation functions, Tchebycheff in Zhang and Li (2007) and achievement scalarizing function in Deb and Jain (2012c), were proposed for fitness assignment. The resulting fitness values are applied to select individuals instead of Pareto-dominance. Therefore, this method can be easily extended to solve MaOPs.

The last class is the grid-based method. From Yang *et al.* (2013), a grid can reflect the status quo of convergence and diversity simultaneously. Grid-based evolutionary algorithm (GrEA) [Yang *et al.* (2013)] exploits the potential of the grid-based approach to strengthen the selection pressure toward the optimal direction, while maintaining an extensive and uniform distribution among solutions. Territory defining multiobjective evolutionary algorithm (TDEA) [di Pierro *et al.* (2007)] defines a territory around each individual to prevent crowdness in any region. The hyperbox of TDEA is related to individuals rather than predefined by decision maker.

Although numerous MOEAs were proposed for many-objective optimization problems, by the No Free Lunch Theorem [Wolpert and Macready (1997)], any algorithm's elevated performance over one class of problems is exactly paid for in loss over another class. Therefore, comparative studies are often conducted to reveal advantages and weaknesses of the underlying MOEAs and at determining the best performance pertaining to specific class of problem characteristics [Li *et al.* (2013)]. Recently, multiple comprehensive comparisons between latest improvements on NSGA-II and MOEA/D for many-objective optimization problems have been made in Deb and Jain (2012a,b); Saxena *et al.* (2011). In those experiments, only single performance metric is used therein. However, in absence of any established comparison criteria, none of the claims based on heuristically chosen performance metrics for the Pareto-optimal solutions generated can be made convincingly. In literature, when an MOEA is proposed, a number of benchmark problems are often selected to quantify the performance. Since these are artificially crafted benchmark functions, their corresponding Pareto fronts can be made available to measure the performance. Based on a set of heuristically chosen performance metrics, the proposed MOEA and some competitive representatives are evaluated statistically given a large number of independent trials. The conclusion, if any is drawn, is often indecisive and reveals no additional insight pertaining to the specific problem characteristics that the proposed MOEA would perform the best [Tessema and Yen (2009); Yen and Lu (2003)].

Zitzler *et al.* (2000) proposed three optimization goals to be measured: the distance of the resulting non-dominated set to the Pareto-optimal front should be minimized, a good (in most cases uniform) distribution of the solutions found in objective space is desirable, and the extent of the obtained non-dominated front should be maximized. In literature, there are many *unary* performance metrics used to compare MOEAs. These metrics can be broadly divided into five categories according to the optimization goals. Each category mainly evaluates the quality of a Pareto-optimal set in one aspect only. The first category involves metrics assessing the number of Pareto optimal solutions in the set: *Ratio of non-dominated individuals* (RNI) [Tan *et al.* (2002)] measures the proportion of the non-dominated solutions found with respect to the population size; *error ratio* (ER) [Van Veldhuizen (1999)] checks the proportion of non true Pareto points in the approximation front over the population size; *overall non-dominated vector generation* (ONVG) [Van Veldhuizen (1999)] simply counts the number of distinct non-dominated individuals found; and the n-ary performance metric, *Pareto dominance indicator* (NR) [Goh and Tan (2009)], measures the ratio of non-dominated solutions contributed by a particular approximation front to the non-dominated solutions provided collectively by all approximation fronts. Within the second category, metrics measuring the closeness of the solutions to the theoretical Pareto front are given: *Generational distance* (GD) [Van Veldhuizen (1999)] measures how far the evolved solution set is from the true Pareto front; a complementary metric of GD called *inverted generational distance* (IGD) [Czyzzak and Jaszkiewicz (1998)] concerns how well is the Pareto-optimal front represented by the obtained solution set; and *maximum Pareto front error* (MPFE) [Van Veldhuizen (1999)] focuses on the largest distance between the individual in the theoretical Pareto front and the points in the approximation front. In the third category, metrics are relating on distribution of the solutions: *Uniform distribution* (UD) [Tan *et al.* (2002)] quantifies the distribution of an approximation front under a pre-defined parameter; *spacing* [Schott (1995)] measures how evenly the evolved solutions distribute themselves; and *number of distinct choices* (NDC_m) [Wu and Azarm (2001)] identifies solutions that are sufficiently distinct for a special value m. Fourth, metrics concerning spread of the solutions are included: *Maximum spread* (MS) [Zitzler *et al.* (2000)] measures how well the true Pareto front is covered by the approximation set. In the last category, metrics consider both closeness and diversity at the same time: *Hyperarea and ratio* (or hypervolume indicator) [Van Veldhuizen (1999); Zitzler and Thiele (1998)] calculates the volume covered by the approximation front with respect to a properly chosen reference point.

Furthermore, there are some *binary* performance metrics used to compare a pair of algorithms. I_ε [Zitzler *et al.* (2003)] defines an ε-dominant relation between algorithms, enclosing hypercube indicator and coverage difference metrics (D-metric) [Zitzler (1999)]. The C-metric, or set coverage, considers the domination relations between two algorithms, i.e., how good an approximation front obtained from one algorithm dominates an approximation front obtained by another algorithm and vice versa [Hansen and Jaszkiewicz (1998)].

However, no single metric alone can faithfully measure MOEA performance. Every metric can provide some specific, but incomplete, quantifications of performance and can only be used effectively under specified conditions. For example, UD does a poor job when the Pareto front is discontinuous, while hypervolume indicator can be misleading if the Pareto optimal front is non-convex [Van Veldhuizen (1999)]. This implies that one metric alone cannot entirely evaluate MOEAs under various conditions. Every metric focuses on some problem-specific characteristics while neglects information in others. Every *carefully crafted* metric has its unique attribute; no metrics alone can substitute others completely. Therefore, a single metric alone cannot provide a comprehensive measure for MOEAs. For a specific test problem, we cannot ascertain which metrics should be applied in order to faithfully quantify the performance of MOEAs. Common practice is to exploit various metrics to determine which combination is a better choice. Apparently, this process adds a heavy computational cost.

To overcome these deficiencies and arrive at a fair evaluation of MOEAs, performance metrics ensemble is proposed in this research work. The ensemble method uses multiple metrics collectively to obtain a better assessment than what could be obtained from any of single performance metric alone. Metrics ensemble not only can give a comprehensive comparison between different algorithms, but avoid the choosing process and can be directly used to assessing MOEAs.

In literature, the ensemble approaches can be found in statistics and machine learning. Supervised learning algorithms search through a space to find a suitable hypothesis that will make good predictions for a given problem. Ensemble methods combine multiple hypotheses to form a better one than could be obtained from any of the constituent models [Polikar (2006)]. It always combines many weak learners in an attempt to produce a strong one. Furthermore, ensembles tend to yield better results when there is a significant diversity among the models [Kuncheva and Whitaker (2003)]. There are some well-regarded designs: bagging [Breiman (1996)], boosting [Schapire *et al.* (1998)], Bayesian model averaging [Hoeting *et al.* (1999)], stacked generalization [Wolpert (1992)] and the random subspace method [Ho (1998)]. The application of multiple performance metrics is first introduced by Zitzler *et al.* (2010). They discuss how to use hierarchies of metrics so that each metric is a refinement of the preference structure detected by a lower-level metric. However, there exists no publication in literature, to our best knowledge, regarding performance metrics ensemble. Without any reference information, MOEAs are evaluated and compared based on a single metric at a time. In this chapter, we propose a double elimination tournament selection operator to compare approximation fronts obtained from different MOEAs in a statistically meaningful way. In every competition, one metric is chosen randomly to compare. After the whole process, every metric could be selected multiple times and a final winning algorithm is to be identified. This final winner would have been compared under all the metrics considered so that we can make a fair conclusion based on an overall assessment.

The remaining sections complete the presentation of this chapter. Section 13.2 provides the consolidated literature review on the performance metrics proposed in literature and Section 13.3 introduces MOEAs designed specifically for MaOPs proposed in literature. Section 13.4 describes the proposed performance metrics ensemble approach in detail, including the double elimination tournament selection operator. In Section 13.5, we elaborate on the experiment results for selected benchmark problems. Finally, a conclusion is drawn in Section 13.6 along with pertinent observations.

13.2 Literature Review on Performance Metrics

Selected performance metrics will be briefly reviewed according to the way how they are classified in this chapter.

13.2.1 *Metrics assessing the number of Pareto optimal solutions in the set*

1) Ratio of Non-dominated Individuals (RNI) [Tan *et al.* (2002)]:

The performance measure of an approximation front X is:

$$\text{RNI} = \frac{|\overline{X}|}{n} \tag{13.1}$$

where $\overline{X}$ denotes the set of non-dominated individuals in population X whose size is n. Clearly RNI $\in [0, 1]$, the larger is the better. When RNI $= 1$, it implies all the individuals in X are non-dominated. When RNI $= 0$, it implies none of the individuals in X is non-dominated. RNI is a significant measure in that it checks the proportion of non-dominated individuals in population, X.

2) Error Ratio (ER) [Van Veldhuizen (1999)]:

It is defined as the proportion of non-true Pareto points:

$$\mathrm{ER} = \frac{\sum_{i=1}^{n} e(x_i)}{n} \tag{13.2}$$

where x_i denotes an individual in an approximation front X and n is the number of individuals in X. $e(x_i) = 0$, if $x_i \in \mathrm{PF}_{true}$. Otherwise, $e(x_i) = 1$. PF_{true} denotes the true Pareto set. This metric does assume the PF_{true} is made available. Lower values of ER refer to smaller proportions of non-true Pareto points in X and represent better non-dominated sets.

3) Overall Non-dominated Vector Generation (ONVG) [Van Veldhuizen (1999)]:

It measures the number of non-dominated individuals found in an approximation front during MOEA evolution. It is defined as

$$\mathrm{ONVG} = |\mathrm{PF}_{known}| \tag{13.3}$$

where PF_{known} represents the obtained approximation front. From Van Veldhuizen and Lamont (2000), too few individuals in PF_{known} make the front's representation poor and too many vectors may overwhelm the decision maker. Also, Knowles and Corne (2002) proves that algorithm A outperforms B on this metric does not necessarily imply algorithm A is clearly better than B.

4) Pareto Dominance Indicator (NR) [Goh and Tan (2009)]:

Considering the approximation fronts, $A_1, A_2, \ldots, A_m$ obtained by different algorithms, this n-ary metric measures the ratio of non-dominated solutions that is contributed by a particular solution set A_1 to the non-dominated solutions provided by all algorithms:

$$\mathrm{NR}(A_1, A_2, \ldots, A_m) = \frac{|A_1 \cap B|}{|B|} \tag{13.4}$$

where $B = \{b_i | \forall b_i, \neg\exists a_j \in (A_1 \cup A_2 \cup \cdots \cup A_m) \prec b_i\}$, and $a_j \prec b_i$ implies that a_j dominates b_i. A_1 is the set under evaluation.

13.2.2 *Metrics measuring the closeness of the solutions to the true Pareto front*

1) Final Generational Distance (GD) [Van Veldhuizen (1999)]:

$$\mathrm{GD} = \frac{\sqrt{\sum_{i=1}^{n} d_i^2}}{n} \tag{13.5}$$

where $d_i = \min_j \|f(x_i) - \mathrm{PF}_{true}(x_j)\|$ refers to the distance in objective space between individual x_i and the nearest member in the true Pareto front, and n is the number of individuals in the approximation front. This metric, assuming PF_{true} is readily available, is a measure representing how "far" the approximation front is from the true Pareto front. Lower value of GD represents a better performance.

2) Inverted Generational Distance (IGD) [Czyzzak and Jaszkiewicz (1998)]:

This metric measures both convergence and diversity. Let PF_{true} is a set of uniformly distributed solutions in true Pareto front. X is the set of non-dominated solutions in the approximation front PF_{known}:

$$\mathrm{IGD} = \frac{\sum_{v \in \mathrm{PF}_{true}} d\,(v, X)}{|\mathrm{PF}_{true}|} \tag{13.6}$$

$d\,(v, X)$ denotes the minimum Euclidean distance between v and the points in X. To have a low value of IGD, the set X should be close to PF_{true} and cannot miss any part of the whole PF_{true}.

3) Maximum Pareto Front Error (MPFE) [Van Veldhuizen (1999)]:

It measures a worst case scenario in term of the largest distance in the objective space between any individual in the approximation front and the corresponding closest vector in the true Pareto front.

$$\mathrm{MPFE} = \max_i d_i. \tag{13.7}$$

d_i, defined earlier, is referred to as the distance in objective space between individual x_i and the nearest member in the true Pareto front. From Van Veldhuizen and Lamont (2000), for a non-dominated set, a good performance in MPFE does not ensure it is better than another one with a much worse MPFE.

13.2.3 *Metrics focusing on distribution of the solutions*

1) Uniform Distribution (UD) [Tan *et al.* (2002)]:

It measures the distribution of non-dominated individuals on the found trade-off surface. For a given set of non-dominated individuals $\bar{X}$ in a population X:

$$\mathrm{UD} = \frac{1}{1 + S_{nc}} \tag{13.8}$$

where $S_{nc} = \sqrt{\frac{\sum_{i=1}^{N_{\bar{X}}} \left(nc(\bar{x}_i) - \overline{nc}(\bar{X})\right)^2}{N_{\bar{X}} - 1}}$ is the standard deviation of niche count of the overall set of non-dominated individuals in $\bar{X}$, $N_{\bar{X}}$ is the size of the set $\bar{X}$, and $\overline{nc}(\bar{X})$ is the mean value of niche counts, $nc(\bar{x}_i)$, $\forall i = 1, 2, \ldots, N_{\bar{X}}$. Specifically, niche count of individual x_i is defined as

$$nc(\bar{x}_i) = \sum_{j=1, j\neq i}^{N_{\bar{X}}} \mathrm{Sh}(x_i, x_j) \text{ and } \mathrm{Sh}(x_i, x_j) = \begin{cases} 1, & \text{if } d(x_i, x_j) < \sigma_{share} \\ 0, & \text{otherwise.} \end{cases}$$

$d(x_i, x_j)$ is the distance between individuals x_i and x_j in the objective space, and σ_{share} is a user-defined parameter to quantify the closeness.

2) Spacing [Schott (1995)]:

This metric is a value measuring how evenly the non-dominated solutions are distributed along the approximation front,

$$S = \sqrt{\frac{1}{\bar{n}} \sum_{i=1}^{\bar{n}} (d_i - \bar{d})^2} \tag{13.9}$$

where d_i is the Euclidean distance in objective space between individual x_i and the nearest member in the true Pareto front, and $\bar{n}$ is the number of individuals in the approximation front. This metric requires low computational overhead and can be generalized to more than two dimensions.

3) Number of Distinct Choices (NDC_m) [Wu and Azarm (2001)]:

In this metric, only those solutions that are sufficiently distinct from one another should be accounted for as useful design options. Let m, $(0 < \mu < 1)$, be a user specified parameter which can be used to divide an m-dimensional objective space into $1/\mu^m$ number of small grids. Each of the grids refers to indifference region $T_\mu(q)$ wherein any two solutions within the same grid are considered similar to one another. The quality $\text{NT}_\mu(q, P)$ indicates whether or not there is any individual $p_k \in P$ that falls into the region $T_\mu(q)$. Specifically

$$\text{NT}_\mu(q, P) = \begin{cases} 1, & \exists p_k \in P, \ p_k \in T_\mu(q), \\ 0, & \forall p_k \in P, \ p_k \notin T_\mu(q). \end{cases}$$

$\text{NDC}_\mu(P)$ defines the number of distinct choices for a pre-specified value of m:

$$\text{NDC}_\mu(P) = \sum_{l_m=0}^{(1/\mu)-1} \cdots \sum_{l_2=0}^{(1/\mu)-1} \sum_{l_1=0}^{(1/\mu)-1} \text{NT}_\mu(q, P) \tag{13.10}$$

From Bader and Zitzler (2011), for a pre-specified value of m, an observed Pareto solution set with a higher value of the quantity $\text{NDC}_\mu(P)$ is preferred to a set with a lower value.

13.2.4 *Metrics concerning spread of the solutions*

1) Maximum Spread (MS) [Zitzler *et al.* (2000)]:

It addresses the range of objective function values and takes into account the proximity to the true Pareto front, assuming available. This metric is applied to measure how well the PF_{true} is covered by the PF_{known}

$$\text{MS} = \sqrt{\frac{1}{M} \sum_{i=1}^{M} \left[\frac{\min(\text{PF}_{known,i}^{\max}, \text{PF}_{true,i}^{\max}) - \max(\text{PF}_{known,i}^{\min}, \text{PF}_{true,i}^{\min})}{\text{PF}_{true,i}^{\max} - \text{PF}_{true,i}^{\min}} \right]^2} \tag{13.11}$$

where $\text{PF}_{known,i}^{\max}$ and $\text{PF}_{known,i}^{\min}$ are the maximum and minimum of the ith objective in PF_{known}, respectively; and $\text{PF}_{true,i}^{\max}$ and $\text{PF}_{true,i}^{\min}$ are the maximum and minimum of the ith objective in PF_{true}, respectively. M denotes the number of objectives considered. A higher value of MS reflects that a larger area of the PF_{true} is covered by the PF_{known}.

13.2.5 *Metrics considering both closeness and diversity*

1) Hyperarea and Ratio (*Hypervolume Indicator*) [Van Veldhuizen (1999); Zitzler and Thiele (1998)]:

It calculates the hypervolume of the multi-dimensional objective space enclosed by approximation front PF_{known} and a reference point. For example, an individual x_i in PF_{known} for a two-dimensional MOP defines a rectangle area, $a(x_i)$, bounded by an origin and $f(x_i)$. The union of such rectangle areas is referred to as hyperarea of PF_{known},

$$H(\text{PF}_{known}) = \left\{ \bigcup_i a(x_i) \middle| \forall x_i \in \text{PF}_{known} \right\}. \tag{13.12}$$

As pointed out in Zitzler and Thiele (1998), this metric requires defining a reference point of the region and could be misleading if PF_{known} is nonconvex. In Auger *et al.* (2009), suggestion is given as how to properly choose a reference point. In Van Veldhuizen (1999), Veldhuizen also propose a hyperarea ratio metric defined as:

$$\text{HR} = \frac{H(\text{PF}_{known})}{H(\text{PF}_{true})}. \tag{13.13}$$

Apparently, PF_{true} is given as a reference. In the proposed performance metrics ensemble to be presented in the next section, we adopt the hyperarea ratio metric.

13.3 Literature Review on Many-Objective Evolutionary Algorithms

In this study, six state-of-the-art MOEAs are chosen for competition. They are FD-NSGA-II [He and Yen (2013)], MOEA/D [Zhang and Li (2007)], e-MOEA [Deb *et al.* (2005)], MO-NSGA-II [Deb and Jain (2012c)], GrEA [Yang *et al.* (2013)], and HypE [Bader and Zitzler (2011)]. A brief overview of each chosen MOEA is given below.

FD-NSGA-II [He and Yen (2013)] is the improved NSGA-II by adopting the fuzzy Pareto dominance relations and the corresponding fuzzy fitness assignment process. In the proposed design, fuzzy Pareto dominance relation is applied to determine the rank value of each individual instead of Pareto dominance in the original NSGA-II. After the rank value is determined, the same crowding-distance is used as the original design of NSGA-II. The fuzzy fitness assignment process ensures one individual is fuzzy non-dominated with respect to others in the same rank.

MOEA/D [Zhang and Li (2007)] decomposes a MOP into a number of scalar optimization subproblems and optimizes them simultaneously. Each subproblem has a different weight vector and a single solution. For each subproblem, a certain number of the nearest subproblems are defined as its neighbors based on the Euclidean distance between their weight vectors. Each subproblem is optimized by only using information from its several neighboring subproblems. For each subproblem, a new solution is generated by current solutions in its neighboring subproblems and is compared with current solutions in the neighboring subproblems.

e-MOEA [Deb *et al.* (2005)] is a steady-state algorithm based on the e-dominance relation. It divides the objective space into hyperboxes by a size of e. Each hyperbox is assigned at most a single solution on the basis of e-dominance. From Deb *et al.* (2005). e-MOEA provides a tradeoff among convergence, diversity, and computational time. Furthermore, it could be made interactive with a decision-maker which implies e can be chosen by decision-maker according to user's preference.

MO-NSGA-II [Deb and Jain (2012c)] is a hybrid NSGA-II with similar framework as the original NSGA-II except a modified selection mechanism is designed for handling many-objective optimization problems. It uses multiple predefined reference points to guide the search in the evolution process. Then, population members are projected on a hyper-plane and a clustering operation is performed on the hyper-plane to select a desired number of clusters (user-defined). Based on the diversity of the population, either a local search operation on a random cluster member is used to move the solution closer to the Pareto-optimal front or a diversity enhancement operator is used to choose population members from all clusters.

Grid-based evolutionary algorithm (GrEA) [Yang *et al.* (2013)] exploits the potential of the grid-based approach to strengthen the selection pressure towards the global Pareto front while maintaining an extensive and uniform distribution among solutions. Two concepts, grid dominance and grid difference, were introduced to determine the mutual relationship of individuals in a grid environment. Then, three grid-based criteria, grid ranking, grid crowding distance, and grid coordinate point distance, are incorporated into the fitness of individuals to distinguish them in both the mating and environmental selection processes. GrEA uses the basic framework of NSGA-II while modifying three main steps of evolution process: fitness assignment, mating selection, and environmental selection.

HypE [Bader and Zitzler (2011)] is a hypervolume-based evolutionary many objective optimization algorithm. It applies Monte Carlo simulation to approximate the exact hypervolume value, and assigns ranks of solutions induced by the hypervolume indicator. These ranks of solutions can be used in fitness evaluation, mating selection, and environmental selection. Overall, it balances the accuracy of the estimates and the computation cost of the hypervolume calculation.

From literature, because these MOEAs apply different method to handle the difficulty of increasing objectives, each of them has some distinct advantages and disadvantages compared with others. FD-NSGA-II can ensure its convergence no matter how large the dimension of objective space but it is time consuming. MOEA/D can quickly generate a converged and well-distributed approximation front, but requiring a huge number of target vectors. The grid methods, such as e-MOEA and GrEA, can effectively deal with convergence and diversity simultaneously, but the choice of hyperbox size and grid parameter is made *ad hoc*. In HypE, hypervolume indicator is strictly monotonic with Pareto dominance [Bader and Zitzler (2011)] but the approximation of its value is not easy and influenced by the number of sampling points. From above discussions, a comprehensive comparison is necessary to arrive at a fair evaluation among these state-of-the-art MOEAs designed specifically for MaOPs.

13.4 Performance Metrics Ensemble

13.4.1 *The proposed framework*

Fig. 13.1 shows the process of Performance Metrics Ensemble proposed. The final output from the performance metrics ensemble is a ranking order of all MOEAs considered. Please note the proposed design does not provide a quantifiable performance measure for a given MOEA. Instead it attempts to rank the selected MOEAs comprehensively through a collection of performance metrics. A number of MOEAs are presented as input. Given the same initial population, each of MOEAs considered generates an approximation front. Among these approximation fronts, a winning front is selected according to a randomly chosen performance metric. To arrive at a statistically meaningful conclusion, 50 independent trials are conducted. This

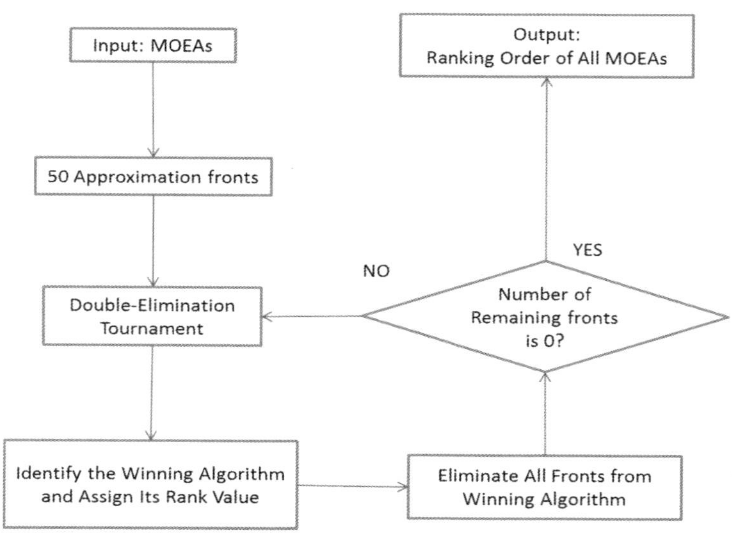

Fig. 13.1 The proposed framework for performance metrics ensemble

process results into 50 approximation fronts deriving from the MOEAs considered. A double elimination tournament selection is applied to these 50 approximation fronts and one ultimate winning approximation front will be identified. The MOEA which is responsible to this approximation front will be assigned with ranking one. This MOEA is regarded as the winning algorithm among all MOEAs participated. The approximation fronts which are generated by this winning MOEA will be removed from 50 approximation fronts. The remaining approximation fronts will then go through another round of double elimination tournament to identify the second winning MOEA with ranking two. The process will repeat until the complete ranking order of all MOEAs considered is assigned.

13.4.2 *Double elimination tournament*

The proposed double elimination tournament down selects an approximation front (as the winning front) out of all approximation fronts available using a series of binary tournament selections. In each tournament selection, a performance metric from metrics ensemble is randomly chosen for comparison.

Fig. 13.2 depicts the process of double elimination tournament in a general setting. Suppose the tournament has a pool size of N approximation fronts to begin with. The $N/2$ "qualifier" binary tournaments are held as normal, and the whole pool is divided into two parts: winner bracket contains $N/2$ winners and loser bracket $N/2$ losers. Then, in each of the bracket, $N/4$ binary tournament

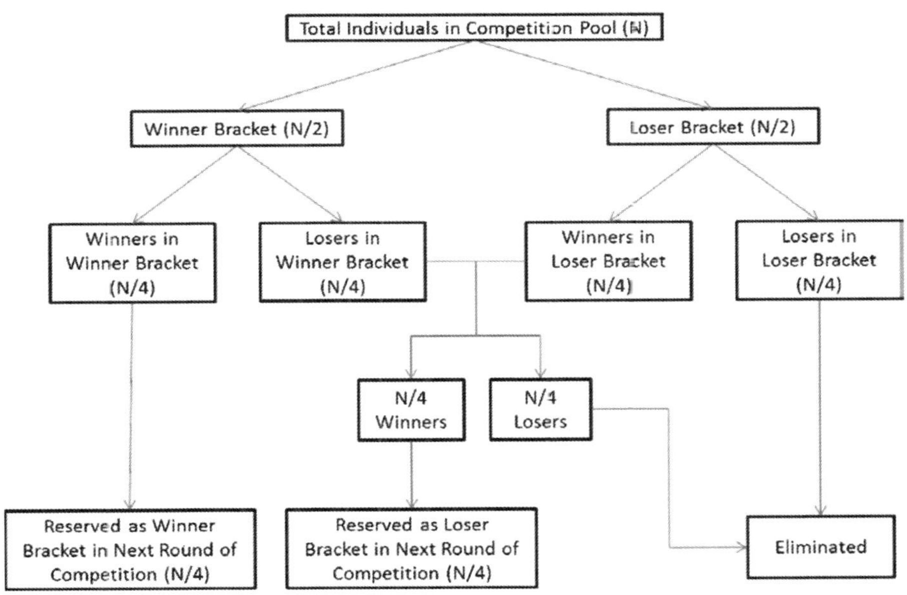

Fig. 13.2 The process of double elimination tournament

selections are competed so that each part is further divided again. In both parts, there are $N/4$ new winners and $N/4$ new losers. The $N/4$ losers from loser bracket will lose twice and be eliminated from further consideration. The $N/4$ winner from winner bracket will be reserved in winner bracket for the next round of competition. Additionally, $N/4$ losers from winner bracket and $N/4$ winners from loser bracket will be paired for binary tournaments. Specifically, one approximation front from winner bracket and one from loser bracket will be matched for a binary tournament. Afterward, we obtain $N/4$ winners which will be placed in the loser bracket for the next round of competition. Those $N/4$ losers lost twice and will be eliminated from the pool. This process reduces the total number of approximation fronts in the pool from N to $N/2$ (i.e., $N/4$ in winner bracket and $N/4$ in loser bracket). Repeat the same process; the number of candidate approximation fronts will be trimmed down from $N/2$ to $N/4$, $N/4$ to $N/8$, and eventually down to 2. The remaining two will then compete given a randomly chosen performance metric. If the one from winner bracket wins, it will be declared as the final winner. If the one from loser bracket wins, one more round of competition will be held to decide the ultimate winner. Please note if N is an odd number to begin with the double elimination tournament process, one approximation front randomly chosen will be held back and $(N-1)/2$ binary tournaments will be called. After competitions, the one that was held back will be added into both the winner bracket and loser bracket to assure it will be fully considered in the competition process.

The motivation for applying the double elimination tournament is that it gives every individual approximation front at least two chances to take part in the competition. This design would be helpful to preserve good approximation fronts. Because of the stochastic process, one quality approximation front may lose the competition if a biased performance metric is chosen. For example, for a benchmark problem with discontinuous Pareto front, performance metric UD will not offer a fair assessment. If this occurs in the single elimination tournament, a quality front could be lost forever. However, in the double elimination tournament, even an approximation front loses once; it still has an opportunity to compete and to win it all. Double elimination design allows a characteristically poor performance of a quality MOEA under the special environment still be able to win it all.

Specifically, each competing MOEA will produce an approximation front given the same initial population. One will be donned as the winner using a randomly chosen performance metric. Out of 50 independent runs, 50 approximation fronts will be resulted. Some may come from the same MOEA. It is also possible that an MOEA has no representation in the 50 approximation fronts. Out of such a large number of competitions, most likely every performance metric will be chosen multiple times to compete.

In Fig. 13.3, 25 pairs of binary tournaments will be held to result 25 winners in winner bracket and 25 losers in loser bracket. In every competition, a randomly chosen performance metric from metric ensemble will be used. In each bracket, one

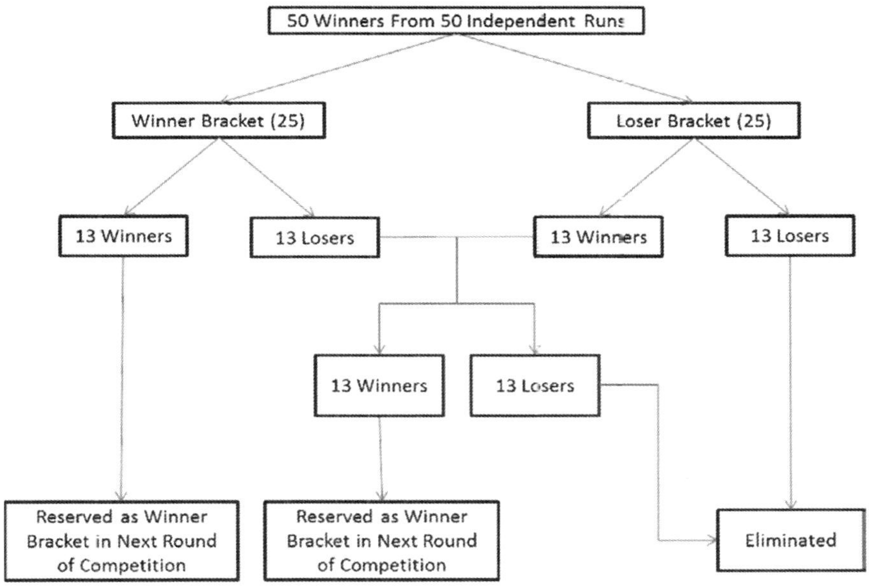

Fig. 13.3 From 50 individuals down to 26 individuals

approximation front will be randomly chosen. The remaining 24 approximation fronts will form 12 pairs of binary tournaments. The one that was held back will join both winner bracket and loser bracket to result into 13 winners and 13 losers in each bracket. Those 13 winners from winner bracket will be reserved as winners for the next round of double elimination tournament. Those 13 losers from loser bracket, which lost twice already, will be eliminated from the candidate pool. 13 losers from winner bracket and 13 winners from loser bracket are then paired to compete. 13 winners will be reserved as losers in the next round of double elimination tournament, while 13 losers which lost twice each and will be eliminated from further consideration.

In Fig. 13.4, 13 remaining approximation fronts are in the winner bracket while 13 are in the loser bracket. A similar process of double elimination tournament continues to trim down the number of approximation fronts to 7 winners and 7 losers. 14 more approximation fronts will be eliminated from the pool. In Fig. 13.5, 7 remaining approximation fronts are in the winner bracket and 7 are in the loser bracket. A similar process repeats to cut down the number of approximation fronts to 4 winners and 4 losers. 8 more approximation fronts will be eliminated from the pool. In Fig. 13.6, the process takes one more step to down select the number of approximation fronts to 2 winners and 2 losers, while in Fig. 13.7, one more step results into one winner and one loser. In Fig. 13.8, the remaining two will compete based on a randomly chosen performance metric. If the one from winner bracket wins, it will be declared as the final winner. If the one from loser bracket wins, one more round of competition will be held to decide the ultimate winner. The MOEA which is responsible to the ultimate winning approximation front will be honored as the winning MOEA with ranking order one. Please note 101 binary tournaments will be called to decide the overall winning MOEA beginning with 50 approximation fronts.

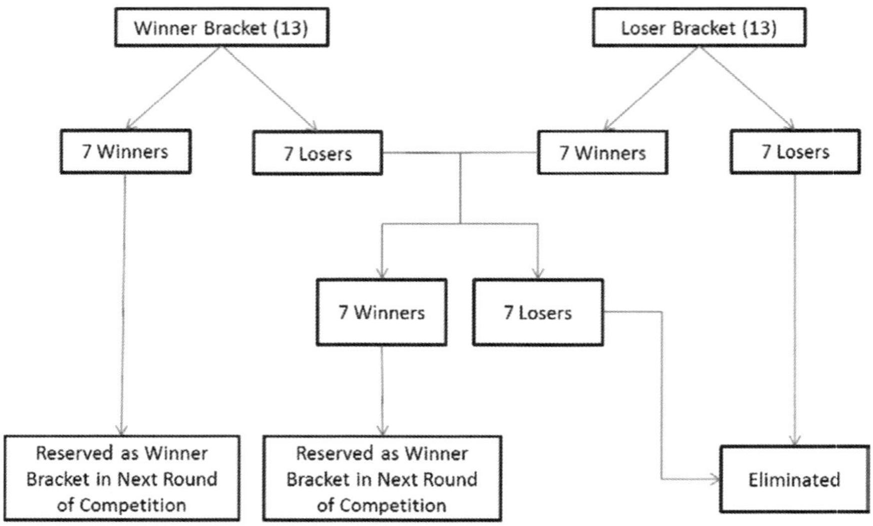

Fig. 13.4 From 26 individuals down to 14 individuals

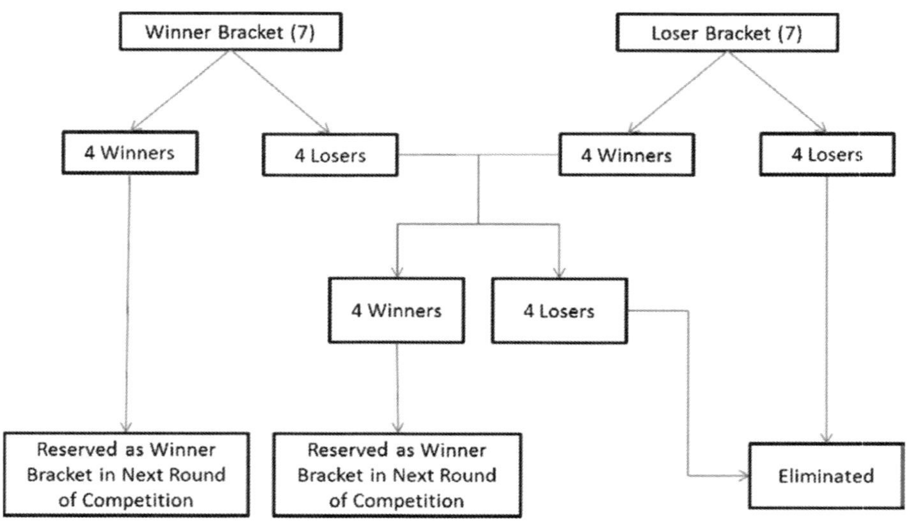

Fig. 13.5 From 14 individuals down to 8 individuals

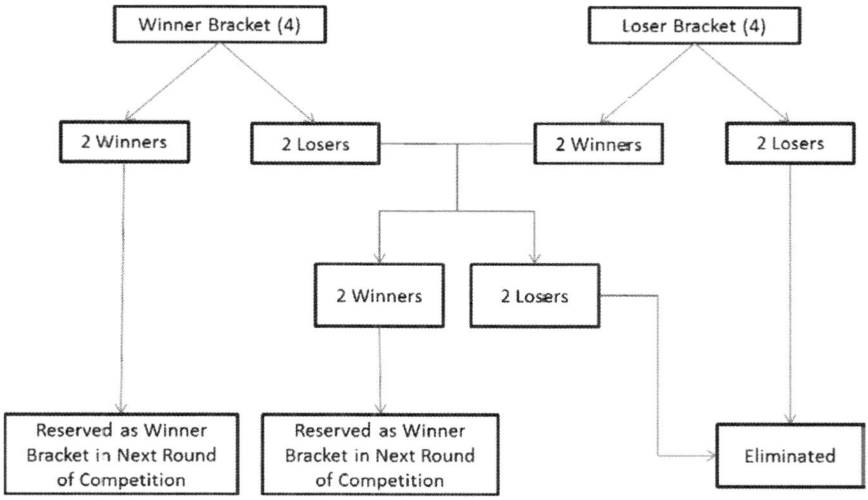

Fig. 13.6 From 8 individuals down to 4 individuals

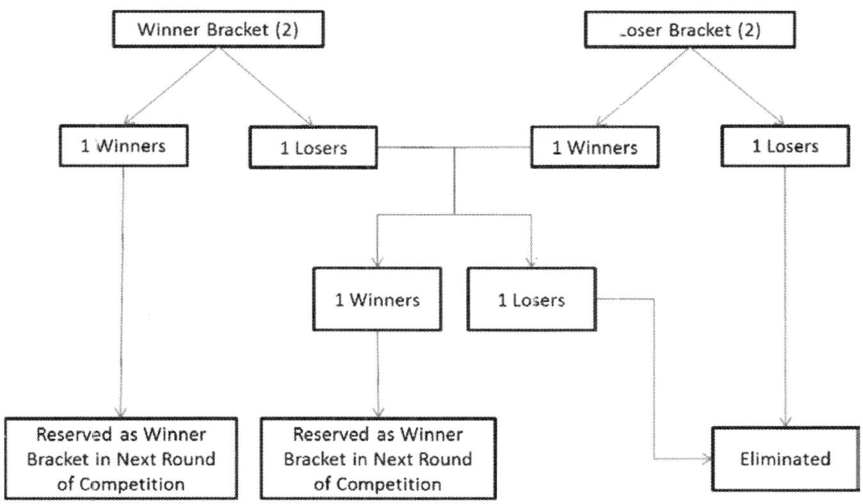

Fig. 13.7 From 4 individuals down to 2 individuals

Those approximation fronts out of original 50, if generated from the winning MOEA, will be removed from the candidate pool. The double elimination tournament process repeats until a winning approximation front is found and the MOEA which is responsible to this winning approximation front will be declared as the second winning MOEA with ranking order two. The process repeats until all MOEAs are ranked.

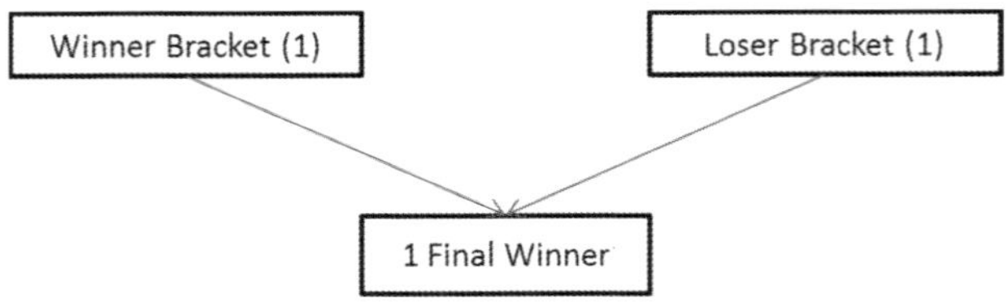

Fig. 13.8 From 2 individuals down to 1 winner

13.5 Experimental Results

13.5.1 *Selected benchmark test problems*

Two sets of widely used scalable many-objective benchmark problems are chosen to evaluate the performance of the MOEAs considered. The first set includes DTLZ1, DTLZ2, DTLZ3, DTLZ4, DTLZ5, DTLZ6, and DTLZ7 [Deb *et al.* (2002b)]; and the second has WFG1, WFG2, WFG3, WFG4, WFG5, WFG6, WFG7, WFG8, and WFG9 [Huband *et al.* (2006)]. In this experiment, chosen MOEAs are tested in five-dimension and ten-dimension objective spaces of these benchmark problems.

The DTLZ test suite contains a variety of problem characteristics that present various degrees of complications for the underlying MOEAs. These problem characteristics are summarized in Table 13.1 [Pierro *et al.* (2007)]. Specifically, "multimodal" implies there are many local fronts in the search space; "bias" means the Pareto optimal solutions are non-uniformly distributed along the global Pareto front; "disconnect" refers to the disconnected global Pareto front; and "many-to-one mapping" shows that several different variables in decision space are mapped to the same solutions in objective space.

Table 13.1 Problem characteristics of the DTLZ test suite [Huband *et al.* (2006)]

MOP	Multimodal	Bias	Disconnect	Many-to-One Mapping
DTLZ1	Yes	No	No	Yes
DTLZ2	No	No	No	Yes
DTLZ3	Yes	No	No	Yes
DTLZ4	No	Yes	No	Yes
DTLZ5	No	No	Unknown	Yes
DTLZ6	No	Yes	Unknown	Yes
DTLZ7	Yes	No	Yes	No

According to Deb (1999), DTLZ1, DTLZ3, and DTLZ6 introduce a large number of local Pareto fronts and DTLZ2 presents a spherical Pareto front so as to test MOEAs ability to converge to the global Pareto front. DTLZ4 generates a nonuniform distribution of points along the Pareto front, and therefore it challenges MOEA's ability to maintain a good distribution of solutions. In DTLZ5, its Pareto front is a degenerated hypersurface. DTLZ6 has disconnected Pareto-optimal regions and it will test an algorithm's ability to maintain subpopulations in disconnected portions of the objective space. DTLZ7 is constructed with the constrained surface approach. The Pareto front of DTLZ7 is the intersection of a straight line and a hyperplane. The tested MOEA may find it difficult to converge to the Pareto front and to maintain a good distribution of solutions along it.

WFG toolkit is flexible for creating multi-objective, real-valued, unconstrained problems. It provides a true means of assessing the performance of optimization algorithms on a wide range of problem characteristics, including nonseparable problems, deceptive problems, a truly degenerate problem, a mixed shape Pareto front problem, problems scalable in the number of position related parameters, and problems with dependencies between position- and distance-related parameters [Pierro *et al.* (2007)]. These problem features are summarized in Table 13.2 [Pierro *et al.* (2007)]. Here, "multimodal" and "bias" have the same meaning as in the Table 13.1; "separate objectives" means each objective is not related with others and every parameter of objective should be separable, and "deceptive" means the problem has at least two optima: a true optima and deceptive optima [Deb (1999)].

Table 13.2 Problem characteristics of the WFG test suite [Pierro *et al.* (2007)]

MOP	Multimodal	Bias	Separate Objectives	Deceptive
WFG1	No	Yes	Yes	No
WFG2	Yes	No	No	No
WFG3	No	No	No	No
WFG4	Yes	No	Yes	No
WFG5	No	No	Yes	Yes
WFG6	No	No	No	No
WFG7	No	Yes	Yes	No
WFG8	No	Yes	No	No
WFG9	Yes	Yes	No	Yes

According to Pierro *et al.* (2007), WFG1 employs dissimilar weights in its weighted sum reduction, making it both separable and unimodal, which is similar to WFG7. There are five problems, WFG2, WFG3, WFG6, WFG8, and WFG9, with the nonseparable reduction. The position-related parameters of WFG9 are dependent on its distance-related parameters and other position-related parameters. On the other hand, the distance-related parameters of WFG8 are dependent on its position-related parameters and other distance-related parameters and as a result the problem is nonseparable. Furthermore, Both WFG4 and WFG9 involve the multimodality. The deceptiveness of WFG5 in both position- and distance-related parameters is more difficult than that of WFG9 which is only deceptive on its position parameters.

13.5.2 *Selected performance metrics*

In this experiment, four different performance metrics are chosen. inverted generational distance (IGD) concerns how well is the Pareto-optimal front represented by the obtained approximation front [Czyzzak and Jaszkiewicz (1998)]; spacing measures how evenly the evolved solutions distribute themselves [Schott (1995)]; maximum spread (MS) measures how well the true Pareto front is covered by the approximation set [Zitzler *et al.* (2000)], and hypervolume indicator (also called S-metric) considers both closeness and diversity of obtained approximation front by calculating hypervolume [Van Veldhuizen (1999); Zitzler and Thiele (1998)].

The less the IGD and spacing values, the better the algorithm's performance; the more the MS and hyperarea values, the better the algorithm's performance. Please note binary performance metrics, such as ε-indicator or C-metric, can be easily adopted into the proposed design since binary tournament is used here as a baseline.

13.5.3 *Parameter setting in experiment*

The population size in all six MOEAs is set to be 100 for all test instances. The stopping criterion is set at 200 generations. Initial populations are generated by uniformly random sampling from the search space in all MOEAs considered.

The simulated binary crossover (SBX) and polynomial mutation are used. The crossover operator generates one offspring, which is then modified by the mutation operator. Following the practice in Breiman (1996), the distribution indexes in SBX and the polynomial mutation are set to be 20. The crossover rate is 1.00, while the mutation rate is $1/m$ and m is the number of decision variables. For GrEA and ε-MOEA, the parameter setting follows Yang *et al.* (2013). In MOEA/D [Zhang and Li (2007)], the number of the weight vectors in the neighborhood of each weight vector T is set to be 10. For HypE, according to Bader and Zitzler (2011), 10,000 sampling points are used.

In S-metric, we define the reference point according to Auger *et al.* (2009) for DTLZ1-DTLZ7. That is, the reference point with the integer value 110 for each objective. For WFG1-WFG9, choose the reference point with its value slightly larger than the worst value of all the obtained solution sets in each objective.

13.5.4 *Experiment results*

In this subsection, we demonstrate the whole process of performance metric ensemble through benchmark function DTLZ1 in five objectives.

1) **Preliminary Iteration**
   This step generates 50 approximation fronts as the initial population for double elimination tournament. In these 50 winning fronts, FD-NSGA-II wins 10 times, MOEA/D wins 11 times, ε-MOEA wins once, MO-NSGA-II wins 10

times, GrEA wins 17 times, and HypE wins once. During the competitions, performance metric is randomly chosen from four available metrics. In summary, IGD is used 12 times, Spacing 12 times, S-metric 12 times, and MS 14 times.

2) **Iteration 1**
This is the first round in the double elimination tournament that 50 fronts are competed to survive 26 in the candidate pool. In these 26 fronts, FD-NSGA-II wins six times, MOEA/D wins seven times, MO-NSGA-II wins seven times, and GrEA wins six times. Note that ε-MOEA and HypE are completely eliminated without any hope to win the overall competition.
The 50 approximation fronts are first paired into 25 binary tournaments. In this process, IGD is chosen four times, Spacing eight times, S-metric eight times, and MS five times, respectively.
The remaining process involves 12 binary tournaments to generate 13 winners reserved in a winner bracket for the next iteration, 12 binary tournaments to generate 13 losers to be eliminated from further consideration, and 13 binary tournaments to generate 13 fronts reserved in a loser bracket for the next iteration. Altogether 37 binary tournaments are held using 37 randomly chosen performance metrics: IGD is used nine times, Spacing 12 times, S-metric seven times, and MS nine times.

3) **Iteration 2**
In the second round of double elimination tournament, 26 remaining approximation fronts are trimmed down to 14 in the candidate pool. In these remaining 14 fronts, FD-NSGA-II wins twice, MOEA/D wins five times, MO-NSGA-II wins four times, and GrEA wins three times.
Out of 19 binary tournaments held in this iteration, performance metric of IGD is used three times, Spacing four times, S-metric eight times, and MS four times.

4) **Iteration 3**
In the third round of double elimination tournament, 14 approximation fronts remained in the pool are once again down select to eight fronts. In these eight fronts, FD-NSGA-II wins once, MOEA/D wins three times, MO-NSGA-II wins three times, and GrEA wins once.
Out of 10 binary tournaments called in this iteration, IGD is chosen four times, Spacing four times, S-metric once, and MS once.

5) **Iteration 4**
It is the fourth round of double elimination tournament; eight approximation fronts will be further reduced down to four survivors. In these four fronts, FD-NSGA-II wins once, MOEA/D wins once, and MO-NSGA-II wins twice. Note that GrEA is eliminated in this iteration to be the ultimate winner.
Out of six binary tournaments held in this iteration, Spacing three times, S-metric twice, and MS once. Here, IGD has never been chosen.

6) **Iteration 5**
   In the fifth round of double elimination tournament, four approximation fronts are trimmed one more time to two in the candidate pool. In these two remaining fronts, MO-NSGA-II wins once and MOEA/D wins once. Here, FD-NSGA-II is eliminated in this iteration to be the ultimate winner.
   Out of three competitions, performance metrics Spacing is selected once, S-metric once, and MS once.
7) **Iteration 6**
   In the final round of double elimination tournament, the ultimate winner is to be identified. The final winner is MOEA/D and performance metric, S-metric, is chosen to compare.
8) **Iteration 7**
   Removing all the fronts (i.e., 11) generated by MOEA/D out of 50 approximation fronts, the remaining 39 fronts continue through double elimination tournaments. MO-NSGA-II is identified as the second winner with respect to the benchmark function, five-dimension DTLZ1. FD-NSGA-II is the third winner. When the proposed framework complete, the resulting ranking order shows: rank 1-MOEA/D, rank 2-MO-NSGA-II, rank 3-FD-NSGA-II, rank 4-GrEA, rank 5-ε-MOEA, and rank 6-HypE for the benchmark function five-dimension DTLZ1.

Table 13.3 provides a complete ranking order for all benchmark functions considered in five-dimension and ten-dimension. Please note each ranking order is derived from multiple, independent runs and the ranking order has always been consistent.

It is interesting to note that the number of times each MOEA wins in the preliminary iteration is not necessarily related to its final ranking order. Even one MOEA wins many times here, it may not be the final winner at the end. For example, in 5-objective WFG1, although FD-NSGA-II wins 18 times in the preliminary round, which is much more than others; it loses in the final round to ε-MOEA. Also, in 10-objective WFG2, although MOEA/D wins 18 times in the preliminary iteration, which is much more than that of FD-NSGA-II, it loses in the final round of competition to FD-NSGA-II. Moreover, in 10-objective WFG6, although MOEA/D wins 10 times in the preliminary iteration, it has the worst ranking order at the end. On the other hand, even one MOEA wins very few times in the preliminary round, it can also be the final winner. For instance, in 10-objective DTLZ4, despite GrEA only win twice in preliminary iteration, it wins all at the final stage. Similarly, in 5-objective DTLZ6, MOEA/D only wins twice in preliminary round, but wins all at the final stage.

It is also worthy of noting that if an inappropriate performance metric is chosen alone to judge the performance of MOEAs, the result will not be fair. For example, in 10-objective DTLZ1, if IGD was used to compare GrEA and MOEA/D, MOEA/D would win the close competition. However, if the other metric (e.g., Spacing, S-metric, or MS) was used instead, GrEA will be favored as oppose to

Table 13.3 Ranking orderings for all benchmark problems

	Rank 1	Rank 2	Rank 3	Rank 4	Rank 5	Rank 6
DTLZ1 (5-D)	MOEA/D	MO-NSGA-II	FD-NSGA-II	GrEA	ε-MOEA	HypE
DTLZ1(10-D)	FD-NSGA-II	GrEA	MO-NSGA-II	MOEA/D	ε-MOEA	HypE
DTLZ2(5-D)	MO-NSGA-II	FD-NSGA-II	GrEA	ε-MOEA	MOEA/D	HypE
DTLZ2(10-D)	FD-NSGA-II	GrEA	ε-MOEA	MO-NSGA-II	MOEA/D	HypE
DTLZ3(5-D)	MO-NSGA-II	GrEA	FD-NSGA-II	ε-MOEA	MOEA/D	HypE
DTLZ3(10-D)	GrEA	FD-NSGA-II	MOEA/D	MO-NSGA-II	HypE	ε-MOEA
DTLZ4(5-D)	GrEA	MOEA/D	FD-NSGA-II	ε-MOEA	MO-NSGA-II	HypE
DTLZ4(10-D)	GrEA	MOEA/D	FD-NSGA-II	ε-MOEA	MO-NSGA-II	HypE
DTLZ5(5-D)	ε-MOEA	MO-NSGA-II	GrEA	MOEA/D	FD-NSGA-II	HypE
DTLZ5(10-D)	ε-MOEA	FD-NSGA-II	MO-NSGA-II	MOEA/D	HypE	GrEA
DTLZ6(5-D)	MOEA/D	ε-MOEA	FD-NSGA-II	MO-NSGA-II	GrEA	HypE
DTLZ6(10-D)	FD-NSGA-II	ε-MOEA	MOEA/D	MO-NSGA-II	GrEA	HypE
DTLZ7(5-D)	MOEA/D	FD-NSGA-II	GrEA	ε-MOEA	HypE	MO-NSGA-II
DTLZ7(10-D)	GrEA	MO-NSGA-II	FD-NSGA-II	HypE	ε-MOEA	MOEA/D
WFG1 (5-D)	ε-MOEA	FD-NSGA-II	HypE	MOEA/D	MO-NSGA-II	GrEA
WFG1(10-D)	ε-MOEA	FD-NSGA-II	GrEA	MO-NSGA-II	MOEA/D	HypE
WFG2(5-D)	MOEA/D	MO-NSGA-II	GrEA	FD-NSGA-II	HypE	ε-MOEA
WFG2(10-D)	FD-NSGA-II	MO-NSGA-II	MOEA/D	ε-MOEA	HypE	GrEA
WFG3(5-D)	MOEA/D	GrEA	FD-NSGA-II	HypE	MO-NSGA-II	ε-MOEA
WFG3(10-D)	GrEA	ε-MOEA	FD-NSGA-II	MOEA/D	MO-NSGA-II	HypE
WFG4(5-D)	MOEA/D	FD-NSGA-II	MO-NSGA-II	HypE	ε-MOEA	GrEA
WFG4(10-D)	FD-NSGA-II	HypE	GrEA	MO-NSGA-II	MOEA/D	ε-MOEA
WFG5(5-D)	MO-NSGA-II	MOEA/D	HypE	GrEA	ε-MOEA	FD-NSGA-II
WFG5(10-D)	HypE	FD-NSGA-II	MOEA/D	GrEA	ε-MOEA	MO-NSGA-II
WFG6(5-D)	FD-NSGA-II	GrEA	ε-MOEA	MOEA/D	HypE	MO-NSGA-II
WFG6(10-D)	GrEA	FD-NSGA-II	HypE	ε-MOEA	MO-NSGA-II	MOEA/D
WFG7(5-D)	MO-NSGA-II	FD-NSGA-II	MOEA/D	GrEA	ε-MOEA	HypE
WFG7(10-D)	HypE	GrEA	MO-NSGA-II	FD-NSGA-II	MOEA/D	ε-MOEA
WFG8(5-D)	MO-NSGA-II	HypE	MOEA/D	ε-MOEA	GrEA	FD-NSGA-II
WFG8(10-D)	FD-NSGA-II	GrEA	MO-NSGA-II	HypE	ε-MOEA	MOEA/D
WFG9(5-D)	GrEA	HypE	MO-NSGA-II	MOEA/D	ε-MOEA	FD-NSGA-II
WFG9(10-D)	ε-MOEA	GrEA	FD-NSGA-II	MOEA/D	MO-NSGA-II	HypE

MOEA/D. According to Yen and He (2014), when the test function is more complex, the proposed metric ensemble provides a much more rational result than that of any metric alone. This is simply due to the drawback that any individual metric alone can only quantify one aspect of performance measure.

Although we have exact ranking order for each benchmark problems, multiple MOEAs may have very similar performance in some problems. For example, in 5-objective DTLZ2, both GrEA and FD-NSGA-II have nearly equal performance with slight edge goes to FD-NSGA-II. Moreover, in 5-objective DTLZ3, GrEA, FD-NSGA-II, and ε-MOEA have very similar performance. They are not easy to be differentiated under each of these four chosen metric individually. Only metrics ensemble and double elimination method can offer a clear distinction among them. Furthermore, in 10-objective DTLZ7, all MOEAs share nearly equal performance; there is no significant difference among their approximation fronts by any of the single metric alone. The proposal performance metrics ensemble provides a clear judgment about their ranking order. Finally, in 10-objective WFG8, from both results of preliminary iteration and double elimination process, three MOEAs, ε-MOEA, MOEA/D, and HypE, share very close performance. On the other hand, one or two MOEAs perform much better or worse than others in some problems. For instance, in 10-objective DTLZ3, ε-MOEA performs much worse than other MOEAs. This is also confirmed by reference [Yang *et al.* (2013)]. Then, in 5-objective DTLZ7, MO-NSGA-II is much worse than others in this problem. In 10-objective WFG1, ε-MOEA and FD-NSGA-II perform much better than others, which is confirmed in both preliminary iteration and final ranking result. Finally, in 5-objective WFG3, MOEA/D performs much better than others, especially in the double elimination process.

13.5.5 *Observations and insights*

1) Observation from characteristics of benchmark problems

First, for benchmark problems with multimodal Pareto fronts, e.g., DTLZ1, DTLZ3, DTLZ7, WFG2, WFG4, WFG5, and WFG9, the two decomposition-based methods, MO-NSGA-II and MOEA/D, show better performance in nearly all of these five-dimensional multimodal problems except WFG9. However, both MOEAs cannot get very good result in ten-dimensional multimodal problems. On the other hand, FD-NSGA-II, which is based on fuzzy Pareto dominance, performs better in many of ten-dimensional problems, but gives poor performance in some of five-dimensional problems. One thing should be paid attention is that FD-NSGA-II does bad in five-dimensional WFG5 and WFG9, both of which involves deceptive characteristics; therefore, FD-NSGA-II cannot handle problems combining difficulties with both multimodal and deceptive. Also, in five-dimensional WFG2 and WFG9 with non-separate objectives, FD-NSGA-II cannot give the satisfactory results. Then, in low-dimension problems, the correlation between each objective will impact on the performance of FD-NSGA-II to deal with multimodal.

Among above multimodal problems, there are two specific problems with disconnected Pareto front: DTLZ7 and WFG2. Here, HypE does very poorly in both of them while it shows quality performance in other problems. Therefore, HypE cannot handle problems with multimodality in disconnected condition.

The second characteristic of benchmark problems is bias in the Pareto front. There are six problems with this type of difficulty: DTLZ4 DTLZ6, WFG1, WFG7, WFG8, and WFG9. FD-NSGA-II has a good performance in DTLZ6, WFG1, WFG7, and WFG8; all of these four problems have a common point: there are unimodal. However, FD-NSGA-II does poorly in five-dimensional WFG8 and WFG9 with non-separate objectives. Therefore, when FD-NSGA-II is used to solve bias problems, it will have better performance in unimodal condition, while show worse for problems with non-separate objectives. On the other hand, HypE performs better in the same non-separate objectives problems, WFG8 and WFG9, while it does a poor job in other problems.

For the third characteristic, geometry of Pareto front, there are convex, linear, and concave. Convex front exists in both WFG1 and WFG2. For both problems, FD-NSGA-II always has a good performance while HypE always performs poorly. The linear front is in another two problems: DTLZ1 and WFG3. Again, HypE gives bad result in both problems. For concave problems, there is no useful observation we can reveal.

2) Observation from characteristics of MOEAs

First, during the evolution process of ε-MOEA, both grid method and relax form of Pareto dominance are used. Based on both mechanisms, ε-MOEA performs very well in DTLZ5, DTLZ6, and WFG1, where DTLZ5 tests an MOEA's ability to converge to a degenerated curve while DTLZ6 and WFG1 test an MOEA's ability to maintain subpopulation in different disconnected Pareto-optimal regions. On the other hand, we have found that GrEA which is also based on grid approach has a poor performance in both DTLZ5 and DTLZ6. FD-NSGA-II which is also a relax form of Pareto dominance, however, performs well in both problems. Therefore, it appears that the relax form of Pareto dominance can help to solve problems with disconnected Pareto-optimal regions and degenerate curve.

Second, GrEA, which is a grid based method, shows a good performance in both DTLZ3 and DTLZ4, where DTLZ3 has many local Pareto-optimal fronts while DTLZ4 investigates an MOEA's ability to maintain a good distribution of solutions. Also, GrEA performs well in both ten-dimension DTLZ1 and DTLZ7, but poor in five-dimensional problems. On the other hand, another grid based method, ε-MOEA, has very poor performance in all these problems. Based on the above discussions, grid based method and its extending selection approach (as in the selection process of GrEA instead of Pareto dominance) help to solve MOPs with multiple modalities.

Finally, both grid based MOEAs have very dissimilar performance in problems with bias. Here, GrEA performs better in bias problems with concave shape while ε-MOEA is better for other bias problems.

FD-NSGA-II, which is based on fuzzy Pareto dominance, performs better in ten-dimension problems, but poorly in five-dimension problems, except for DTLZ7 and WFG7. This implies the fuzzy fitness assignment is more effective when the number of objective is large.

On the other hand, the two decomposition-based methods, MO-NSGA-II and MOEA/D, show better performance in five-dimension problems than in ten-dimension problems, except for MO-NSGA-II in DTLZ7 and WFG6, and MOEA/D in DTLZ3. Therefore, the decomposition method is better suited in lower dimensional MOPs. Furthermore, the running time of MOEA/D is much less than others. If MOEA/D has similar performance with the other algorithms, it should be chosen favorably.

Once again, our observations confirmed with No Free Lunch Theorem [Wolpert and Macready (1997)]: any algorithm's elevated performance over one class of problems is exactly paid for in loss over another class.

For the same benchmark problem, e.g., DTLZ7 and WFG4, the ranking order in a five-dimension problem is not necessary the same as the order of the corresponding ten-dimension one. It seems to imply that the performance of an MOEA to solve many-objective optimization problem depends on two critical aspects: the ability of MOEA to tackle the specific problem characteristics and the ability of MOEA to handle high-dimensional objective space. This explains why MOEAs using similar design principle show different performances given the same optimization problem. For example, in DTLZ4 and WFG5, the performance of MOEA/D and MO-NSGA-II are fairly different.

13.6 Conclusions

In literature, we have witnessed a growing number of studies devoted for MOEA. When an MOEA was proposed, a number of benchmark problems are often chosen ad hoc to quantify the performance. Given a set of heuristically chosen performance metrics, the proposed MOEA and some state-of-the-art competitors are evaluated statistically given a large number of independent trials. The conclusion, if any been drawn, is often indecisive and reveals no additional insight pertaining to the specific problem characteristics that the proposed MOEA would perform the best. On the other end, when an MOP application with real-world complications arises, we often have no clue which MOEA should be chosen to attain the best opportunity to be successful. When an MOEA was proposed in literature, no insight in this regard has even been offered.

To address this concern, an ensemble method on performance metrics is proposed in this chapter, knowing no single metric alone can faithfully quantify the performance of a given design under real-world scenarios. A collection of performance metrics, measuring the spread across the Pareto-optimal front and the ability to attain the global trade-off surface closeness, could be incorporated into the ensemble approach.

A double elimination tournament selection operator is proposed to compare approximation fronts obtained from different MOEAs in a statistically meaningful way. The double elimination design avoids the risk of a quality MOEA from been easily eliminated due to an unfair assessment from a performance metric chosen. This design allows a comprehensive measure and more importantly reveals additional insight pertaining to specific problem characteristics that the underlying MOEA could perform the best. For a given real-world problem if we know its problem characteristics (e.g., a Pareto front with a number of disconnected segments and a high number of local optima), we may make an educated judgment to choose the specific MOEA for its superior performance given the problem characteristics. Please note the proposed design does not provide an independent, quantifiable performance measure for a given problem. Instead, it attempts to rank the selected MOEAs comprehensively through a collective performance metrics.

In summary, this study is based on the observations that "indecisive" or "inconclusive" findings are often produced when MOEAs are compared on specific test functions and those insights into how to match MOEAs to problems for which they are most suitable are thus lacking from the literature. The authors understand that appreciable progresses have been made in recent years to better understand some of the properties of performance metrics qualitatively and quantitatively. However, it is also commonly agreed upon that this process of fundamental works will take years to mature. Meanwhile, a compromising and empirical strategy is proposed here to gain additional insights and to move forward in solving difficult problems with real-world complications.

References

Auger, A., Bader, J., Brockhoff, D. and Zzitzler, E. (2009). Theory of the hypervolume indicator: Optimal μ-distributions and the choice of the reference point, in *Proceedings of ACM SIGEVO Workshop on Foundations of Genetic Algorithms* (Orlando, USA), pp. 87–102.

Bader, J. and Zitzler, E. (2011). Hype: An algorithm for fast hypervolume based many-objective optimization, *Evolutionary Computation* **19**, 1, pp. 45–76.

Batista, L., Campelo, F., Guimaraes, F. and Ramirez, J. (2010). A comparison of dominance criteria in many-objective optimization problems, in *Proceedings of the IEEE Congress on Evolutionary Computation* (Barcelona, Spain), pp. 2359–2366.

Beume, N., Naujoks, B. and Emmerich, M. (2007). SMS-EMOA: Multiobjective selection based on dominated hypervolume, *European Journal of Operational Research* **181**, 3, pp. 1653–1669.

Breiman, L. (1996). Bagging predictors, *Machine Learning* **24**, 2, pp. 123–140.

Czyzzak, P. and Jaszkiewicz, A. (1998). Pareto simulated annealing-a metaheuristic technique for multiple-objective combinatorial optimization, *Journal of Multi-Criteria Decision Analysis* **7**, pp. 34–47.

Deb, K. (1999). Multi-objective genetic algorithms: Problem difficulties and construction of test problems, *Evolutionary Computation* **7**, 3, pp. 205–230.

Deb, K. and Jain, H. (2012a). An improved NSGA-II procedure for many-objective optimization, part I: Solving problems with box constraints, Technical Report KanGAL-Report 2012009, Indian Institute of Technology, Kanpur, India.

Deb, K. and Jain, H. (2012b). An improved NSGA-II procedure for many-objective optimization, part II: Handling constraints and extending to an adaptive approach, Technical Report KanGAL-Report 2012010, Indian Institute of Technology, Kanpur, India.

Deb, K. and Jain, H. (2012c). Handling many-objective problems using an improved NSGA-II procedure, in *Proceedings of the IEEE Congress on Evolutionary Computation* (Brisbane, Australia), pp. 1–8.

Deb, K., Mohan, M. and Mishra, S. (2005). Evaluating the ε-domination based multi-objective evolutionary algorithm for a quick computation of Pareto-optimal solutions, *Evolutionary Computation* **13**, 4, pp. 501–525.

Deb, K., Pratap, A., Agrawal, S. and Meyarivan, T. (2002a). A fast and elitist multiobjective genetic algorithm: NSGA-II, *IEEE Transactions on Evolutionary Computation* **6**, 2, pp. 182–197.

Deb, K., Thiele, L., Laumanns, M. and Zitzler, E. (2002b). Scalable multiobjective optimization test problems, in *Proceedings of the IEEE Congress on Evolutionary Computation* (Honolulu, USA), pp. 825–830.

di Pierro, F., Khu, S. T. and Savić, D. A. (2007). An investigation on preference order ranking scheme for multiobjective evolutionary optimization, *IEEE Transactions on Evolutionary Computation* **11**, 1, pp. 17–45.

Goh, C. K. and Tan, K. C. (2009). A competitive-cooperative coevolutionary paradigm for dynamic multiobjective optimization, *IEEE Transactions on Evolutionary Computation* **13**, 1, pp. 103–127.

Hadka, D. and Reed, P. (2012). Diagnostic assessment of search controls and failure modes in many-objective evolutionary optimization, *Evolutionary Computation* **20**, 3, pp. 423–452.

Hansen, M. P. and Jaszkiewicz, A. (1998). Evaluating the quality of approximations to the nondominated set, Technical Report IMM-REP-1998-7, Technical University of Denmark, Kongens Lyngby, Denmark.

He, Z., Yen, G. G. and Zhang, J. (2013). Fuzzy-based Pareto optimality for many-objective evolutionary algorithms, *IEEE Transactions on Evolutionary Computation* doi: 10.1109/TEVC.2013.2258025.

Ho, T. K. (1998). The random subspace method for constructing decision forests, *IEEE Transactions on Pattern Analysis and Machine Intelligence* **20**, 8, pp. 832–844.

Hoeting, J. A., Madigan, D., Raftery, A. E. and Volinsky, C. T. (1999). Bayesian model averaging: A tutorial, *Statistical Science* **14**, 4, pp. 382–401.

Huband, S., Hingston, P., Barone, L. and While, L. (2006). A review of multiobjective test problems and a scalable test problem toolkit, *IEEE Transactions on Evolutionary Computation* **10**, 5, pp. 477–506.

Hughes, E. (2008). Fitness assignment methods for many-objective problems, in J. Knowles, D. Corne, and K. Deb (eds.), *Multiobjective Problem Solving from Nature: From Concepts to Applications* (Springer, Berlin), pp. 307–329.

Ishibuchi, H., Tsukamoto, N. and Nojima, Y. (2008) Evolutionary many-objective optimization: A short review, in *Proceedings of the IEEE Congress on Evolutionary Computation* (Hong Kong, China), pp. 2419–2426.

Knowles, J. and Corne, D. (2002). On metrics for comparing nondominated sets, in *Proceedings of the IEEE Congress on Evolutionary Computation* (Honolulu, USA), pp. 711–716.

Kuncheva, L. and Whitaker, C. (2003). Measures of diversity in classifier ensembles and their relationship with the ensemble accuracy, *Machine Learning* **51**, 2, pp. 181–207.

Li, M., Yang, S., Liu, X. and Shen, R. (2013). A comparative study on evolutionary algorithms for many-objective optimization, in *Proceedings of the International Conference on Evolutionary Multi-Criterion Optimization* (Sheffield, UK), pp. 261–275.

Pierro, F., Khu, S. and Savic, D. (2007). An investigation on preference order ranking scheme for multiobjective evolutionary optimization, *IEEE Transactions on Evolutionary Computation* **11**, 1, pp. 17–45.

Polikar, R. (2006). Ensemble based systems in decision making, *IEEE Circuits and Systems Magazine* **6**, 3, pp. 21–45.

Saxena, D.K., Zhang, Q., Duro, J.A. and Tiwari, A. (2011). Framework for many-objective test problems with both simple and complicated Pareto-set shapes, in *Proceedings of International Conference on Evolutionary Multi-Criterion Optimization* (Ouro Preto, Brazil), pp. 197–211.

Schapire, R. E., Freund, Y., Barlett, P. and Lee, W. S. (1998). Boosting the margin: A new explanation for the effectiveness of voting methods, *The Annals of Statistics* **26**, 5, pp. 1651–1686.

Schott, J. R. (1995). *Fault Tolerant Design Using Single and Multicriteria Genetic Algorithm Optimization*, Master's Thesis, Massachusetts Institute of Technology, USA.

Singh, H., Isaacs, A. and Ray, T. (2011). A Pareto corner search evolutionary algorithm and dimensionality reduction in many-objective optimization problems, *IEEE Transactions on Evolutionary Computation* **15**, 4, pp. 539–553.

Tan, K. C., Lee, T. H. and Khor, E. F. (2002). Evolutionary algorithms for multi-objective optimization: Performance assessments and comparisons, *Artificial Intelligence Review* **17**, 4, pp. 253–290.

Tessema, B. and Yen, G. G. (2009). An adaptive penalty formulation for constrained evolutionary optimization, *IEEE Transactions on Systems, Man and Cybernetics, Part A: Systems and Humans* **39**, 3, pp. 565–578.

Van Veldhuizen, D. A. (1999). *Multiobjective Evolutionary Algorithms: Classifications, Analyses, and New Innovations*, Ph.D. Dissertation, Air Force Institute of Technology, Wright-Patterson AFB, USA.

Van Veldhuizen, D. A. and Lamont, G. B. (2000). On measuring multiobjective evolutionary algorithm performance, in *Proceedings of the IEEE Congress on Evolutionary Computation* (La Jolla, USA), pp. 204–211.

Wolpert, D. H. (1992). Stacked generalization, *Neural Networks* **5**, 2, pp. 241–259.

Wolpert, D. H. and Macready, W. G. (1997). No free lunch theorems for optimization, *IEEE Transactions on Evolutionary Computation* **1**, 1, pp. 67–82.

Wu, J. and Azarm, S. (2001). Metrics for quality assessment of a multiobjective design optimization solution set, *Journal of Mechanical Design* **123**, 1, pp. 18–25.

Yang, S., Li, M., Liu, X. and Zheng, J. (2013). A grid-based evolutionary algorithm for many-objective optimization, *IEEE Transactions on Evolutionary Computation* **17**, 5, pp. 721–736.

Yen, G. G. and He, Z. (2014). Performance metric ensemble for multiobjective evolutionary algorithms, *IEEE Transactions on Evolutionary Computation* **18**, 1, pp. 131–144.

Yen, G. G. and Lu, H. (2003). Dynamic multiobjective evolutionary algorithm: Adaptive cell-based rank and density estimation, *IEEE Transactions on Evolutionary Computations* **7**, 3, pp. 253–274.

Zhang, Q. and Li, H. (2007). MOEA/D: A multiobjective evolutionary algorithm based on decomposition, *IEEE Transactions on Evolutionary Computation* **11**, 6, pp. 712–731.

Zitzler, E. (1999). *Evolutionary Algorithms for Multiobjective Optimization: Methods and Applications*, Ph.D. Dissertation, Swiss Federal Institute of Technology, Zurich, Switzerland.

Zitzler, E., Laumanns, M. and Thiele, L. (2001). SPEA2: Improving the strength Pareto evolutionary algorithm, Technical Report TIK-Report 103, Swiss Federal Institute of Technology, Zurich, Switzerland.

Zitzler, E. and Simon, K. (2004). Indicator-based selection in multiobjective search, in *Proceedings of International Conference on Parallel Problem Solving form Nature* (Birmingham, UK), pp. 832–842.

Zitzler, E. and Thiele, L. (1998). Multiobjective optimization using evolutionary algorithms-a comparative case study, in *Proceedings of the International Conference on Parallel Problem Solving from Nature* (Amsterdam, Netherlands), pp. 292–301.

Zitzler, E., Thiele, L. and Bader, J. (2010). On set-based multiobjective optimization, *IEEE Transactions on Evolutionary Computation* **14**, 1, pp. 58–79.

Zitzler, E., Deb, K. and Thiele, L. (2000). Comparison of multiobjective evolutionary algorithms: Empirical results, *Evolutionary Computation* **8**, 2, pp. 173–195.

Zitzler, E., Thiele, L., Laumanns, M., Foneseca, C. M. and Da Fonseca, V. G. (2003). Performance assessment of multiobjective optimizers: An analysis and review, *IEEE Transactions on Evolutionary Computation* **7**, 2, pp. 117–132.

Chapter 14

Synchronization Control of Memristive Chaotic Circuits and Their Applications in Image Encryptions

Shiping Wen and Zhigang Zeng

Abstract This chapter investigates the synchronization control problem of memristive chaotic circuits and their applications in image encryptions. First, several memristive chaotic circuits such as the memrisitve Chua's circuit, and memristive Lorenz circuit are designed. Then, considering the state-dependent properties of the memristor, a new fuzzy model employing parallel distributed compensation (PDC) gives a new way to analyze the complicated memristive circuits with only two subsystems. Several examples are also given to illustrate the effectiveness and potential applications in image encryption of the results.

14.1 Introduction

Memristor is considered to be the missing 4th passive circuit element, as Prof. Chua postulated in 1971 [Chua (1971)]. And it took scientists almost 40 years to invent such a practical device until a team at Hewlett-Packard Labs announced the development of a memristor [Strukov *et al.* (2008)], by titanium dioxide thin films. Memristor takes its place along other three existing elements such as the resistor, capacitor, and inductor, and shares many properties of resistors such as the unit of measurement—ohms. Many attentions have been attracted to this novel device for its resistance upon turning off the power source, in other words, it depends on the integral of its entire past current waveform. Therefore, memristor is nonlinear and shows many special properties, and can be described by a nonlinear constitutive relation [Itoh and Chua (2008)].

$$v = M(q)i, \quad \text{or} \quad i = W(\varphi)v,$$

between the device terminal voltage v and terminal current i, where $\varphi = \int v\mathrm{d}t$

$$M(q) = \frac{\mathrm{d}\varphi(q)}{\mathrm{d}q}, \text{ or } W(\varphi) = \frac{\mathrm{d}q(\varphi)}{\mathrm{d}\varphi}.$$

$M(q)$ and $W(\varphi)$ are called the memristance and memductance, respectively, and the piece-wise linear functions $q(\varphi)$, $W(\varphi)$ are given by

$$q(\varphi) = b\varphi + 0.5(a-b)(|\varphi+1| + |\varphi-1|)$$
$$W(\varphi) = \frac{\mathrm{d}q(\varphi)}{\mathrm{d}\varphi} = \begin{cases} a, & |\varphi| < 1, \\ b, & |\varphi| > 1. \end{cases}$$

At present, research on circuits based on memristor becomes a focal topic for research. Chaotic system based on memristor also attracts attention recently [Bao *et al.* (2010)]. And memristive system has been applied in image encryption [Lin and Wang (2010)].

Many approaches have been studied for the synchronization of chaotic systems such as impulsive control, feedback control and so on. Most are based on the exactly knowing of the system structure and parameters, and in practice, some or all the system's parameters may be not available to the designer of synchronization device. There are a lot of works been preceded to solve this problem using adaptive synchronization [Chen *et al.* (2012); Lee *et al.* (2010); Lin *et al.* (2012); Sudheer and Sabir (2011); Vincent and Guo (2011)]. However, there are few works about the synchronization of memristive chaotic systems, so it is necessary to investigate the synchronization control of the memristive chaotic systems with unmatched parameters to shorten such gap.

Motivated by above discussion, a chaotic system based on memristor is proposed, which has complex dynamical behaviors that possess some special features like other chaotic systems such as being extremely sensitive to tiny variations of initial conditions, having bounded trajectories in the phase space, and so on. In addition, this section presents the adaptive synchronization of memristive Chua's circuit. This approach can synchronize the chaotic system with unmatched parameters, and the adaptive controller can also be used to synchronize the general chaotic systems with unmature parameters. Moreover, the controller design method is further extended to more general cases, where the physical plant contains parameter uncertainties, represented in either polytopic or structured frameworks. And this method can also be used in the common chaotic system without memristor or the memductance of the memristor equals a constant.

14.2 Synchronization of Memristive Chua's Circuits

In this section, a memristive Chua's system is presented, and the chaotic behavior of this system is demonstrated by phase portraits. This section also deals with the problem of adaptive synchronization control of this chaotic system using the drive-response concept, and presents an adaptive control scheme for the synchronization of memristive Chua's circuit, when the parameters of the drive system are fully unknown and different with those of the response system. The sufficient condition for the adaptive synchronization has been analyzed. Moreover, the controller design method is further extended to more general cases, where the physical plant contains parameter uncertainties, represented in either polytopic or structured frameworks. Numerical simulations are used to demonstrate these results.

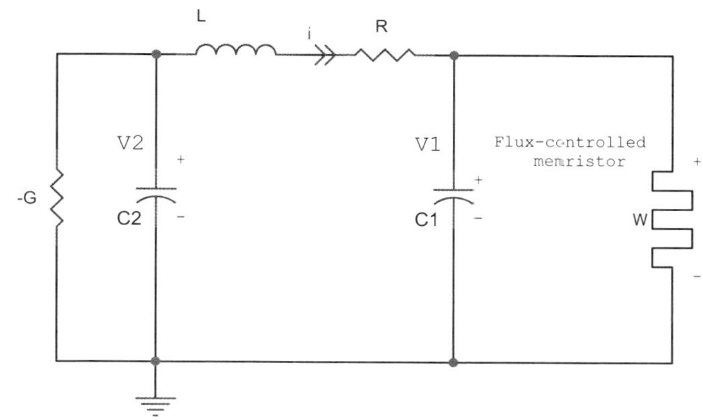

Fig. 14.1 Chua's circuit with memristor

14.2.1 *Memristive Chua's circuits*

M. Itoh and L.O. Chua proposed the possible nonlinear circuits, with a memristor which replaces the Chua's diode in 2008 [Itoh and Chua (2008)]. Fig. 14.1 shows a memristive Chua's circuit which is derived from Chua's circuit using a PWL memristor.

From Fig. 14.1, we can obtain following equations:

$$\begin{cases} \dot{V_1} = \frac{1}{C_1}[i - W(\varphi)V_1], \\ \dot{V_2} = \frac{1}{C_2}(GV_2 - i), \\ \dot{i} = \frac{1}{L}(V_2 - V_1 - iR), \\ \dot{\varphi} = V_1. \end{cases} \tag{14.1}$$

Let $x_1 = V_1,\ x_2 = V_2,\ x_3 = i,\ x_4 = \varphi,\ \frac{1}{C_1} = 5,\ L = 1$, then

$$\begin{cases} \dot{x_1} = 5[x_3 - W(x_4)x_1], \\ \dot{x_2} = \alpha_1 x_2 - \alpha_2 x_3, \\ \dot{x_3} = x_2 - x_1 - \alpha_3 x_3, \\ \dot{x_4} = x_1, \end{cases} \tag{14.2}$$

where $\alpha_1 = \frac{G}{C_2}$, $\alpha_2 = \frac{1}{C_2}$, $\alpha_3 = R$, and the piece-wise linear function $W(x_4)$ is given by

$$W(x_4) = \frac{\mathrm{d}q(x_4)}{\mathrm{d}x_4} = \begin{cases} a, & |x_4| < 1; \\ b, & |x_4| > 1. \end{cases}$$

So memristive Chua's circuit is a state-dependent switching system. If we set $\alpha_1 = 0.7$, $\alpha_2 = 1$, $\alpha_3 = 0.1$, $a = 0.1$ and $b = 18$, the initial values are $(10^{-1}, 0, 0, 0)$, then system (14.2) is chaotic with a positive Lyapunov exponent LE $= 0.4474$ and 2-scroll attractor as shown in Fig. 14.2.

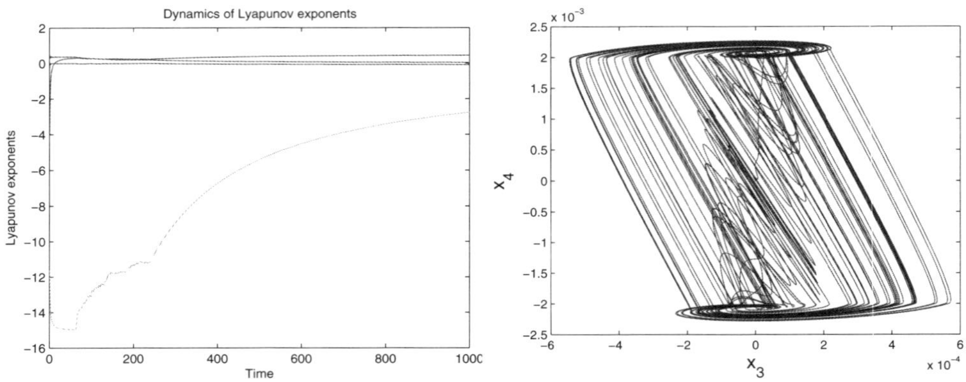

(a) (b)

Fig. 14.2 (a) Lyapunov exponents of system (14.2) (b) Memristor attractor: 2D projection x_3 vs x_4 of system (14.2)

14.2.2 *Adaptive synchronization control of memristive Chua's circuits*

Let system (14.2) be the drive system. Suppose that the parameters of system (14.2) are unknown or uncertain, then the response system is as follows:

$$\begin{cases} \dot{\hat{x}}_1 = 5[\hat{x}_3 - W(\hat{x}_4)\hat{x}_1] - u_1, \\ \dot{\hat{x}}_2 = \hat{\alpha}_1\hat{x}_2 - \hat{\alpha}_2\hat{x}_3 - u_2, \\ \dot{\hat{x}}_3 = \hat{x}_2 - \hat{x}_1 - \hat{\alpha}_3\hat{x}_3 - u_3, \\ \dot{\hat{x}}_4 = \hat{x}_1 - u_4, \end{cases} \tag{14.3}$$

where $\hat{\alpha}_1$, $\hat{\alpha}_2$, $\hat{\alpha}_3$, are parameters of the response system which need to be estimated, and the piece-wise linear function $W(\hat{x}_4)$ is given by

$$W(\hat{x}_4) = \begin{cases} a, & |\hat{x}_4| > 1; \\ b, & |\hat{x}_4| < 1. \end{cases}$$

Supposed that

$$\begin{cases} u_1 = k_1 e_{x1} - 5\big(W(\hat{x}_4) - W(x_4)\big)x_1, \\ u_2 = k_2 e_{x2}, \\ u_3 = k_3 e_{x3}, \\ u_4 = k_4 e_{x4}, \end{cases} \tag{14.4}$$

where

$$e_{x1} = \hat{x}_1 - x_1,\ e_{x2} = \hat{x}_2 - x_2,\ e_{x3} = \hat{x}_3 - x_3,\ e_{x4} = \hat{x}_4 - x_4, \tag{14.5}$$

$k_1, k_2, k_3, k_4 \geq 0$, and

$$\dot{\hat{\alpha}}_1 = -\theta \hat{x}_2 e_{x_1},\ \dot{\hat{\alpha}}_2 = \beta \hat{x}_1 e_{x_2},\ \dot{\hat{\alpha}}_3 = \tau \hat{x}_3 e_{x_3}, \tag{14.6}$$

$\theta, \beta, \tau > 0$ are constants.

Theorem 14.1. *When $k_1, k_2, k_3, k_4 \geq 0$ are properly chosen, so that the following matrix inequality holds:*

$$\Pi = \begin{bmatrix} 5\bar{a} + k_1 & 0 & -2 & -\frac{1}{2} \\ 0 & k_2 - \alpha_1 & -\frac{1}{2}(\alpha_2 + 1) & 0 \\ -2 & -\frac{1}{2}(\alpha_2 + 1) & \alpha_3 + k_3 & 0 \\ -\frac{1}{2} & 0 & 0 & k_4 \end{bmatrix} > 0, \tag{14.7}$$

where $\bar{a} = \min(a, b)$, then memristive Chua's systems (14.2) *and* (14.3) *can be synchronized under the adaptive control of* (14.4) *and* (14.6).

Proof. It is easy to see from systems (14.2) and (14.3), the error system is

$$\begin{cases} \dot{e}_{x1} = 5\hat{x}_3 - 5x_3 - 5W(\hat{x}_4)\hat{x}_1 + 5W(x_4)x_1 - u_1, \\ \dot{e}_{x2} = \hat{\alpha}_1 \hat{x}_2 - \alpha_1 x_2 - \hat{\alpha}_2 \hat{x}_3 + \alpha_2 x_3 - u_2, \\ \dot{e}_{x3} = \hat{x}_2 - x_2 - \hat{x}_1 + x_1 - \hat{\alpha}_3 \hat{x}_3 + \alpha_3 x_3 - u_3, \\ \dot{e}_{x4} = \hat{x}_1 - x_1 - u_4. \end{cases} \tag{14.8}$$

Let

$$e_1 = \hat{\alpha}_1 - \alpha_1,\ e_2 = \hat{\alpha}_2 - \alpha_2,\ e_3 = \hat{\alpha}_3 - \alpha_3,$$

then

$$\dot{e}_1 = \dot{\hat{\alpha}}_1,\ \dot{e}_2 = \dot{\hat{\alpha}}_2,\ \dot{e}_3 = \dot{\hat{\alpha}}_3. \tag{14.9}$$

Choose the Lyapunov function as follows:

$$V(t) = \frac{1}{2}\Big(e_{x1}^2 + e_{x2}^2 + e_{x3}^2 + e_{x4}^2 + \frac{1}{\theta}e_1^2 + \frac{1}{\beta}e_2^2 + \frac{1}{\tau}e_3^2\Big) \tag{14.10}$$

Then, the differentiation of V along the trajectories of (14.8) is

$$\begin{aligned} \dot{V}(t) &= e_{x1}\dot{e}_{x1} + e_{x2}\dot{e}_{x2} + e_{x3}\dot{e}_{x3} + e_{x4}\dot{e}_{x4} + \frac{1}{\theta}e_1\dot{e}_1 + \frac{1}{\beta}e_2\dot{e}_2 + \frac{1}{\tau}e_3\dot{e}_3) \\ &= e_{x1}\Big(5\hat{x}_3 - 5x_3 - 5W(\hat{x}_4)\hat{x}_1 + 5W(x_4)x_1 - u_1\Big) + e_{x2}\Big(\hat{\alpha}_1\hat{x}_2 - \alpha_1 x_2 \\ &\quad - \hat{\alpha}_2\hat{x}_3 + \alpha_2 x_3 - u_2\Big) + e_{x3}\Big(\hat{x}_2 - x_2 - \hat{x}_1 + x_1 - \hat{\alpha}_3\hat{x}_3 + \alpha_3 x_3 - u_3\Big) \\ &\quad + e_{x4}\Big(\hat{x}_1 - x_1 - u_4\Big) + \frac{1}{\theta}e_1\dot{e}_1 + \frac{1}{\beta}e_2\dot{e}_2 + \frac{1}{\tau}e_3\dot{e}_3 \\ &= e_{x1}\Big[5e_{x3} + 5\Big(-W(\hat{x}_4)\hat{x}_1 + W(\hat{x}_4)x_1 - W(\hat{x}_4)x_1 + W(x_4)x_1\Big) - k_1 e_{x1} \end{aligned}$$

$$
\begin{aligned}
&+\Big(W(\hat{x}_4)-W(x_4)\Big)x_1\Big]+e_{x2}\Big[\Big(\hat{\alpha}_1\hat{x}_2-\alpha_1\hat{x}_2+\alpha_1\hat{x}_2-\alpha_1x_2\Big)\\
&+\Big(-\hat{\alpha}_2\hat{x}_3+\alpha_2\hat{x}_3-\alpha_2\hat{x}_3+\alpha_2x_3\Big)-k_2e_{x2}\Big]+e_{x3}\Big[\Big(\hat{x}_2-x_2\Big)\\
&+\Big(-\hat{x}_1+x_1\Big)+\Big(-\hat{\alpha}_3\hat{x}_3+\alpha_3\hat{x}_3-\alpha_3\hat{x}_3+\alpha_3x_3\Big)-k_3e_{x3}\Big]\\
&+e_{x4}\Big[\Big(\hat{x}_1-x_1\Big)-k_4e_{x4}\Big]+\frac{1}{\theta}e_1\dot{e}_1+\frac{1}{\beta}e_2\dot{e}_2+\frac{1}{\tau}e_3\dot{e}_3\\
=&\,e_{x1}\Big[5e_{x3}-5W(\hat{x}_4)e_{x1}-k_1e_{x1}\Big]+e_{x2}\Big[e_1\hat{x}_2+\alpha_1e_{x2}-e_2\hat{x}_3-\alpha_2e_{x3}\\
&-k_2e_{x2}\Big]+e_{x3}\Big[e_{x2}-e_{x1}-e_3\hat{x}_3-\alpha_3e_{x3}-k_3e_{x3}\Big]\\
&+e_{x4}\Big[e_{x1}-k_4e_{x4}\Big]+\frac{1}{\theta}e_1\dot{e}_1+\frac{1}{\beta}e_2\dot{e}_2+\frac{1}{\tau}e_3\dot{e}_3\\
=&-(5W(\hat{x}_4)+k_1)e_{x1}^2-(k_2-\alpha_1)e_{x2}^2-(\alpha_3+k_3)e_{x3}^2-k_4e_{x4}^2\\
&+4e_{x1}e_{x3}+e_{x1}e_{x4}+(-\alpha_2+1)e_{x2}e_{x3}+\Big(\frac{1}{\theta}\dot{e}_1+\hat{x}_2e_{x2}\Big)e_1\\
&+\Big(\frac{1}{\beta}\dot{e}_2-\hat{x}_3e_{x2}\Big)e_2+\Big(\frac{1}{\tau}\dot{e}_3-\hat{x}_3e_{x3}\Big)e_3\\
=&-y^{\mathsf{T}}\Pi y<0
\end{aligned}
$$

where $y=(|e_{x1}|,|e_{x2}|,|e_{x3}|,|e_{x4}|)^{\mathsf{T}}$, Π is shown in (14.7). Since $V(t)$ is positive definite and the differentiation of $V(t)$ is negative definite, which implies that the equilibrium point

$$
\begin{cases}
e_{x1}=e_{x2}=e_{x3}=e_{x4}=0,\\
e_1=e_2=e_3=0,
\end{cases}
$$

of systems (14.8) and (14.9) are asymptotically stable, which in turn implies

$$
\begin{cases}
e_{x1}(t)\to 0,\ e_{x2}\to 0,\ e_{x3}\to 0,\ e_{x4}(t)\to 0,\\
e_1(t)\to 0,\ e_2(t)\to 0,\ e_3(t)\to 0,
\end{cases}
$$

as $t\to\infty$. Thus, in the closed-loop system

$$
\begin{cases}
\hat{x}_1\to x_1,\ \hat{x}_2\to x_2,\ \hat{x}_3\to x_3,\ \hat{x}_4\to x_4,\\
\hat{\alpha}_1\to \alpha_1,\ \hat{\alpha}_2\to \alpha_2,\ \hat{\alpha}_3\to \alpha_3,\ \hat{\alpha}_4\to \alpha_4,
\end{cases}
\tag{14.11}
$$

as $t\to\infty$. This implies that the response system (14.3) is synchronization with the drive system (14.2). □

Polytopic-type uncertainties are an important class of uncertainties because they can be used to represent uncertainties described in terms of interval matrices and have been widely investigated [Mathalagan *et al.* (2012); Phat and Nam (2010); Zhou *et al.* (2009)].

Assume that the parameters $\alpha_1,\alpha_2,\alpha_3,a,b$ in system (14.2) contain partially unknown parameters, and $\Psi\triangleq(\alpha_1,\alpha_2,\alpha_3,a,b)\in\mathcal{R}$, where $\mathcal{R}$ is a given convex-bounded polyhedral domain described by $\mathcal{S}$ vertices

$$\mathcal{R} = \left\{\Psi(\delta)\middle|\Psi(\delta) = \sum_{i=1}^{\mathcal{S}} \delta_i \Psi_i;\ \sum_{i=1}^{\mathcal{S}} \delta_i = 1, \delta_i \geq 0\right\} \tag{14.12}$$

where $\Psi_i = (\alpha_{1i}, \alpha_{2i}, \alpha_{3i}, a_i, b_i)$ denotes the vertices of the polytope. And the main result can be directly used for the robust synchronization control problem on the basis of quadratic stability notion. Then, we obtain the following corollaries without proof.

Corollary 14.1. *There exist the adaptive controllers* (14.4) *and* (14.6)*, such that the memristive Lorenz systems* (14.2) *and* (14.3) *can be synchronized, if there exist* $k_{1i}, k_{2i}, k_{3i}, k_{4i} \geq 0$*, such that LMI* (14.7) *holds, where* $\alpha_1, \alpha_2, \alpha_3, a, b$ *are taken with* $\alpha_{1i}, \alpha_{2i}, \alpha_{3i}, a_i, b_i$*, respectively.*

When the switching properties of the memristor are not under consider, such as $\varphi = a = b$, the following results can be derived

Corollary 14.2. *There exist adaptive controllers* (14.4) *and* (14.6)*, such that the memristive Lorenz systems* (14.2) *and* (14.3) *can be synchronized, if there exist* $k_{1i}, k_{2i}, k_{3i}, k_{4i} \geq 0$*, such that LMI* (14.7) *holds, where* $\alpha_1, \alpha_2, \alpha_3, \varphi$ *are taken with* $\alpha_{1i}, \alpha_{2i}, \alpha_{3i}, \varphi_i$*, respectively.*

Remark 14.1. This adaptive control scheme for the synchronization of the switched Chua's system can be used in the chaotic system without the memristor as well as the parameters of the drive system are known or the same as the response system.

14.2.3 *Numerical simulations*

Example 14.1. The numerical simulations are carried out using the fourth-order Rune-Kutta method. The initial states of drive and response systems are $(10^{-1}, 0, 0, 0)$ and $(0, 0, 0, 0)$. The parameters of the drive system are $\alpha_1 = 0.7, \alpha_2 = 1, \alpha_3 = 0.1$ and $a = 0.1$, $b = 18$. Using MATLAB to solve LMI (14.7), we can get that $k_1 = 0.5$, $k_2 = 1.7$, $k_3 = 10.9$, $k_4 = 1$ which satisfy (14.7). Choose $\theta = \beta = \tau = 2$ and the initial values of the parameters $\hat{\alpha}_1$, $\hat{\alpha}_2$, $\hat{\alpha}_3$ are all 0, and the response system synchronizes with the drive system as shown in Fig. 14.3 and the changing parameters of $\hat{\alpha}_1$, $\hat{\alpha}_2$, $\hat{\alpha}_3$ are shown in Fig. 14.4.

When the control parameters increase, for example, $\theta = \beta = \tau = 20$, the simulation results are shown in Fig. 14.5 and Fig. 14.6.

Remark 14.2. The synchronization errors will converge faster to zero, when $\gamma, \theta, \beta, \tau$ are given larger values, so the rate of synchronization response can be improved distinctly.

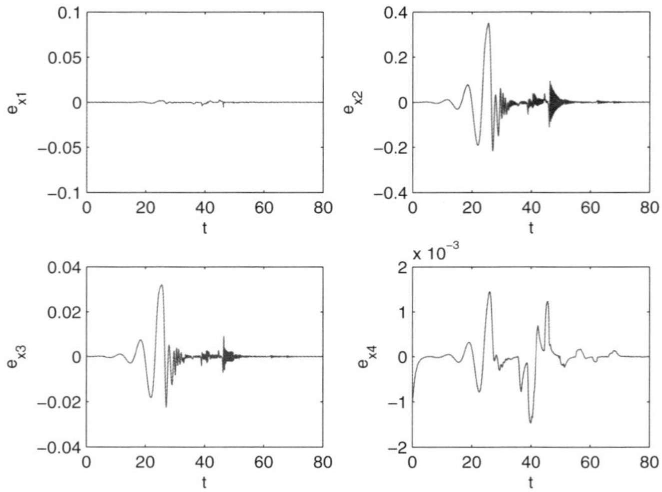

Fig. 14.3 Synchronization errors e_{x1}, e_{x2}, e_{x3}, e_{x4} at $\theta = \beta = \tau = 2$

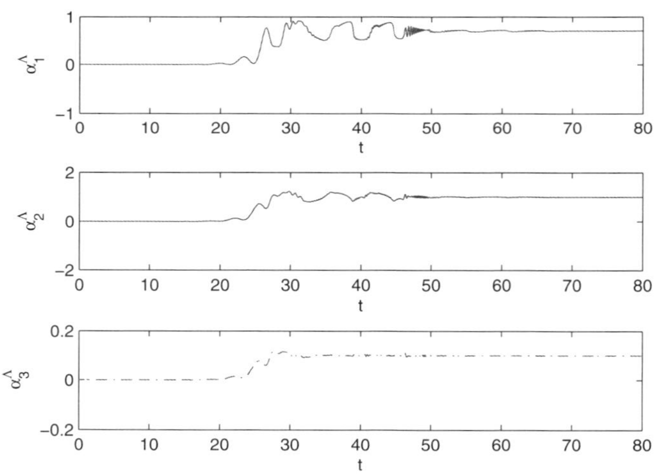

Fig. 14.4 Changing parameters $\hat{\alpha}_1, \hat{\alpha}_2, \hat{\alpha}_3$ at $\theta = \beta = \tau = 2$

Example 14.2. Based on Example 14.1, the obtained results can be applied in the field of digital signal processing, and the algorithm is presented for a color picture F with a size $m \times n \times 3$, as follows:

1. Separating color image F into three gray ones with red, green and blue, respectively, and via sort function to rearrange the pixels in each gray image, therefore, three new ordered pixel series are obtained as $R(i,j)$, $G(i,j)$, $B(i,j)$, $i \in \{1,\ldots,m\}$, $j \in \{1,\ldots,n\}$.

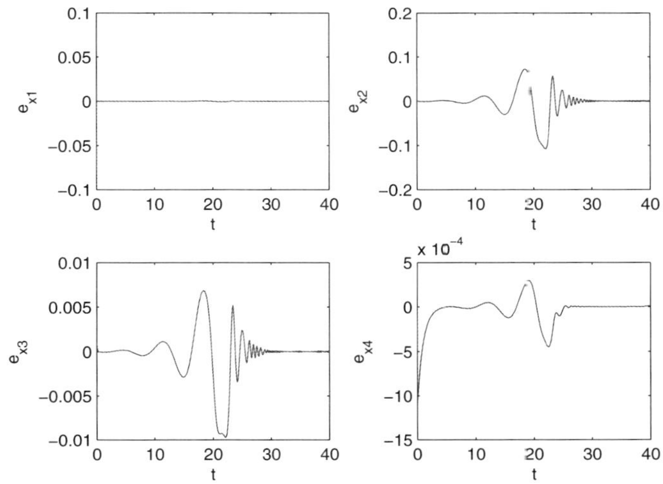

Fig. 14.5 Synchronization errors e_{x1}, e_{x2}, e_{x3}, e_{x4} at $\theta = \beta = \tau = 20$

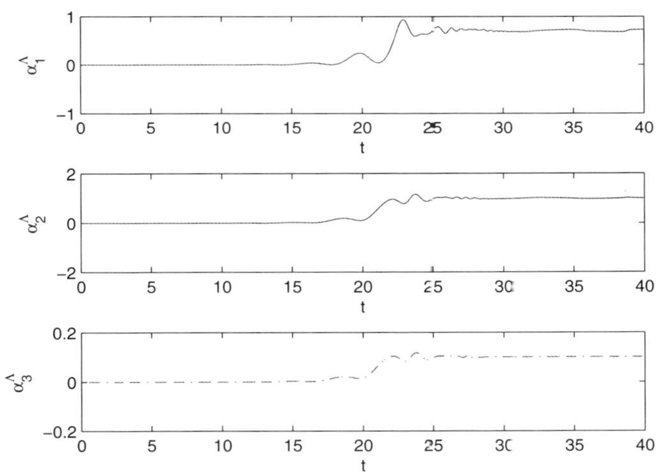

Fig. 14.6 Changing parameters $\hat{\alpha}_1, \hat{\alpha}_2, \hat{\alpha}_3$ at $\theta = \beta = \tau = 20$

2. Through drive system, two groups of time-series chaotic signals obtained as $z_l(i,j) = x_l(k),\ k \in \{1, \ldots, mn\},\ l \in \{1, 2\}$. As there are three gray images needed to encrypted, the third chaotic signal can be set as $z_3(k) = 0.5(x_1(k) + x_2(k))$.
3. Based on the proceeded chaotic signals and gray images, the encrypted gray images can be obtained as the following operation:

$$R(i,j) \leftarrow \mathrm{mod}\Big(R(i,j) * 1000, 256\Big) \oplus z_1(i,j);$$

$$G(i,j) \leftarrow \mathrm{mod}\Big(G(i,j) * 1000, 256\Big) \oplus z_2(i,j);$$

$$B(i,j) \leftarrow \mathrm{mod}\Big(G(i,j) * 1000, 256\Big) \oplus z_3(i,j).$$

4. Reorganizing $R(i,j)$, $G(i,j)$, $B(i,j)$, we can get the encrypted color image.

Although the decryption process is the same as the encryption process, there is omitted. And a simulation about gray image encryption has been provided in Fig. 14.7 and Fig. 14.8, which illustrate the application potential of the lag synchronization of memristive chaotic circuits in signal encryption.

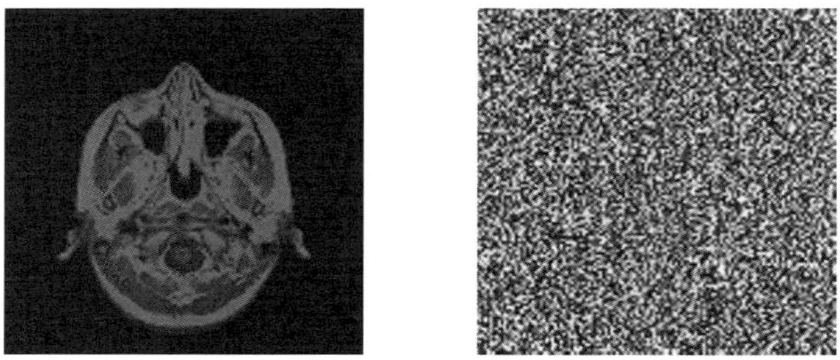

Fig. 14.7 The original image and the encrypted image

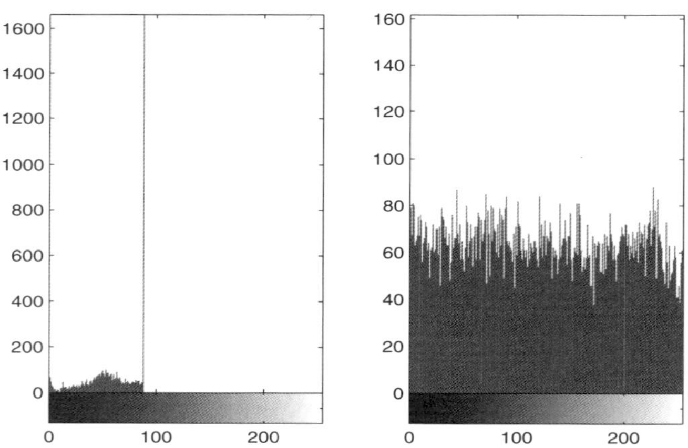

Fig. 14.8 The histograms of the original gray image and encrypted gray image

14.2.4 *Conclusions*

This section investigates the memristive Chua's circuit which is a state-dependent switching system and exhibits complex dynamics characteristics. This system can be used in many fields such as secure communications. And this section also addresses adaptive synchronization of memristive Chua's chaotic circuit with different parameters. The sufficient condition of the control parameters for the synchronization is obtained. The performances of the control scheme are demonstrated by numerical simulations. The controller design method was further extended to more general cases, where the physical plant contains parameter uncertainties, represented in either polytopic or structured frameworks. And this method can also be used in the common chaotic system without memristor or the memductance of the memristor equals a constant. Finally numerical examples have been carried out to show the effectiveness of the proposed method.

14.3 Synchronization of Memristive Lorenz Circuits

This section is concerned with the problem of fuzzy modeling and synchronization of memristive Lorenz circuits with memristive Chua's circuits. In this section, a memristive Lorenz circuit is set up, and illustrated by phase portraits and Lyapunov exponents. Furthermore, a new fuzzy model of memristive Lorenz circuit is presented to simulate and synchronize with memristive Chua's circuit. Through this new fuzzy model, two main advantages can be obtained as: 1) only two linear subsystems are needed; 2) fuzzy synchronization of these two different chaotic circuits with different numbers of nonlinear terms can be achieved with only two sets of gain K. Finally, numerical simulations are used to illustrate the effectiveness of these obtained results.

14.3.1 *Modeling and fuzzy synchronization of memristive Lorenz circuits*

In this subsection, three issues are concerned such as design of memristive Lorenz circuits, fuzzy modeling of memristive Lorenz circuits, synchronization of memristive Lorenz circuits with memristive Chua's circuits.

(1) *Design of memristive Lorenz circuits*

Several new circuits about the memristor have been proposed and investigate in Bao *et al.* (2010); Ventra *et al.* (2009). Based on these results, a new kind of memristive Lorenz circuits is presented in Fig. 14.9. We can obtain the following equations

$$\begin{cases} -R_5C_1\dot{X}(t) = -\frac{R_4}{R_1}Y(t) + (1+\frac{R_4}{R_1})\frac{R_3}{R_2+R_3}X(t) + R_5W(\varphi(t))X(t), \\ -R_{12}C_2\dot{Y}(t) = -\frac{R_{11}}{R_{10}}X(t) + \frac{R_{11}}{R_9}(\frac{R_8}{R_6}Y(t) + \frac{R_8}{R_7}X(t)Z(t)), \\ -R_{17}C_3\dot{Z}(t) = -\frac{R_{16}}{R_{13}}X(t)Y(t) + (\frac{1}{R_{16}} + \frac{1}{R_{13}})\frac{R_{15}R_{16}}{R_{14}+R_{15}}Z(t), \\ \qquad \dot{\varphi}(t) = -X(t). \end{cases} \tag{14.13}$$

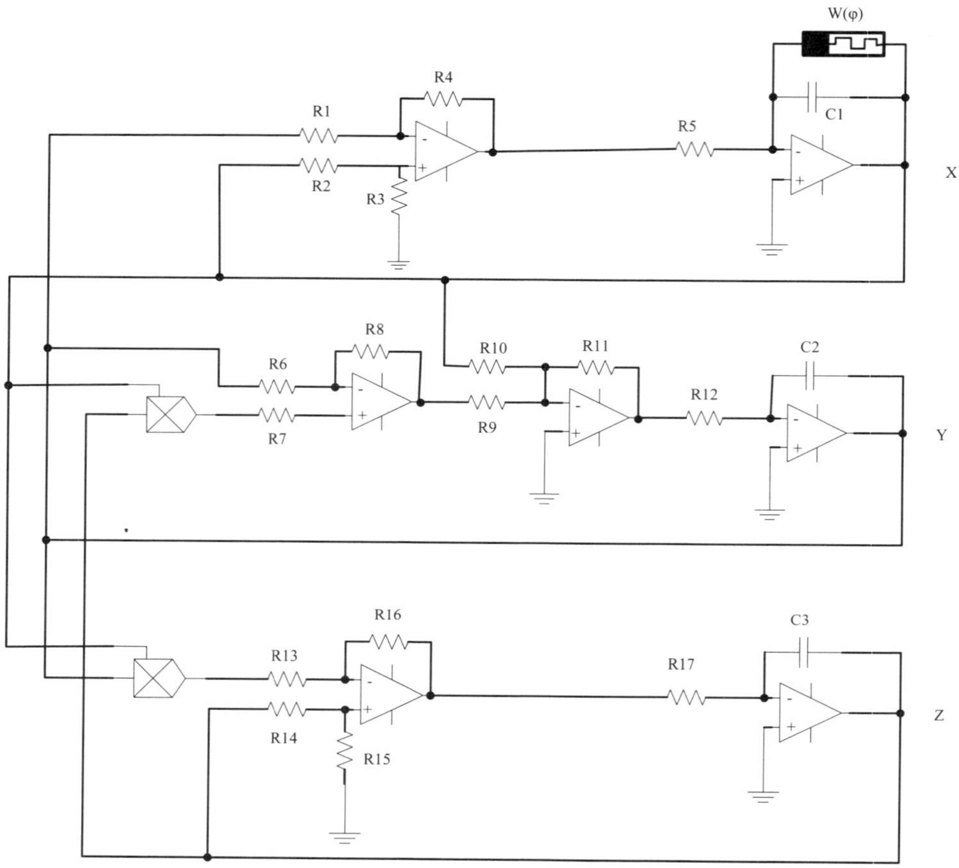

Fig. 14.9 Memristive Lorenz circuits

For the cause of convenience, denote $x(t) = [x_1(t), x_2(t), x_3(t), x_4(t)]^{\mathsf{T}} = [X(t), Y(t), Z(t), \varphi(t)]^{\mathsf{T}}$, $(1+\frac{R_4}{R_1})\frac{R_3}{R_2+R_3}\frac{1}{R_5C_1} = \sigma_1$, $\frac{R_4}{R_1}\frac{1}{R_5C_1} = \sigma_2$, $\frac{R_{11}}{R_{10}}\frac{1}{R_{12}C_2} = \sigma_3$, $(\frac{1}{R_{16}} + \frac{1}{R_{13}})\frac{R_{15}R_{16}}{R_{14}+R_{15}}\frac{1}{R_{17}C_3} = \sigma_4$, $C_1 = 1$, $\frac{R_{11}}{R_9}\frac{R_8}{R_6}\frac{1}{R_{12}C_2} = 1$, $\frac{R_{11}}{R_9}\frac{R_8}{R_7}\frac{1}{R_{12}C_2} = 1$, $\frac{R_{16}}{R_{13}}\frac{1}{R_{17}C_3} = 1$. Therefore, system (14.13) can be rewritten as

$$\begin{cases} \dot{x}_1(t) = -\sigma_1 x_1(t) - W(x_4(t))x_1(t) + \sigma_2 x_2(t), \\ \dot{x}_2(t) = \sigma_3 x_1(t) - x_2(t) - x_1(t)x_3(t), \\ \dot{x}_3(t) = x_1(t)x_2(t) - \sigma_4 x_3(t), \\ \dot{x}_4(t) = -x_1(t), \end{cases} \tag{14.14}$$

where the piece-wise linear function $W(x_4(t))$ is given by

$$W(x_4(t)) = \begin{cases} a, & |x_4(t)| \le 1; \\ b, & |x_4(t)| > 1. \end{cases}$$

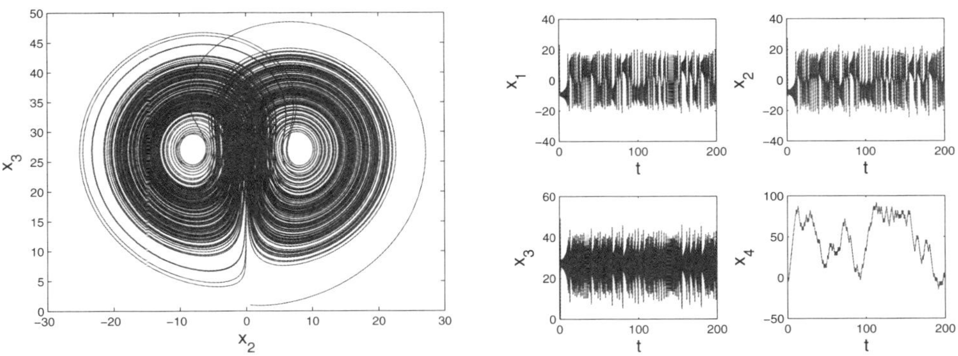

Fig. 14.10 (a) Memristor attractor: 2D projection x_2 vs x_3 of system (14.14) (b) Dynamical behaviors of x_1, x_2, x_3, x_4 of system (14.14)

When $\sigma_1 = 8$, $\sigma_2 = 15$, $\sigma_3 = 28$, $\sigma_4 = 8/3$, $a = 5$ and $b = 8$, system (14.14) is chaotic with a positive Lyapunov exponent LE = 13.2334 and 2-scroll attractor under the initial values of $(10^{-4}, 0, 0, 0)$, as shown in Fig. 14.10.

(2) *Fuzzy modeling of memristive Lorenz circuits*

Suppose that $x_1(t) \in [-\zeta, \zeta]$ and $\zeta > 0$, memristive Lorenz system can be represented with a T-S fuzzy model as follows:

For $\dot{x}_1(t) = -\sigma_1 x_1(t) - W(x_4(t))x_1(t) + \sigma_2 x_2(t)$,

Rule 1: IF $x(t)$ is H_{11}, THEN

$$\dot{x}_1(t) = -\sigma_1 x_1(t) - a x_1(t) + \sigma_2 x_2(t); \tag{14.15}$$

Rule 2: IF $x(t)$ is H_{12}, THEN

$$\dot{x}_1(t) = -\sigma_1 x_1(t) - b x_1(t) + \sigma_2 x_2(t), \tag{14.16}$$

where H_{11} is $|x_4(t)| \leq 1$, H_{12} is $|x_4(t)| > 1$, then

$$M_{11} = \begin{cases} 1, & |x_4(t)| \leq 1, \\ 0, & |x_4(t)| > 1, \end{cases} \quad M_{12} = \begin{cases} 0, & |x_4(t)| \leq 1, \\ 1, & |x_4(t)| > 1; \end{cases}$$

For $\dot{x}_2 = \sigma_3 x_1(t) - x_2(t) - x_1(t)x_3(t)$ and $\dot{x}_3 = x_1(t)x_2(t) - \sigma_4 x_3(t)$

Rule 1: IF $x(t)$ is M_{21}, THEN

$$\dot{x}_2(t) = \sigma_3 x_1(t) - x_2(t) - \zeta x_3(t), \tag{14.17}$$

$$\dot{x}_3(t) = \zeta x_2(t) - \sigma_4 x_3(t). \tag{14.18}$$

Rule 2: IF $x(t)$ is M_{22}, THEN

$$\dot{x}_2(t) = \sigma_3 x_1(t) - x_2(t) + \zeta x_3(t), \tag{14.19}$$

$$\dot{x}_3(t) = -\zeta x_2(t) - \sigma_4 x_3(t), \tag{14.20}$$

where

$$M_{21} = \frac{1}{2}\left(1 + \frac{x_1(t)}{\zeta}\right), \; M_{22} = \frac{1}{2}\left(1 - \frac{x_1(t)}{\zeta}\right).$$

Through calculation by Matlab, we can get $-16.8722 \leq x_1(t) \leq 23.3035$, therefore, we can set $\zeta = 24$. M_{ij}, $i, j = 1, 2$ are fuzzy sets of the first to third equations of memristive Lorenz system (14.14), where $M_{11} + M_{12} = M_{21} + M_{22} = 1$.

Then, (14.15), (14.17), (14.18) are chosen as the members of the first linear subsystem under fuzzy rules. Therefore, the first linear system can be obtained as follows:

$$\begin{cases} \dot{x}_1(t) = -\sigma_1 x_1(t) - a x_1(t) + \sigma_2 x_2(t), \\ \dot{x}_2(t) = \sigma_3 x_1(t) - x_2(t) - \zeta x_3(t), \\ \dot{x}_3(t) = \zeta x_2(t) - \sigma_4 x_3(t), \\ \dot{x}_4(t) = -x_1(t). \end{cases} \tag{14.21}$$

By the same way, (14.16), (14.19), (14.20) are chosen as the members of the second linear subsystem under fuzzy rules. Therefore, the second linear system can be obtained as follows:

$$\begin{cases} \dot{x}_1(t) = -\sigma_1 x_1(t) - b x_1(t) + \sigma_2 x_2(t), \\ \dot{x}_2(t) = \sigma_3 x_1(t) - x_2(t) + \zeta x_3(t), \\ \dot{x}_3(t) = -\zeta x_2(t) - \sigma_4 x_3(t), \\ \dot{x}_4(t) = -x_1(t). \end{cases} \tag{14.22}$$

The final output of the fuzzy memristive Lorenz system is represented as follows:

$$\begin{aligned} \dot{x}(t) = {} & \begin{bmatrix} M_{11}\, 0 \;\; 0 & 0\; 0 \;\; M_{21}\; 0\; 0 \\ 0 \;\; 0\; M_{21} & 0 \\ 0 \;\; 0 \;\; 0 & M_{11} \end{bmatrix} \begin{bmatrix} -\sigma_1 x_1(t) - a x_1 + \sigma_2 x_2(t) \\ \sigma_3 x_1(t) - x_2(t) - \zeta x_3(t) \\ \zeta x_2(t) - \sigma_4 x_3(t) \\ -x_1(t) \end{bmatrix} \\ & + \begin{bmatrix} M_{12} & 0 & 0 & 0 \\ 0 & M_{22} & 0 & 0 \\ 0 & 0 & M_{22} & 0 \\ 0 & 0 & 0 & M_{12} \end{bmatrix} \begin{bmatrix} -\sigma_1 x_1(t) - b x_1 + \sigma_2 x_2(t) \\ \sigma_3 x_1(t) - x_2(t) + \zeta x_3(t) \\ -\zeta x_2(t) - \sigma_4 x_3(t) \\ -x_1(t) \end{bmatrix}, \end{aligned} \tag{14.23}$$

which can be rewritten as $\dot{x}(t) = \sum_{i=1}^{2} \Theta_i A_i x(t)$, where Θ_i are diagonal matrices as

$$\Theta_i = \operatorname{diag}\left[M_{1i}\; M_{2i}\; M_{2i}\; M_{1i} \right],$$

and

$$A_1 = \begin{bmatrix} -\sigma_1 - a & \sigma_2 & 0 & 0 \\ \sigma_3 & -1 & -\zeta & 0 \\ 0 & \zeta & -\sigma_4 & 0 \\ -1 & 0 & 0 & 0 \end{bmatrix}, \quad A_2 = \begin{bmatrix} -\sigma_1 - b & \sigma_2 & 0 & 0 \\ \sigma_3 & -1 & \zeta & 0 \\ 0 & -\zeta & -\sigma_4 & 0 \\ -1 & 0 & 0 & 0 \end{bmatrix}. \tag{14.24}$$

Employing this fuzzy model, the memristive Lorenz system can be represented by only two linear subsystems to express complex chaotic behaviors. With a center-average defuzzier, the over fuzzy system is represented as

$$\dot{x}(t) = \sum_{i=1}^{2} \beta_i A_i x(t), \tag{14.25}$$

where $\beta_i = \Theta_i(\Theta_1 + \Theta_2)^{-1}$.

By the same way, the memristive Chua's system in Phat and Nam (2010) is considered

$$\begin{cases} \dot{y}_1(t) = 5[y_3(t) - W(y_4(t))y_1(t)], \\ \dot{y}_2(t) = \alpha_1 y_2(t) - \alpha_2 y_3(t), \\ \dot{y}_3(t) = y_2(t) - y_1(t) - \alpha_3 y_3(t), \\ \dot{y}_4(t) = y_1(t), \end{cases} \tag{14.26}$$

where

$$W(y_4(t)) = \begin{cases} c, & |y_4(t)| \le 1; \\ d, & |y_4(t)| > 1. \end{cases}$$

Under the initial values of $(0.1, 0.1, 0.1, 0.1)^{\mathrm{T}}$, the dynamics behaviors of this circuit is shown in Fig. 14.11. And memristive Chua's system can be exactly represented by the fuzzy model as follows:

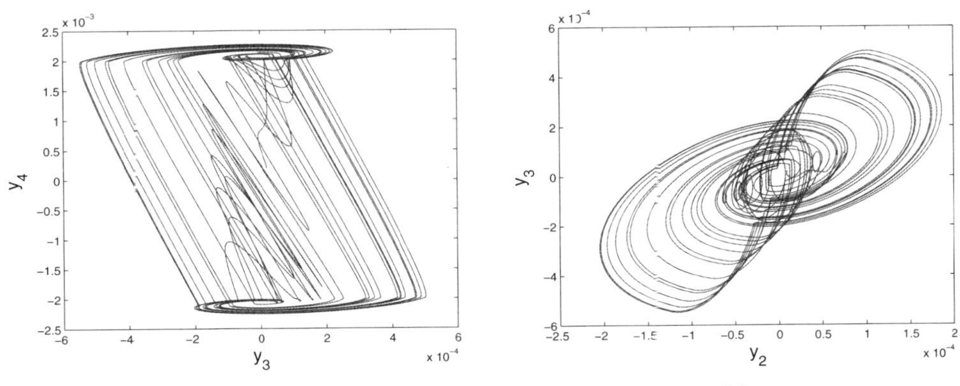

Fig. 14.11 (a) Dynamical behaviors of y_3, y_4 of system (14.26) (b) Dynamical behaviors of y_2, y_3 of system (14.26)

Rule 1: IF $y(t)$ is N_1, THEN

$$\dot{y}(t) = C_1 y(t), \tag{14.27}$$

Rule 2: IF $y(t)$ is N_2, THEN

$$\dot{y}(t) = C_2 y(t), \tag{14.28}$$

where $y(t) = \begin{bmatrix} y_1(t) \; y_2(t) \; y_3(t) \; y_4(t) \end{bmatrix}^{\mathsf{T}}$,

$$C_1 = \begin{bmatrix} -5c & 0 & 5 & 0 \\ 0 & \alpha_1 & -\alpha_2 & 0 \\ -1 & 1 & -\alpha_3 & 0 \\ 1 & 0 & 0 & 0 \end{bmatrix}, \quad C_2 = \begin{bmatrix} -5d & 0 & 5 & 0 \\ 0 & \alpha_1 & -\alpha_2 & 0 \\ -1 & 1 & -\alpha_3 & 0 \\ 1 & 0 & 0 & 0 \end{bmatrix},$$

N_1 is $|y_4(t)| \leq 1$, N_2 is $|y_4(t)| > 1$. With a center-average defuzzier, the over fuzzy system is represented as $\dot{y}(t) = \sum_{i=1}^{2} \eta_i C_i y(t)$, where

$$\eta_1 = \begin{cases} 1, & |y_4(t)| \leq 1, \\ 0, & |y_4(t)| > 1, \end{cases} \quad \eta_2 = \begin{cases} 0, & |y_4(t)| \leq 1, \\ 1, & |y_4(t)| > 1. \end{cases}$$

14.3.2 *Synchronization of fuzzy memristive Lorenz circuits with memristive Chua's circuits*

Let memristive Lorenz circuit as the drive system, and memristive Chua's circuit as the response system, then the response system with feedback control is given as follows

$$\dot{y}(t) = \sum_{i=1}^{2} \eta_i C_i y(t) + Bu(t), \tag{14.29}$$

where $u(t) = \begin{bmatrix} u_1(t) \; u_2(t) \; u_3(t) \; u_4(t) \end{bmatrix}^{\mathsf{T}}$ is the control input, and B is a diagram matrix. Define the error signal as $e(t) = x(t) - y(t)$, then

$$\dot{e}(t) = \dot{x}(t) - \dot{y}(t) = \sum_{i=1}^{2} \beta_i A_i x(t) - \sum_{i=1}^{2} \eta_i C_i y(t) - Bu(t). \tag{14.30}$$

The fuzzy controllers can be designed as follows:

$$u(t) = \sum_{i=1}^{2} \beta_i \Upsilon_i x(t) - \sum_{i=1}^{2} \eta_i \Psi_i y(t), \tag{14.31}$$

such that $\|e(t)\| \to 0$ as $t \to +\infty$. The design is to determine the feedback gains Υ_i and Ψ_i. Substituting (14.31), we can obtain

$$\dot{e}(t) = \sum_{i=1}^{2} \beta_i \{(A_i - B\Upsilon_i)x(t)\} - \sum_{i=1}^{2} \eta_i \{(C_i - B\Psi_i)y(t)\}. \tag{14.32}$$

Theorem 14.2. *The zero solution of error system* (14.32) *is asymptotically stable, and the response system in* (14.29) *can synchronize with the drive system* (14.14) *under the fuzzy controller* (14.31)*, if the following conditions can be satisfied:*

$$F = (A_i - B\Upsilon_i) = (C_i - B\Psi_i) < 0, \ i = 1, 2. \tag{14.33}$$

Proof. The errors dynamics system (14.32) can be linearized via the fuzzy controllers (14.31), if there exist feedback gains Υ_i, Ψ_i such that equation (14.33) exists, therefore, the overall control system is linearized as

$$\dot{e}(t) = Fe(t). \tag{14.34}$$

Consequently, the zero solution of error system (14.34) is asymptotically stable. And the feedback gains Υ_i, Ψ_i can be determined by follows:

$$\begin{cases} \Upsilon_i = B^{-1}(A_i - F), \\ \Psi_i = B^{-1}(C_i - F). \end{cases} \tag{14.35}$$

The proof is completed. □

Remark 14.3. At present, several works have been done about the problem of chaos control and synchronization for memristive chaotic systems [Phat and Nam (2010)]. Generally, there are 2^n linear subsystems (according to 2^n fuzzy rules) and $m \times 2^n$ equations in the T-S fuzzy system, where n is the number of minimum nonlinear terms, and m is the order of the system. If n is large, the number of linear subsystems in the T-S fuzzy system is huge. Through a new fuzzy model, the memristive Lorenz circuit with two nonlinear terms is linearized to a fuzzy system with only two linear subsystems.

Remark 14.4. As the network becomes complex, the number of the memristive chaotic circuits within a complex networks increases, the synchronization of this kind of networks will become difficult as the number of nonlinear terms becomes huge, which decide the number of the linear subsystems. Then, how to linearize memristive Lorenz networks deserves further investigation in future.

14.3.3 *Simulation results*

The numerical simulations are carried out using the fourth-order Rune-Kutta method. The initial stats and parameters of the drive and response systems are set as before, and $F = -I$, $B = I$. Thus, the feedback gains $\Upsilon_1, \Upsilon_2, \Psi_1, \Psi_2$ can be determined as

$$\Upsilon_1 = \begin{bmatrix} -12 & 15 & 0 & 0 \\ 28 & 1 & -24 & 0 \\ 0 & 24 & -5/3 & 0 \\ -1 & 0 & 0 & 1 \end{bmatrix}, \quad \Upsilon_2 = \begin{bmatrix} -15 & 15 & 0 & 0 \\ 28 & 1 & 24 & 0 \\ 0 & -24 & -5/3 & 0 \\ -1 & 0 & 0 & 1 \end{bmatrix},$$

$$\Psi_1 = \begin{bmatrix} 0.5 & 0 & 5 & 0 \\ 0 & 1.7 & -1 & 0 \\ -1 & 1 & 0.9 & 0 \\ 1 & 0 & 0 & 1 \end{bmatrix}, \quad \Psi_2 = \begin{bmatrix} -89 & 0 & 5 & 0 \\ 0 & 1.7 & -1 & 0 \\ -1 & 1 & 0.9 & 0 \\ 1 & 0 & 0 & 1 \end{bmatrix}.$$

The dynamics of the drive memristive Lorenz circuit and response memristive Chua's circuit, as well as the synchronization errors are shown in Fig. 14.12, which illustrate the effectiveness of the proposed method.

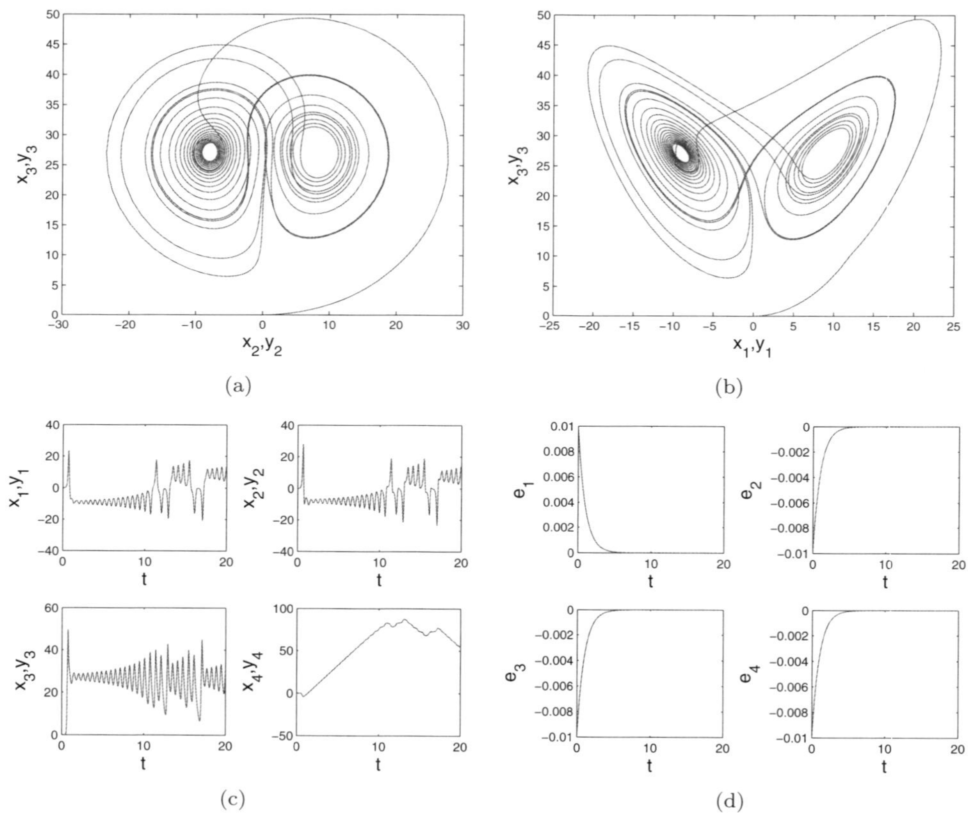

Fig. 14.12 (a) Dynamical behaviors of x_2, x_3, y_2, y_3 of system (14.14) and system (14.29) (b) Dynamical behaviors of x_1, x_3, y_1, y_3, of system (14.14) and system (14.29) (c) State evolution of systems (14.14) and (14.29) (d) Synchronization errors e_1, e_2, e_3, e_4

The encryption/decryption process is the same as in the above section, it is omitted. A simulation is given in Fig. 14.13 and Fig. 14.14, which illustrate the application potential of the lag synchronization of memristive chaotic circuits in signal encryption.

Fig. 14.13 The original image and the encrypted image

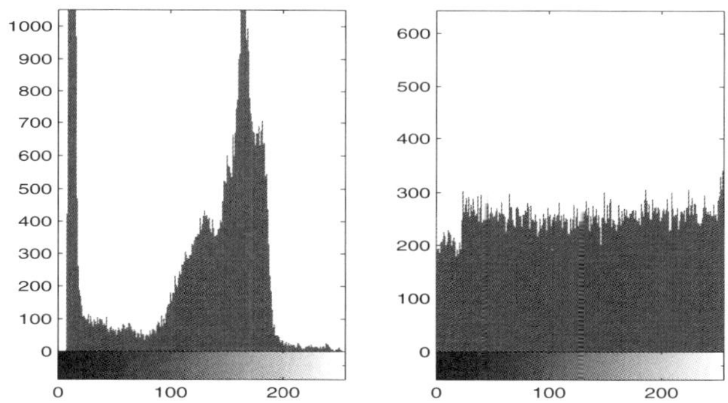

Fig. 14.14 The histograms of the original gray image and encrypted gray image

14.3.4 *Conclusions*

In this section, the problems of fuzzy modeling and synchronization of memristive Lorenz circuits with Chua's circuits, have been concerned. First, a memristive Lorenz circuit is set up, and a new fuzzy model of memristive Lorenz circuit is presented to simulate and synchronize with the memristive Chua's circuit. Furthermore, the synchronization problem of the memristive Lorenz networks is also discussed for the further study. Finally, the effectiveness of the obtained results are illustrated by numerical simulations.

Acknowledgement

The authors thank the Editor and the anonymous referees for their constructive comments and valuable suggestions, which helped improve the quality of this chapter. This work was supported by the Natural Science Foundation of China under Grant 61125303, National Basic Research Program of China (973 Program) under Grant 2011CB710606, the Excellent Youth Foundation of Hubei Province of China under Grant 2010CDA081, the Specialized Research Fund for the Doctoral Program of Higher Education of China under Grant 20100142110021.

References

Bao, B., Liu, Z. and Xu, J. (2010). Transient chaos in smooth memristor oscillator, *Chinese Physics B* **19**, 030510.

Bao, B., Xu, J. and Liu, Z. (2010). Initial state dependent dynamical behaviors in memristor based chaotic circuit, *Chinese Physics Letters* **27**, 070504.

Chen, L., Chai, Y., Wu, R. Sun, J. and Ma, T. (2012). Cluster synchronization in fractional-order complex dynamical networks, *Physics Letters A* **376**, 35, pp. 2381–2388.

Chua, L. (1971). Memristor-the missing circuit element, *IEEE Transactions on Circuit Theory* **18**, 5, pp. 507–519.

Di Ventra, M., Pershin, Y. and Chua, L. (2009). Circuit elements with memory: Memristors, memcapacitors, and meminductors, *Proceedings of the IEEE* **97**, 10, pp. 1717–1724.

Itoh, M. and Chua, L. (2008). Memristor oscillator, *International Journal of Bifurcation and Chaos* **18**, pp. 3183–3206.

Lee, D., Yoo, W. and Won, S. (2010). An integral control for synchronization of a class of unknown nonautonomous chaotic systems, *Physics Letters A* **374**, 41, pp. 4231–4237.

Lin, C., Kuo, C., Hsu, T., Jan, H., Han, S., Ho, M. and Jiang, I. (2012). Experimental observation of chaotic phase synchronization of a periodically pump-modulated multimode microchip Nd: YVO_4 laser, *Physics Letters A* **376**, 15, pp. 1295–1299.

Lin, L. and Wang, H. (2010). Efficient image encryption using a chaos-based PWL memristor, *IETE Technical Review* **27**, 4, pp. 318–325.

Mathiyalagan, K., Sakthivel, R. and Anthoni, S. (2012). Exponential stability result for discrete-time stochastic uncertain neural networks, *Physics Letters A* **376**, 8–9, pp. 901–912.

Phat, N. and Nam, P. (2010). Exponential stability of delayed Hopfield neural networks with various activation functions and polytopic uncertainties, *Physics Letters A* **374**, 25, pp. 2527–2533.

Strukov, D., Snider, G., Stewart, D. and Williams, R. (2008). The missing memristor found, *Nature* **453**, pp. 80–83.

Sudheer, K. and Sabir, M. (2011). Adaptive modified function projective synchronization of multiple time-delayed chaotic Rossler system, *Physics Letters A* **375**, 8, pp. 1176–1178.

Vincent, U. and Guo, R. (2011). Finite-time synchronization for a class of chaotic and hyperchaotic systems via adaptive feedback controller, *Physics Letters A* **375**, 24, pp. 2322–2326.

Zhou, Q., Xu, S., Chen, B., Li, H. and Chu, Y. (2009). Stability analysis of delayed genetic regulatory networks with stochastic disturbances, *Physics Letters A* **373**, 41, pp. 3715–3723.

Chapter 15

Graph Embedded Total Margin Twin Support Vector Machine and Its Applications

Xiaobo Chen and Jian Yang

Abstract Twin support vector machine (TSVM) was proposed recently as a novel binary classifier which aims to seek a pair of nonparallel planes such that each one is closest to the samples of its own class and is at least one distance far from the samples of the other class. However, the samples of the other class with distances to such plane larger than one are discarded, thus implying TSVM fails to exploit the data information residing in those samples. Based on this observation, we extend the original TSVM and propose a novel graph embedded total margin twin support vector machine (GTM-TSVM) in this chapter. The central idea of GTM-TSVM is the plane of one class is required to be far away from overall samples of the other class. Moreover, the intra-class and inter-class graphs which respectively characterize the proximity relationships between samples of within and between classes are embedded into GTM-TSVM formulation so as to exploit the underlying manifold structure of data. The nonlinear classification with kernels is also studied. The experimental results on several publicly available benchmark data sets confirm the feasibility and effectiveness of the proposed method.

15.1 Introduction

Support vector machine (SVM), introduced by Vapnik and colleagues [Cortes and Vapnik (1995); Vapnik (1998)], is a principled and powerful tool for binary data classification. Compared with conventional alternatives, e.g., artificial neural networks [Ripley (2008)] which aim at reducing empirical risk, SVM implements the structural risk minimization principle [Vapnik (1998)], which minimizes the upper bound of the generalization error. Geometrically speaking, SVM seeks a plane to separate samples of different classes by maximizing the margin between them. Such separating plane, i.e., the solution of SVM, follows from solving a quadratic programming problem (QPP) [Boyd and Vandenberghe (2004)]. Moreover, SVM is also a sparse classification algorithm in the sense that the optimal separating plane is determined by a small fraction of training samples, known as support vectors.

Although SVM derives better generalization performance compared with many other machine learning methods, finding the optimal separating plane using naive QPP solver usually leads to high computational complexity. This shortcoming blocks the applications of SVM for large scale training samples to some extent, like text categorization, bioinformatics, etc. To deal with this problem, many novel models were put forward in the literature.

Least squares SVM (LSSVM) [Chen *et al.* (2012, 2011a); Suykens and Vandewalle (1999)] has been proposed as a way to replace inequality constraints with equality constraints in the formulation of SVM. Proximal SVM (PSVM) [Fung and Mangasarian (2005); Mangasarian and Wild (2001)] further maximizes the margin consisting of both the normal vector and bias of the separating plane together. In such a way, LSSVM and PSVM can derive an explicit analytical solution directly from solving a set of linear equations instead of QPP and thus produce extremely fast training speed. The geometric meaning of PSVM is to seek two parallel fitting planes for each class, around which the samples of the corresponding class get clustered and simultaneously the two planes are pushed as far apart as possible. Subsequently, through replacing the parallel constraints with nonparallel ones, Mangasarian and Wild (2006) improved PSVM and proposed PSVM via generalized eigenvalues (GEPSVM). Different from the geometric interpretation of PSVM, GEPSVM aims at searching for a pair of nonparallel planes such that each one is as close as possible to the samples of its own class and at the same time is as far away as possible from the samples of the other class. The resulting plane can be found by solving two related generalized eigenvalue problems (GEPs). In comparison with SVM that classifies samples to one of two disjoint half-space, GEPSVM does this by assigning them to the closest planes [Yang *et al.* (2009)]. Based on GEPSVM, proximal SVM using local information (LIPSVM) was further developed by taking into account the proximity relation of samples [Yang *et al.* (2009)] that may be important for classification performance.

Recently, Jayadeva *et al.* (2007) proposed twin support vector machine (TSVM) classifier which is in the light of GEPSVM. However, instead of solving two GEPs, TSVM solves two related SVM-type problems to obtain the planes. This is achieved by constraining the plane of one class is at least one distance far from the samples of the other class. As stated in Jayadeva *et al.* (2007), the strategy of solving two smaller QPPs rather than a single large one makes the training speed of TSVM approximately four times faster than that of a classic SVM in theory. Moreover, in terms of generalization, TSVM compares favorably with SVM and GEPSVM on many real-world problems. During the past five years, TSVM has become a hot spot in the field of SVM-related machine learning algorithms. In order to accelerate the computation of TSVM further, Kumar and Gopal (2009) developed a least squares algorithm for TSVM by following a similar strategy as in LSSVM. As a result, LSTSVM can generate two planes by solving a pair of small linear systems of equations. By doing so, LSTSVM produces faster training speed than TSVM whereas retains comparable classification performance. Taking motivation from LIPSVM, Ye *et al.* (2011) introduced local similarity information between samples into the formulations of LSTSVM and developed LCTSVM. Besides those aforementioned, many other extensions to the TSVM have been developed in the literatures, including v-TSVM [Peng (2010)], twin bounded SVM [Shao *et al.* (2011)], etc. Note that all the above works are based on the assumption that samples of each class are distributed around planes. However, when it is not the case in practice, we can resort to kernel trick [Cortes and Vapnik (1995)] or use other geometry models, e.g., subspace [Chen *et al.* (2011b)] and hypersphere [Peng and Xu (2013)], to represent data.

Although TSVM has shown its superiority over both SVM and GEPSVM in many experiments, its optimization criterion still has some potential drawbacks. This is because only a small fraction of samples in one class, i.e., support vectors (SVs), are made to be at least one distance from the plane of the other class. As a consequence, TSVM may suffer the problem that the information contained by other samples is lost. In order to mitigate the above deficiency, in this chapter, we extend TSVM to the so-called graph embedded total margin twin support vector machine (GTM-TSVM). Specifically, we first introduce the notion of surplus variables [Yoon *et al.* (2003)] for those non-SVs such that their distances from the fitting plane are made as large as possible. Simultaneously, we put weights [Yang *et al.* (2009); Ye *et al.* (2011)] on fitting errors, slack variables and surplus variables in order to distinguish the importance of different samples. Besides, a rule to determine the weight of each sample is developed, which incorporates proximity relationship between data points. In addition, the proposed method is further extended for nonlinear classification by applying kernel trick. In what follows, we summarize the favorable and attractive characteristics of the proposed algorithms:

1. The information of overall samples is taken into consideration which means not only SVs but also non-SVs in typical TSVM contribute to the construction of fitting plane of each class. Therefore, GTM-TSVM is expected to yield better classification performance than TSVM.
2. To distinguish the importance of different samples, a weighting processing is applied such that fitting errors, slack variables and surplus variables are associated with different weighting factors to reflect relative impact of samples on the fitting plane.
3. A concrete weighting scheme for GTM-TSVM is developed by considering the proximity information between samples. As in many manifold-based feature extraction algorithms [Xu *et al.* (2007); Yan *et al.* (2007)], local information has been proved to be beneficial for classification performance [Yang *et al.* (2009)].

The rest of this chapter is organized as follows: we briefly review twin support vector machine in Section 15.2. In Section 15.3, we propose GTM-TSVM and devise two different weighting schemes. Subsequently, in Section 15.4, we report the experimental results on synthetic and benchmark data sets. Finally, we conclude this chapter in Section 15.5.

15.2 A Brief Review on TSVM

Consider a binary classification problem in n dimensional real space $\mathbb{R}^n$ with m training samples represented by $\{(x_{ji}, y_{ji}) | j = 1, 2;\ i = 1, 2, \ldots, m_j\}$ where x_{ji} denotes the i-th training sample belonging to class j and $m_1 + m_2 = m$, $y_{ji} \in \{-1, 1\}$. We further organize the m_1 samples of class 1 by a $m_1 \times n$ matrix A and m_2 samples of class 2 by a $m_2 \times n$ matrix B. For the sake of simplicity, we also call class 1 as positive class while class 2 as negative class.

The goal of TSVM [Jayadeva *et al.* (2007)] is similar to that of GEPSVM [Mangasarian and Wild (2006)] in spirit, which is to seek two nonparallel planes to proximate samples of distinct class, respectively. Concretely, TSVM finds the following pair of nonparallel positive and negative planes

$$w_1^{\mathsf{T}} x + b_1 = 0 \qquad \text{and} \qquad w_2^{\mathsf{T}} x + b_2 = 0 \tag{15.1}$$

such that each plane is closest to one of two classes and at the same time is at least one distance far away from the other class. Mathematically speaking, TSVM achieves this goal by solving the following pair of SVM-type problems

$$\begin{aligned} \text{(TSVM1)} \qquad & \min_{w_1, b_1, \xi} \frac{1}{2} \sum_{i=1}^{m_1} (w_1^{\mathsf{T}} x_{1i} + b_1)^2 + C_1 \sum_{j=1}^{m_2} \xi_j \\ & \text{s.t.} \qquad -(w_1^{\mathsf{T}} x_{2j} + b_1) \geq 1 - \xi_j \\ & \qquad\qquad \xi_j \geq 0,\ j = 1, 2, \ldots, m_1 \end{aligned} \tag{15.2}$$

and

$$\begin{aligned} \text{(TSVM2)} \qquad & \min_{w_2, b_2, \eta} \frac{1}{2} \sum_{i=1}^{m_2} (w_2^{\mathsf{T}} x_{2i} + b_2)^2 + C_2 \sum_{j=1}^{m_1} \eta_j \\ & \text{s.t.} \qquad w_2^{\mathsf{T}} x_{1j} + b_2 \geq 1 - \eta_j \\ & \qquad\qquad \eta_j \geq 0,\ j = 1, 2, \ldots, m_2 \end{aligned} \tag{15.3}$$

where C_1 and C_2 are tradeoff hyperparameters, ξ_j and η_j are slack variables for negative and positive samples respectively. Taking TSVM1 (15.2) for instance, the positive plane is determined by simultaneously minimizing a sum of squared distances from positive samples to such plane and a hinge loss of distances from negative samples to such plane. Parameter C_1 is used to balance their relative importance. By introducing Lagrangian multiplier vectors α_1 and α_2, we obtain the following dual QPPs [Boyd and Vandenberghe (2004)] for (15.2) and (15.3)

$$\max_{\alpha_1} -\frac{1}{2}\alpha_1^\top H(G^\top G)^{-1}H^\top \alpha_1 + e_2^\top \alpha_1 \quad \text{s.t.} \quad C_1 e_2 \geq \alpha_1 \geq 0 \tag{15.4}$$

and

$$\max_{\alpha_2} -\frac{1}{2}\alpha_2^\top G(H^\top H)^{-1}G^\top \alpha_2 + e_1^\top \alpha_2 \quad \text{s.t.} \quad C_2 e_1 \geq \alpha_2 \geq 0 \tag{15.5}$$

where e_1, e_2 are vectors of ones of appropriate dimension and $G = [A\ e_1]$, $H = [B\ e_2]$. After solving the above two QPPs, the augmented vectors of two planes are given by

$$u_1 = \begin{bmatrix} w_1 \\ b_1 \end{bmatrix} = -(G^\top G)^{-1}H^\top \alpha_1 \quad \text{and} \quad u_2 = \begin{bmatrix} w_2 \\ b_2 \end{bmatrix} = (H^\top H)^{-1}G^\top \alpha_2. \tag{15.6}$$

To obtain more deep insights to TSVM, we rewrite the above u_1 and u_2 as

$$u_1 = -\left(\sum_{i=1}^{m_1} z_{1i}z_{1i}^\top\right)^{-1}\left(\sum_{j=1}^{m_2} z_{2j}\alpha_{1j}\right) \quad \text{and} \quad u_2 = \left(\sum_{i=1}^{m_2} z_{2i}z_{2i}^\top\right)^{-1}\left(\sum_{j=1}^{m_1} z_{1j}\alpha_{2j}\right) \tag{15.7}$$

where $z_{1i} = \left[x_{1i}^\top\ 1\right]^\top$ and $z_{2i} = \left[x_{2i}^\top\ 1\right]^\top$.

From Karush-Kuhn-Tucker (KKT) complementarity conditions [Boyd and Vandenberghe (2004)], we know the optimal values of α_1 and α_2 in (15.4) and (15.5) must be sparse. In fact, taking α_1 for example, only those samples of class 2 whose distances from positive plane are less than or equal to 1 will have positive values and thus they are called support vectors (SVs) with respect to class 1. An intuitive illustration of TSVM1 is shown in Fig. 15.1 where the samples of class 1 and 2 are denoted by circles and squares, respectively. The samples of class 2 are further classified into SVs (black squares) and non-SVs (white squares) depending on whether

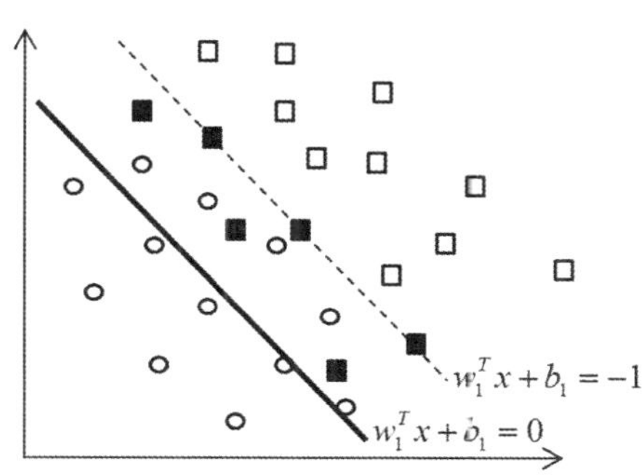

Fig. 15.1 Illustration of the geometric meaning of TSVM1

or not the associated ξ_j is zero. The fitting plane for class 1 and its shifted function which form a margin are denoted by black line and dotted line respectively.

From (15.7), it is obvious that only those SVs (black squares) have an effect on the final plane of class 1, whereas the remaining non-SVs (white squares) can be discarded. This conclusion holds for negative plane as well. Observing Fig. 15.1, we can further see that SVs play an important role in discriminating different classes while non-SVs can provide information on the distribution of classes. However, TSVM ignores the information contained by non-SVs and thus may result in sub-optimal solution.

15.3 Graph Embedded Total Margin TSVM

In this section, we first present a novel framework coined graph embedded total margin twin support vector machine (GTM-TSVM) in detail. Then a concrete model of GTM-TSVM is presented according to its specific weighting scheme.

15.3.1 *Model formulation of GTM-TSVM*

GTM-TSVM aims at looking for two nonparallel planes $f_1(x) = w_1^\top x + b_1$ and $f_2(x) = w_2^\top x + b_2$ such that each one is closest to the samples of its own class and meantime is far away from the samples of other class. However, not only SVs but also the other non-SVs are involved in determining the planes. This is achieved by introducing surplus variables for those non-SVs. In addition, fitting errors, slack variables and surplus variables are all weighted by prescribed factors so as to take into account different effects of data samples. Therefore, GTM-TSVM determines two planes by solving the following problems

$$\begin{aligned}
\text{(GTM-TSVM1)} \qquad \min_{w_1,b_1,\xi,\delta} \ & \frac{1}{2}\sum_{i=1}^{m_1} d_{1i}(w_1^\top x_{1i}+b_1)^2 + C_1\Bigg(\sum_{j=1}^{m_2} f_{2j}\xi_j \\
& -(1-\lambda_1)\sum_{j=1}^{m_2} p_{2j}\delta_j\Bigg) \qquad (15.8) \\
\text{s.t.} \quad & -(w_1^\top x_{2j}+b_1) \geq 1-\xi_j+\delta_j \\
& \xi_j \geq 0,\ \delta_j \geq 0,\ j=1,2,\ldots,m_1
\end{aligned}$$

and

$$\begin{aligned}
\text{(GTM-TSVM2)} \qquad \min_{w_2,b_2,\eta,\zeta} \ & \frac{1}{2}\sum_{i=1}^{m_2} d_{2i}(w_2^\top x_{2i}+b_2)^2 + C_2\Bigg(\sum_{j=1}^{m_1} f_{1j}\eta_j \\
& -(1-\lambda_2)\sum_{j=1}^{m_1} p_{1j}\zeta_j\Bigg) \qquad (15.9) \\
\text{s.t.} \quad & w_2^\top x_{1j}+b_2 \geq 1-\eta_j+\zeta_j \\
& \eta_j \geq 0,\ \zeta_j \geq 0,\ j=1,2,\ldots,m_2.
\end{aligned}$$

Taking GTM-TSVM1 (15.8) for example, ξ_j and δ_j are respectively slack variables and surplus variables corresponding to those SVs and non-SVs of negative samples. λ_1 is a parameter controlling the importance of surplus variables with respect to slack variables. The large λ_1 leads to δ_j having little impact on the fitting plane. d_{1i}, f_{2j} and p_{2j} are all nonnegative weighting coefficients. Now, we give a detailed explanation of the optimization problem (15.8). The first term in the objective function of (15.8) is the weighted sum of squared distances from the positive plane to positive samples. Minimizing it tends to give a robust estimation of plane such that it approximates positive samples [Yang *et al.* (2009); Ye *et al.* (2011)]. For those negative samples whose distances from positive plane are less than or equal to 1, we introduce slack variables ξ_j and minimize them attempting to reduce the misclassification. For those negative samples whose distances from positive plane are greater than 1, we introduce surplus variables δ_j and maximize δ_j intending to make them as far away as possible from positive plane. The geometrical relationship between the slack variables and surplus variables is shown in Fig. 15.2. We can also obtain similar insights for problem (15.9) since it has a similar formulation as (15.8). Therefore, we will focus on (15.8) hereafter for convenience.

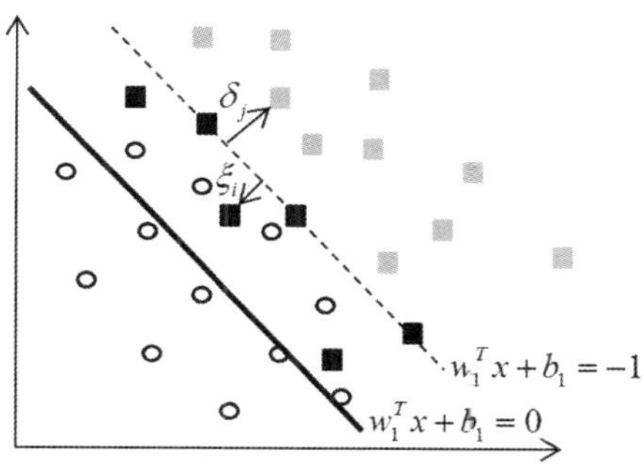

Fig. 15.2 Illustration of slack variables and surplus variables for the positive plane

In order to ensure the minimization problem (15.8) of GTM-TSVM1 to be meaningful, we give the following proposition.

Lemma 15.1. *Given $d_{1i} \geq 0$, $f_{2j} \geq 0$ and $p_{2j} \geq 0$, the objective function of problem (15.8) has a lower bound if $f_{2j} \geq (1 - \lambda_1)p_{2j}$.*

Proof. Suppose there exists some index k $(1 \leq k \leq m_2)$ such that $f_{2j} < (1 - \lambda_1)p_{2j}$ and in such case the optimal solutions for (15.8) are denoted by w_1^*, b_1^*, ξ^* and δ^*. Then, we have $-((w_1^*)^{\mathsf{T}} + b_1^*) \geq 1 - \xi_j^* + \delta_j^*$. We first increase the value of variables ξ_j^* and δ_j^* by a positive increment $\Delta(\Delta \geq 0)$ respectively, and the constraints are still satisfied since $1 - (\xi_j^* + \Delta) + (\delta_j^* + \Delta) = 1 - \xi_j^* + \delta_j^*$. That is to say, w_1^*, b_1^*, $\xi_j^* + \Delta$ and $\delta_j^* + \Delta$ are also feasible solutions. However, the value of objective function of (15.8) can be reduced at this solution. Therefore, we must have $f_{2j} \geq (1-\lambda_1)p_{2j}, \forall j$ to guarantee that the objective function of (15.8) has a lower bound. □

A similar conclusion also holds for GTM-TSVM2 (15.9). This proposition tells us a necessary condition for optimal solution of (15.8) to exist which is critical to guiding the design of weighting factors f_{2j} and p_{2j}.

15.3.2 *Algorithm derivation*

To solve (15.8) efficiently, we introduce the following Lagrangian function

$$\begin{aligned} L = &\frac{1}{2}\sum_{i=1}^{m_1} d_{1i}(w_1^\mathsf{T} x_{1i} + b_1)^2 + C_1\bigg(\sum_{j=1}^{m_2} f_{2j}\xi_j - (1-\lambda_1)\sum_{j=1}^{m_2} p_{2j}\delta_j\bigg) \\ &- \sum_{j=1}^{m_2}\alpha_{1j}\big(-w_1^\mathsf{T} x_{2j} - b_1 - 1 + \xi_j - \delta_j\big) - \sum_{j=1}^{m_2}\alpha_{2j}\xi_j - \sum_{j=1}^{m_2}\alpha_{3j}\delta_j \end{aligned} \tag{15.10}$$

where α_{1j}, α_{2j} and α_{3j} are all nonnegative Lagrangian multipliers. To minimize (15.8), we have to find the saddle point of L, implying that we minimize over the primal variables w_1, b_1, ξ, δ and maximize over the dual variables α_{1j}, α_{2j} and α_{3j}. Setting the derivatives with respect to the primal variables equal to zero yields the following equations

$$\frac{\partial L}{\partial w_1} = \sum_{i=1}^{m_1} d_{1i}(w_1^\mathsf{T} x_{1i} + b_1)x_{1i}^\mathsf{T} + \sum_{j=1}^{m_2}\alpha_{1j}x_{2j}^\mathsf{T} = 0, \tag{15.11}$$

$$\frac{\partial L}{\partial b_1} = \sum_{i=1}^{m_1} d_{1i}(w_1^\mathsf{T} x_{1i} + b_1) + \sum_{j=1}^{m_2}\alpha_{1j} = 0, \tag{15.12}$$

$$\frac{\partial L}{\partial \xi_j} = C_1 f_{2j} - \alpha_{1j} - \alpha_{2j} = 0, \tag{15.13}$$

$$\frac{\partial L}{\partial \delta_j} = -(1-\lambda_1)C_1 p_{2j} + \alpha_{1j} - \alpha_{3j} = 0. \tag{15.14}$$

The Karush-Kuhn-Tucker (KKT) complementarity conditions are given by

$$\alpha_{1j}(-w_1^\mathsf{T} x_{2i} - b_1 - 1 + \xi_j - \delta_j) = 0, \tag{15.15}$$

$$(C_1 f_{2j} - \alpha_{1j})\xi_j = 0, \tag{15.16}$$

$$(-(1-\lambda_1)C_1 p_{2j} + \alpha_{1j})\delta_j = 0. \tag{15.17}$$

To simplify the above equations, we define

$$\begin{aligned} D_1 &= \mathrm{diag}(d_{11}, d_{12}, \ldots, d_{1m_1}), \\ p_2 &= [p_{21}, p_{22}, \ldots, p_{2m_2}]^\mathsf{T}, \\ f_2 &= [f_{21}, f_{22}, \ldots, f_{2m_2}]^\mathsf{T}. \end{aligned} \tag{15.18}$$

Then, the above equations can be rewritten in matrix form as

$$G^\top D_1 G u_1 + H^\top \alpha_1 = 0 \to u_1 = -(G^\top D_1 G) H^\top \alpha_1, \tag{15.19}$$

$$C_1 f_2 - \alpha_1 - \alpha_2 = 0 \to \alpha_2 = C_1 f_2 - \alpha_1 \geq 0, \tag{15.20}$$

$$-(1-\lambda_1)C_1 p_2 + \alpha_1 - \alpha_3 = 0 \to \alpha_3 = -(1-\lambda_1)C_1 p_2 + \alpha_1 \geq 0 \tag{15.21}$$

where $G = [A\ e_1]$, $H = [B\ e_2]$, $u_1 = \begin{bmatrix} w_1 \\ b_1 \end{bmatrix}$ is the augmented weight vector for positive plane. Substituting (15.19)–(15.21) into the Lagrangian function (15.10) finally yields the following dual problem

$$\begin{aligned} \max_{\alpha_1} & -\frac{1}{2}\alpha_1^\top H(G^\top D_1 G)^{-1} H^\top \alpha_1 + e_2^\top \alpha_1 \\ \text{s.t.} \quad & C_1 f_2 \geq \alpha_1 \geq (1-\lambda_1) C_1 p_2. \end{aligned} \tag{15.22}$$

Similarly, the dual problem of GTM-TSVM2 (15.9) can be derived as

$$\begin{aligned} \max_{\alpha_2} & -\frac{1}{2}\alpha_2^\top G(H^\top D_2 H)^{-1} G^\top \alpha_2 + e_1^\top \alpha_2 \\ \text{s.t.} \quad & C_2 f_1 \geq \alpha_2 \geq (1-\lambda_2) C_2 p_1. \end{aligned} \tag{15.23}$$

After solving (15.23) to obtain α_2, the augmented vector for negative plane is given by

$$u_2 = (H^\top D_2 H)^{-1} G^\top \alpha_2 \tag{15.24}$$

where the definitions of D_2, p_1, f_1 are similar to (15.18). Once augmented vectors u_1 and u_2 are figured out from the above equations, the two planes $f_1(x) = w_1^\top x + b_1$ and $f_2(x) = w_2^\top x + b_2$ are known. A new coming sample x is then assigned to class 1 or 2, depending on which of the two planes it lies closest to, that is, $\text{label}(x) = \arg\min_{i=1,2} |w_i^\top x + b_i|$.

Lemma 15.2. *Under the assumption of Lemma 15.1, when the optimal solution* $(w_1^*, b_1^*, \xi^*, \delta^*)$ *for (15.8) is reached, we must have* $\xi_j^* \delta_j^* = 0$.

Proof. We consider this problem from the viewpoint of dual problem (15.22). Supposing the optimal solution for α_1 is denoted by α_1^*, we need to distinguish the three different cases concerning α_1^*, that is, (*i*) $C_1 f_{2j} > \alpha_{1j}^* > (1-\lambda_1)C_1 p_{2j}$, (*ii*) $\alpha_{1j}^* = C_1 f_{2j}$ and (*iii*) $\alpha_{1j}^* = (1-\lambda_1)C_1 p_{2j}$. In the first case, from (15.18) and (15.19), it follows that $\xi_j^* = 0$ and $\delta_j^* = 0$. In other words, those negative samples with $C_1 f_{2j} > \alpha_{1j}^* > (1-\lambda_1)C_1 p_{2j}$ lie exactly on the shifted positive plane $w_1^\top x_{2j} + b_1 = -1$. In the second case, from (15.17) and Lemma 15.1, we have $\xi_j^* > 0$ and $\delta_j^* = 0$. That is to say, the distance between the negative sample satisfying $\alpha_{1j}^* = C_1 f_{2j}$ and positive plane is less than 1. In the third case, from (15.16) and Lemma 15.1, we have $\xi_j^* = 0$ and $\delta_j^* > 0$, indicating the distance between the negative sample satisfying $\alpha_{1j}^* = (1-\lambda_1)C_1 p_{2j}$ and positive plane is great than 1. In conclusion, we always have $\xi_j^* \delta_j^* = 0,\ \forall j$, $\square$

This proposition tells us either slack error ξ_j or surplus error δ_j corresponding to arbitrary sample must be zero when the optimal solution is reached.

In order to extend the above model for nonlinear classification, we explicitly map the samples into the empirical feature space [Jayadeva *et al.* (2007); Ye *et al.* (2011)]. In such case, we seek two hyperplanes in the feature space rather than planes in the input space. To achieve this goal, we express the hyperplanes as follows

$$w_1^\top K(x) + b_1 = 0 \qquad \text{and} \qquad w_2^\top K(x) + b_2 = 0 \tag{15.25}$$

where $K(x) = [k(x_1, x), k(x_2, x), \ldots, k(x_m, x)]^\top$ and k stands for an arbitrary kernel [Cortes and Vapnik (1995)]. Based on the geometric meaning of GTM-TSVM, we can obtain a pair of similar primal QPPs for this nonlinear case. For example, the nonlinear version of GTM-TSVM1 (15.8) can be expressed as

$$\text{(NGTM-TSVM1)} \quad \begin{aligned} \min_{w_1, b_1, \xi, \delta} \ & \frac{1}{2}\sum_{i=1}^{m_1} d_{1i}(w_1^\top K(x_{1i}) + b_1)^2 + C_1\Bigg(\sum_{j=1}^{m_2} f_{2j}\xi_j \\ & -(1-\lambda_1)\sum_{j=1}^{m_2} p_{2j}\delta_j\Bigg) \\ \text{s.t.} \quad & -(w_1^\top K(x_{2j}) + b_1) \geq 1 - \xi_j + \delta_j \\ & \xi_j \geq 0,\ \delta_j \geq 0,\ j = 1, 2, \ldots, m_1 \end{aligned} \tag{15.26}$$

where the involved parameters and variables are defined as in the linear case. The dual problem for (15.26) is formulated as follows by using a similar process to the linear case

$$\begin{aligned} \max_{\alpha_1} \ & -\frac{1}{2}\alpha_1^\top H(G^\top D_1 G)^{-1} H^\top \alpha_1 + e_2^\top \alpha_1 \\ \text{s.t.} \quad & C_1 f_2 \geq \alpha_1 \geq (1-\lambda_1)C_1 p_2 \end{aligned} \tag{15.27}$$

where $G = [K(A) \quad e_1]$, $H = [K(B) \quad e_2]$. After solving (15.27) to obtain the Lagrangian multiplier vector α_1, the parameter for hyperplane of class 1 is then given by $\begin{bmatrix} w_1 \\ b_1 \end{bmatrix} = -(G^\top D_1 G)^{-1} H^\top \alpha_1$.

15.3.3 *Weighting factors for GTM-TSVM*

We introduce local geometry information [Yang *et al.* (2009); Ye *et al.* (2011)] embedded in samples to put different weights on different samples in (15.8) and (15.9). To describe local structure between samples, two kinds of graphs which reflect the k-nearest neighbor relations lurking in the samples need to be constructed in advance.

Taking class 1 for instance, we first construct an intra-class graph S_1 to model intra-class similarity of class 1. Each vertex in S_1 corresponds to a sample of class 1 and an edge between a vertex pair is added when the corresponding sample pair is each others k_1-nearest neighbor (k_1-NN). Let $Ne^+_{k_1}(x_{1i})$ be the k_1-NN set of sample x_{1i} in the same class of x_{1i}. Then, the weight matrix of S_1 is defined as follows

$$\text{if} \quad x_{1j} \in Ne^+_{k_1}(x_{1i}) \quad \text{or} \quad x_{1i} \in Ne^+_{k_1}(x_{1j}) \quad \text{then} \quad W^{S_1}_{ij} = \theta_{ij} \qquad \text{else} \quad W^{S_1}_{ij} = 0. \tag{15.28}$$

When $W^{S_1}_{ij} > 0$, an edge between x_{1i} and x_{1j} is added to graph S_1. Intuitively, the samples in high density regions of class 1 have more chance to become nonzero-degree vertexes and are referred to as interior points (IPs for short), whereas the samples in low density regions, for example, outliers, become more likely isolated vertexes (zero-degree) [Yang *et al.* (2009)].

Then, we construct an inter-class graph R_1 to describe relative separability from class 2 to class 1. The vertex pair in R_1, whose corresponding samples come from different classes, is connected when one of the pair is a k_2-nearest neighbor (k_2-NN) of the other. Let k_2-NN set in the different class of x_{1i} be denoted by $Ne^-_{k_2}(x_{1i})$. Then, the weight matrix of R_1 is defined as follows

$$\text{if} \quad x_{2j} \in Ne^-_{k_2}(x_{1i}) \quad \text{then} \quad W^{R_1}_{ij} = \rho_{ij} \qquad \text{else} \quad W^{R_1}_{ij} = 0. \tag{15.29}$$

When $W^{R_1}_{ij} > 0$, an edge between x_{1i} and x_{2j} is added to graph R_1. It can be observed that the samples of class 2 in the marginal regions, which means they are closer to class 1 than the remaining samples of class 2, have more possibility to become nonzero-degree vertexes. Therefore, such samples of class 2 which may contain important discriminative information are termed as relative marginal points (RMPs). Accordingly, the remaining samples which may contain valuable information about the distribution of class 2 are referred to as relative interior points (RIPs). A geometrical illustration of IPs, RMPs and RIPs is depicted in Fig. 15.3.

Based on the geometrical intuition of intra-class graph S_1 and inter-class graph R_1, and simultaneously considering the necessary condition $f_{2j} \geq (1 - \lambda_1)p_{2j}$ as stated in Lemma 15.1, we present an scheme to set weighting factors for GTM-TSVM1 (15.8) as follows

$$\text{if} \quad \exists j, W^{S_1}_{ij} > 0 \quad \text{then} \quad d_{1i} = 1 \quad \text{else} \quad d_{1i} = 0. \tag{15.30}$$

$$\text{if} \quad \exists i, W^{R_1}_{ij} > 0 \quad \text{then} \quad f_{2j} = 1 \quad \text{else} \quad f_{2j} = 1 - \lambda_1. \tag{15.31}$$

$$\text{if} \quad \exists i, W^{R_1}_{ij} > 0 \quad \text{then} \quad p_{2j} = 0 \quad \text{else} \quad p_{2j} = 1. \tag{15.32}$$

An intuitive explanation for the above weighting formula is elaborated as follows. For the intra-class case, the fitting errors associated with IPs of class 1 are minimized to train the corresponding plane, whereas the other non-IPs are removed since they become more likely outliers. For the inter-class case, the slack variables associated

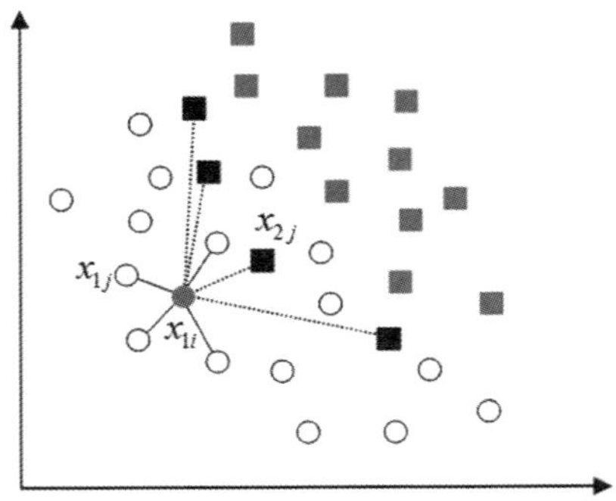

Fig. 15.3 An illustration of IPs (gray circle) of Class 1, RMPs (black square) and RIPs (gray square) of Class 2 with respect to Class 1

with RMPs of class 2 should be punished more severely than those associated with RIPs. At the same time, the surplus variable associated with RIPs should be maximized to ensure they are as far away from the plane of class 1 as possible. Following an analogous idea, we construct intra-class graph S_2 to model intra-class similarity of class 2 and inter-class graph R_2 to describe its relative separability from class 1 to class 2. Then, the weighting factors for GTM-TSVM2 (15.9) can be derived similarly.

15.4 Experimental Results and Analysis

To verify the effectiveness of the proposed TM-TSVM and GTM-TSVM, we investigate their performance and computational cost on the UCI [Blake and Merz (1998)], USPS and MIT CBCL real-world benchmark data sets. In experiments, we compare the performances of GEPSVM, LIPSVM, TSVM, LSTSVM, LCTSVM, and our algorithm. Specifically, we first present a synthetic data set for intuitively demonstrating the behavior of the proposed methods. Then several publicly available benchmark databases are used to check their performance.

15.4.1 *Experimental specification*

All the classification algorithms are implemented in MATLAB (7.8.0) R2009a environment on a PC with Intel Core 2 Duo processor, 3GB RAM. Mosek optimization toolbox for MATLAB was used to solve all the dual QPPs in TSVM, and GTM-TSVM. In the experiments, we follow the methodology in most TSVM-related literatures to compare the involved methods through five-fold cross-validation. Namely, each dataset is partitioned into five disjoint subsets of approximately equal size. The selected algorithm is trained on all the subsets except for one, and the testing accuracy is measured on the subset left out. This procedure is repeated for five times and each time uses a different subset for testing. The optimal model parameters for each algorithm are selected from the corresponding set by grid search. Specifically, to simplify parameter selection, let $C_1 = C_2 = C$ and C be selected from the set $\{10^{-3}, 10^{-2}, \dots, 10^2, 10^3\}$, $\lambda_1 = \lambda_2 = \lambda$ and λ be selected from $\{0.0, 0.2, \dots, 0.8, 1.0\}$, while k of NN in GTM-TSVM, LIPSVM, LCTSVM is selected from the set $\{1, 3, 5, 7, 9\}$. In addition, all samples are normalized such that the features are located in $[0, 1]$. To compare effectiveness and efficiency, average testing accuracy and corresponding standard deviation as well as training CPU time (s) are reported. It should be noticed that the cost of kNN search is also included in the training time for GTM-TSVM, LIPSVM and LCTSVM.

15.4.2 *A toy example*

In this subsection, we construct an artificial 2D dataset in order to intuitively illustrate the behavior of the proposed GTM-TSVM as well as the original TSVM. The dataset is a typical XOR problem which is linearly inseparable. The dataset which comprises of samples belonging to two classes is shown in Fig. 15.4(a). Each class contains 100 training samples and 20 testing samples. We learnt the planes for TSVM and GTM-TSVM on the training set, and evaluated their performance on the testing set. The optimal value of C is tuned to 0.1 for the three methods, k and λ are respectively 3 and 0.0 for GTM-TSVM. Fig. 15.4(a) and Fig. 15.5(a) respectively show the training set as well as the optimal planes of TSVM and GTM-TSVM. In addition, Fig. 15.4(b) and Fig. 15.5(b) show the two-dimensional scatter plots on the training set by plotting samples with coordinates (d_{1i}, d_{2i}), where d_{1i} and d_{2i} are distances of a sample x_i to the plane of class 1 and class 2, i.e., $d_{1i} = |w_1^\top x_i + b_1|$ and $d_{2i} = |w_2^\top x_i + b_2|$, respectively. Furthermore, Fig. 15.6 and Fig. 15.7 show the corresponding results on the testing set for the two methods. The recognition accuracies on the training set (training accuracy) are respectively 88% and 96% for the two methods, whereas the testing accuracies are 92.50% and 97.50%.

From the experimental results, we observe that, for the two methods, the majority of points are clustered around its corresponding plane. The cluster results obtained by a certain multi-plane algorithm also indicate how well the algorithm is able to discriminate between two classes. In fact, according to the scatter plots shown in Fig. 15.6(b) and Fig. 15.7(b), we find that, compared with TSVM, the overlapping of the different classes is reduced in GTM-TSVM and thus achieves better separation. This may because our method requires all samples of other classes are far away from the plane of a certain class, and leads to the distances from the two planes to most samples obtained by GTM-TSVM methods are larger than those by TSVM.

15.4.3 *Evaluations on UCI benchmark data sets*

For further evaluation, we test the involved methods on UCI benchmark datasets. Table 15.1 and 15.2 report the results of the involved algorithms with linear and

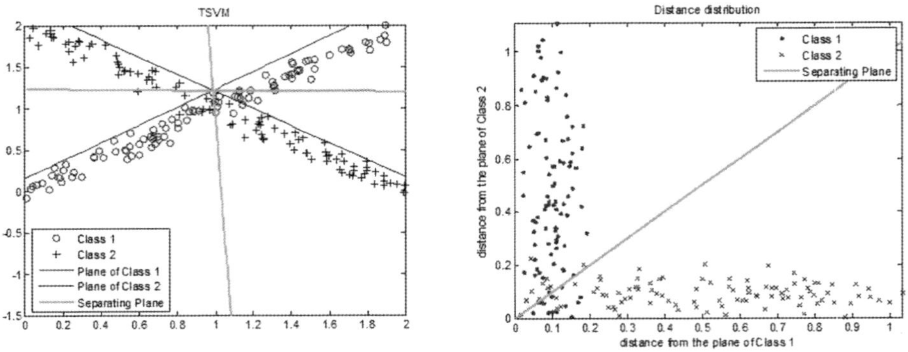

(a) (b)

Fig. 15.4 Classification results on training data of TSVM

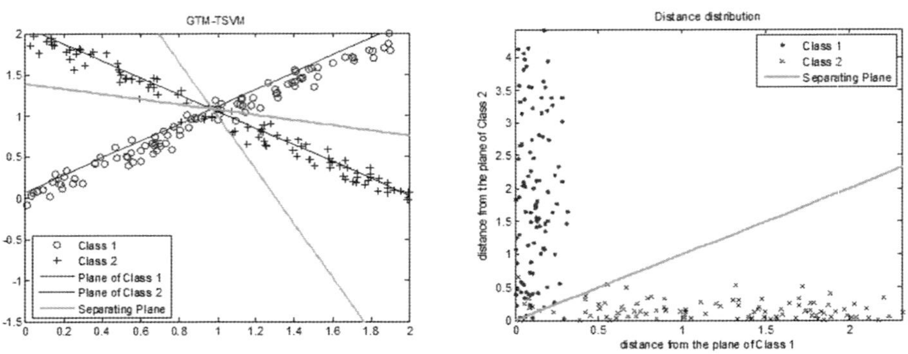

(a) (b)

Fig. 15.5 Classification results on training data of GTM-TSVM

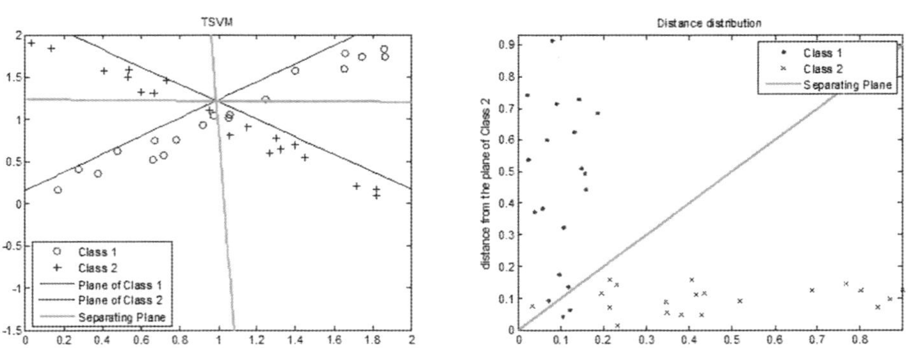

(a) (b)

Fig. 15.6 Classification results on testing data of TSVM

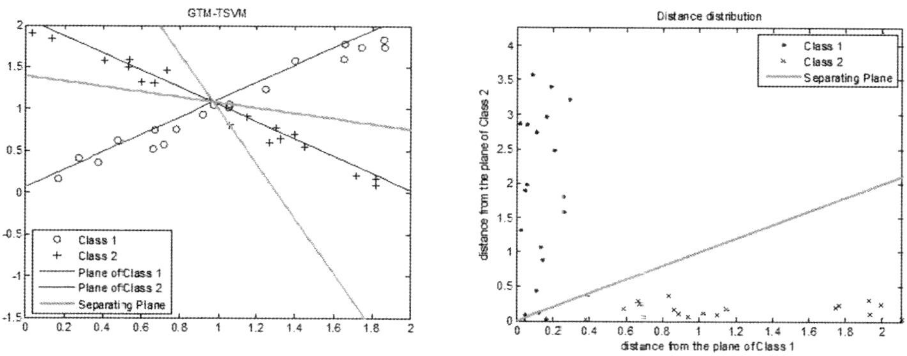

(a) (b)

Fig. 15.7 Classification results on testing data of GTM-TSVM

nonlinear kernel, respectively. For the nonlinear case, Gaussian kernel is employed since it has found wide applications in pattern recognition field and the kernel parameter is chosen from the set $\{2^{-5}, 2^{-4}, \ldots, 2^4, 2^5\}$. From Table 15.1 we can remark that, in comparison with the other methods, the proposed GTM-TSVM usually performs better in terms of classification performance. To show the statistical significance of the results, we performed paired t-tests by comparing GTM-TSVM to the other methods using five-fold testing accuracies, and reported the p-value in Table 15.1. The significance level is set to 0.05. When p-value < 0.05, it implies great performance difference between GTM-TSVM and the compared method exists. Note that bold values in Table 1 denote the best classification performance among all algorithms and an asterisk $(*)$ indicates a significant difference from GTM-TSVM based on p-value < 0.05. Actually, it is easy to see that LCTSVM and GTM-TSVM are among the most competitive methods which take local proximity information into account. A win-loss-tie (W-L-T) summarization based on average testing accuracy is also listed at the bottom of Table 15.1. These W-L-T results verify that GTM-TSVM, on the whole, obtains better accuracies compared with the other methods. Based on the above observations, we may draw the conclusion that the information contained by RMPs and RIPs are both beneficial to the performance of classifier. Table 15.1 also reports the training time (CPU time) of different methods which includes five-fold training. We can observe the computational cost of GTM-TSVM is slightly slower than TSVM since it has to use extra neighbor search procedure to construct proximity graphs. GEPSVM, LIPSVM, LSTSVM and LCTSVM often run faster than the other methods in linear case since they only need to solve small-scale GEPs or linear systems of equations when the dimension of features is not very high. For nonlinear case, the experimental results in Table 15.2 also indicate some similar conclusions except that the training speed of GEPSVM is much slower than the other methods. This is because it has to solve two GEPs of higher dimension. It is also consistent with results in.

Table 15.1 Performance on UCI benchmark data sets with linear kernel

Dataset m×p	GEPSVM Test±Std CPU time p-value	TSVM Test±Std CPU time p-value	LSTSVM Test±Std CPU time p-value	LIPSVM Test±Std CPU time p-value	LCTSVM Test±Std CPU time p-value	GTM-TSVM Test±Std CPU time
Heart-statlog	76.67±1.89*	84.44±1.89	83.33±3.10*	78.89±4.77*	85.19±4.83	**87.04±3.51**
270×13	0.0094	0.1924	0.0043	0.0710	0.0660	0.2061
	0.0038	0.1347	0.0217	0.0375	0.4456	
Hepatitis	81.32±4.05*	85.84±3.08*	85.86±4.57*	81.30±4.66*	**89.69±3.68**	**89.69±3.68**
155×19	0.0093	0.1466	0.0050	0.0343	0.0262	0.1346
	0.0027	0.0334	0.0320	0.0186	-	
Wpbc	79.39±1.51*	81.49±2.65*	81.48±2.79	78.36±0.98*	85.60±3.36	**86.15±4.56**
194×33	0.0126	0.1616	0.0075	0.0640	0.0414	0.1764
	0.0257	0.0221	0.1216	0.0291	0.7881	
Sonar	79.34±2.27	77.92±6.48*	79.37±5.89	76.41±3.08	79.84±5.30	**80.82±6.68**
208×60	0.1126	0.1751	0.0088	0.0935	0.0532	0.1977
	0.6387	0.0321	0.0705	0.2264	0.4655	
Haberman	75.48±2.14*	77.11±3.01	73.86±1.80*	76.13±7.12	76.48±2.97	**78.75±1.86**
306×3	0.0044	0.2463	0.0032	0.1243	0.0859	0.2343
	0.0032	0.0890	0.0006	0.3977	0.1330	
Pima	75.91±5.24	77.48±3.53*	77.74±3.28	63.28±2.18*	77.87±3.35	**78.26±3.63**
768×8	0.0057	1.4519	0.0050	0.7784	0.6311	1.3120
	0.0573	0.0040	0.2425	0.0004	0.3731	
Monk3	82.49±3.88*	83.95±2.43	78.16±3.82*	69.85±3.44*	81.22±4.36*	**85.57±2.89**
554×6	0.0059	0.7643	0.0041	0.3825	0.3120	6.6595
	0.0105	0.1669	0.0007	0.0014	0.0218	
CMC	68.23±2.25*	68.77±1.82	68.02±2.35*	61.17±2.69*	68.77±2.09	**69.58±2.23**
1473×9	0.0110	5.7673	0.0065	3.0801	2.6876	0.3741
	0.0341	0.0606	0.0168	0.0060	0.1177	
Germ	72.10±1.39*	77.70±1.54	78.20±0.75*	68.70±1.29*	78.60±0.97	**78.90±0.97**]
1000×24	0.0261	3.0640	0.0079	1.5605	1.3602	2.9478
	0.0009	0.0802	0.0249	0.0001	0.3739	
Tic-Tac-Toe	64.93±2.34*	67.75±2.27	67.33±2.70*	63.98±2.13*	69.11±2.41*	**70.89±2.37**
958×9	0.0055	1.9438	0.0096	1.1805	1.0546	2.5140
	0.0039	0.0448	0.0238	0.0180	0.0027	
W-T-L	0-0-10	0-0-10	0-0-10	0-0-10	0-1-9	

Table 15.2 Performance on UCI benchmark data sets with Gaussian kernel

Dataset m×p	GEPSVM Test±Std CPU time *p*-value	TSVM Test±Std CPU time *p*-value	LSTSVM Test±Std CPU time *p*-value	LIPSVM Test±Std CPU time *p*-value	LCTSVM Test±Std CPU time *p*-value	GTM-TSVM Test±Std CPU time
Heart-statlog	72.96±6.69*	86.30±2.51	86.67±2.16	77.41±2.16*	87.41±2.46	**88.15±1.48**
270×13	2.0460	0.3476	0.1163	0.6823	0.1924	0.4044
	0.0170	0.1419	0.0993	0.0001	0.3739	
Hepatitis	83.20±2.60*	89.67±3.77	87.74±4.76*	86.45±1.31	89.65±4.39	**90.94±4.81**
155×19	0.4766	0.1961	0.0434	0.1857	0.0718	0.1789
	0.0240	0.1779	0.0316	0.1094	0.3738	
Wpbc	78.83±2.23*	84.02±3.81	85.56±2.11	82.99±1.20*	**86.62±2.88**	**86.62±2.86**
194×33	0.7734	0.2537	0.0650	0.3890	0.1200	0.2493
	0.0141	0.0885	0.5693	0.0246	0.9996	
Sonar	84.65±3.41	89.39±3.38	88.96±4.08	85.13±2.59*	88.96±3.77	**90.38±3.04**
208×60	0.9535	0.2984	0.0825	0.3066	0.1399	0.2737
	0.1171	0.5329	0.3073	0.0285	0.2044	
Haberman	79.73±2.29	78.44±1.79*	78.43±2.37*	78.43±1.25*	80.07±2.94	**82.67±1.75**
306×3	2.5661	0.4844	0.1753	1.1837	0.2864	0.5117
	0.0538	0.0470	0.0399	0.0026	0.1941	
Pima	75.91±3.94*	78.39±4.48	78.26±4.07	75.00±3.15*	79.56±4.50	**79.82±3.68**
768×8	51.8091	3.7979	1.3693	13.9723	2.3876	4.2270
	0.0062	0.1613	0.0000	0.0165	0.6893	
Wdbc	92.45±1.59*	97.19±1.01	96.49±1.44	95.96±0.69*	97.19±1.01	**97.37±0.94**
569×30	20.3604	1.9231	0.6970	2.5265	1.1615	1.9475
	0.0003	0.3739	0.0894	0.0031	0.3739	
Liver	66.09±3.73*	**78.55±4.14**	75.65±4.24*	68.41±2.96*	77.97±4.62	**78.55±3.48**
345×6	4.6204	0.5921	0.2136	1.8381	0.3629	0.6979
	0.0005	1.0000	0.0217	0.0036	0.5415	
Blood	78.21±1.82*	84.62±4.21	81.82±3.44	81.95±2.60*	84.62±3.70	**85.96±1.21**
748×4	42.4563	4.4438	1.3227	7.3024	2.3645	4.2840
	0.0060	0.5094	0.0847	0.0156	0.4244	
Australia	61.88±2.48*	79.71±3.50*	81.16±3.62	80.14±1.99	81.59±3.19	**82.03±3.29**
690×14	33.0955	3.2485	1.0761	9.6884	1.8162	3.4443
	0.0005	0.0030	0.2356	0.4624	0.5279	
W-T-L	0-0-10	0-1-9	0-0-10	0-0-10	0-1-9	

15.4.4 *Experiments on USPS and MIT CBCL database*

To further evaluate the performance of these methods, we run them on the well-known USPS and MIT CBCL database. The USPS database comprises of 8-bit grayscale handwritten digit images from 0 to 9 and 1100 images of each digit. The size of each sample image is 16×16 pixels with 256 gray levels. In the experiments, we choose eight pairwise digits of varying difficulty for odd vs. even digit recognition as in [Peng and Xu (2013)]. For each pair, 50 samples per class are randomly selected for training whereas the remaining for testing. The MIT CBCL database consists of 2429 face images and 4548 non-face images of dimension 19×19. A subset consisting of 500 face images and 500 non-face images is used in our experiments. Randomly selected $n(n = 10, 20, \ldots, 50)$ samples per class are used for training and the rest for testing. Principle component analysis (PCA) is conducted a prior to reduce the dimension of each sample. Fig. 15.8 and Fig. 15.9 show some typical sample images from USPS and CBCL database. The above evaluation repeats 10 times and average testing accuracies are recorded. The p-value calculated by performing paired t-test on the 10 classification accuracies is also used to show the statistical significance of GTM-TSVM. The experimental results including average testing accuracies and p-values of each method are reported in Tables 15.3 and 15.4. From the experimental results, we can see again that the proposed GTM-TSVM is superior to the other algorithms in most cases. Moreover, the experiments also show that the methods which make use of local proximity information between samples, i.e., LIPSVM, LCTSVM and GTM-TSVM generally work better than the original methods which ignore such kind of information. In short, the improved performance of our proposed algorithm shows it is able to enhance TSVM by appropriately selecting data samples and at the same time making full use of the associated neighborhood information.

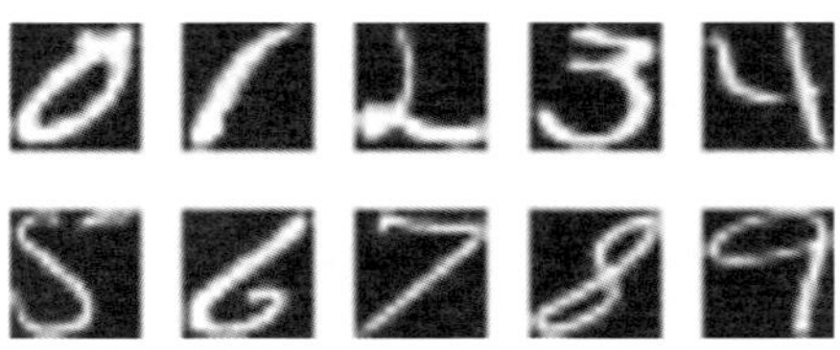

Fig. 15.8 An illustration of 10 subjects in the USPS database

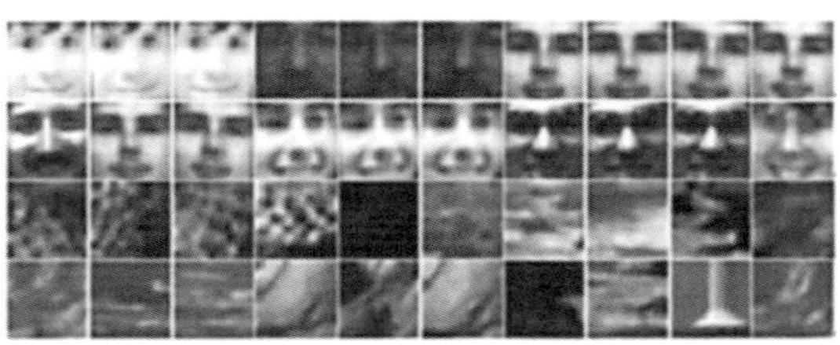

Fig. 15.9 An illustration of face and non-face images in MIT CBCL database

Table 15.3 Performance on USPS benchmark data sets

Case	GEPSVM Test±Std p-value	TSVM Test±Std p-value	LSTSVM Test±Std p-value	LIPSVM Test±Std p-value	LCTSVM Test±Std p-value	GTM-TSVM Test±Std
0 vs. 1	98.21±0.55*	98.88±0.14	98.84±0.11	98.57±0.32*	**98.96±0.06**	98.91±0.13
	0.0023	0.2977	0.2163	0.0092	0.2620	
1 vs. 2	93.63±2.25*	96.99±0.33*	96.68±0.68*	95.22±1.31*	96.98±0.62	**97.23±0.41**
	0.0007	0.0019	0.0174	0.0009	0.1397	
1 vs. 7	91.16±3.71*	94.18±2.83	93.95±2.93*	92.80±3.36	94.32±2.79	**94.41±2.70**
	0.0025	0.1253	0.0062	0.0594	0.3481	
2 vs. 3	94.16±1.33*	95.39±0.86*	95.39±0.94*	94.84±1.06*	**95.89±0.84**	95.84±0.80
	0.0007	0.0027	0.0009	0.0002	0.5323	
2 vs. 7	97.49±0.93*	98.60±0.47*	98.35±0.45*	97.69±0.52*	98.65±0.44	**98.74±0.40**
	0.0087	0.0282	0.0020	0.0006	0.1093	
3 vs. 8	93.69±2.56*	95.17±1.24*	94.80±1.57*	93.98±2.26*	95.01±1.62*	**95.38±1.24**
	0.0076	0.0277	0.0083	0.0066	0.0436	
4 vs. 7	98.68±0.24*	98.74±0.38	98.60±0.32*	98.67±0.31	98.78±0.28	**98.79±0.37**
	0.0887	0.1022	0.0048	0.2190	0.7995	
5 vs. 6	97.08±0.61*	97.98±0.41*	97.93±0.50*	97.74±0.64*	98.11±0.35	**98.15±0.41**
	0.0000	0.0053	0.0061	0.0275	0.6592	
W-T-L	0-0-8	0-0-8	0-0-8	0-0-8	2-0-6	

Table 15.4 Performance on CBCL benchmark data sets

#Samples	GEPSVM Test±Std p-value	TSVM Test±Std p-value	LSTSVM Test±Std p-value	LIPSVM Test±Std p-value	LCTSVM Test±Std p-value	GTM-TSVM Test±Std
10	92.86±2.21	93.13±2.44	93.51±1.95	93.13±2.09	**94.08±1.58**	93.69±2.02
	0.1601	0.0776	0.6600	0.2277	0.1827	
20	94.63±1.36*	96.27±0.88	96.08±0.95*	95.59±1.28*	96.41±1.00	**96.51±1.03**
	0.0001	0.0511	0.0237	0.0078	0.4199	
30	95.70±0.80*	96.77±0.70	96.33±0.69	95.27±1.60*	96.84±0.76	**96.87±0.96**
	0.0002	0.6203	0.0811	0.0045	0.8154	
40	96.08±0.48*	96.71±0.76	96.45±0.75	96.27±1.21	96.61±0.94	**96.80±0.84**
	0.0228	0.5478	0.1465	0.0537	0.0576	
50	96.38±0.59*	96.81±0.51	96.59±0.53*	95.98±1.04*	96.92±0.46	**97.08±0.51**
	0.0089	0.0944	0.0053	0.0039	0.3217	
W-T-L	0-0-5	0-0-5	0-0-5	0-0-5	1-0-4	

15.5 Conclusions

In this chapter, by taking into account the information contained by all samples, we proposed a novel supervised learning algorithm coined graph embedded total margin twin support vector machine (GTM-TSVM). It attempts to combine the merits of total margin and graph embedding into a unified optimization model. Furthermore, by assigning different weighting factors, we develop a concrete algorithm based on the proximity information between data samples. The experiments carried out on many benchmark datasets validate the effectiveness of the proposed methods. However, each method has its own advantages and disadvantages. For the proposed algorithm, the training speed is not fast enough in its current implementation. In the future works, we will try to accelerate the proposed GTM-TSVM using other optimization techniques, such as coordinate descent, geometric algorithm etc, thus facilitating its applications to large sized datasets. In addition, how to impose sparsity for the solution of GTM-TSVM so as to reduce testing time is another interesting problem which deserves further investigation.

Acknowledgement

This work was partially supported by the National Science Fund for Distinguished Young Scholars under Grant Nos. 61125305, 61233011 and 61203244, the Key Project of Chinese Ministry of Education under Grant No. 313030.

References

Blake, C. and Merz, C. (1998). UCI repository of machine learning databases, `http://www.ics.uci.edu/mlearn/MLRepository.html`.

Boyd, S. and Vandenberghe, L. (2004). *Convex Optimization* (Cambridge University Press).

Chen, X., Yang, J. and Liang, J. (2011). Optimal locality regularized least squares support vector machine via alternating optimization, *Neural Processing Letters* **33**, 3, pp. 301–315.

Chen, X., Yang, J., Liang, J. and Ye, Q. (2012). Recursive robust least squares support vector regression based on maximum correntropy criterion, *Neurocomputing* **97**, pp. 63–73.

Chen, X., Yang. J., Ye, Q. and Liang, J. (2011). Recursive projection twin support vector machine via within-class variance minimization, *Pattern Recognition* **44**, 10–11, pp. 2643–2655.

Cortes, C. and Vapnik, V. (1995). Support-vector networks, *Machine Learning* **20**, 3, pp. 273–297.

Fung, G. and Mangasarian, O. (2005). Multicategory proximal support vector machine classifiers, *Machine Learning* **59**, 1–2, pp. 77–97.

Jayadeva, Khemchandani, R. and Chandra, S. (2007). Twin support vector machines for pattern classification, *IEEE Transactions on Pattern Analysis and Machine Intelligence* **29**, 5, pp. 905–910.

Kuma, M. A. and Gopal, M. (2009). Least squares twin support vector machines for pattern classification, *Expert Systems with Applications* **36**, 4, pp. 7535–7543.

Mangasarian, O. and Wild, E. (2001). Proximal support vector machine classifiers, in *International Conference on Knowledge Discovery and Data Mining* (San Francisco, USA), pp. 77–86.

Mangasarian, O. and Wild, E. (2006). Multisurface proximal support vector machine classification via generalized eigenvalues, *IEEE Transactions on Pattern Analysis and Machine Intelligence* **28**, 1, pp. 69–74.

Peng, X. (2010). A ν-twin support vector machine (ν-TSVM) classifier and its geometric algorithms, *Information Sciences* **180**, pp. 3863–3875.

Peng, X. and Xu, D. (2013). A twin-hypersphere support vector machine classifier and the fast learning algorithm, *Information Sciences* **221**, pp. 12–27.

Ripley, B. (2008). *Pattern Recognition and Neural Networks* (Cambridge University Press).

Shao, Y., Zhang. C., Wang, X. and Deng, N. (2011). Improvements on twin support vector machines, *IEEE Transactions on Neural Networks* **22**, 16, pp. 962–968.

Suykens, J. and Vandewalle, J. (1999). Least squares support vector machine classifiers, *Neural Processing Letters* **9**, 3, pp. 293–300.

Vapnik, V. (1998). *Statistical Learning Theory* (Wiley).

Xu. D., Yan, S., Tao, D., Lin, S. and Zhang, H. (2007). Marginal fisher analysis and its variants for human gait recognition and content-based image retrieval, *IEEE Transactions on Image Processing* **16**, 11, pp. 2811–2821.

Yan, S., Xu. D., Zhang, B., Zhang, H., Yang, Q. and Lin, S. (2007). Graph embedding and extensions: A general framework for dimensionality reduction, *IEEE Transactions on Pattern Analysis and Machine Intelligence* **29**, 1, pp. 40–51.

Yang, X., Chen, S., Chen, B. and Pan, Z. (2009). Proximal support vector machine using local information, *Neurocomputing* **73**, 1–3, pp. 357–365.

Ye, Q., Zhao. C., Ye, N. and Chen, X. (2011). Localized twin SVM via convex minimization, *Neurocomputing* **74**, 4, pp. 580–587.

Yoon, M., Yun. Y. and Nakayama, H. (2003). A role of total margin in support vector machines, in *Proceedings of the International Joint Conference on Neural Networks* (Portland, USA), pp. 2049–2053.

Chapter 16

Regularized Covariance Matrix Estimation Based on MDL Principle

Xiuling Zhou, Ping Guo and C. L. Philip Chen

Abstract When Gaussian Mixture Model (GMM) is used for classification for small sample problem with high dimension, the estimation of the sample covariance matrix will be singular, which can lead to lower classification accuracy or can not achieve credible results. KLIM is a regularization method proposed for solving the problem. However, the regularization parameter in the KLIM is single and fixed which makes it suitable for a certain class of problems. In this chapter, under the framework of minimum description length (MDL) principle, the extensions of KLIM are investigated. One is the covariance matrix estimation with multi-regularization parameters, and the other is the covariance matrix estimation with variable regularization parameters. In the developed methods, based on MDL principle, regularization parameters are selected by the criterion of minimization the KL divergence and approximated efficiently by second-order Taylor expansion. The feasibility and effectiveness of developed methods are validated with experiment studies.

16.1 Introduction

Gaussian mixture model (GMM) has been widely used in pattern recognition problem for clustering and classification, where the maximum likelihood is adopted to estimate the model parameters with the training samples [Bishop (2007); Everitt and Hand (1981); McLachlan and Basford (1988)]. To obtain reasonably accurate estimation of parameters (such as prior probability, mean and covariance matrix for each class) in Gaussian classifier, it is often assumed that there are enough training samples available. However, in the real world, often only a small sample size data set can be obtained for high dimensional data, because gathering these training samples is difficult and expensive. In this case, for d-dimensional data, if less than $d+1$ training samples is available for each class, the sample covariance matrix estimate in Gaussian classifier is singular, and therefore unstable. Since inaccurate estimation of the covariance matrix can lead to lower classification accuracy, it is an obstacle to classify high dimensional data by the Gaussian classifier when there are too few training samples.

There are two ways to solve this kind of problem. One is dimensionality reduction by feature extraction or feature selection [Jolliffe (1996)], the other is regularization of sample covariance matrix [Friedman (1989); Friedman *et al.* (2008); Hoffbeck and Landgrebe (1996); Srivastava *et al.* (2007)]. Regularization is the procedure of allowing parameters bias toward what are thought to be more plausible values, which will reduce the variance of the estimates at the cost of introducing bias.

Many regularized classification techniques have been proposed. Shrinkage and regularized covariance estimators are examples of such techniques. Shrinkage estimator is a widely used class of estimator, which regularizes the covariance matrix by shrinking it toward some positive definite target structures, such as the identity matrix or the diagonal of the sample covariance [Daniels and Kass (2001); Friedman (1989); Hoffbeck and Landgrebe (1996); Kuo and Landgrebe (2002); Ledoit and Wolf (2004); Schafer and Strimmer (2005); Srivastava *et al.* (2007)]. More recently, a number of methods have been proposed for regularizing the covariance estimate by constraining the estimate of the covariance or its inverse to be sparse [Bickel and Levina (2008); Cao *et al.* (2009, 2011); Friedman *et al.* (2008)].

The above regularization methods mainly focus on various mixture of sample covariance matrix, common covariance matrix and identity matrix or constraint the estimate of the covariance or its inverse to be sparse. In these methods, the regularization parameters are required to be determined by cross validation technique. Although the regularization methods have been successfully applied for classifying small size data with some heuristic approximations [Friedman (1989); Hoffbeck and Landgrebe (1996)], the selection of regularization parameters by cross validation technique is computationally expensive. In some cases, cross-validation does not always perform well in the selection of linear models [Rivals and Personnaz (1999)].

Originally proposed as an estimation criterion by Rissanen [Barron*et al.* (1998); Rissanen (1978)], the central idea of the minimum description length (MDL) principle is to represent an entire class of probability distributions as models by a single universal representative model, such that it would be able to imitate the behavior of any model in the class. The best model class for a set of observed data is the one whose representative permits the shortest coding of the data. The MDL estimates of both the parameters and their total number are consistent; i.e., the estimates converge and the limit specifies the data generating model. The codelength (probability distribution or a model) criterion of MDL involves in the Kullback-Leibler (KL) divergence [Barron*et al.* (1998)].

Recently, a covariance matrix estimation method named as KLIM (Kullback-Leibler information measure) estimator is developed based on MDL principle for small size samples with high dimension data [Guo *et al.* (2008)]. The KLIM is derived theoretically by minimizing KL (Kullback-Leibler) divergence. And a formula for fast estimation the regularization parameter is derived. However, since multi-parameters optimization is more difficult than single parameter optimization, only a special case where the regularization parameters are taken the same value for all dimensions is considered in KLIM. Though estimation of regularization parameter becomes simple, ignoring the local difference in each dimension decreases the accuracy of covariance matrix estimation. Consequently, that leads to decrease the classification accuracy of Gaussian classifier. On the other hand, the regularization parameter in KLIM is fixed and taken the same value for all the classes. That is, it is irrelevant to the class. This fixed regularization parameter approach may result in undersmooth in regions with only sparse samples while oversmooth in other regions, whereas varying the regularization parameter gives flexibility to reduce the variance of the estimators in regions with few samples, and to reduce the bias of the estimators in regions with lots of samples. Therefore, the regularization parameter should be variable according to the class in such cases.

In this chapter, the covariance matrix estimation with multi-regularization parameters and the covariance matrix estimation with variable regularization parameters are investigated. Based on MDL principle, new covariance matrix estimators named KLIM_L, KLIM_V and KLIM_V_L are derived for the small sample size problem with high dimension settings. KLIM_L is an extension of KLIM where local difference in each dimension is considered. KLIM_V and KLIM_V_L are extensions of KLIM where regularization parameters are variable according to the class. Regularization parameters in the developed methods are selected by the criteria of minimizing the KL divergence and efficiently estimated under the approximation of two-order Taylor expansion. The feasibility and efficiency of the developed methods are shown by the experiments.

16.2 Theoretical Backgrounds

In this section, some theoretical backgrounds are introduced including Gaussian mixture model, some classical discriminant analysis and KLIM method.

16.2.1 *Gaussian mixture model classifier*

Given a data set $D = \{\mathbf{x}_i\}_{i=1}^{N}$ which will be classified. Assume that the data point in $\mathbf{D}$ is sampled from a Gaussian mixture model which has k components:

$$p(\mathbf{x}, \Theta) = \sum_{j=1}^{k} \alpha_j G(\mathbf{x}, \mathbf{m}_j, \mathbf{\Sigma}_j) \tag{16.1}$$

with $\alpha_j \geq 0$ and $\sum_{j=1}^{k} \alpha_j = 1$, where

$$G(\mathbf{x}, \mathbf{m}_j, \mathbf{\Sigma}_j) = \frac{1}{(2\pi)^{d/2}|\mathbf{\Sigma}_j|^{1/2}} \exp\left\{-\frac{1}{2}(\mathbf{x}-\mathbf{m}_j)^\top \mathbf{\Sigma}_j{}^{-1}(\mathbf{x}-\mathbf{m}_j)\right\} \tag{16.2}$$

is a general multivariate Gaussian function; $\mathbf{x}$ is a random vector, the dimension of $\mathbf{x}$ is d and $\Theta=\{\alpha_j, \mathbf{m}_j, \mathbf{\Sigma}_j\}_{j=1}^{k}$ is a parameter vector set of Gaussian mixture model. Here α_j denotes the prior probability, $\mathbf{m}_j$ denotes the mean vector, and $\mathbf{\Sigma}_j$ denotes the covariance matrix of the j-th component.

In the case of Gaussian mixture model, the Bayesian decision rule

$$j^* = \arg\max_j p(j|\mathbf{x}, \Theta) \tag{16.3}$$

is used to classify the vector $\mathbf{x}$ into class j^* with the largest posterior probability $p(j|\mathbf{x}, \Theta)$. The posterior probability functions $p(j|\mathbf{x}, \Theta)$ are usually unknown and have to be estimated from the training samples. After model parameters Θ estimated by the maximum likelihood (ML) method with expectation-maximization (EM) algorithm [Redner and Walker (1984)], the posterior probability can be written in the form:

$$p(j|\mathbf{x}, \hat{\Theta}) = \frac{\hat{\alpha}_j G(\mathbf{x}, \hat{\mathbf{m}}_j, \hat{\mathbf{\Sigma}}_j)}{p(\mathbf{x}, \hat{\Theta})}, \quad j = 1, 2, \ldots, k. \tag{16.4}$$

By taking the logarithm of (16.3) and ignoring the common factors of the classes, the classification rule becomes:

$$j^* = \arg\min_j d_j(\mathbf{x}), \quad j = 1, 2, \ldots, k, \tag{16.5}$$

where

$$d_j(\mathbf{x}) = (\mathbf{x} - \hat{\mathbf{m}}_j)^\top \hat{\mathbf{\Sigma}}_j^{-1}(\mathbf{x} - \hat{\mathbf{m}}_j) + \ln|\hat{\mathbf{\Sigma}}_j| - 2\ln\hat{\alpha}_j. \tag{16.6}$$

(16.5) is called the discriminant function for the class j [Aeberhard *et al.* (1994)]. Furthermore, if the prior probability $\hat{\alpha}_j$ is the same for all classes, the term $2\ln\hat{\alpha}_j$ can be ignored and the discriminant function reduces to a simpler form.

16.2.2 *Classical discriminant analysis*

In this section, four classical discriminant analysis methods, QDA (Quadratic discriminant analysis), LDA (Linear discriminant analysis), RDA (Regularized discriminant analysis), and LOOC (Leave-one-out likelihood) are reviewed. Except QDA, the remainders are all regularization methods. The main difference among these methods is the estimation formula of covariance matrix.

(1) QDA

The method using the classification rules of (16.5) and (16.6) is called QDA, since it separates the disjoint regions of the input space corresponding to each class assignment by quadratic boundaries [Aeberhard *et al.* (1994); Webb (1999)]. It is widely used in pattern recognition problem. In QDA, the parameters in (16.6) can be estimated with traditional maximum likelihood estimator shown as follows:

$$\hat{\mathbf{m}}_j = \frac{1}{n_j}\sum_{i=1}^{n_j}\mathbf{x}_i \tag{16.7}$$

$$\hat{\boldsymbol{\Sigma}}_j = \frac{1}{n_j}\sum_{i=1}^{n_j}(\mathbf{x}_i - \hat{\mathbf{m}}_j)(\mathbf{x}_i - \hat{\mathbf{m}}_j)^{\mathsf{T}} \tag{16.8}$$

where $\mathbf{x}_i$ is a sample from class j with probability one, and n_j is the number of training sample of class j. Here $\hat{\boldsymbol{\Sigma}}_\mathbf{j}$ is called sample covariance matrix.

QDA often suffers poorly-posed problem and ill-posed problem. Roughly speaking, a problem is poorly posed if the number of parameters estimated is comparable to the number of training sample. Worse, the problem is ill-posed if the number of parameters estimated exceeds the size of training sample. In these cases the parameter estimates can be highly unstable, giving rise to high variance in classification accuracy. By employing a method of regularization, the estimates will be improved by biasing them away from their sample based values towards values that are deemed to be more "physically plausible" [Friedman (1989)].

(2) LDA

LDA [Aeberhard *et al.* (1994)] can be seen as a regularization method to deal with the poorly-posed problem. In LDA, the covariance matrix in (16.6) is taken the following common covariance matrix:

$$\hat{\boldsymbol{\Sigma}} = \frac{1}{N}\sum_{j=1}^{k} n_j\hat{\boldsymbol{\Sigma}}_j. \tag{16.9}$$

In this case, the quadratic terms in (16.5) and (16.6) are canceled, resulting in linear decision boundaries.

(3) RDA

RDA proposed by Friedman [Friedman (1989)] is designed for poorly-posed or ill-posed problem. In RDA, the covariance matrix in (16.6) is taken the following form:

$$\hat{\boldsymbol{\Sigma}}_j(\lambda,\gamma) = (1-\gamma)\hat{\boldsymbol{\Sigma}}_j(\lambda) + \gamma\frac{\mathrm{trace}(\hat{\boldsymbol{\Sigma}}_j(\lambda))}{d}\mathbf{I}_d \tag{16.10}$$

$$\hat{\boldsymbol{\Sigma}}_j(\lambda) = \frac{(1-\lambda)n_j\hat{\boldsymbol{\Sigma}}_j + \lambda N\hat{\boldsymbol{\Sigma}}}{(1-\lambda)n_j + \lambda N}. \tag{16.11}$$

The two parameters λ and γ, which are restricted to the range from 0 to 1, are common regularization parameters for each class. The two parameters are selected according to leave one out cross validation technique by maximizing the classification accuracy. Parameter γ controls the degree of shrinkage of the individual class covariance matrix $\hat{\mathbf{\Sigma}}_\mathbf{j}$ towards the common matrix $\hat{\mathbf{\Sigma}}$, while γ controls the shrinkage towards a multiple of the identity matrix. QDA and LDA are two special cases of RDA. The values $\lambda = 0$ and $\gamma = 0$ gives QDA, whereas $\lambda = 1$ and $\gamma = 0$ yields LDA.

(4) LOOC

LOOC is another regularization method for covariance matrix estimation proposed by Hoffbeck and Landgrebe [Hoffbeck and Landgrebe (1996)]. In LOOC, the diagonal sample covariance matrix, the diagonal common covariance matrix, and some pair-wise mixtures of those matrices are considered. The proposed estimator has the following form:

$$\hat{\mathbf{\Sigma}}_j(\xi_j) = \xi_{j1}\mathrm{diag}(\hat{\mathbf{\Sigma}}_j) + \xi_{j2}\hat{\mathbf{\Sigma}}_j + \xi_{j3}\hat{\mathbf{\Sigma}} + \xi_{j4}\mathrm{diag}(\hat{\mathbf{\Sigma}}). \tag{16.12}$$

The elements of the separate regularization parameter $\xi_j = (\xi_{j1}, \xi_{j2}, \xi_{j3}, \xi_{j4})^\mathsf{T}$ for each class are required to sum to unity: $\sum_{i=1}^{4} \xi_{ji} = 1$. In order to reduce the computation cost, three cases are considered: $\xi_{j3} = \xi_{j4} = 0$, $\xi_{j1} = \xi_{j4} = 0$ and $\xi_{j1} = \xi_{j2} = 0$. The regularization parameters are determined by leave-one-out cross validation method. The regularization parameters of LOOC require only one-dimension optimization and consider only three special cases, while the regularization parameters of RDA involve two-dimension optimization. So LOOC requires less computation time than RDA.

16.2.3 *KLIM method*

Different from RDA and LOOC which are heuristically proposed, KLIM is derived by Ping Guo to estimate the covariance matrix and the regularization parameter under the framework of MDL principle [Guo *et al.* (2008)].

When GMM is used for clustering, prior probability and posterior probability require to be estimated. If these probabilities are known, it becomes a classification problem. So clustering is more general than classification in the mixture model analysis. Let us consider the general case first.

Now we consider that given sample data set $D = \{\mathbf{x}_i\}_{i=1}^N$ generated from an unknown density $p(\mathbf{X})$, it can be modeled by a finite Gaussian mixture density $p(\mathbf{x}, \Theta)$, where Θ is the parameter set. In the absence of knowledge of $p(\mathbf{X})$, it may be estimated by an empirical kernel density estimate $p_\mathbf{W}(\mathbf{X})$ obtained from the data set. Because these two probability densities describe the same unknown density $p(\mathbf{X})$, they should be best matched with proper mixture parameters and smoothing parameters. According to MDL principle, the model parameters should be estimated with minimized KL divergence $KL(\mathbf{W}, \Theta)$ based on the given data drawn from the unknown density $p(\mathbf{X})$. With the constraint of the given data set, the "distance" of these two probability densities can be measured with the following KL divergence function [Kullback (1959)]

$$KL(\mathbf{W}, \Theta) = \int p_{\mathbf{W}}(\mathbf{X}) \ln \frac{p_{\mathbf{W}}(\mathbf{X})}{p(\mathbf{X}, \Theta)} \mathrm{d}\mathbf{x} \tag{16.13}$$

with

$$\begin{aligned} p_{\mathbf{W}}(\mathbf{X}) &= \frac{1}{N} \sum_{i=1}^{N} G(\mathbf{X}, \mathbf{X}_i, \mathbf{W}) \\ &= \frac{1}{N(2\pi)^{d/2} |\mathbf{W}|^{1/2}} \sum_{i=1}^{N} \exp\left\{ -\frac{1}{2} (\mathbf{X} - \mathbf{X}_i)^{\mathsf{T}} \mathbf{W}^{-1} (\mathbf{X} - \mathbf{X}_i) \right\} \end{aligned} \tag{16.14}$$

is assigned as the Gaussian kernel density for the given samples. Here $\mathbf{W}$ is called the smoothing matrix (or regularization parameter matrix) which is symmetric and positive definite. Here $\mathbf{W}$ is a $d \times d$ dimensional diagonal matrix with the following form:

$$\mathbf{W} = \mathrm{diag}(h_1, h_2, \ldots h_d) \tag{16.15}$$

where $h_i,\ i = 1, \ldots, d$ are smoothing parameters (or regularization parameters) in the nonparametric kernel density. (16.13) equals to zero if and only if $p_{\mathbf{W}}(\mathbf{X}) = p(\mathbf{X}\,\Theta)$.

If $\mathbf{W} \to 0$, the kernel density function $p_{\mathbf{W}}(\mathbf{X})$ becomes a δ function, then, (16.13) reduces to the negative log likelihood function. So the ordinary EM algorithm can be re-derived based on the minimization of this KL divergence function with$\mathbf{W} \to 0$ [Redner and Walker (1984); Webb (1999)]. The ML-estimated parameters are shown as follows:

$$\begin{aligned} n_j &= \sum_{i=1}^{N} p(j|\mathbf{x}_i, \hat{\Theta}), \\ \hat{\alpha}_j &= \frac{n_j}{N}, \\ \hat{\mathbf{m}}_j &= \frac{1}{n_j} \sum_{i=1}^{N} p(j|\mathbf{x}_i, \hat{\Theta}) \mathbf{x}_i, \\ \hat{\mathbf{\Sigma}}_j &= \frac{1}{n_j} \sum_{i=1}^{N} p(j|\mathbf{x}_i, \hat{\Theta}) (\mathbf{x}_i - \hat{\mathbf{m}}_j)(\mathbf{x}_i - \hat{\mathbf{m}}_j)^{\mathsf{T}}. \end{aligned} \tag{16.16}$$

The covariance matrix estimation for the limit $\mathbf{W} \not\to 0$ is shown as follows. By minimizing (16.13) with respect to $\mathbf{\Sigma}_j$, i.e., setting $\partial KL(\mathbf{W}, \Theta)/\partial \mathbf{\Sigma}_j = 0$, the following covariance matrix estimation formula can be obtained:

$$\hat{\mathbf{\Sigma}}_j(\mathbf{W}) = \frac{\int p_{\mathbf{W}}(\mathbf{x}) p(j|\mathbf{x}, \hat{\Theta})(\mathbf{x} - \hat{\mathbf{m}}_j)(\mathbf{x} - \hat{\mathbf{m}}_j)^{\mathsf{T}} \mathrm{d}\mathbf{x}}{\int p_{\mathbf{W}}(\mathbf{x}) p(j|\mathbf{x}, \hat{\Theta}) \mathrm{d}\mathbf{x}}. \tag{16.17}$$

In this case, the Taylor expansion is used for $p(j|\mathbf{x},\Theta)$ at $\mathbf{x}=\mathbf{x}_i$ with respect to $\mathbf{x}$ and it is expanded to first order approximation:

$$p(j|\mathbf{x},\hat{\Theta}) \approx p(j|\mathbf{x}_i,\hat{\Theta}) + (\mathbf{x}-\mathbf{x}_i)^{\mathsf{T}}\nabla_{\mathbf{x}}p(j|\mathbf{x}_i,\hat{\Theta}) \tag{16.18}$$

where

$$\nabla_{\mathbf{x}}p(j|\mathbf{x}_i,\hat{\Theta}) = \nabla_{\mathbf{x}}p(j|\mathbf{x},\hat{\Theta})|_{\mathbf{x}=\mathbf{x}_i}. \tag{16.19}$$

By substituting (16.18) into (16.17) and according to the properties of probability density function, the following approximation is finally derived:

$$\mathbf{\Sigma}_j(\mathbf{W}) = \mathbf{W} + \hat{\mathbf{\Sigma}}_j \tag{16.20}$$

where $\hat{\mathbf{\Sigma}}_j$ is the ML estimation when $\mathbf{W}\to\mathbf{0}$, computed from (16.16).

In Guo *et al.* (2008), a special case of the regularization parameter matrix is considered, that is

$$\mathbf{W} = h\mathbf{I}_d. \tag{16.21}$$

Now, the corresponding covariance matrix estimation is

$$\mathbf{\Sigma}_j(h) = h\mathbf{I}_d + \hat{\mathbf{\Sigma}}_j. \tag{16.22}$$

The estimation in (16.22) is called as KLIM.

The regularization parameter can be estimated based on the model selection criteria derived under the MDL principle with all samples:

$$h = \mathrm{trace}(\hat{\mathbf{\Sigma}})/\mathrm{d} \tag{16.23}$$

where $\hat{\mathbf{\Sigma}}$ is the common covariance matrix. The details of parameter estimation can be referred to Guo *et al.* (2008).

16.3 Covariance Matrix Estimation with Multi-Regularization Parameters

16.3.1 *KLIM_L method with multi-regularization parameters*

Since multi-parameters optimization is more difficult than single parameter optimization, only a special case is considered in KLIM, that is, the regularization parameters are taken the same value for all dimensions. Though estimation of regularization parameter becomes simple, ignoring the local difference in each dimension decreases the accuracy of covariance matrix estimation. Consequently, that leads to decrease the classification accuracy of Gaussian classifier.

In this section we study the covariance matrix estimation with multi-regularization parameters. If the local difference in each dimension is considered, the regularization parameter matrix is taken the general form as in (16.15).

$$\mathbf{W} = \mathrm{diag}(h_1, h_2, \ldots, h_d).$$

The corresponding covariance matrix estimation is shown as in (16.20)

$$\mathbf{\Sigma}_j(\mathbf{W}) = \mathbf{W} + \hat{\mathbf{\Sigma}}_j.$$

The estimation in (16.20) is called as KLIM_L in the paper, where multi-regularization parameters are considered. By (16.20), the main problem is that how to determine the multi-regularization parameters in regularization matrix $\mathbf{W}$.

In the following, method of multi-regularization parameters estimation will be developed.

16.3.2 *Multi-regularization parameters estimation*

(*i*) *Selection of multi-regularization parameters*

The regularization parameter set $\bar{h} = \{h_i\}_{i=1}^d$ in the Gaussian kernel density plays an important role in estimating the mixture model parameters. Different $\bar{h}$ will generate different models. So selecting the regularization parameters $\bar{h}$ is a model selection problem.

According to the analysis above, the goal of selecting the regularization parameters is to produce a model for the probability kernel density $p_{\mathbf{W}}(\mathbf{X})$ which is as close as possible to the probability density $p(\mathbf{X}, \Theta)$. For empirical kernel density estimate, it is known that when sample number $N \to \infty$, $p_{\mathbf{W}}(\mathbf{X})$ will best describe the unknown density with $\mathbf{W} \to 0$. When N becomes small, the value of multi-regularization parameters $\bar{h}$ in $\mathbf{W}$ should be increased in order to get a plausibly true density estimate. However, if $\bar{h}$ is too big, the estimate will become too smooth and far away from the true density. So there exists an optimal $\bar{h}$ value, such that the probability kernel density $p_{\mathbf{W}}(\mathbf{X})$ matches the probability density $p(\mathbf{X}, \Theta)$ well with training samples.

It is very difficult to select an optimal $\bar{h}$ for regularized estimation. In RDA and in LOOC, the regularization parameters are selected by using the statistical cross-validation technique based on criterion of classification accuracy with training samples. The cross-validation methods used in RDA and LOOC are computationally expensive. In our work the similar method as in Guo *et al.* (2008) is used based on MDL principle to estimate the regularization parameters in (16.20). However, in Guo *et al.* (2008), only single regularization parameter is considered, here the challenge problem we deal with is multi-regularization parameter estimation.

According to the MDL principle, it should be with the shortest codelength to select a model. When $\mathbf{W} \not\to 0$, the regularization parameters $\mathbf{W}$ can be estimated with the minimized KL divergence regarding $\mathbf{W}$ with ML estimated parameter,

$$\mathbf{W}^* = \arg\min J(\mathbf{W}), \quad J(\mathbf{W}) = KL(\mathbf{W}, \Theta). \tag{16.24}$$

Now a second order approximation for estimating the regularization parameter $\mathbf{W}$ is used here. Rewrite the $J(\mathbf{W})$ as:

$$J(\mathbf{W}) = J_0(\mathbf{W}) + J_e(\mathbf{W}) \tag{16.25}$$

where

$$\begin{aligned} J_0(\mathbf{W}) &= -\int p_{\mathbf{W}}(\mathbf{x}) \ln p(\mathbf{x}, \Theta) \mathrm{d}\mathbf{x}, \\ J_e(\mathbf{W}) &= \int p_{\mathbf{W}}(\mathbf{x}) \ln p_{\mathbf{W}}(\mathbf{x}) \mathrm{d}\mathbf{x}. \end{aligned} \tag{16.26}$$

Applying the Taylor expansion of $\ln p(\mathbf{x}, \Theta)$ with respect to $\mathbf{x}$ to the function at $\mathbf{x} = \mathbf{x}_i$ and only keeping the second order term,

$$\begin{aligned} \ln p(\mathbf{x}, \Theta) \approx\ & \ln p(\mathbf{x}_i, \Theta) + (\mathbf{x} - \mathbf{x}_i)^{\mathsf{T}} \nabla_{\mathbf{x}} \ln p(\mathbf{x}_i, \Theta) \\ & + \frac{1}{2} (\mathbf{x} - \mathbf{x}_i)^{\mathsf{T}} \nabla_{\mathbf{x}} \nabla_{\mathbf{x}} \ln p(\mathbf{x}_i, \Theta)(\mathbf{x} - \mathbf{x}_i). \end{aligned} \tag{16.27}$$

Replacing $\ln p(\mathbf{x}, \Theta)$ with the second order term of Taylor expansion into the integral of $J_0(\mathbf{W})$ that results the following approximation of $J_0(\mathbf{W})$,

$$J_0(\mathbf{W}) \approx -\frac{1}{N} \sum_{i=1}^{N} \ln p(\mathbf{x}_i, \Theta) - \frac{1}{2N} \sum_{i=1}^{N} \mathrm{trace}(\mathbf{W}(\nabla_{\mathbf{x}} \nabla_{\mathbf{x}} \ln p(\mathbf{x}_i, \Theta))). \tag{16.28}$$

For very sparse data distribution, there is almost no intersection among kernel functions in the kernel density function $p_{\mathbf{W}}(\mathbf{X})$. Therefore, the following approximation can be used:

$$\begin{aligned} p_{\mathbf{W}}(\mathbf{x}) \ln p_{\mathbf{W}}(\mathbf{x}) &\approx \frac{1}{N} \sum_{i=1}^{N} G(\mathbf{x}, \mathbf{x}_i, \mathbf{W}) \ln \frac{1}{N} G(\mathbf{x}, \mathbf{x}_i, \mathbf{W}) \\ &= \frac{1}{N} \sum_{i=1}^{N} G(\mathbf{x}, \mathbf{x}_i, \mathbf{W}) \Big[- \ln N - \frac{d}{2} \ln 2\pi - \frac{1}{2} \ln |\mathbf{W}| \\ &\quad - \frac{1}{2} (\mathbf{x} - \mathbf{x}_i)^{\mathsf{T}} \mathbf{W}^{-1} (\mathbf{x} - \mathbf{x}_i) \Big]. \end{aligned} \tag{16.29}$$

Substituting (16.29) into $J_e(\mathbf{W})$, it can be gotten as

$$J_e(\mathbf{W}) = \int p_{\mathbf{W}}(\mathbf{x}) \ln p_{\mathbf{W}}(\mathbf{x}) \mathrm{d}\mathbf{x} \approx - \ln N - \frac{d}{2} \ln 2\pi - \frac{1}{2} \ln |\mathbf{W}| - \frac{d}{2}. \tag{16.30}$$

So far, the approximation formula of $J(\mathbf{W})$ is obtained:

$$J(\mathbf{W}) \approx -\frac{1}{2N} \sum_{i=1}^{N} \mathrm{trace}(\mathbf{W}(\nabla_{\mathbf{x}} \nabla_{\mathbf{x}} \ln p(\mathbf{x}_i, \Theta))) - \frac{1}{2} \ln |\mathbf{W}| + C \tag{16.31}$$

where C is a constant irrelevant to $\mathbf{W}$. Let $\mathbf{H}_{d \times d} = \nabla_{\mathbf{x}} \nabla_{\mathbf{x}} \ln p(\mathbf{x}_i, \Theta)$. Taking partial derivative of $J(\mathbf{W})$ to $\mathbf{W}$ and letting it be equal to zero, that is, $\partial J(\mathbf{W}) / \partial \mathbf{W} = 0$, the rough approximation formula of is obtained as follows:

$$\mathbf{W}^{-1} = -\frac{1}{N}\sum_{i=1}^{N} \text{diag}(\mathbf{H}). \tag{16.32}$$

Now $\bar{h}$ can be estimated with (16.32).

(ii) Approximation for regularization parameters estimation

Due to complication to compute matrix H, there is a need to take some approximation to make it computable. (16.32) can be rewritten as follows:

$$\mathbf{W}^{-1} = -\frac{1}{N}\sum_{i=1}^{N} \text{diag}(\mathbf{H}) = \frac{1}{N}\text{diag}\left(-\sum_{i=1}^{N} (\nabla_{\mathbf{x}}\nabla_{\mathbf{x}} \ln p(\mathbf{x}_i, \Theta)\right), \tag{16.33}$$

with

$$\begin{aligned}-\sum_{i=1}^{N} \nabla_{\mathbf{x}}\nabla_{\mathbf{x}} \ln p(\mathbf{x_i}, \boldsymbol{\Theta}) = \sum_{i=1}^{N} \Bigg\{ & \sum_{j=1}^{k} p(j|\mathbf{x}_i, \Theta)\Big[\boldsymbol{\Sigma}_\mathbf{j}^{-1} - \boldsymbol{\Sigma}_\mathbf{j}^{-1} \\ & \times (\mathbf{x}_i - \mathbf{m}_j)(\mathbf{x}_i - \mathbf{m}_j)^\top \boldsymbol{\Sigma}_\mathbf{j}^{-1}\Big] \\ & + \sum_{j=1}^{k} p(j|\mathbf{x}_i, \Theta)\boldsymbol{\Sigma}_\mathbf{j}^{-1}(\mathbf{x_i} - \mathbf{m_j}) \\ & \times \sum_{j=1}^{k} p(j|\mathbf{x}_i, \Theta)(\mathbf{x}_i - \mathbf{m}_j)^\top \boldsymbol{\Sigma}_\mathbf{j}^{-1}\Bigg\}.\end{aligned} \tag{16.34}$$

Considering hard-cut case ($p(j|\mathbf{x}_i, \Theta) = 1$ or 0) and using the approximations

$$\sum_{i=1}^{N} p(j|\mathbf{x}_i, \Theta)(\mathbf{x}_i - \mathbf{m}_j)(\mathbf{x}_i - \mathbf{m}_j)^\top \approx n_j \hat{\boldsymbol{\Sigma}}_j,$$

$$\sum_{i=1}^{N} p(j|\mathbf{x}_i, \Theta) \approx n_j, \quad \boldsymbol{\Sigma}_j{}^{-1}\hat{\boldsymbol{\Sigma}}_j = \mathbf{I}, \quad \sum_{j=1}^{k} \alpha_j \boldsymbol{\Sigma}_j{}^{-1} = \boldsymbol{\Sigma}^{-1},$$

it can be obtained:

$$\mathbf{W}^{-1} \approx \text{diag}(\boldsymbol{\Sigma}^{-1}). \tag{16.35}$$

Suppose the cofactor of the common covariance matrix $\boldsymbol{\Sigma}$ is $\boldsymbol{\Sigma}^{\#}$, then there exists:

$$\boldsymbol{\Sigma}^{-1} = \frac{1}{|\boldsymbol{\Sigma}|}\boldsymbol{\Sigma}^{\#}. \tag{16.36}$$

Substituting (16.36) into (16.35) and using the product of diagonal elements to approximate the determinant, the regularization parameters can be finally approximated by the following equation:

$$\mathbf{W} = \text{diag}(\boldsymbol{\Sigma}). \tag{16.37}$$

Substituting (16.37) into (16.20), the estimation can be written as:

$$\mathbf{\Sigma}_j(\mathbf{W}) = \text{diag}(\mathbf{\Sigma}) + \mathbf{\Sigma}_j. \tag{16.38}$$

This is the final form of the KLIM_L estimation.

16.3.3 *Comparison of KLIM_L with regularization methods*

The four methods (KLIM_L, KLIM, RDA and LOOC) are all regularization methods to estimate the covariance matrix for small sample size problem. They all consider ML estimated covariance matrix with the additional extra matrices.

KLIM_L is derived by the similar way as KLIM under the framework of MDL principle. Meanwhile, the estimation of regularization parameters is similar to KLIM based on MDL principle. KLIM_L is a generalization of KLIM. Multi-regularization parameters are included and estimated in KLIM_L while one regularization parameter is estimated in KLIM. For every σ_{ii} $(i = 1, \ldots, d\ )$ in common covariance matrix $\mathbf{\Sigma}$, if it is taken by trace$(\mathbf{\Sigma})/d$, then $\mathbf{W} = \text{trace}(\mathbf{\Sigma})/d\mathbf{I}_d$. It will reduce to the case of $\mathbf{W} = h\mathbf{I}_d$ in KLIM, where $h = \text{trace}(\mathbf{\Sigma})/d$.

KLIM_L is derived based on MDL principle while RDA and LOOC are heuristically proposed. They differ in the criterion used to select the regularization parameters. The criterion minimized in KLIM_L is the KL divergence approximation by two-order Taylor expansion to estimate the same regularization parameters for all the classes. RDA maximizes the leave-one-out classification accuracy to select the two global regularization parameters for all the classes. And the criterion maximized in LOOC is the leave-one-out likelihood which allows a separate regularization parameters to be calculated for each class.

Different computation time costs are required in the four regularization discriminant methods. KLIM costs the same or a little more time than KLIM_L. The estimations of regularization parameters in KLIM_L and in KLIM are all point estimation which the finally approximation value is given directly, whereas the selection of regularization parameters in RDA and in LOOC requires to be optimized by leave-one-out cross validation technique. So RDA and LOOC cost much more computational time than KLIM_L and KLIM. Also, RDA requires more time than LOOC since RDA involves a two dimensional grid optimization while LOOC involves a one dimensional grid optimization for regularization parameters estimation. This will be validated by the following experiments.

16.3.4 *Experiments and discussions*

In this section, the classification accuracy and computational cost of KLIM_L are compared with QDA, LDA, RDA, LOOC and KLIM on synthetic data [Friedman (1989)], COIL-20 object data [Nene *et al.* (1996)] and wine data [Frank and Asuncion (2010)].

(a) *Experiments*

In the experiments, the regularization parameter h of KLIM is estimated by $h = \text{trace}(\mathbf{\Sigma})/\mathbf{d}$. The parameter matrix $\mathbf{W}$ of KLIM_L is estimated by (16.37). In RDA, the values of λ and γ are sampled in a coarse grid $(0.0, 0.25, 0.50, 0.75, 1.0)$, resulting in 25 data points. In LOOC, the four parameters are taken according to the table in Hoffbeck and Landgrebe (1996).

(1) Experiment 1

Experiment 1 is conducted on the synthetic data adopted from Friedman's paper [Friedman (1989)]. The synthetic data is generated under Gaussian distribution and five dimensions (d = 6, 10, 20, 40, 60). The covariance matrices of all three classes are the same and highly elliptical. The diagonal elements of the covariance matrices are defined by $\sigma_i = [9(i-1)/(d-1)+1]^2$ for any $1 \le i \le d$. The mean of the first class is at the origin, the means of class 2 and class 3 are given by

$$m_{2i} = 2.5\sqrt{\sigma_i/d}\,\frac{(d-i)}{(d/2-1)}$$

and $m_{3i} = (-1)^i m_{2i}$ for any $1 \le i \le d$.

In each class, 15 training samples are randomly drawn from three different Gaussian distributions, and the mean and covariance matrix are estimated based on the training samples. Additional 100 independent test samples from each class are generated to verify the classification accuracy. Each experiment runs 100 times, and the mean and standard deviation of classification accuracy are reported as results. The result of experiment 1 is shown in Table 16.1. The notation N/A indicates that the covariance matrix is singular in which case reliable results cannot be obtained.

Table 16.1 Mean classification accuracy for experiment 1

Classfier/dimension	QDA	LDA	RDA	KLIM	KLIM_L
6	87.8(3.4)	93.3(1.7)	92.8(3.2)	76.8(3.0)	93.7(1.4)
10	68.6(5.6)	88.3(2.3)	87.3(3.5)	72.0(2.7)	89.1(1.7)
20	N/A	77.0(4.5)	74.7(5.5)	59.2(3.3)	78.7(2.9)
40	N/A	48.2(6.9)	61.3(5.1)	53.8(3.1)	73.7(2.6)
60	N/A	N/A	60.0(4.5)	51.8(3.2)	71.8(2.6)

It is shown by Table 16.1 that among the three regularization methods (RDA, KLIM, KLIM_L), the performance of KLIM_L is the best, the performance of RDA is the second and the performance of KLIM is the worst. The performance of LDA is degrading greatly with the increasing of feature dimensionality. When the dimension of variable is greater than 10, the covariance matrix estimation in QDA is singular, resulting in incredible classification results.

(2) Experiment 2

Experiment 2 is conducted on COIL-20 object data. COIL-20 is a database of gray-scale images of 20 objects. The objects were placed on a motorized turntable against a black background. The turntable was rotated through 360 degrees to vary object pose with respect to a fix camera. Images of the objects were taken at pose intervals of 5 degrees, which corresponds to 72 images per object. The total number of images is 1440 and the size of each image is 128∗128.

In this experiment, seven objects are randomly drawn from 20 objects. Six images are randomly selected as training samples from each class to estimate the mean and covariance matrix. And the remaining images are employed as testing samples to verify the classification accuracy. Since the dimension of image data is very high of 128∗128, PCA is used to reduce the data dimension. Experiments are performed on four different numbers of dimensions: d = 40, 30, 20, 10. Each experiment runs 25 times, and the mean and standard deviation of classification accuracy are reported as results. The results of experiment 2 are shown in Table 16.2. Computational costs of estimating regularization parameters on COIL-20 object database are shown in Table 16.3.

Table 16.2 Mean classification accuracy on COIL-20 object database

Classfier/dimension	LDA	RDA	LOOC	KLIM	KLIM_L
40	N/A	82.1(4.1)	77.6(5.0)	82.7(3.7)	82.8(3.6)
30	77.7(3.5)	81.3(3.7)	77.5(4.3)	81.8(3.5)	82.1(2.9)
20	82.1(3.4)	80.5(4.6)	77.8(3.2)	81.1(3.8)	82.3(3.6)
10	78.8(3.8)	78.9(3.9)	76.8(4.9)	78.8(3.4)	79.6(2.7)

Table 16.3 Computational cost (in seconds) of estimating regularization parameters on COIL-20 object database

Classfier/dimension	RDA	LOOC	KLIM	KLIM_L
40	2.3124	0.1839	9.4302e-005	4.9365e-005
30	1.5292	0.1511	9.4201e-005	4.8459e-005
20	0.9927	0.1097	9.5560e-005	5.1629e-005
10	0.6461	0.0813	9.4654e-005	4.7553e-005

In the experiment 2, the classification accuracy of KLIM_L is the best among the five compared methods. In the case of dimensions of 30 and 40, the classification accuracy of KLIM is on the second. The classification accuracy of LOOC is the worst among the compared methods except in dimension 40, where the classification accuracy of LOOC is higher than that of LDA. The computational cost of regularization parameters estimation in RDA and LOOC are much more than these of KLIM and KLIM_L.

Table 16.4 Mean classification accuracy on wine data

Classifier/training number	LDA	RDA	LOOC	KLIM	KLIM_L
3	N/A	64.5(9.1)	N/A	64.2(9.1)	90.4(3.5)
4	N/A	67.9(7.3)	N/A	69.2(5.7)	90.9(4.1)
6	78.8(11.2)	73.5(9.3)	92.9(3.6)	71.1(2.3)	93.5(2.7)
8	90.3(4.0)	89.1(5.4)	92.8(9.1)	69.3(5.1)	93.5(2.6)
10	93.3(2.8)	93.4(2.8)	94.5(2.6)	70.9(3.9)	94.3(2.0)

(3) Experiment 3

Experiment 3 is conducted on the "wine" data. The real world "wine" data set is 13-dimensional with three classes. This data set is large with 59, 71 and 48 samples per class. In order to study the performance of regularized methods, 3-10 training samples are randomly drawn from each class, whereas the remaining samples are used to verify classification accuracy. Each experiment runs 25 times, and the mean and standard deviation of classification accuracy are reported as results. The results of experiment 3 are shown in Table 16.4.

On the real world "wine" data, the classification accuracy of each method becomes higher with increasing of the number of training sample. The performance of KLIM_L is the best when the number of training samples is not equal to 10. In this case, the performance of LOOC is the best.

(*b*) *Discussions*

In all experiments, KLIM_L leads to more satisfying results than any other compared methods except in experiment 3, where the classification accuracy of KLIM_L is worse than LOOC when the number of training samples is equal to 10. It can be shown by experiments that KLIM_L covariance matrix estimator is simple and can obtain satisfying results in most of the cases.

By comparing KLIM_L with KLIM, KLIM_L always perform better in term of classification accuracy. KLIM is a special case of KLIM_L. KLIM_L considers the local regularization in each dimension, while KLIM considers the global regularization by average the trace of common matrix in each dimension. So the covariance matrix estimation of KLIM_L is more precise than that of KLIM. However, KLIM_L is more sensitive to outliers than KLIM because average the trace of common matrix in KLIM can decrease the disturbance of noise.

Considering the computational cost of regularization parameters estimating, KLIM_L needs less time to estimate the regularization parameters than KLIM needs, while RDA and LOOC need much more time than KLIM_L needs. The reason is that the regularization parameters in KLIM_L are estimated by determinant point estimation, while the regularization parameters in RDA and LOOC are optimized by cross validation statistical technique. The experimental results are consistent with the theoretical analysis.

16.4 Covariance Matrix Estimation with Variable Regularization Parameters

In KLIM, the regularization parameter is fixed and taken the same value for all the classes. That is, it is irrelevant to the class. This fixed regularization parameter approach may result in undersmooth in regions with only sparse samples while oversmooth in other regions, whereas varying the regularization parameter gives flexibility to reduce the variance of the estimators in regions with few samples, and to reduce the bias of the estimators in regions with lots of samples. Therefore, the regularization parameter should be variable according to the class in such cases.

In this section, two new covariance matrix estimators with variable regularization parameters are developed, which can be regarded as the generalized KLIM estimators.

16.4.1 *Covariance matrix estimation with variable regularization parameters*

For the Gaussian kernel density $p_{\mathbf{W}}(\mathbf{X})$, we consider the sample point regularization parameter matrix:

$$\mathbf{W} = \mathbf{W}(\mathbf{X}_i) = \mathrm{diag}\big(h_1(\mathbf{X}_i), h_2(\mathbf{X}_i), \ldots, h_d(\mathbf{X}_i)\big), \tag{16.39}$$

which is variable and relevant the sample point $\mathbf{X}_i$. Let $\tilde{\mathbf{W}} = \{\mathbf{W}(\mathbf{X}_i)\}_{i=1}^{N}$, then

$$p_{\tilde{\mathbf{W}}}(\mathbf{X}) = \frac{1}{N}\sum_{i=1}^{N} G\big(\mathbf{X}, \mathbf{X}_i, \mathbf{W}(\mathbf{X}_i)\big). \tag{16.40}$$

By the similar way as in KLIM, it can be obtained:

$$\hat{\boldsymbol{\Sigma}}_j(\tilde{\mathbf{W}}) = \frac{\int p_{\tilde{\mathbf{W}}}(\mathbf{X}) p\big(j|\mathbf{X}, \hat{\Theta}\big)(\mathbf{X} - \hat{\mathbf{m}}_j)(\mathbf{X} - \hat{\mathbf{m}}_j)^{\mathsf{T}} \mathrm{d}\mathbf{X}}{\int p_{\tilde{\mathbf{W}}}(\mathbf{X}) p\big(j|\mathbf{X}, \hat{\Theta}\big) \mathrm{d}\mathbf{X}}. \tag{16.41}$$

Applying the Taylor expansion approximation of $p\big(j|\mathbf{X}, \hat{\Theta}\big)$ at $\mathbf{X} = \mathbf{X}_i$ for (16.41),

$$\hat{\boldsymbol{\Sigma}}_j(\tilde{\mathbf{W}}) = \hat{\boldsymbol{\Sigma}}_j + \frac{\sum\limits_{i=1}^{N} p\big(j|\mathbf{X}_i, \hat{\Theta}\big)\mathbf{W}(\mathbf{X}_i)}{\sum\limits_{i=1}^{N} p\big(j|\mathbf{X}_i \hat{\Theta}\big)}. \tag{16.42}$$

It is shown by the second term in (16.42) that the regularization term is only relative to the point $\mathbf{X}_i$ belonging to the class j. For convenience and simplification, it is assumed that $\mathbf{W}(\mathbf{X}_i) = \mathbf{W}(\mathbf{X}_k)$ for any $\mathbf{X}_i \in j$ and $\mathbf{X}_k \in j$. Denoted it as $\mathbf{W}(\mathbf{X}_i \in j)$. So, (16.42) can be rewritten as

$$\hat{\boldsymbol{\Sigma}}_j(\tilde{\mathbf{W}}) = \hat{\boldsymbol{\Sigma}}_j + \mathbf{W}(\mathbf{X}_i \in j). \tag{16.43}$$

This new estimator in (16.43) is called as KLIM_V_L. Similarly, if

$$\mathbf{W} = \mathbf{W}(\mathbf{X}_i) = h(\mathbf{X}_i)\mathrm{I}_d, \tag{16.44}$$

then

$$\hat{\mathbf{\Sigma}}_j(\tilde{\mathbf{W}}) = \hat{\mathbf{\Sigma}}_j + h(\mathbf{X}_i \in j)\mathrm{I}_d, \tag{16.45}$$

which is denoted as KLIM_V. So far, two new covariance estimators, KLIM_V and KLIM_V_L, are derived based on MDL principle.

16.4.2 *Variable regularization matrix estimation*

The regularization parameter matrices $\tilde{\mathbf{W}}$ in the Gaussian kernel density play an important role in estimating the mixture model parameters. Different $\tilde{\mathbf{W}}$ will generate different models. So selecting the regularization parameter matrix is a model selection problem. According to the principle of MDL, it should be with the shortest codelength to select a model. So the regularization parameter matrix $\mathbf{W}(\mathbf{X}_i \in j)$ can be estimated with the minimized KL divergence regarding $\mathbf{W}(\mathbf{X}_i \in j)$ with ML estimated parameter,

$$\mathbf{W}(\mathbf{X}_i \in j)^* = \arg\min J(\tilde{\mathbf{W}}), \quad J(\tilde{\mathbf{W}}) = KL(\tilde{\mathbf{W}}, \Theta). \tag{16.46}$$

Now rewrite the $J(\tilde{\mathbf{W}})$ as:

$$J(\tilde{\mathbf{W}}) = J_0(\tilde{\mathbf{W}}) + J_e(\tilde{\mathbf{W}}) \tag{16.47}$$

where

$$\begin{aligned} J_0(\tilde{\mathbf{W}}) &= -\int p_{\tilde{\mathbf{W}}}(\mathbf{X}) \ln p(\mathbf{X}, \Theta) \mathrm{d}\mathbf{X}, \\ J_e(\tilde{\mathbf{W}}) &= \int p_{\tilde{\mathbf{W}}}(\mathbf{X}) \ln p_{\tilde{\mathbf{W}}}(\mathbf{X}) \mathrm{d}\mathbf{X}. \end{aligned} \tag{16.48}$$

Replacing $\ln p(\mathbf{X}, \Theta)$ with the second order term of Taylor expansion into the integral of $J_0(\tilde{\mathbf{W}})$ and resulting in the following approximation of $J_0(\tilde{\mathbf{W}})$,

$$\begin{aligned} J_0(\tilde{\mathbf{W}}) \approx &-\frac{1}{N}\sum_{i=1}^{N} \ln p(\mathbf{X}_i, \Theta) \\ &-\frac{1}{2N}\sum_{i=1}^{N} \mathrm{trace}\Big(\mathbf{W}(\mathbf{X}_i)\big(\nabla_{\mathbf{X}}\nabla_{\mathbf{X}} \ln p(\mathbf{X}_i, \Theta)\big)\Big). \end{aligned} \tag{16.49}$$

For very sparse data distribution, the following approximation can be used:

$$p_{\tilde{\mathbf{W}}}(\mathbf{X}) \ln p_{\tilde{\mathbf{W}}}(\mathbf{X}) \approx \frac{1}{N}\sum_{i=1}^{N} G(\mathbf{X}, \mathbf{X}_i, \mathbf{W}(\mathbf{X}_i)) \ln \frac{1}{N} G(\mathbf{X}, \mathbf{X}_i, \mathbf{W}(\mathbf{X}_i)). \tag{16.50}$$

Substituting (16.50) into $J_e(\tilde{\mathbf{W}})$, it can be obtained:

$$J_e(\tilde{\mathbf{W}}) \approx -\ln N - \frac{d}{2}\ln 2\pi - \frac{1}{2N}\sum_{i=1}^{N}\ln|\mathbf{W}(\mathbf{X}_i)| - \frac{d}{2}. \tag{16.51}$$

So far, the approximation formula of $J(\tilde{\mathbf{W}})$ is:

$$\begin{aligned} J(\tilde{\mathbf{W}}) \approx &-\frac{1}{2N}\sum_{i=1}^{N}\operatorname{trace}\big(\mathbf{W}(\mathbf{X}_i)(\nabla_{\mathbf{X}}\nabla_{\mathbf{X}}\ln p(\mathbf{X}_i,\Theta))\big) \\ &-\frac{1}{2N}\sum_{i=1}^{N}\ln|\mathbf{W}(\mathbf{X}_i)| + C \\ = &-\frac{1}{2N}\sum_{j=1}^{k}\sum_{i=1}^{n_j}\operatorname{trace}\big(\mathbf{W}(\mathbf{X}_i\in j)(\nabla_{\mathbf{X}}\nabla_{\mathbf{X}}\ln p(\mathbf{X}_i,\Theta))\big) \\ &-\frac{1}{2N}\sum_{j=1}^{k}\sum_{i=1}^{n_j}\ln|\mathbf{W}(\mathbf{X}_i\in j)| + C \end{aligned} \tag{16.52}$$

where $n_j = \sum_{i=1}^{N} p(j|\mathbf{X}_i,\Theta)$, $N = \sum_{j=1}^{N} n_j$, and C is a constant irrelevant to $\tilde{\mathbf{W}}$. Taking partial derivative of $J(\tilde{\mathbf{W}})$ to $\mathbf{W}(\mathbf{X}_i\in j)$ and letting it be equal to zero, the rough approximation formula of $\mathbf{W}(\mathbf{X}_i\in j)$ is obtained as follows:

$$\mathbf{W}(\mathbf{X}_i\in j)^{-1} = -\frac{1}{n_j}\sum_{i=1}^{n_j}\operatorname{diag}\big(\nabla_{\mathbf{X}}\nabla_{\mathbf{X}}\ln p(\mathbf{X}_i,\Theta)\big) \tag{16.53}$$

where

$$\begin{aligned} -\sum_{i=1}^{N}\nabla_{\mathbf{x}}\nabla_{\mathbf{x}}\ln p(\mathbf{x}_i,\Theta) = &\sum_{i=1}^{N}\Bigg\{\sum_{j=1}^{k}p(j|\mathbf{x}_i,\Theta)\Big[\boldsymbol{\Sigma}_j^{-1} - \boldsymbol{\Sigma}_j^{-1} \\ &\times(\mathbf{x}_i-\mathbf{m}_j)(\mathbf{x}_i-\mathbf{m}_j)^{\mathsf{T}}\boldsymbol{\Sigma}_j^{-1}\Big] \\ &+\sum_{j=1}^{k}p(j|\mathbf{x}_i,\Theta)\boldsymbol{\Sigma}_j^{-1}(\mathbf{x}_i-\mathbf{m}_j) \\ &\times\sum_{j=1}^{k}p(j|\mathbf{x}_i,\Theta)(\mathbf{x}_i-\mathbf{m}_j)^{\mathsf{T}}\boldsymbol{\Sigma}_j^{-1}\Bigg\}. \end{aligned} \tag{16.54}$$

Using the approximations

$$\sum_{i=1}^{N}p(j|\mathbf{X}_i,\Theta)(\mathbf{X}_i-\mathbf{m}_j)(\mathbf{X}_i-\mathbf{m_j})^{\mathsf{T}} \approx n_j\hat{\boldsymbol{\Sigma}}_j,$$

$$\sum_{i=1}^{N}p(j|\mathbf{X}_i,\Theta) \approx n_j, \qquad \boldsymbol{\Sigma}_j^{-1}\hat{\boldsymbol{\Sigma}}_j = \mathbf{I},$$

considering hard-cut case ($p(j|\mathbf{X}_i, \Theta) = 1$ or 0), and restricting $\mathbf{X}_i \in j$, it can be obtained:

$$\mathbf{W}(\mathbf{X}_i \in j)^{-1} = \mathrm{diag}(\mathbf{\Sigma}_j{}^{-1}). \tag{16.55}$$

Suppose the cofactor of the covariance matrix $\mathbf{\Sigma}_j$ is $\mathbf{\Sigma}_j^{\#}$, then there exists:

$$\mathbf{\Sigma}_j{}^{-1} = \frac{1}{|\mathbf{\Sigma}_j|}\mathbf{\Sigma}_j^{\#}. \tag{16.56}$$

Substituting (16.56) into (16.55) and using the product of diagonal elements to approximate the determinant, the regularization parameter matrix can be finally approximated by the following equation:

$$\mathbf{W}(\mathbf{X}_i \in j) = \mathrm{diag}(\mathbf{\Sigma_j}). \tag{16.57}$$

Substituting (16.57) into (16.43), it can be gotten

$$\mathbf{\Sigma}_j(\tilde{\mathbf{W}}) = \mathbf{\Sigma}_j + \mathrm{diag}(\mathbf{\Sigma}_j). \tag{16.58}$$

This is the final form of the KLIM_V_L estimation.

By the similar way, the regularization parameter in KLIM_V can be approximated by the following formula:

$$h(\mathbf{X}_i \in j) = \frac{\mathrm{trace}(\mathbf{\Sigma}_j)}{d}. \tag{16.59}$$

Substituting (16.59) into (16.45), it can be gotten

$$\mathbf{\Sigma}_j(\tilde{\mathbf{W}}) = \mathbf{\Sigma}_j + \frac{\mathrm{trace}(\mathbf{\Sigma}_j)}{d}\mathbf{I}_d. \tag{16.60}$$

This is the final form of the KLIM_V estimation.

16.4.3 *Comparisons*

The four estimators (KLIM, KLIM_L, KLIM_V, and KLIM_V_L) are all regularization methods to estimate the covariance matrix for small sample size problem. They are derived under the framework of MDL principle. Meanwhile, the derivation of regularization term is also based on MDL principle They all consider ML estimated covariance matrix with the additional regularization parameter matrices. Since the estimation of elements in regularization parameter matrix is point estimation, the computational speed is very fast.

The elements in regularization parameter matrix of KLIM or KLIM_L is taken the same value for all the classes, while the elements in regularization parameter matrix of KLIM_V or KLIM_V_L are different for each class which is variable according to sample points in each class. KLIM_V is a more general case than KLIM, which can be regarded as an extension of KLIM with the variable regularization parameters in principle. KLIM_V is a special case of KLIM_V_L which considers the local difference of regularization parameters in each direction. For each σ_{ii}^{j} $(i = 1, \ldots, d)$ in $\boldsymbol{\Sigma}_j$, if it is taken by $\text{trace}(\boldsymbol{\Sigma}_j)/d$, then

$$\mathbf{W} = \frac{\text{trace}(\boldsymbol{\Sigma}_j)}{d}\mathbf{I}_d.$$

It will reduce to the case of $\mathbf{W} = h(\mathbf{X}_i \in j)\text{I}_d$ in KLIM_V, where $h(\mathbf{X}_i \in j) = \text{trace}(\boldsymbol{\Sigma}_j)/d$. The relationship between KLIM and KLIM_L is similar as that of KLIM_V and KLIM_V_L.

On considering the difference of regularization parameter matrix, the four estimators can be adopted to meet requirements of different conditions. If the same covariance is taken for all the classes, KLIM or KLIM_L can be used. Otherwise, if the covariance for each class is different, KLIM_V or KLIM_V_L can be applied. This will be validated by the following experiments.

16.4.4 *Experiments and discussions*

For the purpose of testing the feasibility and applicability of the four methods (KLIM, KLIM_L, KLIM_V, and KLIM_V_L), the classification accuracies are compared on synthetic data sets. Also, the classification accuracies of them are compared with those of LDA and RDA.

(a) Experiments

The synthetic data sets are generated under four different Gaussian distributions and five dimension (d = 6, 10, 20, 40, 60) adopted from Friedman's paper [Friedman (1989)]. For all cases there are 3 classes. In each class, 15 training samples are randomly drawn from three different Gaussian distributions, and the mean and covariance matrix are estimated based on the training samples. Additional 100 independent test samples from each class are generated to verify the classification accuracy.

In the experiments, the regularization parameter h of KLIM and the regularization parameter $h(\mathbf{X}_i \in j)$ of KLIM_V are estimated by (16.23) and by (16.60) respectively. The regularization matrix of KLIM_L and the regularization matrix of KLIM_L_V are estimated by (16.38) and by (16.58) respectively. In RDA, the values of and are sampled in a coarse grid, (0.0, 0.25, 0.50, 0.75, 1.0), resulting in 25 points. Each experiment runs 100 times, and the mean and standard deviation of classification accuracy are reported as results.

Experiment 1 is conducted under condition of equal spherical covariance matrices. For all three classes, their covariance matrices are equal to the identity matrix, but each class has a different mean vector. The mean of the first class is at the origin, the mean of second class is 3.0 in the first variable and 0 in the other variables, and the mean of third class is 3.0 in the second variable and 0 in the other variables. The result of experiment 1 is shown in Table 16.5. In the following tables, the value in parentheses represents the standard deviation and the notation N/A indicates that the covariance matrix is singular in which case credible results cant be obtained.

Experiment 2 is conducted under condition of unequal spherical covariance matrices. The covariance matrices of each class is equal to $k\mathbf{I}$, where k is the class number $(1 \le k \le 3)$ and $\mathbf{I}$ is the identity matrix. The mean of the first class is at the origin, the means of class 2 and class 3 are shifted in orthogonal directions, class 2 by a distance of 3.0 and class 3 by a distance of 4.0. The result of experiment 2 is shown in Table 16.6.

Experiment 3 is conducted under condition of equal highly elliptical covariance matrices for all three classes. The diagonal elements of the covariance matrices is defined by $\sigma_i = [9(i-1)/(d-1)+1]^2$ for any $1 \le i \le d$. The mean of the first class is at the origin, the means of class 2 and class 3 are given by

$$m_{2i} = 2.5\sqrt{\sigma_i/d}\frac{(d-i)}{(d/2-1)}$$

and $m_{3i} = (-1)^i m_{2i}$ for any $1 \le i \le d$. The result of experiment 3 is shown in Table 16.7.

Experiment 4 is conducted under condition of unequal highly elliptical covariance matrices. The diagonal elements of the covariance matrices for each class are defined respectively by $\sigma_{1,i} = [9(i-1)/(d-1)+1]^2$, $\sigma_{2,i} = [9(d-i)/(d-1)+1]^2$ and $\sigma_{3,i} = [9(i-(d-1)/2)/(d-1)]^2$ for any $1 \le i \le d$. The mean of the first class is at the origin, the means of class 2 and class 3 are given by $m_{2,i} = 14/\sqrt{d}$ and $m_{3,i} = (-1)^i m_{2,i}$ for any $1 \le i \le d$. The result of experiment 4 is shown in Table 16.8.

Table 16.5 Mean classification accuracy for experiment 1

Classifier/dimension	6	10	20	40	60
LDA	89.1(1.9)	82.3(2.7)	76.6(3.7)	49.2(6.3)	N/A
RDA	89.8(1.6)	86.4(2.4)	85.6(2.5)	80.3(6.9)	79.3(2.6)
KLIM	89.2(1.5)	83.4(2.1)	80.9(2.5)	73.6(2.8)	72.7(2.6)
KLIM_L	89.0(1.6)	82.5(2.6)	79.7 (3.3)	72.3(3.5)	72.4(3.5)
KLIM_V	89.0(1.6)	83.0(2.3)	80.9(2.5)	73.5(2.7)	72.5(2.7)
KLIM_V_L	87.7(1.9)	79.7(3.6)	76.1(3.9)	68.1(4.4)	66.4(4.1)

Table 16.6 Mean classification accuracy for experiment 2

Classifier/dimension	6	10	20	40	60
LDA	82.3(2.6)	78.25(2.8)	73.3(3.6)	48.3(6.4)	N/A
RDA	84.4(2.0)	86.0(2.9)	90.6(3.0)	90.3(3.4)	91.6(3.0)
KLIM	82.5(2.1)	79.7(2.5)	74.2(2.5)	63.0(3.2)	58.7(3.0)
KLIM_L	82.3(2.2)	79.4(2.7)	74.9(3.0)	63.3(3.8)	58.5(3.7)
KLIM_V	82.7(2.2)	82.7(2.6)	84.1(1.9)	84.1(2.2)	88.8(2.1)
KLIM_V_L	81.5(2.5)	81.2(3.2)	84.1(2.6)	81.9(2.6)	85.5(2.8)

Table 16.7 Mean classification accuracy for experiment 3

Classifier/dimension	6	10	20	40	60
LDA	93.5(1.5)	88.1(2.5)	77.3(3.9)	48.8(6.8)	N/A
RDA	93.3(1.7)	87.7(2.6)	75.1(5.1)	62.2(5.1)	60.5(4.1)
KLIM	77.1(2.8)	72.0(3.0)	60.1(3.3)	54.0(3.1)	52.0(3.4)
KLIM_L	93.8(1.2)	89.2(1.5)	79.1(2.5)	73.4(2.8)	71.3(2.8)
KLIM_V	75.1(3.7)	70.0(3.8)	58.9(3.6)	53.6(3.2)	51.8(3.4)
KLIM_V_L	92.8(1.3)	87.7(2.0)	75.8(2.9)	69.8(3.5)	66.3(3.0)

Table 16.8 Mean classification accuracy for experiment 4

Classifier/dimension	6	10	20	40	60
LDA	88.0(2.7)	80.7(2.7)	75.7(3.9)	53.7(6.5)	N/A
RDA	94.7(2.8)	92.6(3.8)	95.5(2.1)	95.6(2.0)	95.4(2.3)
KLIM	92.9(2.2)	90.4(1.7)	91.5(2.0)	82.5(2.4)	75.1(3.2)
KLIM_L	92.7(2.0)	90.1(1.8)	90.7(1.7)	81.9(2.7)	75.0(3.2)
KLIM_V	93.2(1.8)	92.2(1.5)	96.5(1.0)	97.5(0.9)	97.9(0.9)
KLIM_V_L	98.2(0.4)	98.9(0.5)	99.9(0.1)	100.0(0.0)	100.0(0.0)

(*b*) *Discussions*

For the four methods (KLIM, KLIM_L, KLIM_V, and KLIM_V_L) compared in the experiments, the classification accuracy of KLIM (/KLIM_L) is the best under the condition of equal and spherical (/elliptical) covariance matrices, while the classification accuracy of KLIM_V (/KLIM_V_L) is the best under condition of unequal and spherical (/elliptical) covariance matrices. The experimental results are consistent with the theoretical analysis.

The LDA method favors the situation of equal covariance matrices. But its classification accuracy is degraded greatly when the dimension is more than 20. Under the condition of spherical covariance matrices, the classification accuracy of RDA is the best; while under the condition of elliptical covariance matrices, the classification accuracy of one of KLIM-like method is the best. However, RDA needs much more time to estimate the regularization parameters than KLIM-like methods do.

16.5 Conclusions

In this chapter, the KLIM_L covariance matrix estimator is derived based on MDL principle for the small sample size problem with high dimension. Under the framework of MDL principle, multi-regularization parameters are simply and directly estimated by point estimation that is approximated by two-order Taylor expansion. KLIM_L is a generalization of KLIM. With the KL information measure, total samples can be used to estimate the regularization parameters in KLIM_L, and can reduce the computational cost as compared to using leave-one-out cross-validation method in RDA and LOOC. KLIM_L achieves the highest classification accuracy among LDA, RDA, LOOC, KLIM and KLIM_L in all experiments.

Moreover, two new covariance matrix estimators, called KLIM_V and KLIM_V_L, are derived based on MDL principle for the small sample problem with high dimension. The developed estimators can be regarded as extensions of KLIM with variable regularization parameters. The regularization parameters in KLIM_V or KLIM_V_L are variable according to sample points in each class, while the regularization parameters in KLIM or KLIM_L are taken the same value for all the classes. KLIM_V is a special case of KLIM_V_L which considers the local difference of regularization parameters in each direction. It is shown by experiments that different KLIM-like estimator favors different situation.

In the future work, the proposed methods will be further investigated and applied to real world applications. The kernel method combined with these regularization discriminant methods will be studied and the selection of kernel parameters will be investigated under some criterion.

Acknowledgement

The research work described in this chapter was fully supported by the grants from the National Natural Science Foundation of China (Project No.61375045, 60911130513).

References

Aeberhard, S., Coomans, D. and De Vel, O. (1994). Comparative analysis of statistical pattern recognition methods in high dimensional settings, *Pattern Recognition* **27**, 8, pp. 1065–1077.

Barron, A., Rissanen, J. and Yu, B. (1998). The minimum description length principle in coding and modelling, *IEEE Transactions on Information Theory* **44**, 6, pp. 2743–2760.

Bickel, P. J. and Levina, E. (2008). Regularized estimation of large covariance matrices, *Annals of Statistics* **36**, 1, pp. 199–227.

Bishop, C. M. (2007). *Pattern Recognition and Machine Learning* (Springer).

Cao, G., Bachega, L. R. and Bouman, C. A. (2011). The sparse matrix transform for covariance estimation and analysis of high dimensional signals, *IEEE Transactions on Image Processing* **20**, 3, pp. 625–640.

Cao, G., Bouman, C. and Webb, K. (2009). Noniterative map reconstruction using sparse matrix representations, *IEEE Transactions on Image Processing* **18**, 9, pp. 2085–2099.

Daniels, M. J. and Kass, R. E. (2001). Shrinkage estimators for covariance matrices, *Biometrics* **57**, 4, pp. 1173–1184.

Everitt, B. and Hand, D. J. (1981). *Finite Mixture Distributions* (Chapman and Hall).

Frank, A. and Asuncion, A. (2010). UCI machine learning repository, `http://archive.ics.uci.edu/ml`.

Friedman, J. H. (1989). Regularized discriminant analysis, *Journal of the American Statistical Association* **84**, 405, pp. 165–175.

Friedman, J., Hastie, T. and Tibshirani, R. (2008). Sparse inverse covariance estimation with the graphical lasso, *Biostatistics*, **9**, 3, pp. 432–441.

Guo, P., Jia, Y. and Lyu, M. R. (2008). A study of regularized Gaussian classifier in high-dimension small sample set case based on MDL principle with application to spectrum recognition, *Pattern Recognition* **41**, 9, pp. 2842–2854.

Hoffbeck, J. P. and Landgrebe, D. A. (1996). Covariance matrix estimation and classification with limited training data, *IEEE Transactions on Pattern Analysis and Machine Intelligence* **18**, 7, pp. 763–767.

Jolliffe, I. T. (1996). *Principal Component Analysis* (Springer).

Kullback, S. (1959). *Information Theory and Statistics* (Wiley).

Kuo, B. C. and Landgrebe, D. A. (2002). A covariance estimator for small sample size classification problems and its application to feature extraction, *IEEE Transactions on Geoscience and Remote Sensing* **40**, 4, pp. 814–819.

Ledoit, O. and Wolf, M. (2004). A well-conditioned estimator for large-dimensional covariance matrices, *Journal of Multivariate Analysis* **88**, 2, pp. 365–411.

McLachlan, G. J. and Basford, K. E. (1988). *Mixture Models: Inference and Applications to Clustering* (Dekker).

Nene, S. A., Nayar, S. K. and Murase, H. (1996). Columbia object image library (COIL-20), Technical Report CUCS-005-96, Department of Computer Science, Columbia University, USA.

Redner, R. A. and Walker, H. F. (1984). Mixture densities, maximum likelihood and the EM algorithm, *SIAM Review* **26**, 2, pp. 195–239.

Rissanen, J. (1978). Modeling by shortest data description, *Automatica* **14**, 5, pp. 465–471.

Rivals, I. and Personnaz, L. (1999). On cross validation for model selection, *Neural Computation* **11**, 4, pp. 863–870.

Schafer, J. and Strimmer, K. (2005). A shrinkage approach to large-scale covariance matrix estimation and implications for functional genomics, *Statistical Applications in Genetics and Molecular Biology* **4**, 1, pp. 1–28.

Srivastava, S., Gupta, M. R. and Frigyik, B. A. (2007). Bayesian quadratic discriminant analysis, *Journal of Machine Learning Research* **8**, pp. 1277–1305.

Webb, A. R. (1999). *Statistical Pattern Recognition* (Oxford University Press).

Chapter 17

An Evolution of Evolutionary Algorithms with Big Data

Weian Guo, Lei Wang and Qidi Wu

Abstract We use this chapter to illustrate the correlation analysis for big data based on evolutionary algorithms. Since there exists a huge amount of information in big data, traditional methods which process and analyze the database based on causality is powerless in dealing with big data. To exhibit the influences of data volume to methods, numerical simulations with low and high dimensional benchmarks are employed, which points out that the performances of algorithms with small data is not scalable to evaluate the performances in the cases of high-dimensional problems. Besides, to overcome the disadvantage that there is no "one-fits-all" evolutionary algorithm for various problems, we design a framework of hyper evolutionary algorithms and the simulation results validate the feasibility and effectiveness of the framework. An example of business intelligence to conduct the correlation analysis for big data is presented and based on the evolutionary algorithms demonstrates that the evolutionary algorithm is useful and helpful for e-business to make correct judgements for commodities storage, transfer and sales.

17.1 Introduction

In current years, big data gains a popularity since people attaches a great importance to its current and potential values. It has been widely investigated and applied in sociology, informatics, finance, geology, and etc. Other than the traditional data set, big data contains the data sets which are so large and complex that it is difficult for traditional data management tools and applications. Hence, there are increasing challenge problems including data gain, storage, search, sharing, transfer and analysis raised by big data, which people never met before and the conventional methods are powerless to deal with them.

For the definition of big data, IBM presents a popular and simple explanation which involves four critical features: volume, velocity, variety, and veracity, which are abbreviated as "4V". First, as the literal meaning of big data, the data set is always large in volume. Although there is no definite number to qualify the data scale, the data set is so large that it is not easy to handle by most current methods and keeps increasing. The amount of data being collected is so huge that current popular database management methods and tools are becoming overloaded and therefore obsolete, which points out that it is keenly necessary to find novel solutions to support big data analyzing. Second, velocity refers to the speed which the data is coming and the requirement for real time analyze or process. With the increasing velocity at which we are creating a huge amount of data nowadays, a quick response for big data is crucial. To solve this challenge, it needs the suggestions for hardware, software and data management strategies. As we continue to create more data, we also will employ more methods to monitor and manage the information. Third, variety means the kinds of data sources, which may produced by email, video, phone number, climate records, stock prices and much more. The amount is still only increasing. It refers to the data storage, maintain, management and processing. Fourth, veracity is the toughest which points to that if there exits too many errors, redundancy and corruption, the processing for big data means nothing. Therefore, to reduce the accumulation of distrust data is hard but crucial. As data currently stands, it's difficult to know which information is accurate and which is out of date. If organized and used correctly, big data can help us spot business trends, prevent diseases, and combat crime, among other things. As humans continue to create more data in their daily lives, the work of data scientists will become that much more important and useful.

As a successful application of business intelligence, e-business gains a rapid development in current decades. To maximize the benefits, a perfect commodity management which includes the commodity storage, transfer, sale is necessary. By analyzing the browsing records of customers, dealers could make a correct and quick judgement. However, this kind of business intelligence belongs to big data. The correlation analysis is helpful in commodities management. Evolutionary algorithms (EAs) has been developed for several decades and achieve a great success. As a key subfield of artificial intelligence, evolutionary algorithms plays an important role in various areas such as business management, engineering optimization, medical rehabilitation and etc. Considering the 4V feature of big data, evolutionary algorithms is very suitable in dealing with big data problem since it involves metaheuristic and stochastic features. In previous work, the researches on evolutionary algorithms regarding big data are quite few. Meanwhile according to our recent research, the data scale has positive connections to algorithms' performance. Inspired by biological mechanisms of evolution, evolutionary algorithms employs an iterative development in a population towards objectives. Research on optimization has been very active during the last decades in almost every field of science and engineering, ranging from efficiency maximization in job-shop scheduling problems [Zelenka and Kasanicky (2011)] to cost minimization in robotic path planning [Williams and Yang (2006)]. There are also various optimization problems in our daily life, including minimizing charging the cost of electric vehicles (EV) [Kristoffersen *et al.* (2011)], and maximizing profits from investments [Chang and Shi (2011)], etc. In recent years, a variety of evolutionary algorithms have been developed as feasible and effective methods for optimization problems [Akay and Karaboga (2012); Oca *et al.* (2009)], especially for non-deterministic polynomial (NP) problems [Garey and Johnson (1990)]. Evolutionary algorithms have several advantages such as robustness, reliability, global search capability and the fact that there is little or no prior knowledge required [Noman and Iba (2008)]. Many of them [Hansen and Ostermeier (2001); Herrera *et al.* (1998); Holland (1992)], such as evolutionary strategies (ES), genetic algorithms (GA), differential evolution (DE), have been successfully implemented in practical problems.

The main ideas of most EAs are inspired from nature. During the past decades, researchers have proposed novel heuristic algorithms by drawing inspiration from different natural phenomena. Genetic algorithm (GA) mimics the biological process through producing generations of chromosomes [Herrera *et al.* (1998); Holland (1992)]. Simulated Annealing (SA) was inspired by the annealing in metallurgy [Kirkpatrick *et al.* (1983)]. The idea of Particle Swarm Optimization (PSO) comes from the flocking behavior of birds [Kennedy *et al.* (2012)]. Ant colony optimization (ACO) simulates the ecological behavior of ants in foraging [Dorigo and Stutzle (2004)], and artificial bee colony (ABC) was inspired by the behavior of honeybees in collecting nectar [Akay and Karaboga (2012)]. With the development of these algorithms, several hybrid methods have been proposed as well [Yang *et al.* (2008); Yuan *et al.* (2010)]. This shows that nature does not only provide us with resources, but that it may also serve as inspiration for developing new methodologies.

The rest of this chapter is organized as follow. Section 17.2 briefly introduces several classical algorithms and their varieties. We also give examples to illustrate the influences of data to algorithms. According to our numerical analysis, it is obvious that in traditional design of evolutionary algorithm, the algorithms are preferred to consider, while the conclusions may be not correct with a huge volume of big data. Section 17.3 focuses on the varieties of big data, which design a framework for hyper heuristic algorithms, and the numerical simulation results validate the effectiveness of our design. In Section 17.4, we summarize five kinds of correlations for commodities and present the correlation analysis models for business intelligence, which is applied to the e-business. In simulations, evolutionary algorithms is employed to conduct the correlation analysis. Based on the simulation results, evolutionary algorithms are feasible and effective to predict the commodity sales and cam simulate the commodities correlation. We end this chapter in Section 17.5 with conclusions and also proposed several problems which need to be solved in future.

17.2 Design of Hyper-Heuristic Framework for Evolutionary Algorithms

Although evolutionary algorithms are developed for several decades, for different problems there is no a uniform algorithm performs the best than any others, which means the traditional evolutionary algorithms is difficult to solve the variety problem in big data. In 1997, Wolfpert and Macready proposed "No Free Lunch" (NFL) theorem [Wolpert and Macready (1997)] and in 2005, the authors developed their work which indicated that "Any two optimization algorithms are equivalent when their performance is averaged across all possible problems" [Wolpert and Macready (2005)]. Hence, how to choose the suitable algorithm is a problem before computation. To date, some researches about hyper heuristic algorithms are investigated and the objectives are to ask the hyper algorithm to choose suitable algorithm during the computation by itself [Ozcan *et al.* (2008)]. In this section, we design a novel framework for the hyper-heuristic algorithm. Through the time complexity analysis, the novel framework will not increase the time complexity. Besides, Simon [Simon *et al.* (2009)] explored the effect of elitism on performance. The paper showed that elitism can significantly improve the performance in some problems. Also the conclusion is drawn if the counts of elites are too many, convergence will become worse. Elites are excellent individuals in one generation. They have some good features of solution and the next generation can inherit the features from elites probabilistically. Hence, a proper numbers of elites can help algorithms' convergence be better. However, too many elites may result in local optimal. To employ a large proportion of population as elites, in the proposed framework, we split the whole population into several optimizers and select the large number of good solutions to do the second-layer optimization. The elites will be generated in the second-layer optimization and used in the next generation.

17.2.1 *Genetic algorithm*

Genetic algorithm (GA) as the most famous algorithm in evolutionary algorithms has been investigated and applied for many decades [Srinivas and Patnaik (1994)]. It uses the crossover operator and mutation operator to recombine the solutions in a stochastic way. In GA, a population composed by a certain number of candidate solutions termed chromosome is evolved toward better solutions. The solutions are generally represented in binary as strings of 0 and 1, but other encoding ways such as real number encoding are feasible as well. Each bit in the chromosome is called gene. The dimension of an optimization problem could be considered as the genes of the chromosome which could be mutated and recombined. In each generation, the fitness of every chromosome in the population is evaluated by fitness function which is the optimization goal. The best chromosome is selected from the current population and each individual's genome is modified to form a new population for the generation. Generally, the algorithm terminated when the solution met users' requirement or the maximum number of generations is been achieved.

17.2.2 *Particle swarm optimization*

As a popular heuristic algorithm, particle swarm optimization (PSO) has drawn a lot of attentions and implemented in various kinds of areas [Kennedy *et al.* (2012)]. It built mathematical formula based on the birds' flock, which could guide the particles in optimization. Each particle with individual position and velocity presents a candidate solution. The particle moves based on its own best history position and the best position of the whole population. The mathematical formulae are given in (17.1).

$$\begin{aligned} v_{id} &= v_{id}(t-1) + c_1\varphi_1(p_{id} - x_{id}(t-1)) + c_2\varphi_2(p_{gd} - x_{id}(t-1)) \\ x_{id} &= x_{id}(t-1) + v_{id}(t) \end{aligned} \tag{17.1}$$

where $v_{id}(t)$ is the velocity of the d-th dimension of the i-th particle at the t-th generation, $x_{id}(t)$ presents the current position of d-th dimension of the i-th particle at the t-th generation, is the i-th individual's best position found so far at d-th dimension, p_{gd} is the global best position found so far at d-th dimension, φ_1 and φ_2 are two random values in interval $(0, 1)$, c_1 and c_2 are learning factors.

17.2.3 *Biogeography-based optimization*

Compared with GA and PSO, Biogeography-based Optimization (BBO) is a new evolutionary algorithm [Simon (2008); Simon *et al.* (2009) Guo *et al.* (2014)] and in the subsequent simulations several kinds of varieties of BBO are employed, so here we introduce this algorithm in detail. Biogeography-based Optimization is inspired from the science of Biogeography. The habitats in biogeography are analogous to the solutions in optimization problems. The suitability of one island for living is evaluated by High Suitability Index (HSI). The variables that characterize the habitability are called Suitability Index Variables (SIVs). In optimization problems, a good solution is analogous to an island with a high HSI, while a poor solution represents an island with a low HSI. An island with high HSI has more species than an island with low HSI. This means the immigration to a good island is difficult for the aliens. Therefore, the emigration is dominant in good islands. In contrast, since few species live in poor islands, immigration is dominant in poor islands. To sum up, good solutions tend to share their features with poor solutions while poor solutions tend to accept features from good solutions. We use the notation Δt to denote a time period which is so small that during Δt at most one migration occurs. In addition, three assumptions are described as follows,

(1) There were K species at time t and no any immigration or emigration occurred during the period from t to $t + \Delta t$.

(2) There were $K+1$ species at time t and at most only one species emigrate.
(3) There were $K-1$ species at time t and at most one species immigrated.

Considering that the probability P_K denotes the habitat contains exactly K species, P_K updates in each generation from t to $t+\Delta t$ as (17.2),

$$P_K(t+\Delta t) = P_K(t)(1-\lambda_K \Delta t - \mu_K \Delta t) + P_{K-1}\lambda_{K-1}\Delta t + P_{K+1}\mu_{K+1}\Delta t \quad (17.2)$$

where λ_K and μ_K are the immigration rate and emigration rate in the habitat that has K species respectively.

By defining

$$\dot{P}_K = \begin{cases} -(\lambda_K+\mu_K)P_K + \mu_{K+1}P_{K+1}, & K=0 \\ -(\lambda_K+\mu_K)P_K + \lambda_{K-1}P_{K-1} + \mu_{K+1}P_{K+1}, & 1 \le K \le K_{\max}-1 \\ -(\lambda_K+\mu_K)P_K + \lambda_{K-1}P_{K-1} & K = K_{\max}. \end{cases} \quad (17.3)$$

we obtain

$$\dot{P} = AP \quad (17.4)$$

where

$$P = [P_0, ..., P_N]^{\mathsf{T}} \quad (17.5)$$

$$A = \begin{pmatrix} -(\lambda_0+\mu_0) & \mu_1 & 0 & \cdots & 0 \\ \lambda_0 & -(\lambda_1+\mu_1) & \mu_2 & \ddots & \vdots \\ \vdots & \ddots & \ddots & \ddots & \vdots \\ \vdots & \ddots & \lambda_{n-2} & -(\lambda_{n-1}+\mu_{n-1}) & \mu_n \\ 0 & \cdots & 0 & \lambda_{n-1} & -(\lambda_n+\mu_n) \end{pmatrix}. \quad (17.6)$$

Fig. 17.1 illustrates a linear model of species distribution in a single habitat [Simon (2008)]. The researches on migration models have been investigated in [Ma and Simon (2011a)]. In the linear migration model, I is the maximum possible immigration rate and E represents the maximum possible emigration rate. The immigration rate and the emigration rate are functions of the number of species in the habitat. For one island, the immigration rate increases and the emigration rate decreases as the number of species increases. The pseudo-codes of migration are shown in Algorithm 17.1.

Like most other evolutionary algorithms, Biogeography-Based Optimization has the mutation operator which could help algorithms break away local optimum and explore the searching space. In BBO [Simon (2008)], the mutation rate is calculated by (17.7).

$$m(K) = m_{\max}\left(\frac{P_{\max}-P_K}{P_{\max}}\right) \quad (17.7)$$

where $m_{\max}$ is defined by users in advance and $P_{\max}$ and P_K can be obtained in (17.2). The pseudo-codes of mutation are shown in Algorithm 17.2.

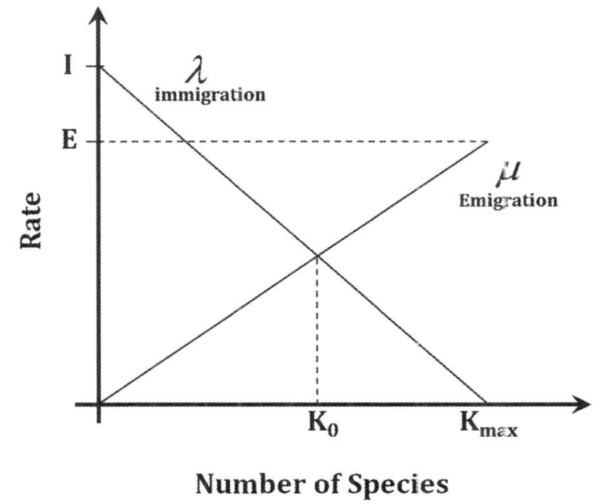

Fig. 17.1 Linear migration model in biogeography-based optimization

---

Algorithm 17.1 Pseudo-codes of Biogeography-Based Optimization Migration

---

Select H_i according to immigration rate λ_i
if H_i is selected **then**
  for $j = 1$ to n **do**
    Select H_j according to emigration rate μ_i
    if H_j is selected **then**
      Replace SIV in H_i with SIV in H_j
    end if
  end for
end if

Algorithm 17.2 Pseudo-codes of Biogeography-Based Optimization Mutation

---

for $i = 1$ to n **do**
  Use λ_i and μ_i to compute the probability P_i and mutation rate m_i
  Select H_i(SIV) with probability m_i
  if H_i(SIV) is selected **then**
    Replace H_i(SIV) with a random value
  end if
end for

17.2.4 *Design of framework of hyper-heuristic algorithm*

In this section, we present a framework of hyper heuristic algorithm to obtain a comprehensive adaption to different algorithms, lots of improvements are investigated and one of them is hyper-heuristics. A hyper-heuristic is a heuristic search method that seeks to automate, often by the incorporation of machine learning techniques, the process of selecting, combining, generating or adapting several simpler heuristics (or components of such heuristics) to efficiently solve computational search problems [Ozcan *et al.* (2008)]. One of the motivations for studying hyper-heuristics is to build systems which can handle classes of problems rather than solving just one problem. It does not use a single algorithm during the whole progress of optimization, but let the algorithm choose a suitable algorithm by itself. Hence, the idea of hyper heuristic algorithm overcomes the disadvantage that a single algorithm only performs the best only for specific problems. Due to its comprehensive adaption to optimization problem, hyper heuristic algorithm has drawn a lot of attentions.

In evolutionary algorithms, elite plays a key role to retain good solutions and share helpful information in the whole population. In generally, only a small proportion of population could be considered as elites, while a large number of elites will make the algorithm achieve local minimal and stagnate the computational progress. In our proposed framework, the whole population will be split into several groups. In different groups, several kinds of algorithms are used to select elites. The number of elites of each algorithm is based on the performance of the algorithm in that generation. The flowchart is given in Fig. 17.2 and the schedule is present as follows.

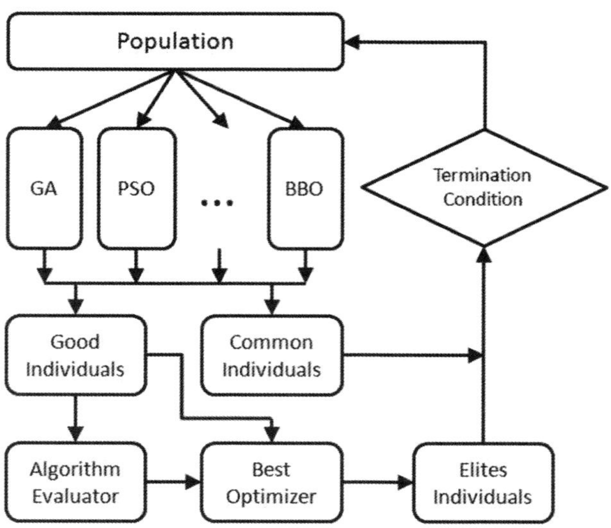

Fig. 17.2 The flowchart of proposed hyper heuristic algorithm

- Step 1. Generate the initial population and split them evenly into several optimizers.
- Step 2. All optimizer run the corresponding algorithm in a parallel way. In each optimizer, select the half of individuals as good solutions. The rest individuals are considered as common individuals.
- Step 3. Algorithm evaluator accesses the performance of each optimizer based on the best solution of each optimizer. The best optimizer uses the good solutions to compute elites.
- Step 4. The elites and the common individuals compose the whole population in the next generation.
- Step 5. If the termination conditions are met, end the algorithms; otherwise, go to Step 1.

Since the parallel computation strategy is employed, the time complexity of the hyper heuristic algorithm is equal to the worst time complexity of all the individual optimizer. Hence, this design could not increase the time cost in computation.

17.2.5 *Simulations for the novel framework of evolutionary algorithms*

In this section, 14 functions are employed as numerical benchmarks to test the performance of the proposed migration operators and the results are compared with previous migration operators. The functions are given in Table 17.1 and the properties of the 14 benchmarks are given in Table 17.2. More details can be found in [Yao *et al.* (1999)].

Table 17.1 Benchmarks functions. The granularity of each domain was 0.1. Except for the quartic Function, which has a granularity of 0.01.

Function	Name	Dimension	Domain
f01	Ackley's Function	20	[-30,30]
f02	Fletcher-Powell	20	[$-\pi$,π]
f03	Generalized Griewank's function	20	[-600,600]
f04	Generalized Penalized function 1	20	[-50,50]
f05	Generalized Penalized function 2	20	[-50,50]
f06	Quartic function	20	[-1.28,1.28]
f07	Generalised Rastrigin's function	20	[-5.12,5.12]
f08	Generalized Rosenbrock's function	20	[-2.048,2.048]
f09	Schwefel' Problem 1.2	20	[-65.535,65.535]
f10	Schwefel' Problem 2.21	20	[-100,100]
f11	Schwefel' Problem 2.22	20	[-10,10]
f12	Schwefel' Problem 2.26	20	[-512,512]
f13	Sphere Model	20	[-5.12,5.12]
f14	Step Function	20	[-200,200]

Table 17.2 Properties of each benchmarks functions

Function Name	Multimodal?	Separable?	Regular?
Ackley's Function	Yes	No	Yes
Fletcher-Powell	Yes	No	No
Generalized Griewank's function	Yes	No	Yes
Generalized Penalized function 1	Yes	No	Yes
Generalized Penalized function 2	Yes	No	Yes
Quartic function	No	Yes	Yes
Generalised Rastrigin's function	Yes	Yes	Yes
Generalized Rosenbrock's function	No	No	Yes
Schwefel' Problem 1.2	No	No	Yes
Schwefel' Problem 2.21	No	No	No
Schwefel' Problem 2.22	Yes	No	No
Schwefel' Problem 2.26	Yes	Yes	No
Sphere Model	No	Yes	Yes
Step Function	No	Yes	No

In simulation, we employ two classic algorithms genetic algorithm and particle swarm optimization, and a novel evolutionary algorithm biogeography-based optimization as the single algorithm. To demonstrate the harm of the large number of elites, Table 17.3 is given to compare the performance of algorithm with few elites and lots of elites. In Table 17.3, the whole population is with a size of 100. In the few-elites strategy, we only adapt 2 elites while in another strategy we employ 50 individuals (half of the whole population) as elites. It is obvious that a large number of elites cannot help algorithm enhance the performance. We also present the simulation results of hyper heuristic algorithm and the results show that the large proportion of elites could help algorithm obtain a good solution in hyper heuristic algorithm.

In simulation, the parameters are set as follows. For GA, roulette wheel selection and uniform crossover are employed. The mutation rate is set as 0.01. For PSO, the two learning factors are both set as 1.296 and the inertial weight is 0.726. For BBO, a linear migration model is employed. The maximum migration rate is 1 and mutation rate is 0.01. For each algorithm, the population size is set as 100 and the maximum generation is 500. For each benchmark, 20 Monte Carlo simulations of each algorithm are conducted to obtain an average performance. PSO-2E means PSO employs 2 elites while PSO-50E means PSO employs 50 elites. The same to GA-2E, GA-50E, BBO-2E, BBO-50E.

According to the results in Table 17.3, compared the performance in pairwise, PSO-2E is superior to PSO-50E, GA-2E outperforms GA-50E and BBO-2E is better than BBO-50E. Hence, it is obvious that the performances of algorithms with large proportion of elites are inferior to those with fewer elites, which means a large number of elites cannot help algorithm enhance their performance. However, the hyper heuristic algorithm performs the best for most benchmarks although it employs a large number of elites, which the so many elites play a positive roble in optimization. Hence, our design is feasible and effective.

Table 17.3 The performances of algorithms with different elitism strategy

Function	PSO-2E	PSO-50E	GA-2E	GA-50E	BBO-2E	BBO-50E	Hyper
F1	1.0326E+01	9.9097E+00	1.2539E+00	2.6602E+00	2.7132E-01	4.7867E-01	1.7751E-01
F2	2.7064E+04	2.9395E+04	1.4506E+03	4.8528E+03	1.2050E+03	1.6040E+03	7.1796E+02
F3	7.8090E+00	8.4455E+00	1.0214E+00	1.0701E+00	1.0027E+00	1.0038E+00	1.0002E+00
F4	1.2383E+03	4.1505E+01	1.7144E-03	6.9510E-03	2.4347E-03	5.8418E-03	1.3090E-04
F5	1.0930E+05	1.5178E+05	2.8683E-02	1.9275E-01	1.7121E-02	4.1205E-02	3.6545E-03
F6	3.8241E-02	4.1181E-02	2.0000E-08	4.0500E-07	1.6000E-08	4.2000E-08	0.0000E+00
F7	4.2363E+01	4.3227E+01	0.0000E+00	8.0872E-01	0.0000E+00	1.0109E-01	0.0000E+00
F8	4.0853E+01	5.4201E+01	6.0126E+00	6.4508E+00	6.7070E+00	7.4567E+00	6.8248E+00
F9	1.1982E+03	1.2632E+03	3.2849E+00	1.4358E+01	8.3695E-01	1.3768E+00	1.7071E-01
F10	8.0366E+02	9.6260E+02	1.6827E+01	1.1809E+02	1.5657E+01	3.4061E+01	5.7964E+00
F11	6.9991E+00	7.8000E+00	9.0000E-02	3.8000E-01	0.0000E+00	1.0000E-02	0.0000E+00
F12	1.9620E+01	1.6928E+01	1.7700E+00	2.5600E+00	1.3900E+00	1.1900E+00	1.0000E+00
F13	2.2753E+00	2.1700E+00	0.0000E+00	1.1086E-02	0.0000E+00	6.0472E-03	0.0000E+00
F14	8.8960E+02	1.0290E+03	0.0000E+00	3.0000E+00	1.0000E-01	1.0000E-01	0.0000E+00

17.3 Performances Analysis of Evolutionary Algorithms for High-Dimensional Benchmarks

In previous work [Guo *et al.* (2014)], the comparisons for different evolutionary algorithms present the performance of different methods. However, there exists differences in dealing with traditional data set and big data. An example is given that for general common benchmarks, the dimension change could results different performance for algorithms. One algorithm could outperforms others for benchmarks with low dimension while it may performs the worst when the dimension is much higher. To exhibit the simulation results, we employed the benchmarks in Table 17.1 and the varieties of BBO which are presented in Section 17.3.1, Section 17.3.2 and Section 17.3.3 to conduct the numerical simulations.

17.3.1 *Uniform blended migration operator*

In [Ma and Simon (2011b)], the blended migration operator employed a non-uniform blended parameter α, which the value of α is set as a const. In the simulation of [Ma and Simon (2011b)], Ma and Simon set the values of α as 0, 0.5 and 0.8, respectively, and the results show that the value of α is influential on BBO's performance. Blended migration operator with $\alpha = 0.5$ is better than that with $\alpha = 0.8$ which shows that when a new solution feature is contributed equally from itself and the selected emigrant, the performance is the best. In this subsection, we extend the work of [Ma and Simon (2011b)] to propose a uniform blended migration operator (UBMO), which set the blended parameter termed $\tilde{\alpha} \in [0, 1]$ as a random value. The uniform blended migration operator is demonstrated as (17.8).

$$H_i(SIV) \leftarrow \tilde{\alpha} H_i(SIV) + (1 - \tilde{\alpha}) H_j(SIV). \tag{17.8}$$

In uniform blended migration operator, the value of $\tilde{\alpha}$ is uniformly distributed in interval $[0, 1]$ and needn't to be set by users in advance. Thereby, the uniform blended migration operator is easy to implement. The pseudo-codes of uniform blended migration operator is given in Algorithm 17.3.

Algorithm 17.3 Pseudo-codes of Uniform Blended Biogeography-Based Optimization Migration

```
Select H_i according to immigration rate λ_i
if H_i is selected then
  for j = 1 to n do
    Select H_j according to emigration rate μ_i
    if H_j is selected then
      Replace SIV in H_i with SIV in H_j
    end if
  end for
end if
```

17.3.2 *Heuristic migration operator*

In genetic algorithm, the heuristic crossover was proposed in [Wright (1991)] and developed by [Michalewicz (1995)]. In this subsection, we import the idea to biogeography-based optimization to propose the heuristic migration operator (HMO). Assuming that H_j(SIV) is selected to immigrate from H_i(SIV), the heuristic migration operator is given as (17.9),

$$H_i(SIV) \leftarrow H_i(SIV) + \gamma\left(H_j(SIV) - H_i(SIV)\right), F(H_j) \geq F(H_i) \qquad (17.9)$$

where $\gamma \in [0, 1]$ and $F(\cdot)$ is a fitness function. In heuristic migration operator, the individual fitness is taken account of the migration operator which is different from the idea of blended migration operator. The better individuals will influence the worse individuals, while the worse individuals cannot affect the better ones. The pseudo-codes of heuristic migration operator is given in Algorithm 17.4.

Algorithm 17.4 Pseudo-codes of Heuristic Migration Operator

```
Select H_i according to immigration rate λ_i
if H_i is selected then
  for j = 1 to i do
    Select H_j according to emigration rate μ_j
    if H_j is selected then
      SIV in H_j migrate to H_i based on Eq. (17.9)
    end if
  end for
end if
```

In (17.9), the value of γ is set as a constant. Like the uniform blended migration operator, we design a uniform heuristic migration operator (UHMO) shown in (17.10).

$$H_i(SIV) \leftarrow H_i(SIV) + \tilde{\gamma}\left(H_j(SIV) - H_i(SIV)\right), F(H_j) \geq F(H_i) \qquad (17.10)$$

$\tilde{\gamma} \in [0, 1]$ is a random value and $F(\cdot)$ is a fitness function. The pseudo-codes of uniform heuristic migration operator is similar with Algorithm 17.4. The difference between blended migration operator (BMO) and heuristic migration operator (HMO) is that the fitness of solutions is considered in HMO, while not in BMO.

17.3.3 *Extended migration operator*

In [Herrera *et al.* (1998); Sharapov (2007)], BLX-ρ crossover operator was proposed for genetic algorithm, which is an extension of flat crossover [Radcliffe (1991)]. By defining x_1 and x_2 are two parents, $x_{1,i}$ and $x_{2,i}$ present the i-th gene and j-th gene in x_1 and x_2, respectively. The offsprings could be located outside the following interval

$$[\min(x_{1,i}, x_{2,i})] - I \cdot \rho, \max(x_{1,i}, x_{2,i}) + I \cdot \rho] \tag{17.11}$$

where $I = \max(x_{1,i}, x_{2,i}) - \min(x_{1,i}, x_{2,i})$ and ρ are defined in advance.

Based on the idea of extended intermediate crossover, we proposed a novel migration operator termed extended migration operator (EMO) in this subsection. By defining H_i being selected to migrate and H_j being emigrant, EMO is described as (17.12),

$$H_i(SIV) = H_i(SIV) + \zeta(H_j(SIV) - H_i(SIV)), F(H_j) \geq F(H_i) \tag{17.12}$$

where $\zeta \in [-0.25, 1.25]$ is a constant and and $F(\cdot)$ is a fitness function. The uniform version of extended migration operator (UEMO) is given as (17.13).

$$H_i(SIV) = H_i(SIV) + \tilde{\zeta}\left(H_j\left(SIV\right) - H_i\left(SIV\right)\right), F(H_j) \geq F(H_i) \tag{17.13}$$

where $\tilde{\zeta} \in [-0.25, 1.25]$ is a random value and $F(\cdot)$ is a fitness function.

17.3.4 *Numerical simulations for high dimensional benchmarks*

In all simulations, the linear migration model [Simon (2008); Guo *et al.* (2014)] is employed in all versions of BBO. We use a population size of 50 and the maximum generation is 1000. For each benchmark the dimension is 20. Since we do not take account of the mutation, the mutation rate is set as 0. For each benchmark, we run 50 times to obtain an average performance. The values in "Mean" row is the average performance in the 50 runs and "Best" is the best performance in the 50 runs.

Since the uniform migration operators are easily implemented, in Table 17.4, we compare the the uniform types of the proposed migration operators. The column of "Original" presents the performance of original migration operator. The column of "UBMO" presents the performance of uniform blended migration operator. The column of "UHMO" presents the performance of uniform heuristic migration operator shown. The column of "UEMO" presents the performance of uniform extended migration operator. To compare the results convenience, for each benchmark, we assign 3 points for the best operator, 2 points for the second best operator, 1 point for the third best operator, and 0 point for the worst one. Then, we add the points for all benchmarks and obtain Table 17.3, where the white column presents the scores for "Mean" value and the black column presents the "Best" value.

First, we compare BBO with different migration operators. For "Mean", original migration operator only wins for f09 and for "Best", original migration operator wins for f07, f11 and f13. For f14, all kinds of migration operator archive the best performance in the column of "Best". For "Mean", "UBMO" only wins for f7 and f10. For "Best", "UBMO" does not perform the best for any benchmark. For f03, "UBMO", "UHMO" and "UEMO" all archive the best performance. For "UHMO", it wins for f10 in the column of "Best" and does not performs the best for any

Table 17.4 Simulation results obtained by original BBO, UBMO, UHMO, UEMO, on 14 functions with dimension=20 over 50 independent runs. The table shows the best and mean values.

Function		BBO	UBMO	UHMO	UEMO
f01	Mean	6.5863E-01	3.3997E-03	1.4071E-03	1.2432E-04
	Best	3.2574E-01	9.6889E-04	4.7880E-04	3.1381E-05
	Stdev	2.1169E-01	8.5456E-04	2.6238E-04	3.3592E-05
f02	Mean	1.3006E+04	2.4833E+04	2.0884E+04	1.2979E+04
	Best	2.0425E+03	1.9552E+03	3.1931E+03	1.7371E+03
	Stdev	6.2198E+03	7.1349E+03	1.5780E+04	5.6596E+03
f03	Mean	1.0168E+00	1.0000E+00	1.0000E+00	1.0000E+00
	Best	1.0047E+00	1.0000E+00	1.0000E+00	1.0000E+00
	Stdev	1.6502E-02	9.9519E-07	2.8271E-07	5.1976E-09
f04	Mean	1.9788E-02	2.2383E-03	2.0216E-06	3.2648E-09
	Best	1.0834E-03	3.2002E-08	5.5094E-09	1.2263E-10
	Stdev	2.2141E-02	1.0582E-06	8.3461E-08	2.7752E-09
f05	Mean	1.0205E-01	1.3052E-04	3.5851E-05	1.1186E-07
	Best	3.4786E-02	3.2683E-07	7.3382E-07	2.4154E-09
	Stdev	3.7674E-02	1.1245E-04	1.7326E-05	2.2910E-07
f06	Mean	5.9660E-07	3.9358E-13	3.6917E-14	1.2961E-15
	Best	3.0000E-08	3.3010E-16	6.4715E-16	1.3731E-18
	Stdev	2.0402E-07	1.0185E-13	9.1092E-14	8.9366E-16
f07	Mean	8.0872E+02	1.2747E-01	1.5291E-01	2.0134E-01
	Best	0.0000E+00	2.8427E-05	2.0774E-05	6.1739E-08
	Stdev	1.1316E+01	2.9723E-01	4.8687E-01	1.9836E-05
f08	Mean	3.5625E+01	1.7770E+01	1.7709E+01	1.7058E+01
	Best	9.3181E+00	1.6840E+01	1.6732E+01	1.2593E+01
	Stdev	2.3533E+01	2.8400E-01	2.8909E-01	8.5342E-01
f09	Mean	4.4812E+00	7.7015E+01	4.0952E+01	4.6887E+00
	Best	1.6951E+00	2.5050E+00	8.9673E-02	2.6094E-04
	Stdev	1.9637E+00	6.3021E+01	6.5889E+01	3.4516E+01
f10	Mean	5.6366E+02	1.6653E+02	1.8878E+02	2.6910E+02
	Best	1.3354E+02	3.7711E+01	3.2155E+01	3.8476E+01
	Stdev	2.8717E+02	4.4016E+01	1.0778E+02	1.2098E+02
f11	Mean	1.3400E-01	2.3695E-03	4.8982E-04	2.7357E-05
	Best	0.0000E+00	5.5519E-04	1.0880E-04	4.6377E-06
	Stdev	7.7460E-02	1.2843E-03	2.9141E-04	2.2767E-05
f12	Mean	6.8460E+00	3.5882E+00	2.9597E+00	1.3942E+00
	Best	3.1000E+00	1.6609E+00	1.1603E+00	4.2796E-01
	Stdev	1.7990E+00	7.3428E-01	1.0527E+00	3.9279E-01
f13	Mean	2.4189E-03	4.1370E-07	7.0628E-08	8.6881E-10
	Best	0.0000E+00	2.0931E-08	3.5954E-09	5.9876E-11
	Stdev	3.0236E-03	3.8354E-07	6.0630E-08	3.6172E-10
f14	Mean	1.5200E+02	6.6000E-01	4.2000E-01	2.2000E-01
	Best	0.0000E+00	0.0000E+00	0.0000E−00	0.0000E+00
	Stdev	1.9621E+00	6.4031E-01	4.5826E-01	8.0000E-01

benchmark in the column of "Mean". For "UEMO", it wins for f01, f02, f04, f05, f06, f08, f11, f12, f13, f14 totally 10 benchmarks, in the column of "Mean", and it also wins f01, f02, f03, f04, f05, f06, f08, f09, f12, totally 9 benchmarks, in the column of "Best". The test results show that the improvements of migration operators are feasible and effective to enhance BBO's performance. In the comparisons of BBO with different migration operators, "UEMO" performs the best, which means to extend the domain of offsprings can help algorithms explore the searching space and pursue a good performance. In addition, by comparing the "UBMO" with "UHMO", it is in general that the "UHMO" is superior to "UBMO". Considering that in "UHMO" only good solutions can migrate to poor solutions, "UHMO" could improve poor solutions by good solutions. Since in BBO the migration occurs in the whole population, which means in any generation, any individual could be affected by any other individuals, the influence of local optimal solutions is weaken.

According to the results in Table 17.3, BBO performs the worst, UEMO performs the best. However, the results of the simulation for high-dimensional benchmarks are not totally same, which are presented in Table 17.5, Table 17.6 and Table 17.7. The comparison results point that the dimension is influential to the performance of evolutionary algorithms. The investigation of evolutionary algorithm for big data is a new field and be fully worthy exploring.

17.4 Correlation Analysis for Business Intelligence Based on Evolutionary Algorithms

In current researches of business intelligence (BI), the data management refers huge potential benefits. An excellent management could help business reduce the cost for storage and transfer and accelerate the flow of funds.

The correlation reagarding various commodities is nonlinear and difficult to present in a determinate way. Some commodities do no have direct correlation. For example, it is difficult to infer correlation of the sales volume of shoes and that of pencils, while some others have obvious correlations. For examples about the positive correlation, for the costumers who buy a toothbrush, it is a high probability for them to buy a toothpaste. For the costumers who buy a cell phone, it is a high probability for them to buy a SIM card. The examples for the negative correlation are given that sales volume of sunglasses and that of umbrella. During rainy days, people would like to buy umbrellas, while during sunshine days, people prefer sunglasses. The above examples are easy to understand. Nevertheless, for most commodities, it is difficult to consider their correlations. Does there have the correlations between water and portable benches? The correlations between the two commodities is not obvious, but with some conditions, they indeed have correlations. An example is given that during the World Expo 2010, Shanghai, since too many visitors are in queues and they have to wait for a long time before visit some hot places, the sales of spring water and benches blowouts. Hence, a good correlation analysis could help dealers make correct decision to maximize their benefits.

Table 17.5 Simulation results obtained by original BBO, UBMO, UHMO, UEMO, on 14 functions with dimension = 100 over 20 independent runs. The table shows the best and mean values.

Function		Original	UBMO	UHMO	UEMO
f01	Mean	1.4463E+00	2.6290E-01	1.1790E-01	1.0787E-02
	Best	1.2384E+00	1.4064E-01	5.2021E-02	5.0160E-03
	Stdev	1.2423E-01	1.3738E-01	8.5726E-02	8.1080E-03
f02	Mean	6.3635E+05	1.0598E+06	8.0707E+05	5.7238E+05
	Best	4.7759E+05	7.2449E+05	3.4629E+05	3.0808E+05
	Stdev	9.6632E+04	2.2838E+05	1.7676E+05	2.0279E+05
f03	Mean	1.2508E+00	1.0082E+00	1.0010E+00	1.0001E+00
	Best	1.1723E+00	1.0010E+00	1.0004E+00	1.0000E+00
	Stdev	5.8018E-02	1.1124E-02	4.3805E-04	5.5579E-05
f04	Mean	6.4926E-02	9.9243E-03	7.9266E-04	1.4362E-04
	Best	3.8130E-02	7.6164E-04	3.0726E-04	4.2027E-05
	Stdev	4.2259E-02	2.2073E-02	3.0836E-04	1.1133E-04
f05	Mean	1.2078E+00	2.2234E-01	6.9362E-02	7.8133E-03
	Best	8.9609E-01	8.0570E-02	3.0045E-02	8.4139E-04
	Stdev	2.0684E-01	1.4024E-01	4.1252E-02	5.7212E-03
f06	Mean	6.5611E-05	9.6532E-07	8.6339E-08	2.9632E-09
	Best	3.0670E-05	6.8751E-08	8.8515E-09	3.5136E-10
	Stdev	1.9893E-05	1.6129E-06	1.6397E-07	3.0616E-09
f07	Mean	3.7277E+00	7.5431E+00	8.3510E+00	7.6193E+00
	Best	2.0218E+00	3.8238E+00	4.1979E+00	3.0414E+00
	Stdev	1.4614E+00	2.1454E+00	2.7038E+00	2.7218E+00
f08	Mean	2.7022E+02	9.7271E+01	9.7192E+01	9.7271E+01
	Best	1.4510E+02	9.6663E+01	9.6798E+01	9.5835E+01
	Stdev	4.1685E+01	2.2715E-01	1.8990E-01	1.0271E+00
f09	Mean	6.8206E+01	2.0917E+03	1.2931E+03	5.0751E+02
	Best	5.4899E+01	1.6214E+03	6.6847E+02	1.1551E+02
	Stdev	7.8995E+00	1.7217E+02	3.1138E+02	2.6721E+02
f10	Mean	3.8215E+04	8.0559E+03	8.3012E+03	2.1688E+04
	Best	2.9573E+04	6.0535E+03	5.5659E+03	1.1099E+04
	Stdev	6.0624E+03	1.2903E+03	1.9187E+03	4.3146E+03
f11	Mean	1.9437E+00	1.6631E-01	8.8361E-02	7.9947E-03
	Best	1.4000E+00	1.0478E-01	5.9975E-02	4.5996E-03
	Stdev	3.9046E-01	8.1251E-02	2.1055E-02	1.8992E-03
f12	Mean	6.2756E+01	1.9868E+01	1.9105E+01	2.1560E+01
	Best	5.0500E+01	1.5551E+01	1.5697E+01	1.7687E+01
	Stdev	6.6209E+00	2.7917E+00	1.6571E+00	2.4698E+00
f13	Mean	7.9369E-02	2.9677E-03	4.4608E-04	1.8459E-05
	Best	0.0000E+00	2.6934E-04	1.2183E-04	8.8291E-06
	Stdev	2.6636E-02	2.9142E-03	3.9130E-04	2.0752E-05
f14	Mean	3.1500E+01	3.2438E+01	9.8125E−00	4.7500E+00
	Best	1.9000E+01	1.3000E+01	3.0000E−00	0.0000E+00
	Stdev	4.1683E+00	1.8439E+01	5.0648E−00	3.2307E+00

Table 17.6 Simulation results obtained by original BBO, UBMO, UHMO, UEMO, on 14 functions with dimension = 500 over 20 independent runs. The table shows the best and mean values.

Function		Original	UBMO	UHMO	UEMO
f01	Mean	1.1972E+01	9.6917E+00	9.2685E+00	1.0019E+01
	Best	1.1305E+01	9.0888E+00	8.7596E+00	9.6286E+00
	Stdev	4.0860E-01	2.5869E-01	3.3681E-01	2.7057E-01
f02	Mean	8.7801E+07	3.7522E+08	2.8780E+08	1.9910E+08
	Best	7.8379E+07	3.3560E+08	2.5908E+08	1.7692E+08
	Stdev	6.1090E+06	1.8481E+07	1.6567E+07	1.1659E+07
f03	Mean	6.1664E+02	3.4835E+02	2.9359E+02	3.8549E+02
	Best	5.0985E+02	3.0369E+02	2.5033E+02	3.1578E+02
	Stdev	5.4517E+01	2.5854E+01	2.1969E+01	3.6544E+01
f04	Mean	3.0844E+07	4.6975E+05	3.0145E+05	3.9630E+06
	Best	5.8528E+06	5.9490E+04	4.6491E+04	8.2623E+05
	Stdev	1.7940E+07	3.6257E+05	3.0111E+05	2.2088E+06
f05	Mean	1.1575E+08	6.9940E+06	7.4245E+06	2.8572E+07
	Best	6.7243E+07	2.7766E+06	2.4911E+06	1.0994E+07
	Stdev	3.6097E+07	2.4387E+06	3.2606E+06	8.6099E+06
f06	Mean	3.9336E+02	9.1383E+01	6.6225E+01	1.6235E+02
	Best	2.6833E+02	6.8418E+01	4.9442E+01	1.2101E+02
	Stdev	8.2379E+01	1.6385E+01	1.0297E+01	3.0395E+01
f07	Mean	1.5597E+03	1.7722E+03	1.4329E+03	1.3227E+03
	Best	1.4695E+03	1.6201E+03	1.2937E+03	1.2263E+03
	Stdev	6.9896E+01	1.0874E+02	1.0201E+02	5.7908E+01
f08	Mean	6.9553E+03	2.3507E+03	2.0597E+03	3.0400E+03
	Best	6.1506E+03	2.1272E+03	1.8443E+03	2.8033E+03
	Stdev	5.5950E+02	1.7906E+02	1.6772E+02	2.2228E+02
f09	Mean	4.5434E+04	1.2381E+05	1.1637E+05	1.0013E+05
	Best	4.2242E+04	1.1855E+05	1.0983E+05	9.5531E+04
	Stdev	1.4501E+03	3.7208E+03	3.5707E+03	3.1807E+03
f10	Mean	1.2460E+06	3.2288E+05	3.4872E+05	9.1917E+05
	Best	9.7522E+05	2.1398E+05	2.0642E+05	4.4147E+05
	Stdev	1.3896E+05	8.9671E+04	8.5792E+04	2.6212E+05
f11	Mean	3.7465E+02	4.3401E+02	4.3979E+02	4.3036E+02
	Best	3.5240E+02	3.4759E+02	3.9906E+02	3.8177E+02
	Stdev	1.3500E+01	3.6731E+01	2.0389E+01	2.4993E+01
f12	Mean	9.6167E+01	3.6865E+01	3.9543E+01	9.7590E+01
	Best	9.4200E+01	3.3147E+01	3.6595E+01	9.6507E+01
	Stdev	1.0044E+00	2.3355E+00	2.2635E+00	6.6959E-01
f13	Mean	1.7663E+02	9.9630E+01	8.4826E+01	1.0707E+02
	Best	1.5159E+02	9.1114E+01	7.8620E+01	9.4933E+01
	Stdev	1.3713E+01	8.4205E+00	5.7356E+00	6.0848E+00
f14	Mean	6.8298E+04	3.8454E+04	3.2724E+04	4.1039E+04
	Best	6.1687E+04	3.4464E+04	2.8033E+04	3.4531E+04
	Stdev	5.3159E+03	2.5500E+03	1.9933E+03	3.1470E+03

Table 17.7 Simulation results obtained by original BBO, UBMO, UHMO, UEMO, on 14 functions with dimension=1000 over 20 independent runs. The table shows the best and mean values.

Function		Original	UBMO	UHMO	UEMO
f01	Mean	1.6184E+01	1.1088E+01	1.1028E+01	1.2414E+01
	Best	1.5894E+01	1.0794E+01	1.0627E+01	1.2134E+01
	Stdev	1.3682E-01	1.4903E-01	1.8803E-01	2.0417E-01
f02	Mean	7.0922E+08	2.7191E+09	2.1616E+09	1.6331E+09
	Best	6.6686E+08	2.4833E+09	2.0036E+09	1.4984E+09
	Stdev	2.4509E+07	1.3048E+08	1.0525E+08	7.6548E+07
f03	Mean	3.8651E+03	1.0145E+03	1.0143E+03	1.5572E+03
	Best	3.4411E+03	9.3735E+02	8.8221E+02	1.3788E+03
	Stdev	2.1594E+02	4.1113E+01	7.5900E+01	7.2328E+01
f04	Mean	9.5004E+08	2.2359E+06	2.7377E+06	3.4181E+07
	Best	7.6014E+08	1.0197E+06	1.1729E+06	1.9613E+07
	Stdev	1.9184E+08	6.8964E+05	8.3529E+05	7.1243E+06
f05	Mean	2.5127E+09	4.2434E+07	3.6677E+07	1.9248E+08
	Best	1.8736E+09	2.2850E+07	2.7876E+07	1.4590E+08
	Stdev	3.2740E+08	1.3238E+07	7.1465E+06	2.9982E+07
f06	Mean	1.1800E+04	6.0961E+02	5.8756E+02	1.7241E+03
	Best	9.8391E+03	4.5997E+02	4.8811E+02	1.3183E+03
	Stdev	1.1951E+03	6.8762E+01	5.9084E+01	1.7542E+02
f07	Mean	5.9574E+03	5.6713E+03	5.0350E+03	4.7436E+03
	Best	5.5998E+03	5.4003E+03	4.8697E+03	4.4830E+03
	Stdev	1.9628E+02	1.2386E+02	1.1337E+02	1.3925E+02
f08	Mean	3.6837E+04	6.6925E+03	6.4067E+03	1.0295E+04
	Best	3.1742E+04	6.1004E+03	5.8808E+03	9.4775E+03
	Stdev	2.8363E+03	3.2766E+02	2.4102E+02	6.1865E+02
f09	Mean	1.6526E+05	3.0820E+05	2.8859E+05	2.6409E+05
	Best	1.5930E+05	3.0309E+05	2.8317E+05	2.5808E+05
	Stdev	3.5687E+03	3.1509E+03	6.1023E+03	3.4268E+03
f10	Mean	4.6403E+06	1.1710E+06	1.5830E+06	3.5392E+06
	Best	3.6308E+06	7.3061E+05	1.0030E+06	2.3679E+06
	Stdev	6.5067E+05	2.5499E+05	6.8016E+05	1.0111E+06
f11	Mean	1.4683E+03	1.2132E+03	1.2755E+03	1.5119E+03
	Best	1.4189E+03	1.0732E+03	1.1519E+03	1.3999E+03
	Stdev	3.2645E+01	7.4374E+01	5.6264E+01	6.5696E+01
f12	Mean	9.8125E+01	4.1314E+01	4.3177E+01	9.9327E+01
	Best	9.7500E+01	3.8071E+01	3.9098E+01	9.8497E+01
	Stdev	3.7165E-01	2.5397E+00	2.1143E+00	3.3094E-01
f13	Mean	1.1645E+03	3.0362E+02	2.8908E+02	4.5984E+02
	Best	1.0347E+03	2.8064E+02	2.6273E+02	4.3685E+02
	Stdev	5.2986E+01	1.4188E+01	1.8566E+01	1.5227E+01
f14	Mean	4.3130E+05	1.1677E+05	1.0971E+05	1.7118E+05
	Best	4.0139E+05	1.0649E+05	1.0266E+05	1.5430E+05
	Stdev	2.3115E+04	5.2517E+03	3.5623E+03	9.2376E+03

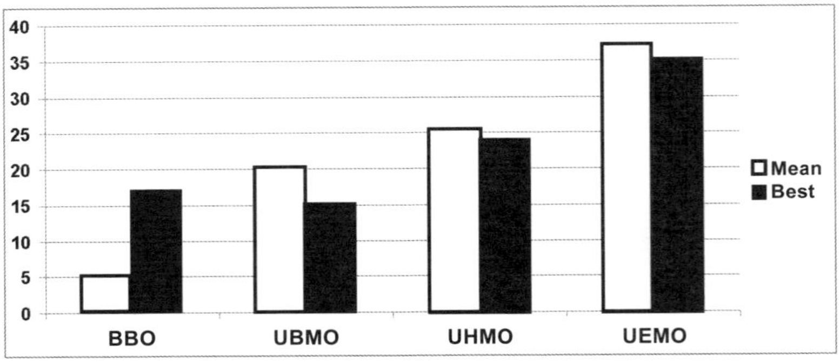

Fig. 17.3 Performances of blended migration operators ($\alpha = 0.2, 0.5, 0.8$) and uniform blended migration operator

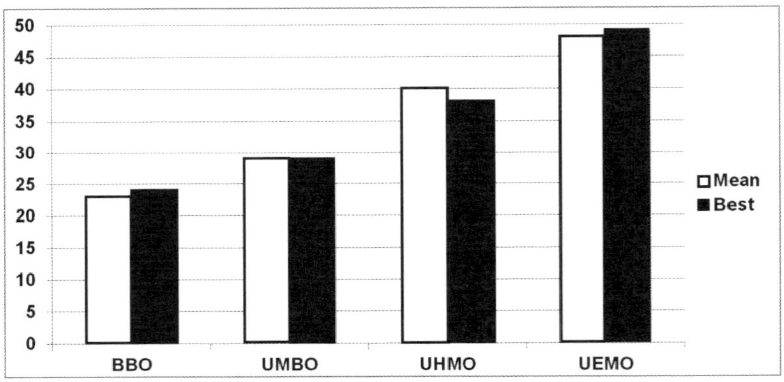

Fig. 17.4 Performances of different migration operators (dimension of benchmarks is 100)

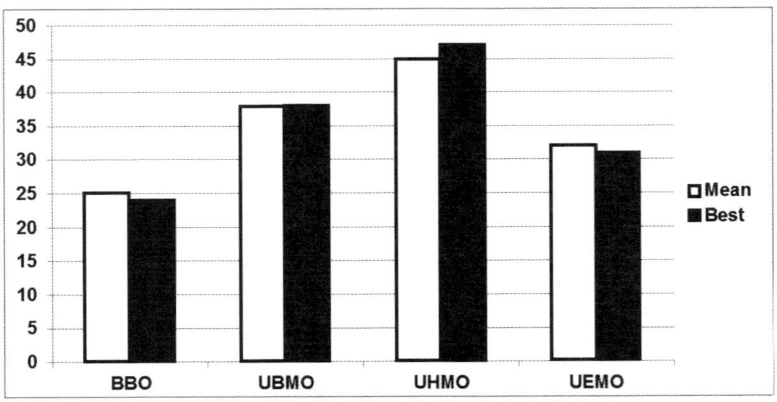

Fig. 17.5 Performances of different migration operators (dimension of benchmarks is 500)

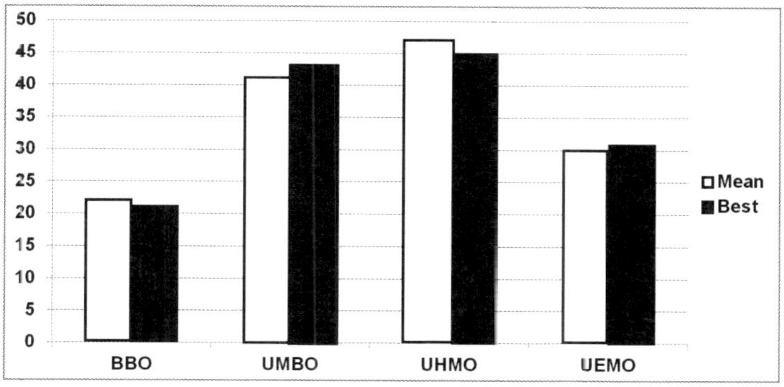

Fig. 17.6 Performances of different migration operators (dimension of benchmarks is 1000)

17.4.1 *Correlation model*

Since the correlation among commodities are nonlinear and implicit, it is difficult to present an exact model for correlation analysis. Besides, the research regarding the big data is correlation of data but not causality of data. The traditional science paid attentions to the causation, which focus on the accuracy. However, for big data, correlation analysis has become the key to investigated since it is impossible to clarify the causality. Meanwhile, it is impossible to present an prediction with an absolute precision, but feasible to make a certain judgement. There are so many cases to illustrate the correlations. For examples, Google employed the search frequency of flu-related symptoms to predict a flu and Alibaba e-commerce predicted the financial crisis by analyzing the declined transactions volume. The two examples are easy to understand, but some implicit correlations cannot be revealed easily. For example, some experts found that there exists correlation between people gaining weight and rental housing, which is not easily to understand although it is indeed true. In general, big data analysis focus on mining correlations between different things and as a basis for people to make all kinds of judgements or decisions.

In addition, differing the traditional analysis method, the correlation analysis for big data is not only dependent on the the algorithms but also the data themselves. The analysis operations could be abstracted into a simple expression that $Y = F(X)$, where $F(\cdot)$ represents the algorithms, X represents the data and Y is on behalf of the results we would like to obtain. In the era of big data, the importance of $F(\cdot)$ has gone down, but a significant increase in the importance of X, as X could influence the performance of $F(\cdot)$. Hence, we do not pursue which algorithms performs the best, but present an idea to analyze the correlation for big data.

To investigate the sales correlation for commodities, we summarize the category of correlations and present the following five kinds of correlations as follows.

- General Correlation: This correlation conforms to the habitats of most people and therefore lasts for a very long time. As mentioned above, toothbrush has a

strong correlation with toothpaste as well as cell phones has a strong correlation with SIM cards. Besides, the strong relationships are common such as raincoats and galoshes, computers and monitors, airplane tickets and hotel reservation, and so forth.
- Event-related Correlation: This kind correlation will be generated by events. During the festivals, the sales of food are better than that in peacetime. A part of events is impossible to predict, such as a break of a road being influent to the sales of vehicles.
- Periodic Correlation: This kind of correlation could be predict by experiences. In summer, the T-shirt is easier to sale than down jacket, while in winter vice versa. Hence, in some cases, the sales of commodities have periodic feature.
- Territorial Correlation: Some commodities are related to the territory. For examples, in South of China, people prefer rice diet than wheaten food, while in North of China, people prefer wheaten food than rice diet.
- Individual Correlation: By recording individual custom, e-business could make a pre-decision before customer checks out. If someone usually buy wines online, e-business could predict the actives that the customer could really buy wines with a high probability. Therefore, the e-business could conduct a logistics management in advance.

The generation of hot spot is dynamic but long term. There are cases that in a sudden dynamic, a hot spot appears and vanishes in a short time, which is termed as a "blink". Hence, although evolutionary algorithms do not perform well in a quick dynamic environment, it happens to be one of the key advantages of the EA, as the investment based on a "blink" is neither rational nor reasonable. The delayed response of evolutionary algorithm helps avoid investment risks.

17.4.2 *Simulation for correlation analysis*

In this subsection, we employ genetic algorithm to conduct the simulation for correlation analysis in business intelligence. First, we assume that there are 1 million commodities denoted by $[x, y]$ for an e-business, where x and y both are integer and $x, y \in [1, 1000]$. For the commodity (i, j), the location in a commodity matrix is on $[x_i, y_j]$. There exist correlations among all the commodities, which include the five kind of correlation described in Section 17.4.1. The correlation matrix is generated in a dynamic random way. The probability to generate a new hot spot marked by "+" is set as 0.00001, and the probability to generate a new spot which is near the hot spot is generated by a probability matrix which is set in advance. We do not consider the probabilities that a hot spot dies out. Fig. 17.7 presents the final location distribution of commodities.

The limitation of hot spot generation is set as 5000. Since the generation of hot spot is dynamic, the evolutionary algorithm runs with the generation. For genetic algorithm, we use a population with size of 100 and employ the half uniform crossover. The mutation rate is set as 0.01. Each individual in the population is encoded by a 1000 by 1000 matrix. For each generation of hot spot, genetic algorithm runs 1000 generations. The simulation results are presented in Fig. 17.8, which is obvious that the cluster regions basically match the ones in Fig. 17.7. The optimization objective is the amount of matching hot spot. Fig. 17.9 is on behalf of the error convergence of genetic algorithms.

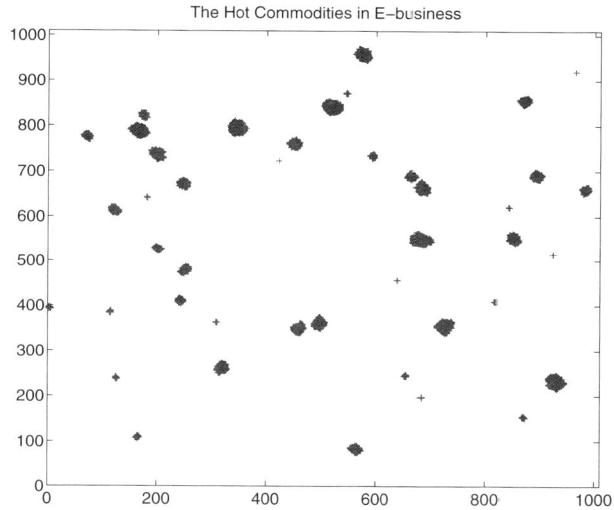

Fig. 17.7 The hot productions in e-business

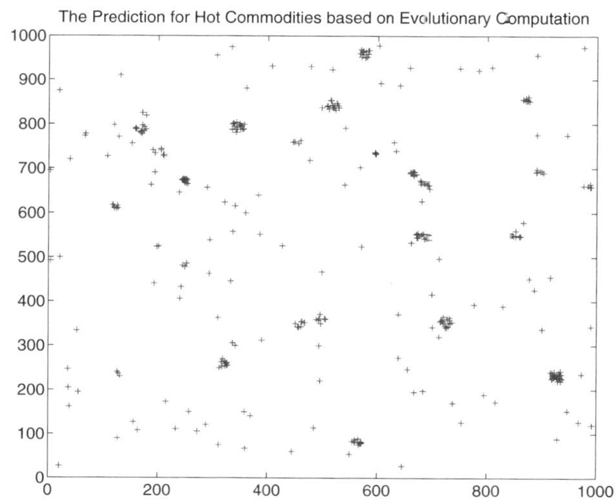

Fig. 17.8 The prediction for hot productions based on evolutionary algorithms

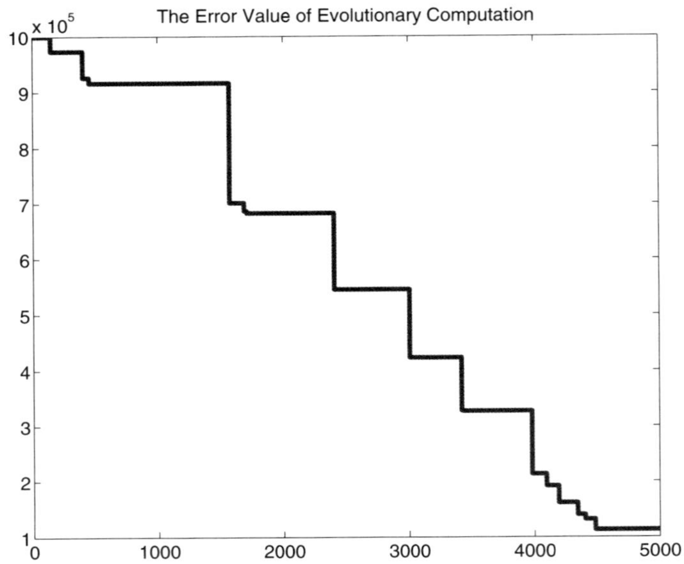

Fig. 17.9 The error values of evolutionary algorithms

As we mentioned above, the correlation analysis is not based on the causality, and therefore it is not necessary and nearly impossible to obtain an accurate results. However, according to the results in Fig. 17.8, the evolutionary algorithms is able to supply a feasible help for e-dealers to predict the hot spots of commodities and help e-dealers to make correct judgements. From the Fig. 17.9, the error converges which means the evolutionary algorithms performs well in predictions.

17.5 Conclusions

Nowadays, with the rapid development of the Internet, big data era has arrived. Big data is increasingly characterized and the value of big data is obvious to explore. Big data is the foundation for creating new chances of business value. For the application of business intelligence, there is irreversible trend that big data will be used in every every aspect of financial and business analysis, especially financial engineering, equity analysis and portfolio management. In this chapter, we investigate the impact of big data to evolutionary algorithms and proposed some suggestions and new design for evolutionary algorithms. Differing from the traditional idea, the processing for big data should fully take account of data itself. The "4V" features are influent to the methods themselves. To make the algorithms performs feasible and effective, novel designs for calculation, storage, management are desperate needs. As the literal meaning of evolutionary algorithm, it faces a urgent evolution to make it adaptive to the era of big data.

References

Akay, B. and Karaboga, D. (2012). A modified artificial bee colony algorithm for realparameter optimization, *Information Sciences* **192**, pp. 120–142.

Chang, J. F. and Shi, P. (2011). Using investment satisfaction capability index based particle swarm optimization to construct a stock portfolio, *Information Sciences* **181**, 14, pp. 2989–2999.

Dorigo, M. and Stützle, T. (2004). *Ant Colony Optimization* (MIT Press).

Garey, M. and Johnson, D. (1990). *Computers and Intractability: A Guide to the Theory of NP-Completeness* (W. H. Freeman).

Guo, W., Wang, L. and Wu, Q. (2014). An analysis of the migration rates of biogeographybased optimization, *Information Sciences* **254**, pp. 111–140.

Hansen, N. and Ostermeier, A. (2001). Completely derandomized self-adaptation in evolution strategies, *Evolutionary Computation* **9**, 2, pp. 159–195.

Herrera, F., Lozano, M. and Verdegay, J. L. (1998). Tackling real-coded genetic algorithms: Operators and tools for behavioural analysis, *Artificial Intelligence Review* **12**, 4, pp. 265–319.

Holland, J. H. (1992). *Adaptation in Natural and Artificial Systems: An Introductory Analysis with Applications to Biology, Control, and Artificial Intelligence* (MIT Press).

Kennedy, J., Eberhart, R. and Shi, Y. (2012). *Swarm Intelligence* (Morgan Kaufmann Publishers).

Kirkpatrick, S., Gelatt, C. and Vecchi, M. (1983). Optimization by simulated annealing, *Science* **220**, 4598, pp. 671–680.

Kristoffersen, T., Capion, K. and Meibom, P. (2011). Optimal charging of electric drive vehicles in a market environment, *Applied Energy* **88**, 5, pp. 1940–1948.

Ma, H. and Simon, D. (2011a). Analysis of migration models of biogeography-based optimization using markov theory, *Engineering Applications of Artificial Intelligence* **24**, 6, pp. 1052–1060.

Ma, H. and Simon, D. (2011b). Blended biogeography-based optimization for constrained optimization, *Engineering Applications of Artificial Intelligence* **24**, 3, pp. 517–525.

Michalewicz, Z. (1995). Genetic algorithms, numerical optimization, and constraints, in *Proceedings of the 6th International Conference on Genetic Algorithms* (Pittsburgh, USA), pp. 151–158.

Noman, N. and Iba, H. (2008). Accelerating differential evolution using an adaptive local search, *IEEE Transactions on Evolutionary Computation* **12**, 1, pp. 107–125.

Oca, M., Stutzle, T., Birattari, M. and Dorigo, M. (2009). Frankensteins pso: A composite particle swarm optimization algorithm, *IEEE Transactions on Evolutionary Computation* **13**, 5, pp. 1120–1132.

Ozcan, E., Bilgin, B. and Korkmaz, E. E. (2008). A comprehensive analysis of hyperheuristics, *Intelligent Data Analysis* **12**, 1, pp. 3–23.

Radcliffe, N. J. (1991). Equivalence class analysis of genetic algorithms, *Complex Systems* **5**, pp. 183–205.

Sharapov, R. R. (2007). Genetic algorithms: Basic ideas, variants and analysis, in G. Obinata and A. Dutta (eds.), *Vision Systems: Segmentation and Pattern Recognition* (I-Tech Education and Publishing), pp. 407–422.

Simon, D. (2008). Biogeography-based optimization, *IEEE Transactions on Evolutionary Computation* **12**, 6, pp. 702–713.

Simon, D., Ergezer, M. and Du, D. (2009). Population distributions in biogeography based optimization algorithms with elitism, in *IEEE International Conference on Systems, Man and Cybernetics* (San Antonio, USA), pp. 991–996.

Srinivas, M. and Patnaik, L. M. (1994). Genetic algorithms: A survey, *Computer* **27**, 6, pp. 17–26.

Williams, A. and Yang, S. (2006). An efficient dynamic system for real-time robot-path planning, *IEEE Transactions on Systems, Man, and Cybernetics, Part B: Cybernetics* **36**, 4, pp. 755–766.

Wolpert, D. H. and Macready, W. G. (1997). No free lunch theorems for optimization, *IEEE Transactions on Evolutionary Computation* **1**, 1, pp. 67–82.

Wolpert, D. H. and Macready, W. G. (2005). Coevolutionary free lunches, *IEEE Transactions on Evolutionary Computation* **9**, 6, pp. 721–735.

Wright, A. (1991). Genetic algorithms for real parameter optimization, in G. J. E. Rawlins (ed.), *Foundations of Genetic Algorithms* (Morgan Kaufmann, USA), pp. 205–218.

Yang, Z., Tang, K. and Yao, X. (2008). Large scale evolutionary optimization using cooperative coevolution, *Information Sciences* **178**, 15, pp. 2985–2999.

Yao, X., Liu, Y. and Lin, G. (1999). Evolutionary programming made faster, *IEEE Transactions on Evolutionary Computation* **3**, 2, pp. 82–102.

Yuan, Q., Qian, F. and Du, W. (2010). A hybrid genetic algorithm with the baldwin effect, *Information Sciences* **180**, 5, pp. 640–652.

Zelenka, J. and Kasanický, T. (2011). Comparison of artificial immune systems with the particle swarm optimization in job-shop scheduling problem, in *IEEE 9th International Symposium on Applied Machine Intelligence and Informatics* (Smolenice, Slovakia), pp. 129–135.

Index

ε-differential privacy, 248
ε-greedy exploration, 113
ε-optimal, 109

action dependent, 84
action neural network, 61
activation function, 60
actor-critic network, 1
admissible control, 29
approximate dynamic programming, 1
approximation front, 343
asymptotically stable, 19
autoregressive, 276
autoregressive coefficients, 296
autoregressive moving average with exogenous input, 277

backpropagation algorithm, 93
baseline scenario, 322
belief space, 88
Bellman equation, 13
Bellman optimality principle, 13
benchmark problem, 348
big data, 431
binary space partitions, 259
biogeography-based optimization, 435
bisection protocol, 253
black-box identification, 206
business intelligence, 446

Cauchy-Schwarz inequality, 65
centripetal-Coriolis matrix, 140
change-detection test, 264
change-time instant, 273
chaotic systems, 366
Chua's circuit, 366
classification performance, 387
coefficient factor, 139
coefficient matrix, 157
cofactor, 425
communication channel, 178
communication complexity, 246
communication constraints, 226
communication network, 154
communication protocol, 246
compact set, 200
completely continuous operator, 58
complex networks, 381
concept drift, 270
constrained robot, 135
continuous-action system, 109
contracting operator, 58
control co-design, 221
control protocol, 225
correlation analysis, 451
cost function, 60
cost-to-go, 156
costate equation, 12
covariance matrices, 235
cross-validation procedure, 268
cross-validation technique, 415
crossover operator, 434
crowding-distance, 344
cryptographic tools, 250
curse of dimensionality, 90
cyclic service, 225

damping matrix, 137
data-driven, 90
data-generating process, 267
datagram protocol, 225
datastream, 268

decay factor, 113
decoding phase, 267
decryption process, 374
delay-dependent, 233
density function, 289
determinism, 287
differential privacy model, 248
directed graphs, 27
discount factor, 87
discounted return criterion, 109
discrete-action system, 109
discriminant analysis, 410
discriminant function, 410
dissection protocol, 258
dissipation function, 137
distributed compensation, 365
distributed function, 248
divide-and-conquer, 259
dual heuristic dynamic programming, 1, 84
dual heuristic programming, 27
dual variables, 204

e-dominance, 345
e-business, 432
elastic deformation, 139
elimination tournament, 347
emigration rate, 436
encoding-decoding machine, 272
encryption process, 374
energy allocation, 325
energy consumption, 312
energy cost, 312
energy distribution, 309
energy efficiency ratio, 313
energy storage, 312
ensemble method, 335
entropy, 287
equality function, 251
error propagation, 315
Euler-Bernoulli beam, 135
event condition, 182
event-generator, 179
event-triggered, 178
evolutionary algorithms, 335, 431
expectation, 87
expectation operator, 180
expectation-maximization, 291
exploration noise, 113, 162
exploration probability, 113
exponential function, 289
extreme learning machine, 193

feature extraction, 271
feature space, 394
feedback linearization methods, 55
feedforward control, 57
final generational distance, 341
fitness function, 444
fitting errors, 395
fixed point, 58
forecast data, 320
forward-in-time, 151
Frobenius norm, 155

Gaussian classifier, 414
Gaussian distribution, 271
Gaussian kernel density, 415
Gaussian mixture density, 412
Gaussian mixture model, 409
Gaussian noise, 277
Gaussian realizations, 272
gradient descent, 36
gradient descent method, 92
gradient-based adaptation, 61
graphical games, 7
greedy policy, 123
grid burden, 312
grid difference, 345
grid dominance, 345

Hamilton-Jacobi-Bellman equations, 1
Hamiltonian, 12
Hamming distance, 250
heat pump, 318
Heaviside function, 287
heuristic crossover, 443
heuristic dynamic programming, 1
high-frequency disturbances, 139
hybrid genetic algorithm, 317
hyper heuristic algorithms, 434

ill-posed problem, 411
immigration rate, 436
implicit function theorem, 57
in-degree matrix, 5
indicator function, 180
infinite horizon rewards, 109
intra-class graph, 395
inverted pendulum, 124

iterative learning, 140

Jacobian linearization, 193
joint probability distribution, 250

k-means, 290
k-nearest neighbor, 394
Karush–Kuhn–Tucker conditions, 204
kernel density function, 413
kinetic energy, 136

Lagrange multiplier, 9
Lagrange operator, 136
Lagrangian function, 392
Lagrangian method, 137
$\mathcal{L}aplace$ distribution, 249
leader node, 5
least additive error, 250
least-squares, 204
leave-one-out likelihood, 410
likelihood function, 289
linear quadratic regulator, 114
linearization residue, 198
linearized error, 55
linearly independent, 160
log-likelihood, 290
Lorenz systems, 371
loss factor, 314
Lyapunov exponent, 368
Lyapunov extension theorem, 70
Lyapunov-Krasovskii-like, 143

machine learning, 89
manifold-based, 387
Markov decision process, 83
maximum likelihood, 410
mean absolute error, 315
memristive Lorenz circuit, 375
memristor, 366
micro-grid level, 309
migration operators, 439
minimum description length, 407
model predictive control, 193
Moore-Penrose generalized inverse, 204
moving-average terms, 276
multi-agent control, 1
multi-dimensional objective, 344
multi-input-multi-output, 49
multi-layered, 315
multi-regularization, 407, 415
mutation operator, 434
mutation rate, 436
mutual information, 286

Nash equilibrium, 9
negative definite, 115
network units, 274
network-induced, 224
networked control system, 151
neurodynamic optimization, 206
niche count, 342
noise-free, 266
non-dominated individuals, 340
non-overlapping windows, 272
nonaffine, 53
nonconvex optimization, 197
nondecreasing functions, 55
nonparallel planes, 388

observation function, 88
offline training, 54
optimal Q function, 109

packet disorder, 224
packet dropout, 224
packet losses, 154
packet-based communication, 221
pair-wise mixtures, 412
Pareto dominance, 344
Pareto front, 341
Pareto set, 341
Pareto-optimal front, 345
Pareto-optimal regions, 359
particle swarm optimization, 435
performance index, 7
performance metrics ensemble, 346
persistently, 159
phase portraits, 366
pinning gain, 6
policy evaluation, 110
policy improvement, 110
policy iteration, 110
policy-mistake count, 121
polyhedral domain, 371
polytopic-type, 370
poorly-posed problem, 411
positive definite, 53
positive semi-definite, 203
posteriori probability, 289
potential energy, 136

pre-specified value, 343
precision degradation, 272
prediction error, 60
prescribed threshold, 60
prior probability, 410
priori information, 273
privacy approximation ratio, 254
privacy-preserving, 252
private protocol, 249
probability density function, 266
projection theorem, 205
projection vector, 267
prosthetics, 283
pseudo-codes, 436

quadratic constraint, 203
quadratic programming, 200

radial basis function, 300
random access, 225
randomized algorithm, 248
reconstruction error, 182
reconstruction residuals, 269
recurrence plots, 286
recurrence points, 288
recurrence quantification analysis, 287
recurrence rate, 287
recurrent machine, 267
recurrent neural network, 205
recursive least-square, 107
regression matrix, 141
rehabilitation, 283
reinforcement-learning, 49
renewable energy, 312
residual error, 161
reward function, 86
Riccati equations, 9
rigid-flexible, 135
rotation matrix, 292

saddle point theorem, 205
Schauder's fixed-point theorem, 58
second-price auction, 247
set-covering, 251
set-disjointness function, 251
set-intersection function, 252
Shannon entropy, 288
shape function, 135
skew-symmetric, 141
slack variables, 387
sliding window, 269
smart grids, 309
smart home, 317
smoothing matrix, 413
smoothing parameters, 413
spanning tree, 5
sparse data, 423
stability criterion, 233
stable matrix, 57
stack-at fault, 272
state aggregation, 121
state-action, 87
stationarity conditions, 30
stiffness matrix, 137
stochastic value function, 162
support vector machine, 283
surface electromyography, 283
surplus variables, 387
switched system, 231
synchronization control, 366
synchronization error, 6

task allocation, 309
thermal regulation, 312
tiling function, 258
time complexity, 434
time-dependent, 160
time-stamped, 228
time-synchronized, 228
tournament selection, 347
transformation matrix, 174
transition probability, 86

uniform distribution, 259, 342
uniform ultimate boundedness, 54
uniformly continuous, 144
uniformly distributed, 442
uniformly ultimately bounded, 160
unmatched parameters, 366
utility function, 18

value iteration algorithms, 1
varying feedback gain, 227
vector-valued functions, 57
vibration theory, 135
virtual states, 91

weighting factors, 387
weighting scheme, 390
well-formed, 7

white noise, 235
winning algorithm, 347

zero-sum games, 84
zeros-order hold, 115